PRODUCTION
PLANNING
AND
INVENTORY
CONTROL

Quantitative Methods and Applied Statistics Series

Barry Render, Consulting Editor
Roy E. Crummer Graduate School of Business, Rollins College

Second Edition

PRODUCTION PLANNING AND INVENTORY CONTROL

Seetharama L. Narasimhan
University of Rhode Island

Dennis W. McLeavey
University of Rhode Island

Peter J. Billington
University of Southern Colorado

PRENTICE HALL, Englewood Cliffs, New Jersey 07632

Library of Congress Cataloging-in-Publication Data
Narasimhan, Seetharama L., 1936–
    Production planning and inventory control / Seetharama L.
Narasimhan, Dennis W. McLeavey, Peter J. Billington.
        p.   cm.
    Revised ed. of: Production planning and inventory control / Dennis
W. McLeavey. 1985.
    Includes index.
    ISBN 0–13–186214–6
    1. Production planning.   2. Inventory control.   I. McLeavey,
Dennis W.   II. Billington, Peter, 1948–   .   III. McLeavey, Dennis
W.   Production planning and inventory control.   1985.   IV. Title.
TS176.N36   1995
658.5—dc20   94–22418
                                                                          CIP

Acquisitions editor: *Rich Wohl*
Project manager: *Tina M. Trautz, PMI*
Cover designer: *Richard Dombrowski*
Production coordinator: *Marie McNamara*

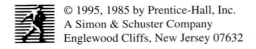
Printed in the United States of America

10   9   8   7   6   5   4   3   2   1

ISBN: 0-13-186214-6

PRENTICE-HALL INTERNATIONAL (UK) LIMITED, LONDON
PRENTICE-HALL OF AUSTRALIA PTY. LIMITED, SYDNEY
PRENTICE-HALL CANADA INC., TORONTO
PRENTICE-HALL HISPANOAMERICANA, S.A., MEXICO
PRENTICE-HALL OF INDIA PRIVATE LIMITED, NEW DELHI
PRENTICE-HALL OF JAPAN, INC., TOKYO
SIMON & SCHUSTER ASIA PTE. LTD., SINGAPORE
EDITORA PRENTICE-HALL DO BRASIL, LTDA., RIO DE JANEIRO

Contents

Chapter 2
Forecasting   25

Chapter 3
Special Topics in Forecasting   66

PART II
MATERIALS MANAGEMENT   87

Chapter 4
Basic Inventory Systems   91

Chapter 8
Distribution Inventory Management   208

PART III
PLANNING ACTIVITIES   251

Chapter 9
Aggregate Planning   255

Chapter 13
High-Volume Production Activity Control and Just-in-Time Systems   436

Chapter 14
Job Shop Production Activity Planning   470

Chapter 21
Factory of the Future   695

Index   707

Preface

The role of production planning and inventory control professionals is to accomplish the organization's mission by using the manufacturing resource of a firm in the most efficient and effective ways possible. In the 1980s American manufacturers realized that they must improve quality, cut costs, and better provide customer service to compete effectively in the global market. *Production Planning and Inventory Control* provides the basics of manufacturing planning and control with the need of the modern manufacturing organizations in mind.

The traditional topics of forecasting, inventory management, aggregate planning, master production planning, material requirements planning, capacity planning and control, and job shop production activity planning, scheduling, and control are all included here. This second edition also deals with important newer topics such as total quality management, just-in-time, manufacturing strategy, theory of constraints, and technology and innovation management.

Separate chapters cover speed to market or so called time-based competition (Chapter 18), multi-item joint inventory replenishment systems (Chapter 5), theory of constraints and synchronous manufacturing (Chapter 16), manufacturing strategies (Chapter 1), technological innovations in manufacturing (Chapter 19), and total quality management (Chapter 20). Areas of expanded coverage included forecasting of slow-moving items (Chapter 3), pragmatic but more complex distribution systems (Chapter 8), newer lot-sizing rules for MRP (Chapter 11), comprehensive coverage of just-in-time (Chapter 13), manpower scheduling for service operations (Chapter 14), and scheduling flexible manufacturing systems (Chapter 15).

Our aim was to cover many emerging and important topics that we as faculty already cover or hope to address in the near future. With suggestions received from students and faculty across the country as well as from international faculty we rewrote many chapters to facilitate easier readings. The new end-of-chapter readings were chosen from varied sources. Articles from *Interfaces* emphasizes that this discipline is for real and is live and well, real-world practitioners are applying these techniques to save real dollars in business and industry and thus increase the quality and productivity of products and services. Some readings were chosen to illustrate the applicability of mathematical programming and networks, even though we do not delve into those topics here. Some readers may not be able to digest an entire article. The implication of these *Interfaces* articles, however, is that the real-world applications help many of our readers understand the concepts behind the techniques. By skipping the mathematical jargon one can appreciate the usefulness of the techniques described in these articles. Thus we hope to motivate our budding POM/OR/MS students to pursue higher level courses and to foster a greater appreciation of these topics with our friends who are not mathematically oriented. Other articles from *Across the Board, Financial World,* and *American Scientist* were selected for their quality, applicability, and usefulness in business and industry. In addition, we have covered many important topics that are considered essential to improve the quality and productivity of manufacturing and services and have guided the reader to where to go for additional information.

SIGNIFICANT FEATURES OF THE BOOK

The most distinctive feature of this book is the heavy use of worked examples at points where concepts are introduced. Tables and figures illustrate difficult concepts. Problems at the end of chapters reinforce the concepts. The end-of-chapter readings illustrate the concepts now in use in business and industry. The concepts are not just academic; they all have day-to-day practical use. An interesting phenomenon is that these new topics are considered important by top management in industry and business. Planning professionals are now important players in the overall strategic planning phase of a firm, whether the strategy is speed of delivery to the customer or increased quality.

ORGANIZATION OF THE BOOK

The book is organized in five parts. Part I starts where most manufacturers are now focusing their attention: customer and demand management. Chapter 1 stresses the importance of manufacturing strategy in the context of traditional production planning and control, Chapter 2 deals with the basic concepts of forecasting, and Chapter 3 presents more specialized forecasting models.

Part II addresses materials management. Chapter 4 presents basic inventory systems, Chapter 5 expands the topic to multi-item joint replenishment, and Chapter 6 introduces the concept of inventory and risk and analyzes the trade-offs between inventory investment and customer service. Chapter 7 introduces aggregate inventory management in the context of risk, and Chapter 8 concludes Part II with distribution requirements planning.

Part III presents planning activities, moving from longer-term aggregate levels in Chapter 9 to shorter term details at the master production level in Chapter 10 to short term requirements for producing individual items in Chapter 11.

Part IV presents control activities. Chapter 12 covers capacity planning, and Chapter 13 presents high-volume control, including just-in-time concepts. Chapters 14 and 15 provide job shop production activity planning and control concepts. Goldratt's theory of constraints and synchronous manufacturing concepts are explained in Chapter 16, and Part IV concludes with project management techniques in Chapter 17.

Part V encompasses chapters on strategy and technology. Chapter 18 presents the emerging concepts in time-based competition that we call speed to market and its connection to the overall competitive advantage of the firm. Technological innovations described in Chapter 19 are playing an increasingly important role in defining the strategy of a firm. Total quality management (Chapter 20) has swept into industry in a dramatic fashion recently, and its impact on production planning and control is discussed. Chapter 21 makes a modest attempt to forecast future directions in manufacturing, which is not an easy task.

AUDIENCE

As with the first edition, we have always wondered about the audience for this book. Are our readers juniors, seniors, or M.B.A.s in the business curriculum, or more quantitative industrial engineering or industry practitioners? We have taken a middle ground and a pragmatic approach rather than a quantitative or a qualitative approach. In the business school, we have used this book for a senior-level, one-semester elective course with an emphasis on forecasting and inventory analysis along with newer topics discussed in other chapters. Using different chapters, we have also taught a junior-level, one-semester elective course on shop floor control, capacity planning, and materials requirements planning. In addition, we have also used this book to teach an introductory production and operations management course for M.B.A. students, and our experience has been very positive and rewarding. The first edition was used by many industrial engineering departments in a one-semester basic course in production planning and control or operations management with a selection of topics. The book was also used for seminar classes in industry. We hope that our readers will find the newer topics very useful and will appreciate the additional clarity provided to other chapters in this edition. Therefore, we feel that this book, with expanded chapters, would be very useful for both academicians and industry practitioners.

In the past, one of the criticisms leveled against these books such as this where various topics are covered, has been that the chapters and topics stand alone and are not cohesively written to motivate the readers. In this edition we have focused our book toward Arthur Young's manufacturing for competitive advantage framework so that readers can see where and how well all these topics fit together.

The topics presented here closely reflect the body of knowledge suggested by the American Production and Inventory Control Society (APICS), with the addition of several important topics such as total quality management, manufacturing technology and strategy. As APICS has changed from the organization often linked to the expansion of material requirements planning use in manufacturing to the educational society for resource management, so has the text changed to represent the expanding field of production planning that is now a key part in the strategy of world-class competitive firms.

USE OF SOFTWARE

We have used software packages such as STORM, AB:POM, OMIS, and Quickquant for solving problems at the end of chapters. For example, STORM was useful for solving open ended problems while AB:POM was helpful for solving specific problems in forecasting. STORM also solved multi-item inventory problems, although not as an aggregate problem as dealt in our book. All these packages were able to solve problems in material requirements planning and project management techniques easily. It is prohibitive for us to provide a comprehensive software package for solving all problems in the book. Many of our management information sciences students have enjoyed writing programs in BASIC or FORTRAN to solve many advanced level problems that we have used in classes in subsequent semesters. Unfortunately, some software still has bugs. Students were also required to use spreadsheets to solve many open ended problems given during take-home examinations.

ACKNOWLEDGMENTS

We hope that our readers will feel free to let us know the mistakes we made in this edition. We have incorporated many suggestions received from faculty and students who used our first edition and we hope to do the same during our next edition. We are indebted to professor Roy D. Harris of the University of Texas at Austin for providing valuable suggestions on the organization of the text into five parts. Chapter 18 on Speed to Market, was conceived during a discussion on the management of technology with Dean Clagett at the University of Rhode Island. He also provided feedback during many rewrites of this chapter, for which we are very grateful. Jack Kanet of Clemson University was gracious to discuss his ideas for the manufacturing technology in Chapter 19. We would like to acknowledge the encouragement of our colleagues, specifically, Russell Koza, Frank Budnick, and Paul Mangiameli at the University of Rhode Island and Stuart Warnock, who provided many insights into the teaching of a course on Production Planning and Inventory Control to business undergraduates at the University of Southern Colorado.

Our sincere apologies to many of our friends who we missed acknowledging in our preface of the first edition. The assistance of our editors, Cary Tengler and Rich Wohl, during this revision was nothing less than world class.

Reviewers Peter Kelle of Louisiana State University–Baton Rouge, Manbir Sodhi of the University of Rhode Island, John Martin of Rochester Institute of Technology, James McKee of Southern Tech, Inder Khosla of the University of Minnesota, Joel Stinson of the University of Syracuse, and John Leschke of the University of Virginia, retained by our publisher for the second edition, provided us with numerous suggestions; we have incorporated most of their suggestions into the text.

We would also like to thank the anonymous reviewers of the first edition; editors Jack Rochester, Jack Peters, Caroline Harris and Bill Burke, who ardently supported our work, as well as our copy editors, who did a magnificent job of keeping us from going insane. The support of editorial assistants Josephine Cashman, Cheryl Ten Eick, Dominique Vachon, and Jennifer Strada kept us going smoothly. We apologize if we inadvertently missed acknowledging our gratitude to any one else. Finally, our families deserve praise for their support and patience as this text was developed. We dedicate this text to Vethanayaki, Priya and Ram Anand; Dennis's loving wife, Janet, and children, Christine and Andy; and Peter's loving wife, Maryann, and sons, Drew and Alex, who always wanted their dad to get off the computer so they could do their writing.

Seetharama Narasimhan, Ph.D, CPIM
Dennis McLeavey, Ph.D, CFPIM
Peter Billington, Ph.D, CPIM

PART ONE

FOUNDATION

This book primarily concentrates on the essentials of manufacturing planning and control. In principle, it follows Arthur Young's *Framework for Manufacturing for Competitive Advantage* as exhibited in Figure I.1. The framework is described more fully in Chapter 1. We refer to this framework frequently during the introductory material for each part of this book.

As shown at the *first level* of Figure I.1, as a part of the long-range planning process management should have a *strategic vision* for a firm to create products and services that a customer needs. These decisions should also be based on a firm's ability to compete in global markets. Chapter 1 and Part V (Chapters 18, 19, 20, and 21) cover many important aspects of strategic planning. The *second level* of Figure I.1 emphasizes the importance of using *world-class manufacturing* techniques to manufacture products using available resources. These resources could consist of possibly the finest technology as described in Chapter 19; commitment to total quality management and respect for people in the organization as described in Chapters 20 and 21; and efficient resource planning and control techniques that are covered in detail in Chapters 2 through 18. In the *third level,* the planning process should adhere to *integrated manufacturing* techniques that consist of integrating supplier networks, after-sale customer service and support distribution, product and process design, manufacturing planning and control, and most important, the production process. Many of these aspects can be classified into three broad categories: (1) customer service and supplier support (covered in Chapters 2 through 8), (2) manufacturing planning and control (covered in Chapters 9 through 12), and (3) production process design (covered in Chapters 12 through 16). In the final analysis, the integrated manufactured process, in turn, must be supported by total quality control, computer-integrated manufacturing, and just-in-time production methods (covered in Chapters 20, 19, and 13, respectively).

1

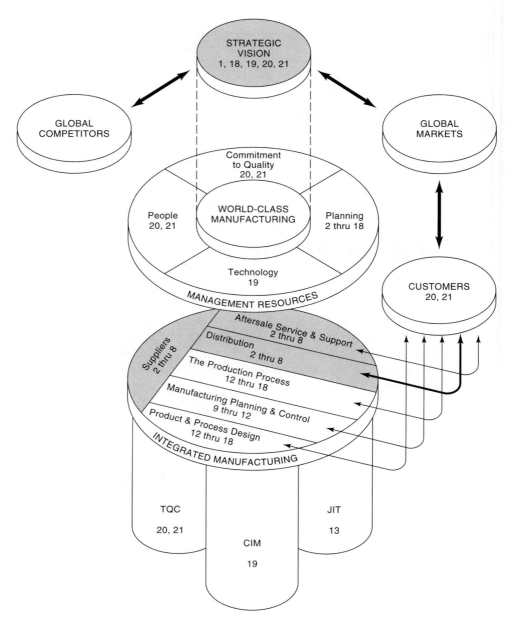

FIGURE I.1 Arthur Young's Manufacturing for competitive advantage framework. (*Source:* Copyright ®️ 1987 by Ernst & Young (formerly Arthur Young, a member of Arthur Young International). Reprinted by permission.) *Note:* numbers on this figure indicates chapter numbers covering these subjects.

This book does not attempt to be a cure-all for all industry ills or a panacea for all productivity problems. We have, however, given comprehensive treatment of many important aspects of manufacturing planning and control. This book is not intended to cover the responsibilities of other functions of business such as marketing, finance, accounting, and management. Many aspects of engineering, such as product and process design, are also beyond the scope of this book.

DEMAND MANAGEMENT

The idea that a manufacturing or service operation could be used for competitive advantage, the premise of Part I, is a relatively new concept. Chapter 1 discusses important advances made in strategic thinking about manufacturing processes. For a firm to be successful, proper manufacturing and service strategies must be integrated into the overall objectives or vision of the firm. The basics of manufacturing strategy are covered in Chapter 1, with an emphasis on showing how these strategies and changes can affect production planning and control activities.

Chapters 2 and 3 cover forecasting techniques, which are a part of demand management. Demand management deals with both customer needs and supplier coordination. Accurate forecasts are valuable for material and resource planning. A firm can release orders for procuring material on time by using material requirements planning techniques described in Chapter 11. By receiving the right quantity of the right material on time, a firm can deliver products to its customers on time. Accurate forecasts are also invaluable for planning resources. These topics are discussed in Parts II, III, and IV.

Chapter 2 starts with an introduction to demand management and then covers short-term and medium-term forecasting techniques in detail. Among the short-term forecasting techniques described are moving averages and smoothing techniques; among medium-term techniques described are regression analysis and the effects of seasonality in demand management. The forecasting techniques are followed by methods for monitoring and making necessary corrections to forecasting models. In addition, Chapter 2 covers some simple techniques such as focus forecasting and pyramid forecasting. Chapter 3 deals with several special topics such as multi-item forecasting and slow-moving item forecasting.

The following chapters are in Part I:

Chapter 1, Manufacturing and Service Strategies

Chapter 2, Forecasting

Chapter 3, Special Topics in Forecasting

Chapter One

Manufacturing and Service Strategies

INTRODUCTION

Using a manufacturing or service operation for competitive advantage is a relatively new concept. Unfortunately, the opposite thinking has been the standard of manufacturing managers for years. Many problems that cause a company to be noncompetitive can be directed at the manufacturing function: poor quality and reliability, late deliveries, high manufacturing costs, and lack of proper inventory at the right place. Top management then thinks that manufacturing is doing its job if manufacturing does not cause these problems to occur. The result is that few managers think of the operations function as providing a competitive advantage; rather, they think of manufacturing as neutral.

It is now obvious that important design aspects can and do have a direct impact on the competitive advantage of companies. A simple example is the home-delivery pizza business. The widespread acceptance of this concept was accomplished by Domino's Pizza. Why was this not done 20 years ago? The answer is that the making and baking process for pizza was not developed around delivery; it was designed for in-store pickup or in-store consumption. Home-delivery pizza restaurants have ovens designed especially for rapid baking, preparation areas that allow a high volume of assembly of pizzas, and a logistics system geared toward rapid delivery. The whole system is designed to make fast delivery possible.

Fast delivery would not be possible using the old process of making pizzas. A company cannot just say that they will deliver quickly; they must be able to do that consistently. Domino's Pizza totally redesigned pizza making for a competitive advantage. This is how operations can be used for competitive advantage and why production plan-

5

ning and inventory control systems must consider the strategy of the firm and be part of the firm's strategy.

This chapter introduces the concept of a manufacturing and service strategy and how it should be integrated into the corporate strategy. Production planning and inventory control processes are an integral part of the infrastructure of manufacturing strategy.

MANUFACTURING AS COMPETITIVE ADVANTAGE

Classical Manufacturing Strategy Factors

The basis for most decisions regarding manufacturing has been cost. A company strategy was formulated, then manufacturing was to find the least-expensive way to deliver the product. Little thought was given to how manufacturing could be used to a competitive advantage other than cost.

Skinner [8] suggests that there are five key decision areas in manufacturing: plant and equipment, production planning and control, labor and staffing, product design, and organization and management. Each area has key decisions with various alternatives. It was once believed that these decisions could be made after the corporate strategy was developed. Now we know that the decisions will lend themselves to competitive advantages.

The area of production planning and inventory control covers a number of key factors, including inventory size and location and the degree of control. Gunn [3, p. 90] proposes a list of classical factors that fits well with Skinner's structure of decision making:

Capacity

Facilities

Technology

Vertical integration

Workforce

Quality

Production planning and material control

Organization

At one time this may have been a sufficient set of factors to consider when designing and managing a competitive manufacturing organization. When manufacturing was part of a stable, static business environment, factors such as these were capable of defining the manufacturing function. These factors and decision areas must still be considered today, but another, more demanding set has emerged. Businesses now see short product life cycles, more demanding customer requirements, higher standards for quality and design, and significant foreign competition.

A New Set of Manufacturing Strategy Factors

A new set of factors is necessary to compete effectively in today's dynamic, global marketplace. These factors are important because the way that manufacturing is used to compete effectively has also changed. No longer do manufacturing firms compete on cost; now they compete on quality, time, service, flexibility, and availability. According to Gunn [3, p. 92], the new strategy factors are:

Shorter new-product lead time

More inventory turns

Shorter manufacturing lead time

Higher quality

Greater flexibility

Better customer service

Less waste

Higher return on assets

These factors all result from a new direction in the use of manufacturing as a basis for competitive advantage. No longer is low cost the only consideration. Returning to the Domino's Pizza example, if Domino's competed not on fast delivery but on low cost, then many customers may find the low cost appealing, but a one-hour wait for delivery may not be satisfactory. Many customers are willing to pay a little more to get fast delivery. Of particular interest is that Domino's can deliver quickly, but there is little cost disadvantage to buying from them. Low cost results from doing many other things effectively, not because it is the driving force of the manufacturing strategy.

Competitive Advantages

How do companies compete in today's market, and how is the manufacturing part of the business important in that competitive advantage? There are several areas for competitive advantage, including cost, time, quality, and products and services. Each of these has a strong manufacturing component required for success.

Cost. Many markets are still driven by cost. Commodity items such as sugar and gasoline and parts of other markets such as IBM-clone personal computers may be very cost sensitive. There is a strong market for low-cost automobiles. Many manufacturers of IBM-clone personal computers compete on price because the product will perform identically to the more expensive IBM models. Gasoline refineries produce at the lowest cost possible, and smaller, high cost products just do not exist. Pricing differences of gasoline are usually linked to location, a service component of the product. The manufacturing part of these businesses must be run efficiently to keep costs down through many different factors: high volume, good design, effective purchasing, component selection, and efficient utilization.

Time. Domino's Pizza illustrates the power of time as a competitive advantage. One-hour photo developing is another consumer product market that has been revolutionized by time as a competitive weapon. There are actually two components to time competition: (1) how fast can the product be delivered and (2) how fast can new products be introduced to the market.

This second factor is important in markets with short product life cycles. Again, consider the personal computer market. Each year there are new advancements in central processing units' (CPU) processing capabilities and in software. Manufacturers must be able to take the new CPU chips and get them into new products very rapidly, or competitors will be in the market ahead of them. Software companies must be able to take advantage of new technology and develop more powerful software programs in a short amount of time, or lose the market to competitors that can.

In this latter area of time advantages, the manufacturing team must work concurrently with the product design team to have the manufacturing processes in place for fast start-up. The older method of consecutive processes will no longer work. In the old system, the design team develops a new product and then hands it to manufacturing personnel, who discovers that the product cannot possibly be manufactured at the estimated cost. A series of back-and-forth volleys of the product results in a long lead time until delivery of the product.

Concurrent engineering is the new way of product and process design. The design team includes product designers and manufacturing professionals who can develop the manufacturing process concurrently with the design of the product. Time to introduction is shortened. In the PC software industry, this concurrent engineering has even extended to external companies. When a software company is developing a new (or upgrade) operating system, early versions are available to other software application developers so that new versions of those products can be developed concurrently and introduced to the market soon after the operating system is introduced.

Fast delivery to the customer is forcing the production planning and inventory control areas of manufacturers to consider new methods. Chapter 16 discusses the new concepts of synchronous manufacturing. Shorter lead times are a key underlying driving force to competitive advantage. Goldratt's theories on synchronous manufacturing suggest smaller transfer batches and better scheduling of key bottleneck resources as the way to speed the flow of product through a manufacturing facility.

Quality. Quality has become *the* management initiative of the 1990s. Quality can be defined from both a design and a conformance consideration. Customers determine how the quality of the product is defined from both aspects. Product design takes into account the performance and features that customers want. Conformance means that the product is produced according to the specifications determined in the design phase. Customers believe that products are of low quality if customers' needs are not satisfied from either a design or conformance aspect.

Many customers will buy a higher-quality designed product and be willing to pay more. Mercedes-Benz is a designed car with features that are deemed of higher quality than competing cars. Mercedes Benz customers are willing to pay more money because of these high-quality features.

Customers are still buying lower-priced cars, however, because many customers believe that a low-cost car is also high quality. In addition, if the car works as designed and gets the customer from point A to B with no problems, then the customer will believe that the car is of high quality. A Mercedes-Benz that has constant mechanical problems will be considered a low-quality car by the disgruntled owner because the car does not meet the specification of high reliability.

Higher quality will require organization-wide activity. Production planning and control activities should be designed around the strategy of quality, if that is how the firm will compete. Just-in-time systems require virtually zero-defect incoming material to be run effectively. A component of poor quality will disrupt the production line because inventory will not be available to cover the variability.

Products and Services.   The products and services approach to competing ties closely with time as a competitive advantage. New products need to be introduced rapidly to beat competitors to the market. New products add complexity to the production planning and inventory control function, however. New products may need to be made on existing production lines with existing products, which adds the complexity of ramp-up (introduction) of the new product into the production process. How fast can the production rate be changed to handle the uncertainty in demand increase? How will that affect the production scheduling of the existing products? Timing of both the new product and the elimination of the old product will also complicate the scheduling process.

In materials requirements planning (MRP), for example, increases in demand for a new product may result in system nervousness, or changes in the production plans for all the subassemblies. This is discussed in more detail in Chapter 11.

No matter in what area a company tries to compete, the key is that the production planning and control function must be matched to that competitive advantage. If a company is trying to compete on fast delivery of custom-made products, the schedule cannot be determined by using large batches to keep costs down: Long lead times will result and the firm will not be competitive.

The next section considers how manufacturing fits within a company's overall strategy.

MANUFACTURING FOR COMPETITIVE ADVANTAGE FRAMEWORK

Figure I.1 showed a framework for manufacturing for a competitive advantage. This framework starts at the top with a strategic vision from the organization that describes the overall business objectives of that organization, which could be a company, a group, division, or a plant. This vision is taken with respect to two aspects: the global markets that exist and the competitors within those global markets. To compete effectively in these global markets, a company must be a world-class manufacturer.

Management Resources
for World-Class Manufacturing

A class A world-class manufacturer must have annual inventory turns of 80 to 100 and fewer than 200 defective units per million produced, and the value-added lead time must be greater than half the total product lead time through the manufacturing facility. Achieving this level will require effective use of four critical management resources.

Commitment to Quality. A commitment to quality must be made by everyone in the organization, from the chief executive officer on down. We now know that quality and low cost are not opposites that need to be traded off. Higher quality in the production process can, in fact, result in lower total cost through reduction of scrap and rework.

Technology. Technology must be appropriate for the task at hand. Many companies have tried robotics as a savior only to find that flexibility is lost due to high setup time. One the other hand, automation may be the most appropriate technology in certain settings.

Planning. Planning is the process of translating the strategy into action. Management must plan so that fire fighting is not an everyday activity. If there is little planning, then the best strategy will not be implemented at the manufacturing level.

People. Human resources are the most important part of any organization. Nothing gets done properly unless people are trained, educated, and motivated to work for success. A world-class manufacturer must have people who can change rapidly in today's rapidly changing world.

Integrated Manufacturing

Management resources are applied to integrated manufacturing to achieve world-class status. Manufacturing includes a host of activities that must be considered together to be successful in the global market. For example, product design cannot develop products without thinking about the customers' needs. The design of the product will also have a significant effect on the ability to repair a product through the proper selection of parts and through accessibility to repair points. Designers must also work with suppliers to ensure that parts and components are available to meet design specifications and after-sale service. Each area of this integrated manufacturing framework interacts with others in many dynamic ways.

We limit our discussion of integrated manufacturing to the planning and control area, although other areas are also important to success.

Manufacturing Planning and Control. Many changes are occurring in planning and control. With the introduction of just-in-time (JIT) systems, there is a greater emphasis on reduction of setup time and cost to afford smaller lot sizes. In addition, it is likely that smaller time buckets will be needed in MRP systems and that the master production schedule will be frozen more because JIT will not accommodate changes in the production rate.

Several important points were made by Gunn [3]. First, the term *integrated manufacturing* is used because all the areas identified at that level in Figure I.1 must be integrated together. Planning and control, then, must consider the design of the product, the production process, and the distribution system, among other areas, when designing a planning and control system that supports the overall world-class manufacturing strategy.

Second, suppliers and customers must be linked with planning and control. Manufacturers are now developing partnerships with suppliers to get better quality and better-designed supplied products and materials. The delivery of materials into the production process can be more closely linked through MRP. Customers are also demanding closer linkages with planning and control functions so that delivery dates and quantities can be more accurately estimated.

Third, quality and JIT systems are the foundation of an integrated manufacturing process.

The Pillars of Integrated Manufacturing

Total quality control (TQC), just-in-time, and computer-integrated manufacturing (CIM) are considered the pillars that support integrated manufacturing. Each is important to ensure that manufacturing is done correctly. As shown in Chapter 20, the quality movement has accelerated in recent years as a competitive initiative. In addition, JIT requires near zero defects in the production process. The combination works together to maintain integrated manufacturing.

Linking Strategy to Production Planning and Inventory Control

The basis for this chapter is that the production planning and inventory control system must be matched with the strategy of the firm so that manufacturing is used for competitive advantage.

Gunn believes that JIT systems are absolutely essential to achieving world-class manufacturing status. If that is the case, then implementation of JIT systems will require shorter setup times, smaller lot sizes, freezing of the master production schedule and MRP plans, reduction of inventory levels, better scheduling and sequencing at bottleneck work centers, change in layout for better flow, group technology and cells for faster flow of material, and faster distribution cycle times.

CORPORATE STRATEGY AND MANUFACTURING STRATEGY

This section presents an overview of corporate strategy development and how it should integrate manufacturing strategy. Many companies have not integrated manufacturing into the strategy process; instead, corporate strategy is developed around marketing issues, and then the strategy is given to manufacturing to implement. Unfortunately, the strengths of manufacturing may not match the marketing strategy, and a mismatch may result in failure. Figure 1.1 provides a framework for integration of manufacturing strategy into corporate strategy.

Development of Corporate Strategy

The first step in developing a corporate strategy is to determine the objectives of the company. Figure 1.1 shows the typical objectives: growth, survival, profit, and return on investment. These objectives are then considered in relation to the external market and strengths of the company to develop the marketing strategy.

The marketing strategy represents the plans of the markets in which the company will compete and the types of products that will be sold in those markets. The marketing strategy also considers whether a company will be a leader or a follower with products.

These marketing strategies consider what the order-winning criteria are in each market. As seen in Figure 1.1, a series of criteria can be used individually or jointly to win orders. The basic questions are, "Why do customers purchase our product?" Is it price, quality, delivery, features, service, or some combination of these and other factors? It is important that the marketing strategy result in proper order-winning criteria.

Corporate Objectives	Marketing Strategy	How Do Products Win Orders in the Marketplace	Manufacturing Strategy	
			Process Choice	Infrastructure
Growth	Product markets	Price	Choice of alterna-	Function support
Survival	and segments	Quality	tive processes	Manufacturing
Profit	Range	Delivery:	Trade-offs em-	planning and
Return on invest-	Mix	speed	bodied in the	control systems
ment	Volumes	reliability	process choice	Quality assurance
Other financial	Standardization	Demand increases	Process position-	and control
measures	versus cus-	Color range	ing	Manufacturing
	tomization	Product range	Capacity;	systems engi-
	Level of innova-	Design leadership	size	neering
	tion	Technical support	timing	Clerical proce-
	Leader versus fol-	being supplied	location	dures
	lower alterna-		Role of inventory	Payment systems
	tives		in the process	Work structuring
			configuration	Organizational
				structure

Note: Although the steps to be followed are given as finite points in a stated procedure, in reality the process will involve statement and restatement, for several of these aspects will impinge on each other.

FIGURE 1.1   Strategy framework. (From Terry Hill, *Manufacturing Strategy*, 2nd ed. Burr Ridge, Ill.: Richard D. Irwin, 1994.)

Manufacturing Strategy in Corporate Strategy Formulation

The order-winning criteria provide the key link between corporate strategy and manufacturing strategy, as shown in Figure 1.2.

Linking Manufacturing Strategy with Order-Winning Criteria.   Once the order-winning criteria are known, manufacturing can develop a strategy that will be successful. The first step is the process choice determination. As seen in Figure 1.1, there are a series of choices regarding the type of process, capacity decisions, inventory, and so forth. We see that Skinner's [8] areas of plant and equipment, labor and staffing, and product design/engineering are all part of the process choice in Figure 1.1.

The second part of the manufacturing strategy is the choice in infrastructure. Figure 1.1 activities such as the planning and control systems, quality assurance, and organizational structure are key components of this. Again, the selection of actions here must be made in relation to the order-winning criteria. Consider the pizza business again. For home delivery, the order-winning criteria is fast delivery of a custom product. Once that is realized, the infrastructure can be designed to complement these criteria. Order entry, production control, customization, and fast production are all infrastructure choices that should be designed in accordance with fast delivery. The success of Domino's Pizza indicates that the infrastructure has been designed with the order-winning criteria in mind.

Many firms already have manufacturing systems in place, so the manufacturing strategy may be well developed. In this case, the processes and infrastructure that are in place should be assessed in relation to the order winners. Notice that the arrows in Figure 1.2 move from the outside into the center, the order-winning criteria, which indicates how the process flows in determining the manufacturing strategy.

If the manufacturing strategy is a good match for the order-winning criteria, then the strategy is appropriate and can be implemented. In many cases, the match may not be good, which results in two possible actions (other than doing nothing):

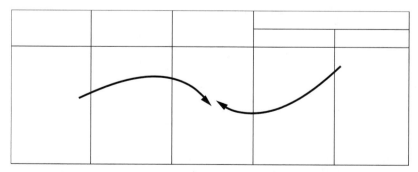

FIGURE 1.2 Linking corporate strategy with manufacturing strategy. (From Terry Hill, *Manufacturing Strategy,* 2nd ed. Burr Ridge, Ill.: Richard D. Irwin, 1994.)

1. Change the order-winning criteria
2. Change the process choice or infrastructure

This results in the circular flow illustrated in Figure 1.3. If the order-winning criteria are changed, the manufacturing strategy is again assessed in relation to the new order-winning criteria. If the process choice or infrastructure is changed, then we must reassess the order-winning criteria again in relation to the changes.

Linking Manufacturing Strategy with Corporate Strategy. In the process described above, the corporate strategy is developed first, then the manufacturing strategy is developed to match that. A more comprehensive process will expand the circular loop to include the manufacturing strategy with the corporate strategy, as shown in Figure 1.4.

This expands the discussion of the manufacturing strategy into the corporate decision-making process. Corporate strategy can now be developed around particular strengths that the manufacturing function has, which may not have previously been considered. For example, manufacturing may now be able to say that the planning process has been redesigned to allow fast and rapid delivery of custom-made products. If the corporate staff did not realize this, they would not include that strength in the development of the corporate and marketing strategy. If that happened, then the marketing strategy would be redesigned to reflect this and to provide a powerful new order-winning criteria.

Production Planning and Inventory Control in Manufacturing Strategy

In Figure 1.1 we see that the manufacturing planning and control function is an important part of the infrastructure part of manufacturing strategy. In the process choice column, we also see that capacity is an important part of the component of strategy. Chapter 12 deals exclusively with the issue of capacity. Long-term capacity planning is an integral part of the corporate strategy discussion. Long-term capacity is usually expensive and represents a major decision that must be included in the wider manufacturing-corporate strategy discussion as shown in Figure 1.4.

The planning and control function within the infrastructure does have a major im-

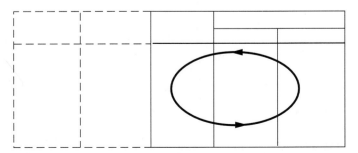

FIGURE 1.3 Assessing the implications for manufacturing processes and infrastructure of order-winning criteria. (From Terry Hill, *Manufacturing Strategy,* 2nd ed. Burr Ridge, Ill.: Richard D. Irwin, 1994.)

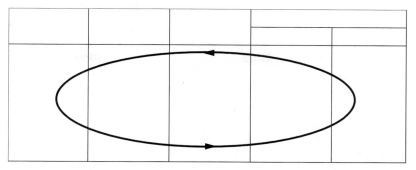

FIGURE 1.4   Manufacturing input into corporate strategy. (From Terry Hill, *Manufacturing Strategy,* 2nd ed. Burr Ridge, Ill.: Richard D. Irwin, 1994.)

pact on overall corporate strategy. MRP, JIT, and TQM have all been implemented with top management commitment and resources. These are seen as systems that will allow a company to compete more effectively in the market. The resulting side benefits can be surprising. Consider that JIT systems were introduced originally to cut inventory and to reduce waste. Successful JIT implementation requires high quality and a reduction of setup times. The reduced setup times allow for more flexibility and greater variety in product families, and for shorter cycle times. The result is that companies now have faster delivery as a competitive advantage.

The original implementations were probably seen as part of the corporate strategy formulation leading into the manufacturing strategy. The surprising results indicate that the full loop (as seen in Figure 1.4) must be completed so that the new corporate and marketing strategy can take full advantage of these new processes.

Consider some additional examples. If the order winner is delivery speed to customer, then the manufacturing strategy must include a planning and control system that will deliver accordingly. If it does not, then either change the planning and control system or change the order-winner criteria. It is also possible that the system is capable of fast delivery but that corporate and marketing strategy did not consider that. Rapid delivery could be done with either inventory or fast cycle time. Can the manufacturing planning and control system do either? For example, if the order winner is a custom product, is the planning and control system capable of handling the variety?

SUMMARY

The development of an effective manufacturing strategy must be done in conjunction with the corporate and marketing strategy. In addition, the manufacturing process choice and infrastructure strengths must be considered when corporate strategy is being developed. This will allow the proper use of the manufacturing strategy for a competitive advantage. The production planning and control system must be designed to complement the corporate strategy and to provide additional competitive advantages.

Each production planning and inventory control method has application in different settings. If an organization is making customer-specified products in small batches, then the methods to use to meet the organization's strategy will be job shop concepts. On the other extreme, if an organization is supplying commodity products with a strategy of low cost in the market, then high volume and just-in-time techniques would be most useful. As you read this text, consider how each of the concepts you study could be applied to an organization to meet the strategy of that organization. Not only should we know how to solve problems and use the techniques, we must be able to determine which techniques should be used in specific types of manufacturing or service organizations.

PROBLEMS

1. What is your strategy for this course?
   a. What is your objective?
   b. How did you arrive at that objective? What factors influenced you?
   c. What do you need to do successfully to meet that objective?
   d. How do you plan to meet that objective?
   e. What are some of the trade-offs that you have to make?
   f. How do you plan your study time? Is it appropriate for your objective for this course?
2. Refer to the Appendix 1A reading regarding Motorola.
   a. Using the strategy process described in Figure 1.1, discuss the strategy of Motorola's Bandit plant.
   b. What aspects of production planning and control are considered important in the strategy of this plant?

REFERENCES AND BIBLIOGRAPHY

1. R. B. Chase and N. J. Aquilano, *Production and Operations Management—6th ed.* (Homewood, Ill.: Richard D. Irwin, 1992).
2. D. A. Garvin, *Operations Strategy—Text and Cases* (Englewood Cliffs, N.J.: Prentice Hall, 1992).
3. A. Young, *Manufacturing for Competitive Advantage: Framework.* (Ernst & Young, formerly Arthur Young, a member of Arthur Young International, reprinted by permission, copyright 1987).
4. R. H. Hayes and S. C. Wheelwright, "Link Manufacturing Process and Product Life Cycles," *Harvard Business Review,* Vol. 57, No. 1 (January–February, 1979), p. 133–140.
5. R. H. Hayes, and S. C. Wheelwright, *Restoring Our Competitive Edge: Competing through Manufacturing* (New York: John Wiley & Sons, 1984).
6. T. Hill. *Manufacturing Strategy, Text and Cases,* 2nd ed. (Burr Ridge, Ill.: Richard D. Irwin, 1994).
7. P. E. Moody, *Strategic Manufacturing: Dynamic New Directions for the 1990's,* (Homewood, Ill.: Dow-Jones Irwin, 1990).
8. W. Skinner, "Manufacturing—Missing Link in Corporate Strategy," *Harvard Business Review,* Vol. 47, No. 3 (May–June 1969), p. 136–145.

APPENDIX 1A

How They Brought Home the Prize*
Tom Inglesby

Florida has gotten a bad reputation lately. *Miami Vice* and other entertainment have skewed the public's perception of this sunbelt state. Ever since the movie *Smokey and the Bandit,* a certain lawlessness and disregard for convention seem to have attached themselves to the Florida image. Perhaps that had something to do with the choice of names for a Motorola automation project, one that in many ways threw convention to the winds and developed new laws of integration. The project and the end product both bear the name: Bandit.

But Bandit, the product, and Bandit, the plant, aren't as fool-hardy as Burt Reynolds' Bandit in the movie. While Burt burned up the highways, Motorola's Bandit is burning up the competition—all overseas built—in a market that is growing quickly throughout the world. They are doing it without "smoke" and mirrors; they're using US know-how, managerial techniques that bring people fully into the project, and a design for manufacturability scheme that has tightened the product while improving its assembly. We aren't privy to the deliberations that went into the selection of Motorola for the Malcom Baldridge Quality Award, but having toured the Bandit works, we think this project may have had something to do with their winning.

So fasten your seatbelt, put on your Stetson and let's take a quick trip to Gatorland where Motorola is making—out like—a Bandit.

SUN, SAND AND SUCCESS

Northerners coming into the Ft. Lauderdale airport in winter are usually surprised at how fast they feel hot and sticky. The humidity in December can be a shock when combined with 70 plus degrees of temperature.

The area boasts a high quality of life rating due partially to its warm weather and maximum sunshine and partially to the relaxed, somewhat laid-back style of the Deep South. Getting people to move here isn't usually a problem, getting the right people can be. In the case of Motorola's Pager Division in Boynton Beach, the bandits first struck by "highjacking" quality people from throughout the company.

The Motorola bandits "stole" more than just the best and the brightest from the organization, they stole ideas. As Russ Strobel, Bandit's engineering manager, recalls, they were after something more than warm bodies.

"We wanted to eliminate a major cause of manufacturing headaches, the *not invented here* syndrome. NIH would kill us, given the chance. In the past, we rewarded the designer who came up with the newest, the greatest, the latest thing. We should have been looking for the designer who took something that basically worked and optimized it.

*Tom Inglesby, *Manufacturing Systems,* Vol. 7, No. 4 (April 1989), p. 25–32. Copyright 1989 by Hitchcock Publishing Company.

"So we wanted to foster the idea of building on past successes and previous designs," he continues. "We have a sign outside the lab that says it all: *Please don't leave the area without leaving us a good idea.* We need ideas from anywhere we can get them to prevent reinventing the product. That's why we called it Bandit."

Strobel and Director of Manufacturing Operations Scott Shamlin might admit they also "stole" a few people they wanted, although they won't use the term. "We had to staff the program," Strobel agrees, "but we didn't steal anybody. Motorola has a program called IOS—internal opportunity system—where you post job openings for anyone in the company to apply. We didn't go up to managers and tell them we were taking someone. We didn't have to, the project was so popular that we got applications from all over. It took us about three months to staff it with 20 people."

And while the staffing was going on, the work was, too. Strobel, Shamlin, "a tooling engineer and a couple of Double-Es" were evaluating pagers and manufacturing approaches. Three attacks were considered: make the new receiver (pager) capable of automated assembly by using surface-mount technology (SMT); adapt an existing receiver that was already close to SMT; or develop something new in both product and process.

"We looked at it and looked at it," remembers Strobel. "When we put it down on paper—in a matrix of cost, performance, space and risk—it was clear that the only way to meet our schedule was to adapt something that was pretty close to SMT already. We just didn't have the time to develop all the SMT components we needed for the other approaches."

The receiver they decided to adapt is called a *five-spot* design. It has both oscillator and preselector in one package—there are four spots on the crystal blank for a four-pole preselector filter and one spot for the oscillator, therefore the name. The design was more efficient in part count, eliminating about 15 components from the standard unit, and all the required integrated circuits (ICs) and support chips were already available. "We hope to fool around with those components some in the next generation," Strobel hints.

As it is, the new design, a hybrid if you will, is so compact that the battery—an AA cell—is the largest component. "We could make more progress in size reduction if we could get a better, smaller battery," comments Strobel, "but a lot of our customers cry that they don't want the pager any smaller."

Because of the tremendous number of variations of the Bandit—millions of combinations are possible—each unit is essentially a one-of-a-kind, lot-size-of-one manufacturing marvel. By the way, although we use the term "Bandit" throughout to identify the pager made in this facility, it is actually a model within the Bravo series. Other Bravo units with slightly different component counts and assembly requirements are made in the same building, but without the advantages of the automation used with Bandit.

STEALING SPACE

The Boynton Beach facility is a large, practically windowless building. The Bandit team needed space, and management found some—in a storage area. "They had a bunch of stuff—scaffolds, stored things—in this area and the space was being used but not utilized," recalls Strobel. "We had a dinky lab back in the bowels of the factory, next to the sprinkler system and fire alarms. We were in pretty close quarters at first."

Buildings don't build product, people and equipment do. The approach the Bandit bunch took was to codevelop the product and the processes, back and forth, back and forth. Again, Strobel reminisces.

"We knew we were going to have to populate the bottom of the board with about seven components. That meant we'd need three robots. We knew how fast the machines were and that dictated how many we'd need to maintain the production schedule. We also a knew the dimensions of the equipment so an industrial engineer used AutoCAD on a Compaq computer to chunk out little designs of how the process would go and how it would look. As we decided on things, the physical layout, he'd redraw the facility.

"We got lucky, I guess. We knew enough about what we wanted that we got pretty close to what we actually ended up with. It would have been nice to use simulation, to be able to make decisions based on running 'what-if' programs, but we didn't simulate—we just went out in the shop and measured it."

FOLLOW THE ROAD MAP

Strobel explained that taking measurements was easy because they have a detailed "road map" of the Bandit factory-in-a-factory posted on the wall outside the actual manufacturing facility. This is up for several reasons, not the least as a motivator to the workers. Bandit is only part of the manufacturing that goes on at this plant, and it has its own climate conditioned room separate from the other production. People not involved in Bandit come by, look through the windows and can track the operations on the "map." It also makes it easier for guides to explain the advanced systems to visitors, especially customers. Good "PR" internally and externally.

What they found was that a customer's order, placed through the Chicago office and stored in an IBM 3090 computer, is downloaded to the division's IBM 4381 where a bill of materials and shop order are generated. Everything is based on the customer input—color of housing, label, with or without a vibrating signal, frequency and other variations—and the options are translated into sequences for the eventual machine operations.

The order goes to a Stratus computer where serial numbers are scheduled and matched against the customer's order. Then to a Model 825 Hewlett-Packard RISC (reduced instruction set computing) computer that supervises five HP1000 computers used as controllers for the Bandit factory. It's here that Strobel confesses to not getting exactly what the team wanted.

"Our computer guys wanted the latest Unix box on the floor, not an older architecture like RTE (Real Time Executive). But we couldn't find a Unix system that would respond to an interrupt. Again, we went to minimizing the risk factor and to building on what we had seen work.

"We prepared a variety of benchmarks—three months from project start we'd have this, six months into it we'd have something more—and then a six month period to make everything work together. It was a phased approach," he explains.

"The very first thing we did," Strobel continues, "was to hook up a couple of robots to the computer to demonstrate they'd respond to commands. The HP1000 had to control Seiko robots in this prototyping exercise. Then we tried it with the HP1000 and a

Panacert machine, then the computer and other equipment. Once we got one cell working, we'd move on to the next."

There were cells that had been "prototyped" in actual production with previous pager models. There were also new ideas and new production that the team had had no exposure to before. According to Strobel, some areas were the toughest. "The housing assembly, putting the chassis into the housing, putting on the back cover and driving the screws, putting on the belt clip, all the mechanical assembly was pretty new to us from the standpoint of automation.

"We hadn't ever used a computer to send instructions to a robot before. That was a real reach-out in our opinion—a high-risk endeavor—but something we thought was critical to the success of a factory building in lot sizes of one."

Throughout the development of the processes, the battle cry was "Make sure you can get English instructions here and machine actions there." The major emphasis was to drive customer orders into robotic actions. But even with this practical approach, the issue of quality was never far from the surface.

"Hewlett-Packard kept trying to focus on SPC (statistical process control) systems, and coming from an engineering background, I was sympathetic," Strobel confides. "Scott (Shamlin) had been disappointed with a computer project and said, 'Don't mess with all that stuff, we can get it later.' Well, we've got it all running now.

"In retrospect, it was probably a good idea that we didn't spend a lot of time trying to figure out how to collect SPC data. Even though, in our black little hearts as computer designers, we knew we were going to have to have it—and even put in a line controller with the knowledge we'd use it to run the SPC software at the line level."

Planning for the future, while installing what was necessary for the present, is a hallmark of the Bandit facility. They found that HP's 1000 series cell control computers, the A400 models, needed more "horse-power." Instead of scrapping the controllers and starting over, they considered changing the configuration to make six cells out of five and add another computer. Strobel says, "It was easy because our software backplane allows us to add and delete cells, move functionality around, talk from computer to computer without paying any attention to which process we're trying to get to. The backplane routes it where it needs to go."

That backplane was codeveloped by Hewlett-Packard's Atlanta Project Center and allows integration of various software developments on multiple vendors' computers, transparently to the user. Using computer aided software engineering (CASE) tools, Motorola can cut its software development from years to months.

THROUGH A WINDOW BRIGHTLY

The Bandit facility is a glass-enclosed structure within the larger pager factory. This means there is no place to hide mistakes. And to make sure, bright red and yellow lights along the production line warn management and technicians alike when something goes wrong.

At the head end of the line, slightly off the actual shop floor, a monitor and control computer system operates under the careful watch of an engineer. Equipped with HP touchscreen terminals, the operator can move through a network of menus with the tip of a finger to access immediate capacity, status, operation, maintenance and other information on any or all of the machines.

Charts for SPC and simulations of where product is in the build cycle can be displayed on color monitors at this and other locations around the line. Workers use the workstation displays for assembly diagrams, troubleshooting information, test and conformance data as well as serial number tracking.

The first three cells, where the boards are stuffed with components, are replicated in Puerto Rico—but the work there is done manually. According to Shamlin, "We run these cells in parallel with our offshore manufacturing. They were the first we set up—starting at the front and working around the line seemed logical. To exercise the new line, we produced some boards for use in the Japanese market, in the standard Bravo pager. It allowed us to tweak the machinery while we were developing and installing the rest of the system."

But there were some problems sharing production facilities with actual products while developing systems for a new line. Shamlin recalls, "In the daytime, Bandit engineers worked on the line, hooking up hoses, plumbing and wiring. At night, the 280 team came in to run production (280 refers to the 280 MHz frequency band of the Japanese pagers). It was somewhat aggravating to come back the next morning and find all of the switches in different positions, reels loaded with different components. But we got some miles on the machinery to prove it out, to get some confidence that we were doing the right things."

The differences between the Bravo—a two-board receiver that is at least partially assembled manually—and the Bandit are mostly skin deep. In designing the Bandit for automated assembly, many components were changed in size or shape, SMT systems were applied and a double-sided single board was used.

Strobel explains, "Bandit is actually ultraconservative in design because we knew we had to achieve a Six-Sigma* process at the end. That's why we went to components on both sides of the board. When I joined the project, the design had all the components crammed onto one side—two mil lines, two mil spaces, custom packages that didn't even exist—and the first thing I told Shamlin was that it wasn't going to work. You can't get Six Sigma out of this state-of-the-art spacing.

"So we ended up with some parts on the bottom of the board and had to add a

---

Six Sigma is defined by Shamlin as a rate of 3.4 defects per million parts. This is an extremely high level of quality. He uses this analogy: The odds of a person getting on an airplane and reaching his or her destination alive is about 6.2 Sigma; the odds of the same person and his or her bags arriving at the same place at the same time is about 4.1 Sigma.

At the time of our visit at the end of 1988, Bandit was approaching Five Sigma, "with fairly traditional technology and approaches," according to Shamlin. He adds, "Five Sigma can be reached with hard work and smart decisions; you can't work hard enough to get to Six Sigma. You have to work at the process level to predict the next defect and then head it off with some preemptive action. That's the way you get to Six Sigma. You have to have a smarter toolbox and that's what we are developing now.

"You have to have the experience to practically smell a defect condition starting. We have to be able to inject a machine controlled or human controlled modification to prevent the condition from developing while the process is still within an acceptable range. We aren't there yet but we're further down that road than when we started. It takes a lot of compute power to predict rather than detect and we're doing some neat things in this area."

whole cell to populate the underside. It was a conservative design strategy because we took things that we knew would work."

Shamlin adds, "We tried to minimize any invention so we could save our resources to concentrate on things we knew we had to develop. We had to create processes where robots would insert components, solder leads, add switches, things that hadn't been done by robots. We even have a robot that tunes the receiver—that's never been done before."

"That's all part of the idea of overcoming the NIH problem," chimes in Strobel. "We concentrated on using good ideas wherever we could find them so we didn't have to reinvent things." Indeed, they go so far as displaying a sign with NIH surrounded by a red circle with the familiar red slash through the center.

MAKING BANDITS PAY

Automation is nice; it can make things no one even wants faster than ever before. But that's only part of the Motorola Bandit story. Using computers to drive cells, control machines and produce unique products one at a time is interesting, but becoming a familiar story. Let's dig a little deeper into the lessons learned from Bandit. Let's find out what the payback has been.

"I tell people we learned four things from Bandit," Shamlin starts. "First, create interdisciplinary teams; then reward risk taking; third is to enforce outrageous goals; and fourth, which might sound strange, is to avoid Japanese management techniques. Oh, yes, there is a fifth: we learned it can be done."

Expanding on some of those points, Shamlin continues. "The part about avoiding Japanese management techniques is important because we are two decidedly different groups of folks. Team building here is different than in Japan. In Japan, they are constantly reinforcing the team as a single entity; in the US, the team is composed of individuals and their roles are emphasized. Both are teams, but the thought of using Japanese management techniques with a team of US individuals doesn't work very well.

"I think we haven't spent enough time and effort considering the strengths we have as a nation, as a culture. In fact, maybe too much time is spent downgrading some cultural aspects of our society and not enough on how we can use them as competitive weapons."

Strobel adds to the discussion on team building by saying, "In the beginning, we didn't know what we couldn't do.

"What we tried was clearly a generation ahead of anybody else—to drive the factory by a bill of materials and produce in lots of one. These things were talked about, but no one had implemented them.

"We were always up-front with our supplier team. We told them what we were up to and what we needed—or thought we needed—to push the envelope a little further. We admitted we didn't know exactly how to do it or whether anyone else had ever done it before. We asked for some unusual accommodations from them and told them that when it was successful, they would share in the success," Strobel goes on.

"We had always heard that it took a young, entrepreneurial company to react quickly, to move back and forth easily. What we found was we got as much entrepreneurial spirit and flexibility from large, well-established companies as from the smaller ones. It was clearly a matter of picking the right companies."

Strobel then relates a story, which sounds vaguely familiar to every manager implementing a new system. "There was one company that, up-front, had us spending more time with their attorneys than with their engineers. That was a signal to us that we had picked the wrong people to work with."

Shamlin picks up on the story, "We were sharing information with our suppliers that was proprietary and they in turn were doing the same with us. We were moving so fast, we were forced into it. We began to lose vendors who weren't willing to participate at our speed, by our rules."

SPEED TRAP

Shamlin, who worked for many years in the NASA space program, puts it in that perspective. "When we were trying to put vehicles on the moon, the old guys would say, 'If you don't like our schedules, talk to God. He's the one who put the planets in their rotation. We have no choice.'

"With Bandit, too, the schedule was our religion. Any supplier candidate who couldn't operate in that environment dropped out."

Meeting schedules is perhaps the hardest part of managing a manufacturing—or any other—operation. At Motorola, the schedule took on a life of its own. Strobel: "We weren't going to miss any of those major prototype dates. When those cells were to be on-line, they would be on-line. When the product had to be in a certain position, it would be. The equipment had to be, the computers had to be, everything had to be. The date was the date and we never missed one."

"Well, we did miss one or two," Shamlin interjects. "But we made them up within a day or two and by the next date we were on schedule."

Throughout the project, quality of the product and the process was important. With a strong team of suppliers on the outside, a strong team of engineers and designers on the inside, Bandit moved along at scheduled speed. At no time was Shamlin or Strobel willing to compromise either quality or schedule.

"Think about it realistically. There is no such thing as slipping a little," Shamlin argues. "There's a mentality that is absolute poison. It says a little bit more won't hurt. Of course it hurts.

"The way you attack a problem really makes a difference," he goes on, getting spirited. "When you're willing to accept less than what you said you were going to accept, that's a matter of integrity. If you build a team where you authorize a disruption in what you expected, that's a much bigger issue than just letting the schedule slip for a day or two. You just won't be able to keep the competition at bay with that type of attitude."

Dealing with suppliers and making them stick to your schedules can be frustrating. In one case, Shamlin got a taste of his own medicine from one of his suppliers. Nancy Ewing of Hewlett-Packard: "Shamlin and Strobel were always saying they wanted partners to provide expertise in various areas. So we approached the project as partners. That meant a two-way street."

"I hate to admit it," Shamlin admits, hating it all the way, "but Nancy would come in and give me performance reviews. She would actually chew me out because we'd make decisions contrary to the stated plan. She was right and we modified our behavior."

MOVING ON AND MOVING UP

Bandit is fact. The Bandit facility is turning out pagers at the planned rate, with the quality expected at this stage of the project. More effort is around the corner, as the program heads for that goal of Six Sigma.

The Bandit team has become something of a cause celebre within Motorola. And that *wasn't* one of the goals. As Shamlin puts it, "We didn't want to build a group of superstars who wanted to get all the hot projects. We built the team to re-establish the level of expectation, put it together to develop a strategy and then dispersed them like seeds that will hopefully have a multiplying effect.

"We're doing that now," he concludes. "Some guys will be involved in the next generation programs, some in propagating the technology we have now, some will continue operating Bandit. We hope that by the end of 1989 we'll be able to point at some who have broken off on their own. Breaking up a team has both positive and negative connotations. Hopefully, this time it's all positive."

Strobel wraps it up by saying, "If we had had 100 percent success at every stage, it would mean we didn't define the project the way our charter was written—to go where no man has gone before. If you go where you think no one has been before and you find beer cans, then you didn't exactly set your criteria properly."

Chapter Two

Forecasting

INTRODUCTION

Many business decisions depend on some sort of forecasting. For example, accountants rely on forecasts of costs and revenues for tax planning; human resource personnel need forecasts for recruiting; marketing teams require forecasts for establishing promotional budgets; financial planners need forecasts for managing cash flows; and production planners require forecasts so that capacity plans, inventory levels, and shop floor planning activities can be established. In addition, forecasts can also measure the variability in demand during lead time that in turn can help carry proper safety stock levels. Appropriate safety stock inventory levels could minimize overall carrying and stockout costs associated with these items. Determination of safety stock levels are covered in inventory management chapters. Trade-offs between more accurate forecasts and the cost associated with obtaining them are covered in a later section of this chapter. This chapter provides an appreciation of the value of forecasting, an understanding of some of the major techniques, and insight into the selection of an appropriate technique for a given situation. This knowledge is necessary for designing a successful materials flow system.

Forecasts are simply statements about the future. Good forecasts can be quite valuable; it would be worth a great deal, for example, to receive tomorrow's *Wall Street Journal* today. Not all forecasts are useful, however. For example, some people have used AFC Super Bowl victories to predict a poor year for the Dow Jones stock averages and NFC victories to predict a good year. This is nonsense, of course. It is clear that we must distinguish between forecasts per se and *good* forecasts.

Forecasting is the art of specifying meaningful information about the future. Long-run planning decisions require consideration of many factors: general economic conditions, industry trends, probable competitor actions, overall political climate, and so on.

Extrinsic forecasts are formulated on external associations—for example, between sales of appliances and disposable personal income or between house sales and mortgage availability. For financial planning, companies need extrinsic forecasts of the year's aggregate sales by product line. These aggregate forecasts are of little use in production planning, however, in which we must plan production quantities of every item in the product line. In some companies with as few as five major product lines, there may be 10,000 individual items to be forecast. Such *item forecasting* requires routine projections of past data. This chapter presents sophisticated methods of projecting data, methods analogous to the old pencil and ruler trick. Such projection forecasts are usually termed *intrinsic forecasts*. Hence, this chapter is about item or intrinsic forecasting.

Item forecasts are required only for end items. Demand for car doors in an automobile assembly plant, for example, can be derived simply by doubling or "exploding" the end item forecast for two-door cars, which is known as dependent demand. The dependent demand is the subject of Chapter 11.

Here we address forecasting approaches for individual items. Chapter 3 deals with forecasting for family of items and other miscellaneous forecasting topics such as slow-moving items.

Demand Management

The purpose of demand management is to coordinate and control all the sources of demand so that production and operations systems can be utilized efficiently. In addition, the right quantity and quality of products will be delivered to customers on time and branch warehouse requirements, interplant shipments, and service parts needs will be met. A firm can take an active role to influence the demand. For example, a firm can increase incentives to the sales force or can wage promotional campaigns to sell more products. On the other hand, demand can be reduced simply through price increases and reduced sales efforts [4]. A firm can also take a passive role and simply respond to the actual demand by projecting past demand patterns to forecast future needs. Our primary interest in this chapter is forecasting for independent items.

Forecasting Period

Forecasts are often classified according to time period and use. In general, short-term (up to one year) forecasts guide current operations. Medium-term (one to three years) and long-term (over five years) forecasts support decisions on plant location and capacity. The item forecasts discussed in this chapter are aimed at the ordering or production lead time, normally a matter of weeks or months. Typically, we want to know the average demand during lead time for inventory control purposes.

Forecast Accuracy

Forecasts are never perfect. Because we are basically dealing with methods that project from past data, our forecasts will be less reliable the further into the future we predict. Causal or explanatory models will generally give the greatest accuracy, especially in predicting turning points, but they do so at a great cost in computing time and data storage. For a product group or family, we can probably get reasonably good forecasts by using explanatory models. A product group forecast will also be more accurate than a single-item forecast, as it is easier to forecast a group of items than a single item.

To appreciate fully the greater accuracy of product group forecasts, we will take a closer look. Suppose that we have ten items in a group or family. All items have the same standard deviation ($\sigma = 2{,}000$) but different means (μ_i, $i = 1, \ldots , 10$). The group mean is $\sum_{i=1}^{10} \mu_i$. Although the demands for individual items in a family might be interdependent, assume for simplicity that they are independent. Then the standard deviation of the group demand would be $\sqrt{10}\,\sigma$, because the variance of the sum is the sum of the variances. For our example, suppose that $\mu_i = 40{,}000$ and $\sum_{i=1}^{10} \mu_i = 400{,}000$. Now examine the coefficient of variation, which is the ratio of a variable's standard deviation to its mean. For the first item, the coefficient of variation is $2000/40{,}000 = 0.05$. The coefficient of variation for the family is $(\sqrt{10} * 2000)/400{,}000 = 0.016$.

Forecasting Approaches

Forecasting approaches vary with the number of items to be forecast and the dollar importance of the decisions. Decisions regarding plant capacity and plant location can be made with long-term, aggregate forecasts, and we would probably be willing to expend a substantial amount of money and computer time to gain accuracy. Economic order quantity (EOQ) decisions for low-value items are based on short-term, individual item forecasts for which we do not want to spend much money.

TABLE 2.1   Forecasting Model Selection According to Problem Type

	Few Time Series, Costly Decisions	*Thousands of Series, Routine Decisions*
Large amount of past data available	Box-Jenkins [1] Econometrics [21]	Exponential smoothing [15] Moving averages
Little past data available	Delphi method [5] Market surveys [5]	Bayes methods [12]

Table 2.1 summarizes the relationship of approaches to problem types. Econometric models, Box-Jenkins methods, and market surveys are expensive but relatively accurate at the aggregate level. When the consequences of the decision, such as new product planning or building a new facility, are costly, these approaches can be justified. Bayes' method is suited when little or no past data are available. It is highly unlikely that a product is launched and produced without a forecast, and hence we will not deal with Bayes methods here. A typical forecasting problem in a production-inventory system, however, includes thousands of individual items. For such problems, the moving averages, exponential smoothing methods and linear regression analysis presented in this chapter have a distinct advantage in computational ease, data storage requirements, and cost.

For production planning purposes, what we consider a "good" forecasting system has the following characteristics:

1. Accuracy
2. Low computer time requirements
3. Low computer storage requirements
4. Low dollar cost of software purchase or development
5. On-line capabilities
6. Ability to link into an existing database management system

Cost versus Accuracy

Two important factors in forecasting are cost and accuracy. In general, the higher the need for accuracy translates to higher costs of developing forecasting models. Therefore, the right questions to ask are: How much money and manpower is budgeted for developing forecasts? What possible benefits are accrued from accurate forecasts? What are possible costs of inaccurate forecasts? The best forecasts are not necessarily the most accurate or the least costly. Factors such as purpose and data availability play important roles in determining the desired accuracy of forecasts. In many instances, very simple forecast models will provide adequate forecasts, and it is difficult even for complex models to improve the accuracy of the forecast significantly [4].

Components of Demand

For systematic analysis of historical data, analysts often use time series analysis. Typically, the production inventory analyst views demand as composed of a base or central tendency, a trend, seasonal variation, cyclical (business cycle) variation, and random variation (noise).

COMMON TIME SERIES
FORECASTING MODELS

The most common and relatively easiest methods for developing a forecast from past data are simple moving averages, weighted moving averages, exponential smoothing, and regression analysis. The calculations in all these methods can be done with a desk calculator or microcomputer.

Simple Moving Average

A moving average is obtained by averaging the demand data from several of the most recent periods. When the demand data do not have rapid growth or seasonal characteristics, the technique can be useful in removing random fluctuations for forecasting. As n (the number of observations to be included in the moving average) increases, the model tends to smooth or dampen out noise. As n grows larger, however, more data are included, and the model becomes less responsive to changes in sales patterns. A simple n-period moving average is defined as follows:

$$\text{Moving average (MA)} = \frac{\text{sum of old demand for last } n \text{ periods}}{\text{number of periods used in the model}}$$

$$= \frac{\sum_{j=1}^{n} D_{t-j+1}}{n} \tag{2.1}$$

$$= \frac{D_t + D_{t-1} + D_{t-2} + \dots + D_{t-n+1}}{n}$$

where t is the index of the current period, j is a general index, and D_j is the demand during period j.

The average moves over time. After each period has elapsed, the demand for the oldest period is removed and the demand for the newest period is added to the next calculation:

$$\text{MA}_t = \text{MA}_{t-1} + \frac{D_t - D_{t-n}}{n} \tag{2.2}$$

Example 2.1

Table 2.2 presents twelve-month demand data for exhaust pipes. Forecasts using three-month and six-month moving averages are also exhibited. The actual demand is quite variable. The three-month moving average, however, is much more stable because

TABLE 2.2 Three- and Six-Month Moving Averages Used as Forecasts

Month	Demand (D_t)	Three-Month Moving Average Forecast* (f_t)	Three-Month Moving Average (MA_t)	Six-Month Moving Average Forecast (f_t)	Six-Month Moving Average (MA_t)
January	450	–	–	–	–
February	440	–	–	–	–
March	460	–	450	–	–
April	510	450	470	–	–
May	520	470	497	–	–
June	495	497	508	–	479
July	475	508	497	479	483
August	560	497	510	483	503
September	510	510	515	503	512
October	520	515	530	512	513
November	540	530	523	513	517
December	550	523	537	517	526

Note: The average at time t becomes a forecast for time $t+1$.

*Using f_t as the forecast for period t, f_t is set equal to the most recently calculated moving average, $f_t = ma_{t-1}$.

the demand for any one month receives only one-third weight. The larger the value of n, the greater the dampening effect.

For the three-month moving average in Table 2.2,

$$\text{MA}_4 = \text{MA}_3 + \frac{D_4 - D_{4-3}}{3}$$

$$= 450 + \frac{510 - 450}{3}$$

$$= 470$$

Weighted Moving Average

The moving average gives equal weight to each observation of past demand used in the average. Sometimes the forecaster wishes to use a moving average but does not want all n periods weighted equally. A weighted moving average allows any desired weights to be placed on old demand. An n-period weighted moving average is defined as follows:

$$\text{Weighted moving average (WMA)} = \sum_{t=1}^{n} C_t D_t \qquad (2.3)$$

where

$$0 \leq C_t \leq 1$$

that is, C_t is a fraction used as a weight for period t, and

$$\sum_{t=1}^{n} C_t = 1$$

In general, more weight is given to the most recent demand and hence the weighted moving average model discounts the value of past information. Thus the forecast tends to be more responsive to genuine changes in demand.

Example 2.2

If n is three periods, we can assign the following weights for the data is Table 2.2: $C_1 = 0.25$, $C_2 = 0.25$, and $C_3 = 0.50$. Thus the weight of 0.50 applies to the most recent observation. We then obtain

$$\text{WMA} = (450 * 0.25) + (440 * 0.25) + (460 * 0.50)$$

$$= 452.5$$

A comparison of three-month moving average and weighted moving average forecasts is given in Table 2.3.

Simple Exponential Smoothing

We begin with a very simple demand process $D_t = \mu + \varepsilon_t$, where ε_t is normally distributed, with mean zero. We would like a model capable of forecasting this process, even when we have an occasional shift in μ, the central tendency. With no shifts, this formula reflects random error around a stable central tendency.

Simple exponential smoothing is a special type of averaging technique that is suitable for forecasting this process. Indeed, Muth [14] showed that the exponential forecast is optimal for such a demand process.

The equation for simple exponential smoothing uses only two pieces of information: (1) actual demand for the most recent period and (2) the most recent forecast. At the end of each period, a new forecast is made. Thus

New exponential average = old exponential average +
fraction (current demand − forecast)

Using the exponential average in one period as a forecast for the next period, we have a process for revising the average upward or downward, depending on forecast error. Current demand minus forecast gives forecast error. Demand that is higher than forecast causes us to revise the average upward, and demand that is lower than forecast causes a downward revision. Simple exponential smoothing has the following equation:

$$F_t = F_{t-1} + \alpha(D_t - F_{t-1})$$

TABLE 2.3   Forecast Comparisons Using Moving Average and Weighted Moving Average

Month	Demand (D_t)	Three-Month Moving Average Forecast (f_t)	Three-Month Moving Average (MA_t)	Three-Month Weighted Moving Average Forecast (f_t)	Three-Month Weighted Moving Average (0.25, 0.25, 0.50), Most Recent (MA_t)
January	450	–	–	–	–
February	440	–	–	–	–
March	460	–	450	–	453
April	510	450	470	453	480
May	520	570	497	480	503
June	495	497	508	503	505
July	475	508	497	505	491
August	560	497	510	491	523
September	510	510	515	523	514
October	520	515	530	514	528
November	540	530	523	528	528
December	550	523	537	528	540

where F_t is the exponential average at time t, D_t is the actual demand in period t, and α is a smoothing constant between zero and one. If we use F_{t-1} as a forecast for D_t, then the error team e_t would be

$$e_t = D_t - F_{t-1}$$

For example, suppose that demand is 100 and that the old average was 90. Using the old average as a forecast, the error term would be $100 - 90 = 10$. Hence, we would revise the average upward a little. If the smoothing factor is 0.2, the new average will be 92, determined as follows:

$$F_t = F_{t-1} + \alpha(D_t - F_{t-1})$$
$$= 90 + 0.2(100 - 90) = 92$$

Slightly rearranging the equation, we obtain the usual form:

$$F_t = \alpha D_t + (1 - \alpha)F_{t-1} \qquad (2.4)$$
$$= 0.2(100) + 0.8(90) = 92$$

Because the simple exponential average will not always be our forecast, we must emphasize the general nature of our revision process. Letting f_t be the forecast of period t sales, we have

$$f_t = F_{t-1}$$

and

$$F_t = \alpha D_t + (1 - \alpha)f_t$$

From equation (2.4), we can see that F_t implicitly captures all past data, even though we use only D_t and F_{t-1} in the calculation at period t. Consider the sequence of forecasts where $t =$ December, and use the data from Table 2.4:

$$F_{t-1} = \alpha D_{t-1} + (1 - \alpha)F_{t-2} = (0.2(540) + 0.8(507)$$
$$F_t = \alpha D_t + (1 - \alpha)F_{t-1} = 0.2(550) + (0.8(514)$$

or

$$F_t = \alpha D_t + (1 - \alpha)[\alpha D_{t-1} + (1 - \alpha)F_{t-2}]$$
$$= 0.2(550) + 0.8[0.2(540) + 0.8(507)]$$

or

$$F_t = \alpha D_t + (1 - \alpha)\alpha D_{t-1} + (1 - \alpha)^2 F_{t-2}$$
$$= 0.2(550) + 0.16(540) + 0.64(507)$$

TABLE 2.4 Simple Exponential Smoothing Forecast

Month	Actual Demand (D_t)	Forecast (f_t)	Old Average (F_{t-1})	New Average (F_t)	Weights[a]
March	460	480	480.00	476.00	0.027
April	510	476	476.00	482.80	0.034
May	520	483	482.80	490.24	0.042
June	495	490	490.24	491.19	0.052
July	475	491	491.19	487.95	0.066
August	560	488	487.95	502.36	0.082
September	510	502	502.36	503.89	0.102
October	520	504	503.89	507.11	0.128
November	540	507	507.11	513.69	0.160
December	550	514	513.69	520.95	0.200

[a]At the end of December, F_{DEC} implicitly applies these weights to the sales from March through December. To see this, calculate $F_{DEC} = 0.2(550) + 0.16(540) + 0.128(520) + \ldots + 0.027(460)$.

In general,

$$F_t = \alpha \sum_{k=0}^{t-1} (1-\alpha)^k D_{t-k} + (1-\alpha)^t F_0$$

where F_0 *is the initial estimate of* μ. For t sufficiently large, $(1 - \alpha)^t$ would approach zero, since α is a fraction. Hence, the initial forecast gets washed out.

Example 2.3

A firm uses simple exponential smoothing, with $\alpha = 0.2$ to forecast demand. The forecast for the month of March was 500 units, whereas the actual demand turned out to be 460 units.

1. Forecast the demand for the month of April.
2. Assume that the actual demand for the month of April turned out to be 480 units. Forecast the demand for the month of May. Continue forecasting for the rest of the year, assuming the subsequent demand as displayed in Table 2.4.
3. Assume that the current average is used as a forecast for the next month.

Solution:

$$f_{APRIL} = F_t = F_{t-1} + \alpha(D_t - F_{t-1})$$
$$= 500 + 0.2 * (460 - 500) = 492 \text{ units}$$
$$f_{MAY} = 492 + 0.2(480 - 492) = 489.6 \text{ units}$$

Selection of the Smoothing Constant

Higher values of the smoothing constant give higher responsiveness to both random fluctuations and shifts in the underlying process. A stable central tendency with large random fluctuation requires a low smoothing constant. A high smoothing constant is more appropriate for small random fluctuations around a somewhat unstable central tendency.

A higher value of the smoothing constant corresponds to fewer months in a moving average. The average age of the data used in a forecasting system can be calculated as $\bar{k} = 0 * C_t + 1 * C_{t-1} + 2 * C_{t-2} + \ldots + n * C_{t-n}$, where C_t is the weight assigned to the data from period t. Brown [2, p. 107] has shown that the average age of the data in a moving average is $(n - 1)/2$. In an exponentially smoothed average, he finds an average age of $(1 - \alpha)/\alpha$. Hence, choosing smoothing constants to give the same average age of the data would produce $\alpha = 2/(n + 1)]$ or $n = (2 - \alpha)/\alpha$.

As a general rule, the smoothing constant for a constant model should be between 0.01 and 0.3. How do we decide on a value for each of the one thousand items to be forecast? First, graphs of sample items are absolutely indispensable. Second, we use trial and error. If we have four years of available data, we might use the first three years as an analysis sample and the last year as a holdout or "trial future" sample. We try various values of α on the analysis sample and choose the one that minimizes a measure, such as sum of squared error. We then forecast the trial future period by period to check out how our system will respond to fresh data.

Winters Trend Model

Winters [20] developed a very popular model for handling both trends and seasons. For explanatory purposes, we will demonstrate his trend calculations first and then add his seasonal factors in the next section. Winters used the Holt trend model, which begins with the usual trend estimation:

$$T_t = \beta(F_t - F_{t-1}) + (1 - \beta)T_{t-1} \tag{2.6}$$

where β is a fraction, T_t is the trend estimate at time t, and F_t is the exponential average at time t. Updating the exponential average requires that we recognize that a forecast now involves the exponential average plus a trend:

$$f_t = F_{t-1} + T_{t-1} \tag{2.7}$$

With this in mind, we can recall our general version of simple exponential smoothing:

$$F_t = \alpha D_t + (1 - \alpha)f_t$$

Substitution for f_t gives

$$F_t = \alpha D_t + (1 - \alpha)(F_{t-1} + T_{t-1}) \tag{2.8}$$

Thus we see that the trend factor becomes part of the old average to be smoothed. This means that F_t no longer behaves as a simple exponentially smoothed average.

Therefore, as a first step, we find F_t using equation (2.8). The second step involves the computation of T_t given by equation (2.6). Finally, we find the trend-adjusted forecast using equation (2.7).

The forecast made at the end of the period t for period $t + 1$ would be

$$f_{t+1} = F_t + T_t$$

Example 2.4

For the data in Example 2.1, calculate forecasts with $\alpha = 0.2$, $\beta = 0.2$, $T_0 = 9$, and $F_0 = 480$ (see Table 2.5, and begin March):

$$F_1 = 0.2(460) + 0.8(480 + 9) = 483.2$$
$$T_1 = 0.2(483.2 - 480) + 0.8(9) = 7.84$$

Seasonally Adjusted Exponential Smoothing

Seasonal demand patterns are characteristic of many demand series, reflecting the Christmas season, the summer doldrums, and the like. It is possible that the seasonal effect could be additive; for example, regardless of the weekly sales rate for the rest of the year, Christmas season sales could be 200 units per week higher. The seasonal effect can also be multiplicative, such as a Christmas weekly sales rate that is double the prevailing weekly rate for the rest of the year.

We can then consider a seasonal demand generation process:

$$D_t = \mu * \delta_t + \varepsilon_t$$

TABLE 2.5   Winters Trend Model

Month	Demand (D_t)	Simple Exponential Average[a]	Winters Exponential Average (F_t)	Trend (T_t)	Winters Forecast (f_t)
			480.00	9.00	
March	460	476.00	483.20	7.84	489.00
April	510	482.80	494.83	8.60	491.04
May	520	490.24	506.74	9.26	503.43
June	495	491.19	511.80	8.42	516.00
July	475	487.95	511.18	6.61	520.22
August	560	502.36	526.23	8.30	517.79
September	510	503.89	529.62	7.32	534.53
October	520	507.11	533.55	6.64	536.94
November	540	513.69	540.16	6.63	540.20
December	550	520.95	547.43	6.76	546.79
January	555	527.76	554.35	6.79	554.19
February	569	536.01	562.72	7.11	561.15

Note: $\alpha = 0.2$; $\beta = 0.2$; $T_0 = 9$; $F_0 = 480$.

[a]Given for comparison purposes. Note how the simple exponential average lags the upward trend.

where μ is the permanent or base sales level and δ_t is a seasonal factor. For example, $\delta_2 = 1.2$ indicates that demand in period 2 is 20% higher than the base level for the year.

Since the simple exponential smoothing average estimates μ, we could estimate δ_t by an index, $I_t = D_t/F_t$. If the simple exponential average is 200 and sales for period 1 were 230, we would calculate I_1 as 1.15. Then next year, with a simple exponential average of 250, we would estimate period 1 sales at $I_1 {}^* F_{t-1} = 1.15 * 250 = 288$.

The seasonal factors allow us to convert back and forth between period sales and the exponential average. If period 1 sales came in at 300, we could deseasonalize these to $300/1.15 = 261$ and use this deseasonalized figure to update the exponential average. Surely the old average of 250 is too low if actual sales were 300 and our seasonalized forecast was 288. The model then becomes

$$F_t = \alpha \frac{D_t}{I_{t-m}} + (1-\alpha)F_{t-1}$$

$$F_t = 0.2\left(\frac{300}{1.15}\right) + (0.8)250 = 252 \tag{2.9}$$

$$= 0.2(261) + 0.8(250) = 252$$

where I_{t-m} is the index calculated $m = 12$ months ago for monthly forecasts or $m = 52$ weeks ago for weekly forecasts. Once we have a new exponential average, we can update the seasonal factor. Winters used

$$I_t = \gamma \frac{D_t}{F_t} + (1-\gamma)I_{t-m} \tag{2.10}$$

where γ is a smoothing constant, preferably set at $\gamma \leqslant 0.05$. Finally, the forecast made at the end of period t for period $t + 1$ becomes

$$f_{t+1} = F_t {}^* I_{t+1-m} \tag{2.11}$$

Further, the full-blown Winters model includes the trend method in the previous section. In that case, F_t includes a smoothed trend factor. The revised equations would be

$$F_t = \alpha \frac{D_t}{I_{t-m}} + (1-\alpha)(F_{t-1} + T_{t-1}) \tag{2.12}$$

$$T_t = \beta(F_t - F_{t-1}) + (1-\beta)T_{t-1} \tag{2.13}$$

$$I_t = \gamma \frac{D_t}{F_t} + (1-\gamma)I_{t-m} \tag{2.14}$$

$$f_{t+1} = (F_t + T_t) {}^* I_{t+1-m} \tag{2.15}$$

TABLE 2.6   Sample Seasonal Index Computation

Month	Demand 1993	Demand 1994	Average Demand[a]	Seasonal Index (I_t)
January	80	100	90	0.957
February	75	85	80	0.851
March	80	90	85	0.904
April	90	110	100	1.064
May	115	131	123	1.309
June	110	120	115	1.223
July	100	110	105	1.117
August	90	110	100	1.064
September	85	95	90	0.957
October	75	85	80	0.851
November	75	85	80	0.851
December	80	80	80	0.851

[a]Average monthly demand: 1128/12 = 94.

For simplicity in presentation, we will use only seasonal components in the following example and will ignore trend calculations. The sample computation in Table 2.6 indicates how initial indices can be calculated from two years of past monthly data. Calculate the average demand for each month and then average these to get average monthly demand. Then divide each month's average by average monthly demand to get initial seasonal factors.

Table 2.7 indicates the forecast computations as we move into the year 1995 with

TABLE 2.7   Computation of Seasonalized Forecasts

Month	Demand (D_t) 1995	Deseasonalized Demand $(D_t I_{t-12})$	Average (F_t) $F_0 = 94$	Forecast (f_t)	Old Seasonal Factor (I_{t-12})	New Seasonal Factor (I_t)
January	95	99.27	94.50	89.96	0.957	0.959
February	75	88.13	93.86	80.42	0.851	0.848
March	90	99.56	94.43	84.85	0.904	etc.
April	105	98.68	94.86	100.47	1.064	
May	120	91.67	94.54	124.17	1.309	
June	117	95.67	94.65	115.62	1.223	
July	102	91.32	94.32	105.72	1.117	
August	98	92.11	94.10	100.36	1.064	
September	95	99.27	94.62	90.05	0.957	
October	75	88.13	93.97	80.52	0.851	
November	85	99.88	94.56	79.97	0.851	
December	75	88.13	93.91	80.47	0.851	

Note: $\alpha = 0.1$; $\gamma = 0.05$; $F_{DEC} = 94$; $f_t = f_{t-1} * I_{t-12}$.

our initial seasonal factors. Set $\alpha = 0.1$, $\gamma = 0.05$ and $F_{DEC} = 94$, the average monthly demand. Then the forecast for January would be

$$f_{JAN} = F_{DEC} * I_{JAN-12} = 94 * 0.957 = 90$$

Using equations (2.9) and (2.10) to update the exponential average and the indices and using Table 2.7 for the demand, we have

$$F_{JAN} = \alpha\left(\frac{D_{JAN}}{I_{JAN-12}}\right) + (1-\alpha)F_{DEC}$$

$$= 0.1\left(\frac{95}{0.957}\right) + 0.9 * 9.4 = 94.5$$

$$I_{JAN} = \gamma\left(\frac{D_{JAN}}{F_{JAN}}\right) + (1-\gamma)I_{JAN-12}$$

$$= 0.05\left(\frac{95}{94.5}\right) + 0.95 * 0.957 = 0.96$$

and

$$f_{FEB} = F_{JAN} * I_{FEB-12} = 94.5 * 0.851 = 80.4$$

Having studied the Winters linear trend and his seasonal factors, we can now combine them in one model, usually referred to simply as the Winters model.

The Complete Winters Model

Given the following data

	1992	*1993*	*1994*
Quarter 1	146	192	272
Quarter 2	96	127	155
Quarter 3	59	79	98
Quarter 4	133	186	219

we will develop forecasts for 1995 using the complete Winters model, with $\alpha = 0.2$, $\beta = 0.1$, and $\gamma = 0.05$.

As developed earlier, the complete model has four main formulas:

$$F_t = \alpha\left(\frac{D_t}{I_{t-m}}\right) + (1-\alpha)(F_{t-1} + T_{t-1}) \qquad (2.12)$$

$$T_t = \beta(F_t - F_{t-1}) + (1-\beta)T_{t-1} \qquad (2.13)$$

and

$$I_t = \gamma\left(\frac{D_t}{F_t}\right) + (1-\gamma)I_{t-m} \tag{2.14}$$

$$f_{t+1} = (F_t + T_t) * I_{t+1-m} \tag{2.15}$$

To being, we see that

$$F_1 = 0.2\left(\frac{146}{?}\right) + 0.8(?+?)$$

involves the unknown values of F_0, T_0, and I_{-3}. What can we do? We need initial values for the intercept and the slope of a trend line (see Figure 2.1).

There are two ways to get these initial estimates. We could regress demand against time, or we could make simple rough estimates. To develop rough estimates, we again use the difference in average sales ($\overline{D}$) between the two years:

$$\overline{D}_{1992} = 108.5$$
$$\overline{D}_{1993} = 146.0$$

Hence, the yearly upward trend is 37.5. On a quarterly basis, the trend would be 9.38 = 37.5/4.

We know, then, that

$$D_t = F_0 + 9.38(\text{quarter number})$$

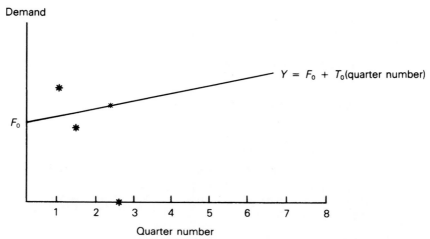

*2.5 = balance point for year. In the case of monthly data, 6.5 is the balance point for the year with six months above and six months below the balance point.

FIGURE 2.1   Simple line fit to demand.

Further, we know that the average for 1992 sales sits right at the midpoint of the year:

$$108.5 = F_0 + 9.38(2.5)(2 \text{ discrete quarters below and 2 above 2.5})$$

or

$$F_0 = 85.05$$

In general,

$$F_0 = \overline{D} - T_0(2.5) \quad \text{for quarterly data}$$

and

$$F_0 = \overline{D} - T_0(6.5) \quad \text{for monthly data.}$$

Using our rough equations, we can now fill in our trend line estimates of sales:

Trend Line Sales Estimates

	1992	*1993*
Q1	94.43	131.95
Q2	103.81	141.33
Q3	113.19	150.71
Q4	122.57	160.09

for example,

1992Q1:   94.43 = 85.05 + 9.38

1993Q2:   103.81 = 85.05 + 2(9.38)

from these trend line estimates, we can develop initial seasonal indices.

$$\text{Index} = \frac{\text{demand}}{\text{trend line estimate}}$$

Seasonal Index Estimates

	1992	*1993*	*Average*
Q1	1.55	1.46	1.51 = (1.55 + 1.46)/2
Q2	0.92	0.90	0.91
Q3	0.52	0.52	0.52
Q4	1.09	1.16	1.13
			4.07

For example,

$$1992Q1: \quad 1.55 = 146/94.43$$

$$1993Q2: \quad 0.92 = 96/103.81$$

These indices add up to 4.07, however, where we would expect them to add to 4. To correct them, we multiply through by 4/4.07:

	Average		*Initial Index*		I_{t-4}		
	$1.51 * (4/4.07)$	$=$	1.48	$= I_{-3}$	I_{1-4}	$=$	I_{-3}
	$0.91 * (4/4.07)$	$=$	0.89	$= I_{-2}$	I_{2-4}	$=$	I_{-2}
	$0.52 * (4/4.07)$	$=$	0.51	$= I_{-1}$	I_{3-4}	$=$	I_{-1}
	$\underline{1.13} * (4/4.07)$	$=$	$\underline{1.11}$	$= I_0$	I_{4-4}	$=$	I_0
	$4.07 * (4/4.07)$	$=$	4.00				

Now we are ready to proceed with Winters's updating equations:

$$F_1 = 0.2\left(\frac{146}{1.48}\right) + 0.8(85.05) + 9.38 = 95.27$$

$$T_1 = 0.1(95.27 - 85.05) + 0.9(9.38) = 9.46$$

$$f_2 = (95.27 + 9.46) * 0.89 = 93.21$$

$$I_1 = 0.05\left(\frac{146}{95.27}\right) + 0.95(1.48) = 1.48$$

$$F_2 = 0.2\left(\frac{96}{0.89}\right) + 0.8(95.27 + 9.46) = 105.36$$

$$T_2 = 0.1(105.36 - 95.27) + 0.9(9.46) = 9.52$$

$$f_3 = (105.36 + 9.52) * 0.51 = 58.59$$

$$I_2 = 0.05\left(\frac{96}{105.36}\right) + 0.95(0.89) = 0.89$$

Continuing in this fashion, we eventually reach F_{12}, T_{12}, and I_9, I_{10}, I_{11}, and I_{12}. Rather than working this out completely by hand, it would be wise to write a simple program. Such an exercise will seal your understanding.

$$F_{12} = 199.92 \qquad I_9 = 1.49$$
$$T_{12} = 9.48 \qquad\quad I_{10} = 0.90$$
$$\phantom{T_{12} = 9.48 \qquad\quad} I_{11} = 0.52$$
$$\phantom{T_{12} = 9.48 \qquad\quad} I_{12} = 1.11$$

To forecast 1994, we then have

1994Q1: $f_{13} = (199.92 + 9.48) * 1.49$ $= 312$

1994Q2: $f_{14} = [199.92 + (2 * 9.48)] * 0.90 = 197$

1994Q3: $f_{15} = [199.92 + (3 * 9.48)] * 0.52 = 119$

1994Q4: $f_{16} = [199.92 + (4 * 9.48)] * 1.11 = 264$

In terms of the actual application of the Winters model, McClain [10] gives us some cause for concern. First, he demonstrates that disturbances picked up in the seasonal factors are long-lived, since these factors are smoothed only once per season. Accordingly, McClain recommends using only very small values of the smoothing constant γ and feels that even zero would be reasonable. Second, he indicates that the choice of α and β must be made very carefully. If values of β are too large relative to α, oscillatory behavior results. The situation for seasonal factors was not studied, but oscillations occurred in the trend model for $\alpha < 4\beta/(1 + \beta)^2$. Since we normally start with a value of α, it is appropriate to give an equation specifying the largest value of β allowed. To prevent oscillations, we must have

$$\beta \leqslant \left(\frac{2}{\alpha} - 1\right) - \sqrt{\left(\frac{2}{\alpha} - 1\right)^2 - 1}$$

Example 2.5

If we want $\alpha = 0.3$ to create a responsive system, we would need $\beta \leqslant 0.089$ to prevent oscillations. In some applications, we have used $\alpha = 0.2$, $\beta = 0.05$, and $\gamma = 0.01$ to cover several hundred demand series. McLeavey implemented a successful forecasting system in which he specified β by making an equality out of the foregoing formula. In this system, one needs to choose only α because β is determined by formula from α. This produces automatic attenuation or dampening of the trend factor.

MONITORING THE FORECASTING SYSTEM

Any forecasting system needs to be regularly monitored for error magnitude and bias. Reasonable errors are to be expected, but any forecaster dreads bias. In this section, we discuss techniques for monitoring forecasts.

Example 2.6

One company had the misfortune to use a computerized forecasting system that continually provided examples of bias and unreasonably large errors. Although it was installed by a well-known consulting firm, the forecasting system was too naive to be practical. To forecast demand for the next week, the procedure began by finding the second highest quarter's sales out of the previous four quarters. Sales for that quarter were then divided by thirteen to put them on a weekly basis, and this figure was used as a forecast. In times of recession, however, the system looked back at the good times of the year before and badly overestimated demand. This positive bias showed up in overproduction in the factories. After some time, when the recession would near its end, the company would adjust to bad times. You can guess the result when the economy would start expanding again—an example of negative bias.

Mean Absolute Deviation

The simplicity of calculation of the mean absolute deviation has made it the most popular technique for monitoring forecast error. We shall define forecast error as actual demand minus the forecast:

$$e_t = D_t - f_t$$

where D_t is demand in period t and f_t is the forecast made at the end of period $t - 1$ for period t. Then we can define the sum of absolute deviations (SAD) and the mean absolute deviation (MAD):

$$\text{SAD} = \sum_{t=1}^{n} |e_t| \qquad \text{MAD} = \frac{\sum_{t=1}^{n} |e_t|}{n}$$

For several error probability distributions, including the normal, Brown [2, p. 282] has shown that MAD is proportional to the standard deviation of forecast errors:

$$\text{MAD} = 0.8\sigma \quad \text{or} \quad \sigma = 1.25\text{MAD}$$

Example 2.7

A firm generates monthly forecasts. For one item, the forecast is 100,000 units, with a MAD of 5000 units. What would be the 66% prediction interval for next month's demand?

$$x = \mu \pm \sigma = 100,000 \pm 1.25 * 5000 \text{(appropriate relationship)}$$
$$= 100,000 \pm 6250$$

Note that $x = \mu \pm 1\sigma = \mu \pm 1.25\text{MAD}$.

Mean Absolute Error

Although MAD is an useful indicator of forecast errors, it is difficult to interpret the meaning of results obtained from the computations. For example, a MAD of 10 with an average demand of 1000 is superior to an average demand of, say, 50. Therefore, a relative error measure becomes more useful compared with indicators reflecting absolute measures. The mean absolute percent error (MAPE) and tracking signal are two such techniques. MAPE relates the average forecast error to the average demand for several consecutive periods.

$$\text{MAPE} = \frac{\Sigma[|e_t|/D_t] * 100}{n}$$

where e_t and D_t represent the forecast error and demand during period t, respectively, as defined earlier.

Tracking Signal

Positive bias should lead to forecasts consistently above the mark and negative bias to forecasts consistently below it. A tracking signal can be created to monitor bias continually. Because a good forecasting system should have equally balancing positive and negative errors, the cumulative or running sum of forecast errors (RSFE) should be close to zero:

$$\text{RSFE} = \sum_{t=1}^{n} e_t \approx 0$$

where $e_t = D_t - f_t$. If the forecasts are consistently too high, we would get a large negative RSFE. A common tracking signal, S, tells us something about the relative size of RSFE.

$$S = \frac{\text{RAFE}}{\text{MAD}} = \frac{\sum e_t / n}{\sum |e_t|/n}$$

where $RAFE$ is the running average forecast error. If positive and negative errors are canceling out, S will be close to zero. If large negative errors occur consistently, S will approach negative unity. We then have $-1 \leq S \leq 1$.

Adaptive Response Systems

Trigg and Leach [17] at Eastman Kodak developed a method of adjusting the responsiveness of the system in the face of bias or large values of S: Simply get α equal to the absolute value of the tracking signal, $\alpha = |S|$. Thus $|S|$ gets large whenever the forecasts are consistently out of line—precisely when we would like to make our system more responsive by raising α.

Although we could use the RAFE and MAD as given, computational convenience and the exponential smoothing spirit suggest that we should use exponential averages:

$$\text{MAD}_t = h|e_t| + (1 - h)\text{MAD}_{t-1}$$
$$\text{RAFE}_t = h(e_t) + (1 - h)\text{RAFE}_{t-1}$$
$$S_t = \text{RAFE}_t/\text{MAD}_t$$

where h is a smoothing constant, often set at $h = 0.05$.

Other adaptive response systems have been proposed, some of which were studied by Whybark [18]. The fundamental principle appears in the Trigg and Leach [17] model. In the forecasting system developed by McLeavey and mentioned in Example 2.5, the trend-smoothing constant β is determined by formula from α and hence automatically adjusts to changes in α. With an upper limit on α of 0.3, this system worked extremely well in practice.

Example 2.8

The actual demand for an item is not behaving as forecasted. Using $h = 0.05$, $\text{RAFE}_0 = 0$, and $\text{MAD}_0 = 0$, calculate the tracking signal (see Table 2.8).

$$\text{RAFE}_1 = 0.05(105 - 100) + 0.95 * (0) = 0.25$$
$$\text{MAD}_1 = 0.05|105 - 100| + 0.95|0| = 0.25$$

Selecting An Appropriate Forecasting Model

In this chapter we have progressed from the simple exponential smoothing model to the Winters model and to adaptive response systems Brown [2] and Montgomery and Johnson [12] give us an even greater variety of models from which to choose. Brown [3], Makridakis and Wheelwright [8], Makridakis et al. [9], and McLeavey et al. [11] discuss on forecast model selection issues. How do we go about choosing a forecasting system to handle thousands of items on a routine basis? Although experience will help to answer this question, we will give you at least a starting point in the form of a series of steps:

1. Graph the data for a random sampling of about thirty items at daily, weekly, or monthly intervals, depending on the company's needs.

TABLE 2.8   Computation of Tracking Signal

Month	Actual	Forecast	RAFE	MAD	S
January	105	100	0.25	0.25	1.00
February	96	100	0.04	0.44	0.09
March	102	100	0.14	0.52	0.27
April	97	100	−0.02	0.64	−0.03
May	121	101	0.98	1.61	0.61
June	118	102	1.73	2.33	0.74
July	119	103	2.44	3.01	0.81
August	123	104	3.27	3.81	0.86
September	121	105	3.91	4.42	0.88

2. If the data exhibit trends and/or seasons, you are probably safe in working with the Winters model. Groff [6] has shown that this model does about as well as most others for a variety of data series. If the data are intermittent (i.e., demand does not occur every period), exponential smoothing will not work very well, and you are probably better off fitting a probability distribution to the data, as in our next chapter. Also, Holt et al. [7] and Brown [2] provide excellent chapters on forecasting with probability distributions.

3. Experiment with smoothing factors according to the procedures outlined earlier for the selection of smoothing constants.

4. Initialize the system with the chosen set of smoothing factors. If you are to forecast next month's demand, you need to bring the system all the way up to last month.

5. Update the system period by period.

CONTROLLING THE FORECASTING SYSTEM

As time passes, older forecasts should be refined in light of recent developments. When the tracking signal or one of the other monitoring measures of error exceeds the specified limits for two or more consecutive periods, it indicates that something is wrong in the forecasting system. Obviously, corrective action is needed. The action to be taken depends upon the reason behind the out-of-control condition [12].

Random Variation

If the out-of-control condition was caused by random variations, obviously no action is required. Such occurrences can be reduced by selecting a sufficiently large value of MAPE or tracking signal.

Inaccurate Parameter Values

Although the forecasts may be satisfactory initially, large errors could emerge as time passes, which would indicate that true parameter values of smoothing constants such as α and β used in the model have changed with time. The model may still be correct, but revised estimates of the model parameters may be required. In these cases, higher values of parameters or smaller values of number of periods in the model can correct the out-of-control situation. Once the forecast appears to be back in control, the smoothing constant can again be revised to its normal values.

Change in Time Series Process

The demand pattern for the product could have changed with time. Terms such as trend, seasonal, and cyclical may need to be added or deleted. Corrective actions should be taken to obtain a better representation of the process. If the demand process appears to be highly autocorrelated [12], then use of different models should be explored.

Change in the Variance of the Demand

Larger smoothing constants can be used to reflect the changed condition of the demand pattern.

In any event, when corrective actions are taken to bring the forecasting system under control, the value of MAD, MAPE, and tracking signal should be calculated by resetting the time periods to prevent the past affecting the new control measures.

FOCUS FORECASTING

We dealt with several quantitative forecasting methods in this chapter. Smith [16] claims that simple forecasting rules can be very effective. The first phase consists of developing such rules of thumb to project forecasts. The second phase conducts a computer simulation on past data to validify the rules.

Some simple and commonsense rules could be: (a) The forecast for the next quarter is the actual demand for the item during the past quarter, (b) the forecast for the next month is 10% more than the actual demand in the corresponding month during the last year, and (c) the forecast for the next month is 110% of the actual demand for the item incurred during the last month.

If some rules of thumb work well for item forecasting, they will be retained. Otherwise, newer rules will be used in the computer simulation for validation. Obviously, whichever newer rules worked well will be used for item forecasting.

PYRAMID FORECASTING

In a large corporation, divisions such as marketing and manufacturing plan their forecasts individually. It is often a problem for these departments to agree on a forecast in a timely manner. One way of reconciling the data and bringing consistency within the firm is called the pyramid principle or pyramid forecasting. As a mechanism to accomplish it, Muir [13] suggests that forecasts be made at the level of detail that a firm needs them, the item-by-location level. The forecasts should also be done at a higher level and then those two forecasts should be tied together. We know that forecasts made for a family of items (major product groups or lines of business) in the aggregate level are more accurate and stable than summing individual item forecasts. To have a companywide forecast, the firm needs a consensus opinion, which can be achieved only by bringing together all concerned parties that might use or influence the forecast. The meetings act as a bridging mechanism to reconcile differences.

THE FORECASTING PROCESS

It is important to follow certain steps in the forecasting process. Wilson and Keating [19] provide a comprehensive treatment of the topic. In this section, we briefly describe their nine steps associated with the process as exhibited in Figure 2.2. The framework will help you to get the most out of the forecasting process.

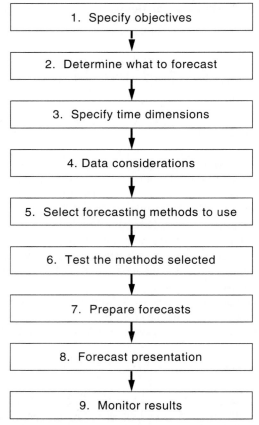

FIGURE 2.2 The forecasting process. From J. H. Wilson and B. Keating, *Business Forecasting* (Homewood, Ill.: Richard D. Irwin, 1990).

Step 1: *Specify objectives.* Management should specify the objectives, such as item forecasting for production planning, shop capacity planning, or marketing planning, as clearly as possible.

Step 2: *What to forecast.* The units of measurement, such as quantities or dollar revenues and the time frame (annual, quarterly, monthly, or weekly), should be clearly stated for the forecast. The quarterly or monthly forecasts are derived for capacity planning or overall inventory planning, whereas the weekly forecasts are more appropriate for materials requirements planning or shop floor planning.

Step 3: *Time dimensions.* Long-term forecasts for resource planning ranges from one to five years. Quarterly forecasts are best suited for observing seasonality, whereas monthly and weekly forecasts are suited for integrating more detailed shop floor operations.

Step 4: *Database considerations.* The kind of data desired depend on their use.

Therefore, any useful units, such as quantity of units per week or dollar volume per week, should be extracted consistently from the database. Any changes in the units of measurements should be clearly recorded. The variability in demand, if necessary, should be measured for the same unit of time and should be clearly labeled.

Step 5:   *Selecting a forecasting model.* Table 2.1 briefly dealt with the selection of a model. For stable demand patterns, which occur during the maturity period of the life cycle of the product, either moving averages or single exponential smoothing techniques are adequate. When the product is in its early growing periods or product phase-out periods, double exponential smoothing could be more suitable. Regression models along with the seasonality indices are found to be effective for medium-range planning purposes.

Step 6:   *Testing the model.* A model should be valid before it can be used for forecasting purposes. Therefore, part of the available data should be used to build the model, whereas the other part of the data should be used to test and validate the model to make sure that the model adequately represents the process.

Step 7:   *Forecast preparation.* One or more models may be adopted simultaneously by management. For example, units of forecast for the next period would be useful for shop floor planning, and a quarterly forecast would be useful for capacity and inventory planning. In such cases, these forecasts should be reconciled wherever and whenever possible.

Step 8:   *Forecast presentation.* People will not use the forecast if they do not have faith in it or if they do not understand how the forecasts were derived. Therefore, forecasts should be presented to the user with explanations as to how the forecast was derived, where the data were found, and what assumptions were implied. It is crucial for users to know the integrity of the data before they can confidently use the data; hence, communication is an important part of the process in any staff activity such as forecasting.

Step 9:   *Monitoring results.* Because forecasts are used in the management decision process, any deviations from the forecasts should be carefully observed using the monitoring techniques described earlier. When the magnitude of deviations is not acceptable, steps should be taken to revise the models as explained in the section on controlling the forecasting system.

SUMMARY

This chapter introduced some practical methods for making routine forecasts for thousands of items. The underlying thread has been the question of responsiveness. We want our system to respond to underlying changes but not to random fluctuations.

In one chapter we can take you only so far, however. For those interested in more advanced work, Montgomery and Johnson [12] will do an excellent job of taking you the next mile.

This chapter developed two fundamental concepts: measuring forecast error and exponential averaging. Exponential averaging assigns declining weights to past observations. In the Winters model, the averaging process is applied to the permanent component or base level, to the trend, and to seasonal indices. After mean absolute deviation and standard deviation have been developed as error measures, the exponential averaging process can be used to calculate the mean absolute deviation.

PROBLEMS

Each of the following problems uses one of the data series in Table 2.9. Visualize a company with hundreds of such series, one for each color, style, size, and so on, of each product. Problems 1 through 7 are based on quarterly data.

1. For data series A, calculate a simple exponential smoothing average. Try $\alpha = 0.1$ and $\alpha = 0.3$, $F_0 = 300$. What is the forecast for quarter 1 of 1995?
2. For series B1, calculate a simple exponential average. Try $\alpha = 0.1$ and $\alpha = 0.3$, $F_0 = 300$. Which is the better α? What is the forecast for quarter 1 of 1995?
3. For series B2, calculate a simple exponential average. Try $\alpha = 0.1$ and $\alpha = 0.3$, $F_0 = 300$. Which is the better α? What is the forecast for quarter 1 of 1995?
4. For series C, calculate a simple exponential average with $\alpha = 0.2$ and $F_0 = 300$. What is the forecast for quarter 1 of 1995?
5. For series C, calculate a Winters trend average with $\alpha = 0.2$, $\beta = 0.05$, $F_0 = 300$, $T_0 = 0$. What is the forecast for quarter 1 of 1995?
6. For series D, calculate a Winters seasonal exponential average with $\alpha = 0.2$, $\gamma = 0.05$, and $F_0 = 300$. What is the forecast for quarter 1 of 1995?
7. For series F, calculate a Winters linear trend and ratio seasonals average. Use $\alpha = 0.2$, $\beta = 0.05$, $\gamma = 0.05$. What is the forecast for quarter 1 of 1995?

For the 1993 data series B2, our forecasting system generated the following forecasts:

	J	F	M	A	M	J	J	A	S	O	N	D
Forecast	100	100	100	100	102	104	106	108	110	112	114	116
Actual	97	93	110	98	130	133	129	138	136	124	139	125

8. For each month, calculate the running average of forecast error, the mean absolute deviation, and the tracking signal. Use $h = 0.05$. How would an adaptive response help the situation? (Assume $RAFE_0 = 1$ and $MAD_0 = 5$.)
9. Recommend a procedure to handle data series F.
10. Which model would you use for series G. Approximately what parameter values would you judge to be reasonable?
11. Which model would you use for series H? Approximately what parameter values would you judge to be reasonable?
12. What smoothing factor (α) is equivalent to six months in a moving average process? Is that a relatively high or low α?
13. What number of months would be equivalent to an α of 0.1? How responsive would such a moving average be?
14. We are presently in December and have been using simple exponential smoothing ($\alpha = 0.3$) for the last twelve months. Hence, we have applied a weight of 0.3 to

TABLE 2.9 Data for Problems

Type of Series	Year	Q1			Q2			Q3			Q4		
		Jan.	Feb.	Mar.	Apr.	May	June	July	Aug.	Sep.	Oct.	Nov.	Dec.
A. Constant	1994	97	93	110	98	104	103	99	108	106	94	109	95
B1. Impulse	1994	97	93	110	138	104	103	99	108	106	94	109	95
B2. Step	1993	97	93	110	98	130	133	129	138	136	124	139	125
	1994	122	127	125	126	139	127	134	128	134	136	132	121
C. Trend	1993	97	96	116	107	116	118	117	129	130	121	139	128
	1994	128	136	137	141	157	148	158	155	164	169	168	160
D. Constant with seasons	1992	145	121	121	88	73	52	50	75	95	103	142	143
	1993	138	126	105	86	76	49	52	69	94	117	133	137
	1994	164	139	117	82	64	53	53	65	95	99	139	141
E. Trend with seasons	1992	146	125	128	96	81	59	59	90	117	133	181	192
	1993	192	177	151	127	110	74	79	109	148	186	218	240
	1994	272	237	202	155	123	97	98	130	182	219	272	299
F. Inter-mittent	1993	85	54	98	90	0	0	92	0	99	0	78	0
	1994	78	0	55	75	87	0	73	0	0	0	0	53
G. Real company data	1992	212	254	291	236	194	229	255	239	303	231	207	219
	1993	201	246	268	176	178	256	199	192	261	161	142	163
	1994	172	143	219	128	157	132	128	170	219	134	108	208
H. Real company data	1992	162	205	265	158	145	171	145	165	251	186	193	183
	1993	192	200	245	149	185	174	159	175	252	172	166	166
	1994	165	199	233	164	158	182	182	139	231	165	173	181

Note: All figures in thousands of units.

51

December sales and 0.21 to November sales in calculating December's exponential average. What weight did we apply to September's and October's sales?

15. Economists differentiate between demand and quantity demanded. In this chapter, we have not preserved that distinction. On the other hand, we have consistently spoken of demand rather than sales. We do this because sales figures do not adequately capture demand. What about lost sales due to stockouts? How would you recommend handling the following?
    a. Sales returns
    b. Orders for future delivery
    c. Keypunch errors in sales figures
    d. Special promotions.

The remaining questions should be solved with the help of a computer.

16. For demand series E, generate monthly forecasts for 1995. Justify your choice of α, β, and γ.
17. For demand series F, generate monthly forecasts for 1995. How meaningful are these forecasts? Can you suggest any sensible way to handle such data? What might cause such lumpiness?
18. For demand series H, choose a model and generate monthly forecasts for 1995. Now generate quarterly forecasts for 1995. Of which forecasts are you more confident?
19. Using a random number generator, generate a monthly demand series with the characteristics

$$D_t = 500 + 10t \pm e_t$$

where e_t is normally distributed with mean $500 + 10t$ and variance 100. Generate six years of data. Now, using the first five years to develop your forecasting system, attempt to forecast year 6, using (a) Winters trend model with seasons and (b) Winters trend model without seasons. Which model gives you the best results?

20. Using a random number generator, generate a monthly demand series with the characteristics

$$D_t = 500 + 10_t \pm e_t \qquad t \leq 24$$
$$D_t = 2000 + 3t \pm e_t \qquad t > 24$$

where e_t is normally distributed with mean $500 \pm 10t$ ($2000 + 3t$ for $t > 24$) and variance 100. Generate six years of data. Now try to forecast year 6, using Winters trend model with seasons and using the trend model only.

21. In problem 20, how good is 1.25MAD as an estimate of the standard deviation of forecast error?

REFERENCES AND BIBLIOGRAPHY

1. G. E. P. Box and G. M. Jenkins, *Time Series Analysis: Forecasting and Control,* rev. ed. (San Francisco: Holden-Day, 1976).
2. R. G. Brown, *Smoothing, Forecasting and Prediction of Discrete Time Series* (Englewood Cliffs, N.J.: Prentice Hall, 1963).

3. R. G. Brown, "The Balance of Effort in Forecasting," *Journal of Forecasting,* Vol. 1, No. 1 (January–March 1982), pp. 49–53.

4. R. B. Chase and N. J. Aquilano, *Production and Operations Management,* 6th ed. (Homewood, Ill.: Richard D. Irwin, 1992).

5. P. E. Green and D. S. Tull, *Research for Marketing Decision,* 4th ed. (Englewood Cliffs, N.J.: Prentice Hall, 1978).

6. G. K. Groff, "Empirical Comparison of Models for Short Range Forecasting," *Management Science,* Vol. 20, No. 1 (September 1973), pp. 22–31.

7. C. C. Holt et al., *Planning Production, Inventories and Work Force* (Englewood Cliffs, N.J.: Prentice Hall, 1960).

8. S. Makridakis and S. C. Wheelwright, *Forecasting Methods and Applications* (New York: John Wiley & Sons, 1978).

9. S. Makridakis et al., "The Accuracy of Extrapolation Methods," *Journal of Forecasting,* Vol. 1, No. 2 (April–June 1982), pp. 111–153.

10. J. O. McClain, "Dynamics of Exponential Smoothing with Trend and Seasonal Terms," *Management Science,* Vol. 20, No. 9 (May 1974), pp. 1300–1304.

11. D. W. McLeavey et al., "An Empirical Evaluation of Individual Item Forecasting Models," *Decision Sciences,* Vol. 12, No. 4 (1981), pp. 708–714.

12. D. C. Montgomery and L. A. Johnson, *Forecasting and Time Series Analysis* (New York: McGraw-Hill, 1976).

13. J. W. Muir, "Forecsting Problems and Their Solutions," in *Understanding Business Forecasting,* C. L. Jain, ed. (Flushing, N.Y.: Graceway Publishing, 1988), pp. 233–237.

14. J. F. Muth, "Optimal Properties of Exponentially Weighted Forecasts," *Journal of the American Statistical Association,* Vol. 55, No. 290 (1960), pp. 299–306.

15. C. C. Pegels, "Exponential Forecasting: Some New Variations," *Management Science,* Vol. 15, No. 5 (January 1969), pp. 311–315.

16. B. T. Smith, *Focus Forecasting* (Boston: CBI Publishing, 1978).

17. D. W. Trigg and A. G. Leach, "Exponential Smoothing with an Adaptive Response Rate," *Operational Research Quarterly,* Vol. 18, No. 1 (1967), pp. 53–59.

18. D. C. Whybark, A Comparison of Adaptive Forecasting Techniques," *Logistics and Transportation Review,* Vol. 8, No. 3 (1972), pp. 13–26.

19. J. H. Wilson and B. Keating, *Business Forecasting* (Homewood, Ill.: Richard D. Irwin, 1990).

20. P. R. Winters, "Forecasting Sales by Exponentially Weighted Moving Averages," *Management Science,* Vol. 6, No. 3 (1960) pp. 324–342.

21. R. J. Wonnacott and T. H. Wonnacott, *Econometrics* (New York: John Wiley & Sons, 1970).

APPENDIX 2A

Double Exponential Smoothing and Trend-Adjusted Exponential Smoothing

Two other forecasting models are quite popular, perhaps because of the robustness of trend-based models. McLeavey et al. [1], for example, found that double exponential smoothing performed well on a variety of demand patterns for forecasts one period ahead and twelve periods ahead. Double exponential smoothing and trend-adjusted exponential smoothing are alternative ways of incorporating trends into exponential smoothing. These two approaches can be shown to be exactly equivalent to Winters trend model (the Holt model) if the proper parameter values are chosen for each model. We present the double exponential smoothing and the trend-adjusted models here only because they have been so popular in the literature. The Winters model is as good.

DOUBLE EXPONENTIAL SMOOTHING

Simple exponential smoothing does not cope very well with a trend. If we were to take equation 2.5 for simple exponential smoothing and to note that $E(S_t) = \mu + \delta t$, we could derive the expected value of the simple exponentially smoothed average. We would discover that we have undershot the mark:

Taking expectations

$$E(F_t) = \alpha \sum_{k=0}^{t-1} (1-\alpha)^k E(D_{t-k}) + (1-\alpha)^t F_0$$

and substituting for $E(D_{t-k})$

$$E(F_t) = \alpha \sum_{k=0}^{t-1} (1-\alpha)^k [\mu + \delta (t-k)] + (1-\alpha)^t F_0$$

As $t \to \infty$, we obtain

$$E(F_t) = [\mu + \delta t]\alpha \sum_{k=0}^{\infty} (1-\alpha)^k - \delta\alpha \sum_{k=0}^{\infty} k(1-\alpha)^k$$

$$E(F_t) = \mu + \delta t - \frac{1-\alpha}{\alpha}\delta$$

$$E(F_t) = E(D_t) - \frac{1-\alpha}{\alpha}\delta$$

Figure 2A.1 portrays the situation. At this juncture, we are helpless in estimating $E(D_t)$, since all we have are the figures F_t, D_t, and $(1-\alpha)/\alpha$. What we need is an estimate of $[(1-\alpha)/\alpha] \delta$, the vertical distance between $E(D_t)$, and our present position F_t.

Double or second-order exponential smoothing uses a trick to estimate $E(D_t)$. We can treat the F_t as a data series to be exponentially smoothed. The second-order smoothed average, $F_2^{(2)}$, will bear the same relationship to F_t as F_t bears to D_t. Specifically,

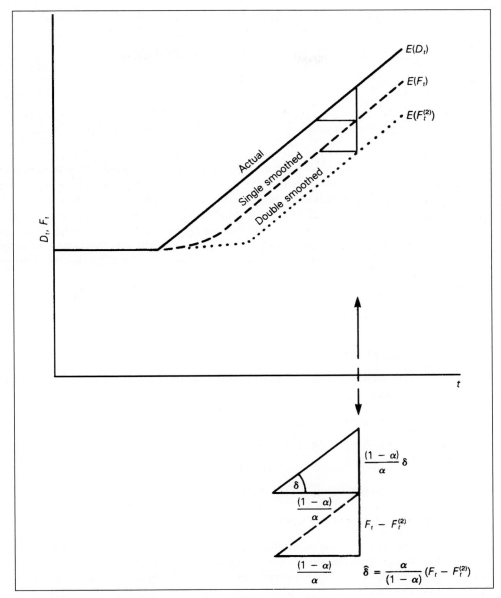

FIGURE 2A.1   The response of simple exponential smoothing to a trend.

$$F_t^{(2)} = \alpha F_t + (1 - \alpha)F_{t-1}^{(2)}$$

$$E(F_t^{(2)}) = E(F_t) - \frac{(1-\alpha)}{\alpha} \delta \, [\text{cf. equation (2.6)}]$$

and

$$E(\text{forecast of single average}) = E(\text{single average}) - \frac{(1-\alpha)}{\alpha} \delta$$

Now we are prepared to estimate $E(D_t)$. From equation (2.6), F_t can be expected to undershoot the mark by $[(1 - \alpha)/\alpha]$ δ. An estimate of $E(D_t)$ would be $F_t + [(1 - \alpha)/\alpha]$ δ. We have just seen, however, that $F_t^{(2)}$ can be expected to undershoot F_t by the same distance $[(1 - \alpha/\alpha]$ δ. Using $F_t - F_t^{(2)}$ to estimate $[(1 - \alpha)/\alpha]\delta$, our estimate of $E(D_t)$ will be $F_t + F_t - F_t^{(2)} = 2F_t - F_t^{(2)}$. To forecast $D_t + j$—that is, to generate a j-period-ahead forecast, we could use $E(D_t) + \text{slope} * j$. Since the height over the base gives us the slope, we take $F_{t-}F_t^{(2)}$ over $(1 - \alpha)/\alpha$ as the slope in Figure 2A.1. At the end of period t, we would forecast demand in period $t + 1$ as

$$f_{t+1} = 2F_t - F_t^{(2)} + \frac{\alpha}{(1-\alpha)}(F_t - F_t^{(2)}) * 1$$

Example 2A.1

Develop a double smoothed forecast for the firm in Example 2.3, using $\alpha = 0.2$, $F_0 = 480$, and $F_0^{(2)} = 480$ (see Table 2A.1).

$$F_1^{(2)} = 0.2 * 476 + 0.8 * 480 = 479.2$$

$$f_2 = 2 * 476 - 479.2 + 0.2/0.8\,(476 - 479.2) = 472$$

$$F_2 = 0.2(510) + 0.8(476) = 482.8$$

$$F_2^{(2)} = 0.2(482.8) + 0.8(479.2) = 479.9$$

TREND-ADJUSTED EXPONENTIAL SMOOTHING

An equivalent method of handling the same trend process relies on a direct estimate of the slope δ. The demand process, $E(D_t)$, will again be estimated by $F_t + [(1 - \alpha)/\alpha]$ δ. To estimate δ, we simply smooth the differences between succeeding exponentially smoothed averages:

$$T_t = \beta(F_t - F_{t-1}) + (1 - \beta)T_{t-1}$$

with $T_0 = 0$. This procedure is justified by the observation that $E(F_t) - E(F_{t-1}) = E(Dt) - E(D_{t-1})$ from equation 2.6, but that $E(D_t) - E(D_{t-1}) = \delta$. Hence, T_t estimates δ, and we can use $F_t + [(1 - \alpha)/\alpha]T_t$ to estimate $E(D_t)$.

TABLE 2A.1   Double Exponentially Smoothed Forecast

Month	Demand (D_t)	Average (F_t) $F_0 = 480$	Average $F_t^{(2)}$	Forecast f_t
March	460	476	479.2	472
April	510	482.8	479.9	472
May	520	490.2	482	486.4
June	495	491.2	483.8	500.5
July	475	488	484.6	500.5
August	560	502.4	488.2	492.3
September	510	503.9	491.3	520.2
October	520	507.1	494.5	519.7
November	540	513.7	498.3	522.9
December	550	521	502.8	533
January				543.8

Our forecast for period $t + 1$ would be

$$f_{t+1} = F_t + \frac{(1-\alpha)}{\alpha} T_t + T_t = F_t + \frac{1}{\alpha} T_t$$

Example 2A.2

Develop a trend-adjusted exponential forecast for the firm in Example 2.3. Assume $T_0 = 0$, $F_0 = 480$, and $\alpha = 0.2$, $\beta = 0.2$ (see Table 2A.2).

$$T_1 = 0.2\,(475 - 480) + 0.8(0) = -0.8$$

$$f_2 = 476 + \frac{1}{0.2}(-0.8) = 472$$

Table 2A.2   Trend-Adjusted Exponentially Weighted Forecast

Month	Demand (D_t)	Average (F_t) $F_0 = 480$	Trend (T_t)	Forecast (f_t)
March	460	476	−0.8	
April	510	482.8	0.72	472
May	520	490.2	2.04	486.4
June	495	491.2	1.83	500.4
July	475	488	0.82	500.4
August	560	502.4	3.54	492.1
September	510	503.9	3.13	520.1
October	520	507.1	3.14	519.6
November	540	513.7	3.83	522.8
December	550	521	4.60	532.9
January				544

REFERENCE

1. D. W. McLeavey et al., "An Empirical Evaluation of Individual Item Forecasting Models," *Decision Sciences,* Vol. 12, No. 4 (1981), pp. 708–714.

APPENDIX 2B

Some Empirical Findings on Short-Term Forecasting: Technique Complexity and Combinations*

Notes and Applications

ABSTRACT

The purpose of this research is to determine if prior findings that favor simple forecasting techniques and technique combinations hold true in a short-term forecasting environment, where demand data can be quite volatile. Twenty-two time series of daily data from a real business setting are used to test one-period ahead forecasts, the epitome of short-term forecasting. The time series vary systematically as to data volatility and forecast difficulty. Forecast accuracy is measured in terms of both mean absolute percentage error (MAPE) and mean percentage error (MPE).

Subject Areas: Decision Support Systems, Forecasting, and Smoothing Techniques.

INTRODUCTION

A large assortment of forecasting techniques has been developed over the last 35 years, which has naturally led to studies comparing their forecasting abilities. The comparisons were often part of a search for the best extrapolation technique but compiled results were mixed and often contradictory. Although no "champion" has been found, two results have emerged with some regularity. First, contrary to the supposition that there is a trade-off between accuracy and cost, greater complexity in extrapolation techniques has not generally improved forecasting ability [1]. A second finding emerges with even more clarity: combining forecasts from two or more techniques (such as simply averaging them) can dramatically improve forecast accuracy [1] [2] [9] [10] [11].

Our empirical study builds on this earlier research in three respects. First, it examines whether the findings on complexity and combinations hold true in situations for which extrapolation techniques are primarily intended—that is, short-term forecasting. Most empirical studies to date deal with macro time series, such as data on a firm, industry, or even nation. Few comparisons are made using micro time series, such as data on an individual product, even though these time series are commonly found in business. A

*Nada R. Sanders and Larry P. Ritzman Reprinted with permission from *Decision Sciences,* Vol. 20, 1989, pp. 635–640.

good example is inventory management, in which hundreds of items need forecasts on a regular basis. Further, the time interval covered by forecasts in prior studies tends to be long, such as with quarterly or yearly data. The shortest time period in the well-known study of 1001 time series is a month, with no weekly or daily data [7] [8]. Daily data (the epitome of short-term forecasting) are used in our study. Such short time intervals are very common in service organizations in which forecasts are often made one period ahead for work force scheduling purposes. Prior findings on complexity and combinations may not generalize to short-term forecasts, particularly since one recent study showed that the relative benefits of combinations diminish with forecast difficulty [5]. Since forecast difficulty in turn is linked with short-term forecasting of micro data [9], the extension of this study helps fill a gap in the knowledge base.

A second feature of this study is that it uses one-step ahead forecasts and the models are refitted after each new observation. The more expedient approach is simply to make all forecasts after the fitting stage is complete. While one-step ahead forecasting with continual refitting is more tedious when doing the research, it gives the techniques every advantage when the next forecast is made.

A third distinguishing feature of this study is that the time series vary systematically according to their inherent variability (which is measured by the coefficient of variation). Prior to the study by Lawrence, Edmunson, and O'Connor [5], scant attention was paid to whether the effects of technique complexity and combinations depend in turn on the characteristics of the time series. Forecasting difficulty seems to be a crucial part of the forecasting situation and therefore is controlled in this study.

THE DATA

The data used in this research come from a national public warehouse and cover a three-year time span. During these three years the warehouse had seven customers. Two functions were provided for each customer: inbound shipment and outbound shipment. The result is 22 time series: 1 for the warehouse as a whole, 7 at the customer level, and 14 at the functional level.

These time series cover a significant range of variability, depending on the customer. As a rough rule, the variability in the data increases with the level of desegregation. For any customer account, the time series of data shipped and received will generally be more variable than the time series of the whole account. If we use the coefficient of variation to measure variability, the values range from 16.7 percent for the more stable to over 200 percent for the highly volatile.

The time series were seasonally adjusted as appropriate. To determine the presence or absence of seasonality, autocorrelation analysis was performed. Only monthly seasonal effects (not day of the week or day of the month) were present in the data, and then for only 5 of the 22 time series.

METHODOLOGY

The five quantitative techniques and their combinations tested here come from a subset of those used recently by Makridakis and Winkler [9]:

1. Single exponential smoothing (SINGLE),
2. Linear trend by regression (LINEAR),
3. Adaptive response rate exponential smoothing (ARRES),
4. Holt's two parameter linear model (HOLT), and
5. Automatic univariate adaptive estimation procedure (AEP).

Several criteria guided the selection of these particular techniques. First, they are among the most common found in the literature. Second, they span a reasonable range from simple to more complex. Third, they can provide forecasts in an expedient fashion, which is particularly important in an environment where forecasts need to be generated on a daily basis. Fourth, will the general patterns from earlier studies hold true in more volatile short-term environments? A sample of representative techniques is sufficient.

Three combination forecasts were randomly selected from these five single techniques, resulting in one two-technique combination and two three-technique combinations. Although more combinations could have been tested, earlier studies [8] [9] [11] show that performance is not significantly affected by the selection of techniques in the combination. Further, rapidly diminishing benefits of additional techniques in any combination beyond two were found in one study [3] and beyond four in another [9]. The following three combinations (COMB 1, 2 and 3) are therefore adequate for our purposes:

1. SINGLE and LINEAR (COMB 1)
2. SINGLE, LINEAR, and ARRES (COMB 2)
3. ARRES, HOLT, and AEP (COMB 3).

The final forecasts for a combination are obtained period by period by taking a simple unweighted average of the forecasts generated by the individual techniques in the combination.

All eight forecasting procedures were executed in an automatic mode using FUTURCAST, an interactive computer package for time series forecasting by Makridakis and Carbone [6]. Two years of data were used for analysis and model start-up purposes, leaving the third year for forecast evaluation. Forecasts generated using these techniques were made for a full year (260 daily observations) on a rolling horizon basis. Performance was evaluated using mean absolute percentage error (MAPE) and mean percentage error (MPE). The mean absolute deviations (MAD) were also collected as a third type of statistic, but our findings are unchanged with their use. For ease in presentation, only MAPE and MPE statistics are presented here.

THE FINDINGS

Table 1 ranks the eight forecasting procedures on their overall performance for the 22 time series using both MAPE and MPE. Shown are the MAPE and MPE scores for each technique, averaged across the 22 time series. It is interesting that the MAPE and MPE rankings are very similar. The one striking exception is LINEAR. It is the worst tech-

TABLE 1 Ranking of forecasting techniques by MAPE and MPE.

	MAPE		MPE	
Rank	Technique	Average MAPE	Technique	Average MPE*
1	COMB 1	61.9	LINEAR	11.7
2	COMB 3	63.0	COMB 1	13.0
3	COMB 2	63.9	COMB 2	16.9
4	SINGLE	70.7	COMB 3	17.1
5	ARRES	72.6	SONIGLE	19.4
6	HOLT	74.1	ARRES	21.1
7	AEP	77.6	HOLT	21.7
8	LINEAR	78.7	AEP	26.0

*For any given time series, a technique's MPE can be either positive or negative. To get meaningful summary data, the absolute values of the MPE statistics were taken prior to averaging across all 22 times series. This same approach is used in Tables 2 and 3.

nique in terms of MAPE but is clearly the best single model in terms of its ability to minimize bias and MPE. If management places primary emphasis on forecasts with minimal bias, serious consideration should be given to the stability offered by LINEAR or just a simple average.

Table 1 sheds considerable light on the main research questions of complexity and combinations. The more complex techniques, HOLT and AEP, rank among the worst performers for both MAPE and MPE. SINGLE and LINEAR (the simplest techniques) rank first on MAPE and MPE. SINGLE and LINEAR (the simplest techniques) rank first on MAPE and MPE, respectively. Adaptive capabilities, which add complexity, do not necessarily improve forecast accuracy. For example, ARRES banks behind SINGLE on both MAPE and MPE. AEP, the other adaptive model, is even worse.

Earlier findings on technique combinations also seem to hold true for short-term forecasting. Table 1 shows that the three combinations are better than all single models in terms of MAPE. They do almost as well in terms of MPE, occupying three of the top four ranks.

The findings in this study are incomplete without testing whether the effects of technique complexity and combinations depend on the degree of forecasting difficulty. The first consideration is how to measure forecast difficulty. As the coefficient of variation of a time series increases, so does the average MAPE produced by the eight techniques. Pearson's simple correlation coefficient was computed as .941, which is highly significant. Such a close correlation suggests that the coefficient of variation is an excellent surrogate for forecast difficulty when using extrapolation techniques for short-term forecasts.

Considering forecasting difficulty, Table 2 carries the complexity question one step further. It contrasts the percent improvement achieved by moving from simple to complex techniques for 11 time series with low variability (coefficients of variation less

TABLE 2   MAPE and MPE Improvements from using more complex techniques, relative to coefficient of variation

	MAPE				⁻MPE		
	Average Coefficient of Variation	Average of Simple Techniques (SINGLE & LINEAR)	Average of Complex Techniques HOLT & AEP)	Percent Improvement	Average of Simple Techniques (SINGLE & LINEAR)	Average of Complex Techniques HOLT & AEP)	Percent Improvement
Average of first 11 series	36.4	33.6	34.2	–1.8	10.0	12.3	–23.8
Average of last 11 series	144.7	115.8	117.5	–1.4	21.2	35.4	–66.6
Average of all 22 series	90.5	74.7	75.9	–1.5	15.6	23.8	–52.9

TABLE 3 MAPE and MPE improvements from combining forecasts, relative to coefficient of variation.

		MAPE					MPE	
	Average Coefficient of Variation	*Average of Individual Techniques*	*Average of Combinations*	*Percent Improvement*	*Average of Individual Techniques*	*Average of Combinations*	*Percent Improvement*
Average of first 11 series	36.4	33.8	25.1	20.6	11.4	7.5	27.4
Average of last 11 series	144.7	115.6	99.9	10.4	28.6	23.8	19.6
Average of all 22 series	90.5	74.7	62.5	15.5	20.0	15.7	23.5

than 60 percent) with 11 time series with high variability. It is interesting that the MAPE percent improvements are quite similar in both cases: −1.8 for low variability and −1.4 for high variability. Complexity actually hurts MAPE performance, although the change is slight. The story is different with bias: the complex techniques are decidedly worse on MPE, particularly at high levels of forecasting difficulty. The more complex techniques seem to have tried to read too much into the data, perceiving changes in patterns when none really existed.

As to how forecasting difficulty affects the benefits of combining forecasts, Table 3 is the counterpart of Table 2. Improvements are evident on almost all time series. The benefit of combinations drops somewhat with higher coefficients of variation, which is consistent with the recent study by Lawrence et al. [5]. The benefit of combining forecasts is statistically significant, regardless of the degree of forecasting difficulty. Turning from MAPE to MPE, the conclusions are almost identical. Combinations help at all levels of data variability, with slightly higher relative improvements for more stable time series. The improvements are statistically significant, however, regardless of the inherent stability in the data.

CONCLUSION

Three main findings emerge from this study. First, complex techniques do not perform any better than simple ones in terms of MAPE, the usual measure of forecast accuracy. Second, complex techniques are definitely worse in terms of forecast bias as measured by MPE. The MPE disadvantage becomes larger with more volatile time series. This disadvantage is statistically significant. Third, combinations of individual techniques yield healthy dividends at all levels of forecasting difficulty. Emerson once write that "to be simple is to be great" [4, p. 145]. His wisdom seems to also apply to forecasting, particularly if you hedge your bets and combine the forecasts of several inexpensive and easily understood methods. [Received: September 3, 1987. Accepted: July 11, 1988.]

REFERENCES

1. Armstrong, J. S. Forecasting by extrapolation: Conclusions from 25 years of research. *Interfaces,* 1984, *14*(4), 52–66.
2. Bates, J. M., & Granger, C. W. J. The combination of forecasts. *Operational Research Quarterly,* 1969, *20*(4), 451–468.
3. Bopp, A. E. On combining forecasts: Some extensions and results. *Management Science,* 1985, *31,* 1492–1498.
4. Emerson, R. W. Self-reliance. *The selected writings of Ralph Waldo Emerson.* New York: Random House, 1968.
5. Lawrence, M. J., Edmundson, R. H., & O'Connor, M. J. The accuracy of combining judgmental and statistical forecasts. *Management Science,* 1986, *32,* 1521–1532.
6. Makridakis, S., & Carbone, R. *An introductory guide to FUTURCAST.* Pittsburgh, PA: Futurion C.M., 1984.
7. Makridakis, S., & Hibon, M. Accuracy of forecasting: An empirical investigation. *Journal of the Royal Statistical Society,* 1979, *142,* 97–144.
8. Makridakis, S., Andersen, A., Carbone, R., Fildes, R., Hibon, M., Lewandowski, R., Newton,

J., Parzen, E., & Winkler, R. The accuracy of extrapolation methods: Results of a forecasting competition. *Journal of Forecasting,* 1982, *1*(2), 111–153.

9. Makridakis, S., & Winkler, R. Averages of forecasts: Some empirical results. *Management Science,* 1983, *29,* 987–996.

10. Newbold, P., & Granger, C. Experience with forecasting univariate time series and the combination of forecasts. *Journal of the Royal Statistical Society,* 1974, *137,* 131–165.

11. Winkler, R., & Makridakis, S. The combination of forecasts. *Journal of the Royal Statistical Society,* 1983, *146,* 150–157.

Chapter Three

Special Topics in Forecasting

INTRODUCTION

We now discuss several interesting special topics in forecasting in production planning and inventory control. The first topic is the extension of the single-item techniques from Chapter 2 to multiple items within a group or family of products. The second topic is the forecasting of slow-moving items.

Chapter 2 presented many techniques for generating individual item forecasts. These techniques can also be used to forecast the demand for a group or family of items. For example, an apparel line may consist of several sizes and colors, or an automotive line may have several choices of an option, such as four-, six-, and eight-cylinder engines. As another example, a specific item may be stocked in several distribution locations, which are known as stock keeping units (SKU). In many of these situations, forecasts of individual item demands are neither economically feasible nor desirable, as will be discussed in Chapter 10. We also know that we can generate a more accurate forecast for a group of items than the individual items in the group. Therefore, in many instances we will resort to forecasting a group, family, or line of items as a whole. Once the total forecast for the entire line is generated, we can generate forecasts for every item in the line. Similarly, the total requirements forecast of a specific item in many distribution locations can be allocated to individual locations.

An accurate forecast of demand for an item at each specific location can improve customer service while keeping inventory and transportation costs in line. A better demand forecast for every item can definitely improve production planning, as exhibited in Figure 3.1. In general, forecasts of families of products assist manufacturing in committing a base plan or a rate of manufacture. Then a detailed forecast of each item is planned for the master production schedule.

Item-level forecasts are the major inputs to the material requirements planning

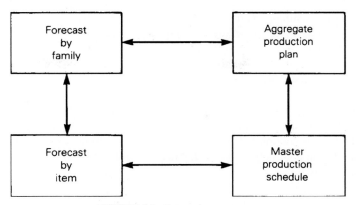

FIGURE 3.1   Demand management.

(MRP) systems via master production schedules. Therefore, these forecasts are important for overall planning in many instances [8]. Chapter 10 focuses on the importance of keeping the number of end items as low as possible on the master production schedule, using a technique known as modularization to accomplish this goal. Several other special situations exist in the industry, however. For example, many seasonal items such as woodstoves, fans and other consumer goods are carried only during a particular season. Many of these seasonal items also reach a peak during the season before the demand slows down. The timing of the peak may vary, but the total demand for the season may be forecast most accurately using regression analysis or other forecasting techniques. Such situations require what is known as multi-item or multidimensional forecasting, whereby forecasts can be generated simultaneously for each item in the group. The first step of such forecasting involves estimating the probability distribution of demand for each item in the group (line or family) or SKU. The second step is to apply one of the models described in this chapter to forecast the demand for an individual item belonging to a group. To utilize these models, we should have a forecast for the group of items as a whole, which can be accomplished by using one of the models described in Chapter 2.

Some products may exhibit low levels of demand or have many periods of no usage. In these situations, the forecasting techniques described in Chapter 2 may not work effectively. The exponential smoothing method with a very small smoothing constant may provide a reasonable forecast for items with demand of less than 100 per year. The vector smoothing method discussed below can also be used for slow demand items. Because the demand is small per period, the number of intervals of demand will be small and can be calculated efficiently.

MULTI-ITEM FORECASTING

Vector Smoothing Method For An Item

Brown [1] illustrates the use of smoothing techniques to find current estimates of the probability distribution of demands. The item demands are grouped into several intervals, and the probability of an item demand falling in an interval is estimated initially with the existing demand data. As new demands become available, the probability values are updated using smoothing techniques. The procedure is known as the vector smoothing method:

$$P_{k,t+1} = \alpha q_{kt} + (1 - \alpha)P_{kt}$$

where $P_{k,t+1}$ represents the updated probability distribution of an item in the interval k for the period $t + 1$, α is the smoothing constant, and q_{kt} is the actual distribution of demand during period t. The vector smoothing technique requires that the time series being forecast be relatively stable or that the demand be slowly shifting as a function of time.

Example 3.1

The following table exhibits the class intervals and the frequency distribution of demand for the past thirty-six days. Assuming that the actual demand during the thirty-seventh day was twenty-one, find the updated probabilities, using $\alpha = 0.20$:

K	*Interval*	*Frequency*	$P_{k,37}$
1	0–9	11	0.306 = 11/36
2	10–19	17	0.472
3	20–29	5	0.139
4	30–40	3	0.083
		36	

Solution: Since the actual demand during the thirty-seventh day was twenty-one, it falls in the third interval. Therefore, $q_{3,37} = 1$, and all other values in the q vector become zeros. Since the value of $P_{k,37}$ is given in the table, we can update the probabilities using the preceding formula. The current estimates of the probability distribution of demand are as follows:

$$P_{k,38} = \alpha q_{k,37} + (1 - \alpha)P_{k,37}$$

$$= (0.2)\begin{bmatrix} 0 \\ 0 \\ 1 \\ 0 \end{bmatrix} + (1 - 0.2)\begin{bmatrix} 0.306 \\ 0.472 \\ 0.139 \\ 0.083 \end{bmatrix}$$

$$= \begin{bmatrix} 0 \\ 0 \\ 0.2 \\ 0 \end{bmatrix} + \begin{bmatrix} 0.245 \\ 0.378 \\ 0.111 \\ 0.066 \end{bmatrix} = \begin{bmatrix} 0.245 \\ 0.378 \\ 0.311 \\ 0.066 \end{bmatrix}$$

Hence, the expected value of the demand for the thirty-eighth day is seventeen, which is obtained by adding the products of medians and their associated probabilities.

Vector Smoothing Method
For Multiple Items

The smoothing technique can also be extended to forecast demand for individual items making up a group of products [10]. The group may represent, for example, different models of bicycles, engines, or hoists. To generate forecasts for individual items, we should have an updated forecast for the total group.

Suppose, for example, that we have K items belonging to a group. Let P_{kt} represent the probability that a demand will occur for item k at time t. The basic law of probabilities requires that individual probabilities of all K items in the group sum to one; that is,

$$\sum_{k=1}^{K} P_{kt} = 1 \quad \text{for all time periods } t.$$

Furthermore, all probabilities in the previous period are assumed to be known. This is true since we can either calculate the probabilities from the actual demands in the past or estimate them from market research activities in the case of a new product. As the cumulative demands D_{kT} for the item k are known at time T, to start the system we can compute an estimate for P_{kt} for item k in the group:

$$P_{kt} = \frac{D_{kT}}{D_T} = \frac{D_{kT}}{\Sigma D_{kT}}$$

where D_T represents the sum of all item demands in the group up to time T. Now the probability mix for the item k can be updated using smoothing formulas discussed in Chapter 2:

$$P_{kt} = \alpha q_{k,t-1} + (1-\alpha)P_{k,t-1}, \quad \text{where } q_{k,t-1} = \frac{D_{k,t-1}}{D_{t-1}}$$

Here $D_{k,t-1}$ and D_{t-1} represent the demand for item k and the total demand for the group, during period $t-1$. Using these most recent probability mixes, we can generate forecasts for individual items:

$$F_{kt} = P_{kt} * F_t$$

where F_{kt} represents the forecast for item k at time t given the total forecast F_t of the group at time t. Notice that the sum of all item forecasts in the group is always equal to the forecast for the group. The only remaining question is what value we should assign to the smoothing constant. The value of α can change over time, and its magnitude depends on the amount of trend in the total demand. The following rule of thumb is suggested:

$$\alpha = \min\left(\frac{D_t}{D}, \ 0.20\right)$$

where D_t represents the group's total demand during period t and D is the total demand for all items in the group last year. Both are known quantities. As is true with all simple exponential smoothing models, the vector smoothing should be used only when the demand is fairly constant or is slowly changing over time.

Example 3.2

Table 3.1 presents six months of data for three different types of exhaust pipes belonging to a group even though the firm forecasts demand only as a group. The November group forecast is twenty-three units.

1. Based on the past demands, generate forecasts for individual exhaust pipes for the month of November.
2. If the actual sales for individual items during November were ten, five, and twelve, whereas the group's forecast was thirty for December, generate individual item forecasts for December.

The total group demand during the previous year was 360 units.

Solution: Based on the six-month demand data, we can compute the probability of selling each item in November. If past demand is not known, we would have to estimate it from experience:

TABLE 3.1 Vector Smoothing Forecast

	Item 1		Item 2		Item 3		Group Total	
Month	Demand	Forecast	Demand	Forecast	Demand	Forecast	Demand	Forecast
May	7		4		8			
June	17		5		25			
July	20		8		15			
August	16		8		17			
September	14		7		16			
October	4		7		6			
Totals	78		39		87			
November	10	8.7	5	4.4	12	9.9	27	23
December		11.4		5.7		12.9		30

Note: Rounded to the nearest digit.

$$P_{1,\text{NOV}} = \frac{78}{204} = 0.38$$

$$P_{2,\text{NOV}} = \frac{39}{204} = 0.19$$

$$P_{3,\text{NOV}} = \frac{87}{204} = 0.43$$

Since the group forecast for November is twenty-three, we can compute item forecasts for November using the probability mix.

$$F_{1,\text{NOV}} = (0.38)(23) = 8.7$$
$$F_{2,\text{NOV}} = (0.19)(23) = 4.4$$
$$F_{3,\text{NOV}} = (0.43)(23) = 9.9$$

Once the actual demand for November is known, we can update the probability mix as follows. The values of α and q_{kt} are computed as

$$\alpha = \min\left(\frac{27}{360}, 0.20\right)$$
$$= 0.075$$

and

$$q_{1,\text{NOV}} = \frac{10}{27} = 0.37$$

$$q_{2,\text{NOV}} = \frac{5}{27} = 0.19$$

$$q_{3,\text{NOV}} = \frac{12}{27} = 0.44$$

Therefore,

$$P_{1,\text{DEC}} = (0.075)(0.37) + (0.925)(0.38) = 0.38$$
$$P_{2,\text{DEC}} = (0.075)(0.19) + (0.925)(0.19) = 0.19$$
$$P_{3,\text{DEC}} = (0.075)(0.44) + (0.925)(0.43) = 0.43$$

The group forecast for December is thirty. We can generate an item forecast (rounded to the nearest digit) as follows:

$$F_{1,\text{DEC}} = (0.38)(30) = 11.4$$

$$F_{2,\text{DEC}} = (0.19)(30) = 5.7$$

$$F_{3,\text{DEC}} = (0.43)(30) = 12.9$$

When the actual December demand for these items becomes known, we can similarly generate forecasts for January and so on.

The vector smoothing method can also be extended to forecast demand for each SKU item used in various distribution locations [8]. In this case, we have K different locations instead of several items in the group. We define P_{kt} as the probability of demand for location k at time t for the SKU instead of for an item. The following example illustrates this concept.

Example 3.3

The following data refer to SKU items in three different locations. The total group forecast for April is 11,500 units. Find April item forecasts.

$$P_{11,\text{MAR}} = 0.2 \qquad \alpha = 0.2 \qquad D_{11,\text{MAR}} = 2010 \qquad q_{1,\text{MAR}} = 0.22$$

$$P_{21,\text{MAR}} = 0.5 \qquad\qquad\qquad D_{21,\text{MAR}} = 4090 \qquad q_{2,\text{MAR}} = 0.44$$

$$P_{31,\text{MAR}} = 0.3 \qquad\qquad\qquad D_{31,\text{MAR}} = 3100 \qquad q_{3,\text{MAR}} = 0.34$$

Solution: Suppose that the probabilities were recomputed for April, using the same formula given in the previous section, as

$$P_{11,\text{APR}} = (0.2)(0.22) + (0.8)(0.20) = 0.20$$

$$P_{21,\text{APR}} = (0.2)(0.44) + (0.8)(0.50) = 0.49$$

$$P_{31,\text{APR}} = (0.2)(0.34) + (0.8)(0.30) = 0.31$$

The forecasts for April, rounded to the nearest integer, would be

$$F_{11,\text{APR}} = (0.20)(11,500) = 2300$$

$$F_{21,\text{APR}} = (0.49)(11,500) = 5635$$

$$F_{31,\text{APR}} = (0.31)(11,500) = 3565$$

Using these forecasts, planners are able to see how many units of the item should be stocked in each location.

The Blending Method

The blending method, developed by Cohen [3], can be used to forecast demands for end items belonging to a group of SKUs of an item located at several distribution points. With this method, a forecast for the entire group is blended with the average demand for each item. Let F be the forecast for the group at time t, and let σ be the corresponding standard error of the forecast. In addition, let $\overline{D}_k$ be the average demand for item k, and let S_k be the standard error associated with the average demand $\overline{D}_k$. According to basic statistics, the sum of all item averages in the group is equal to the group average $\overline{D}$. Therefore,

$$\overline{D} = \Sigma\, \overline{D}_k$$

Assuming that the demand for an item in the group is independent of other items in the group, we can also define the variance of the group average $\overline{D}$ as

$$S_{\overline{D}}^2 = \sum_{k=1}^{K} S_k^2$$

Using the past data, we can now find the forecast F_k for item k:

$$F_k = \overline{D}_k + w_k\, (F - \overline{D})$$

where the weighing (smoothing) factor is defined as

$$w_k = \frac{S_k^2}{(S_{\overline{D}}^2 + \sigma^2)}$$

The item forecast formula consists of two components. The first component represents the average item demand, whereas the second component makes an adjustment to the average. The adjustment is done by multiplying the difference between the group forecast and the demand by a weighing factor w_k that is specific for the item k.

Example 3.4

Using the data given in Example 3.2, find the forecasts for all items by blending methods for the month of November. The variance σ^2 associated with the group forecast $F = 23$ is given as 260.

Solution: Mean and sample standard deviation of each item are calculated as follows for the data given in Table 3.1:

Item	Mean	Sample Variance	Sample S.D.
1	13.0	38.4	6.20
2	6.5	2.7	1.64
3	14.5	46.7	6.83

The mean and variance of group average demand are

$$\overline{D} = \overline{D}_1 + \overline{D}_2 + \overline{D}_3$$

$$= 13 + 6.5 + 14.5 = 34$$

$$S_{\overline{D}}^2 = S_1^2 + S_2^2 + S_3^2$$

$$= 38.4 + 2.7 + 46.7 = 87.8$$

The weights for all individual items are

$$w_1 = \frac{S_1^2}{S_{\overline{D}}^2 + \sigma^2} = \frac{38.4}{87.8 + 260} = 0.110$$

Similarly,

$$w_2 = \frac{2.7}{87.8 + 260} = 0.008$$

$$w_3 = \frac{46.7}{87.8 + 260} = 0.134$$

and the November forecasts rounded to the nearest integer are

$$F_{1,\text{NOV}} = \overline{D}_1 + w_1 (F - \overline{D})$$

$$= 13 + (0.110)(23 - 34)$$

$$= 12$$

Similarly,

$$F_{2,\text{NOV}} = 6.5 + (0.008)(23 - 34) = 6$$

$$F_{3,\text{NOV}} = 14.5 + (0.134)(23 - 34) = 13$$

This example illustrates the use of the blending method for forecasting several items in a group. In a similar fashion, we can solve problems involving multiple stocking locations.

Percentage Done Estimating Method

The percentage done method illustrated by Hartung [4] and by Hertz and Schaffer [7] provides forecasts of the total season's demand for an individual item belonging to a product line or group. It centers on bringing the seasonality factor into the multi-item forecasting problem. This is particularly important in the retailing markets, where an item may be carried only during the season. Examples of such items include woodstoves, fans, clothing, and a variety of consumer goods. Many manufacturers of clothing, for example, also own a chain of retail stores. Therefore, it is very important to take all factors into consideration when determining how much of a particular item or of items in a group will be needed in stock during a season at each location and to devise a proper production planning and inventory control strategy.

In estimating the percentage done, the items are grouped into homogeneous lines. A line is a group of more or less similar merchandise, such as men's shirts or children's shoes. Although a line is usually carried year after year, additions and deletions of items to each line are encountered during consecutive seasons. A season is a portion of the year consisting of several periods. As goods are sold, we need to know how much inventory of what item needs to be replenished—in short, how much more of what item we need to manufacture.

The percentage done method assumes that at any given time of the season in every year, the same percentage of the total season's demand for an item is encountered. The method involves dividing the updated sales by the corresponding percentage to forecast the total demand for the season. From a knowledge of the inventory on hand and total sales to date, we can compute the remainder of the quantity to be produced for the current season. We can also extend this analysis to cover the same item stocked in several locations. The procedure consists of the following steps.

Step 1: Using prior year demand data, find the percentage of cumulative sales of the *group* of items for each time period.

Step 2: Using the current year cumulative demands for each *item* in the group and the corresponding percentage done estimates for the group in the previous year, generate item forecasts for this season's total demand.

Step 3: Calculate the quantity of each item in the group that needs to be produced.

Example 3.5

The cumulative sales and percentage done estimates of the previous year for three items are as follows:

	Cumulative Sales to Week t					
Item (k)	WEEK 1	WEEK 2	WEEK 3	WEEK 4	WEEK 5	WEEK 6
1	7	24	44	60	74	78
2	4	9	17	25	32	39
3	8	33	48	65	81	87
Totals	19	66	109	150	187	204
P_t	9.3	32.4	53.4	73.5	91.7	100

The demands for the items during week 1 were 15, 10, and 18, respectively. The inventories of the items at the start of season were 100, 50, and 120. Based on the results of the first week of demand, how many more of each item should the management expect to produce for this season?

Solution: The demands during week 1 for all items were

$$D_{11} = 15$$
$$D_{21} = 10$$
$$D_{31} = 18$$

Since we know that the percentage done estimates for the group for that period is 9.3%, the total season forecast can be calculated as follows:

$$F_1 = \frac{15}{0.093} = 161$$

$$F_2 = \frac{10}{0.093} = 108$$

$$F_3 = \frac{18}{0.093} = 194$$

From our knowledge of the total season's forecast F_k and the current cumulative production P_k, we can compute the remainder quantity of items R_k expected to be produced for the season:

$$R_1 = F_1 - P_1$$
$$= 161 - 100 = 61$$
$$R_2 = 107 - 50 = 57$$
$$R_3 = 194 - 120 = 74$$

Suppose that at the end of the second period of this season, week 2, we know the cumulative demand for all items to be

$$D_{12} = 55$$
$$D_{22} = 29$$
$$D_{32} = 65$$

We can update the estimates for total seasonal demand and future production requirements using (the percentage done estimate for period two) $P_2 = 0.324$:

$$F_1 = \frac{55}{0.324} = 170$$

$$F_2 = \frac{29}{0.324} = 90$$

$$F_3 = \frac{65}{0.324} = 201$$

and the remainder of the quantity expected to be produced for the season is updated as

$$R_1 = 170 - 100 = 70$$
$$R_2 = 90 - 50 = 40$$
$$R_3 = 201 - 120 = 81$$

Similarly, we can update our estimates during every period for planning production needs of all items.

Percentage of Aggregate Demands Method

The percentage of aggregate demands method was originally developed by Hausman and Sides [5] for forecasting demands of style goods. They dealt with a group of items sold through a catalog during a season consisting of many weeks. Many seasonal items have a slow pickup of demand at the beginning of the season and then reach a peak, after which the demand slows down. This pattern is typical of toys, perfumes, and many other items.

The first step in the method involves consolidating the weeks into intervals consisting of approximately equal total demands. For example, Figure 3.2a exhibits the demand curve for the group during the previous season. We divide the total area into several segments, resulting in unequal interval lengths, as shown in Figure 3.2b. We will update the forecasts periodically at these intervals for all items in the group. In this method, the season total forecast F for the group must be available.

Step 1:   Let D_{kT} represent the cumulative demand for item k at time T, and let D_T be the cumulative demand for the group. The percentage of sales at time $t = T$ for item k is

$$P_{kT} = \frac{D_{kT}}{D_T} = \frac{D_{kT}}{\Sigma D_{kT}}$$

Step 2:   The item forecast F_k for the season is given by

$$F_k = P_{kT} * F$$

We see that two variables affect the item forecast for the season. The group forecast F for the season is assumed to be known. It could change as the season progresses. The probability mix P_{kT} also could vary as changes occur in the demand patterns of various items in the group.

Step 3:   Using the item forecast for the season and up-to-date sales data, we can determine the forecast R_{kT} for the remainder of the season for all items:

$$R_{kT} = F_k - D_{kT}$$

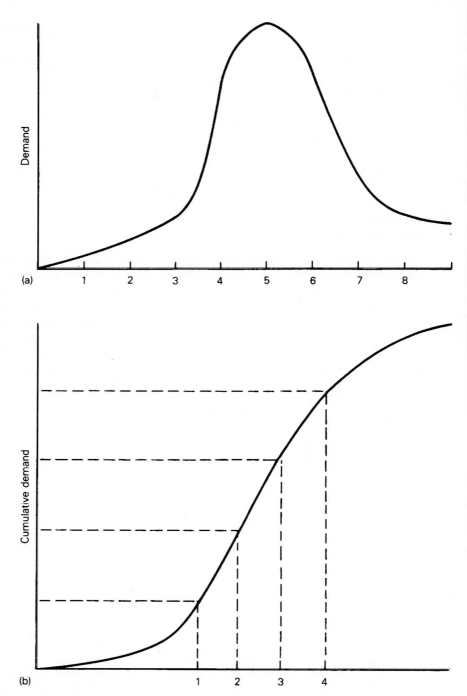

FIGURE 3.2 (a) Real time periods (b) periods for percentage of aggregate demand method.

From our knowledge of available inventory and a forecast for the remainder of the season, we can devise production plans. This concept can be extended for an item located at several distribution points. Although the solution procedure appears to be similar to the percentage done method, the difference should be obvious. The percentage done method forecasts season's demand for all items in the group, and the total group demand for the season is equal to the sum of individual item demands for the season. In the percentage of aggregate demand method, however, the group forecast for the season is assumed to be known, and we allocate the forecast to individual items.

Example 3.6

The demands for three items belonging to a group are twenty-four, nine, and thirty-three, respectively, up to this period. The total group forecast is 204 items for the season. Using the percentage of aggregate demand method, find the individual item forecast for the season. Also find the forecast for the remainder of the season.

Solution: The percentage of aggregate sales (demand) to date for the items is calculated first:

Total sales $D_T = 24 + 9 + 33 = 66$

$$P_{1,1} = \frac{24}{66} = 0.364$$

$$P_{2,1} = \frac{9}{66} = 0.136$$

$$P_{3,1} = \frac{33}{66} = 0.500$$

The season forecasts for individual items are

$F_1 = (204)(0.364) = 74$
$F_2 = (204)(0.136) = 28$
$F_3 = (204)(0.500) = 102$

The forecasts for the remainder of the season are

$R_1 = 74 - 24 = 50$
$R_2 = 28 - 9 = 19$
$R_3 = 102 - 33 = 69$

As mentioned earlier, it is important to keep in mind that the total forecast may vary as the season passes, and the percentage aggregate sales may vary as actual sales data are compiled. They are updated periodically according to the periods displayed in Figure 3.2.

SLOW-MOVING ITEM FORECASTING

Slow-moving items are those that have low demand, or periods of no demand. Peterson and Silver [9] suggest that an item that has a demand of ten or less over the replenishment lead time be classified as slow-moving. The trend toward faster replenishment may make this rule invalid. Hax and Candea [6] suggest an item with annual demand of 100 or lower is slow-moving, although they caution that fifty or lower may be a better threshold of slow-moving.

Lumpy demand may also be an indicator of a slow-moving item, although not all slow-moving items are lumpy nor are all lumpy demands slow-moving. We restrict our discussion in this section to items that are slow-moving and that may have lumpy characteristics, such as extended periods of no usage or very low usage. Brown [2] offers the following procedure to determine if an item has lumpy demand: Use one of the standard forecasting techniques on the historical data. Calculate the standard deviation of the residual differences between the historical data and the best model that could be found. If the standard deviation of these residual differences is greater than the level in the forecast model, than the item exhibits lumpy demand.

Exponential Smoothing

Hax and Candea [6] offer the standard exponential smoothing method as a possible forecasting method for slow-moving items. Because there may be a large amount of noise in the demand pattern, a large smoothing constant would not be appropriate. The standard exponential smoothing model with a very small smoothing constant, α between 0.01 and 0.05, may dampen the instability of the demand with lumpy and low demand items.

Example 3.7

The demand over the last 12 weeks is

Week:	1	2	3	4	5	6	7	8	9	10
Demand:	1	2	0	1	0	4	2	0	2	1

Week	Demand	$\alpha = 0.3$ Forecast	$\alpha = 0.01$ Forecast
1	1		
2	2	1.0	1.0
3	0	1.3	1.01
4	1	0.91	1.00
5	0	0.94	1.00
6	4	0.66	0.99
7	2	1.66	1.02
8	0	1.76	1.03
9	2	1.23	1.02
10	1	1.46	1.03

starting with $F_0 = 1$.

In this example we see that the higher α value forecast is much more erratic than the lower α value forecast. Figure 3.3 shows the effect of a very small α value on the forecast. With an α value that small, there is a very large amount of past history averaged into the forecast. Changes in the demand will have little effect on the forecast. If a trend or significant change in the demand did occur, then the overall demand would no longer be classified as slow-moving and the α value could be adjusted accordingly.

Vector Smoothing

Another method is the vector smoothing method for a single item, described earlier in this chapter.

Example 3.8

Consider a case where the demand has the following type of distribution:

K	Interval	Frequency	$P_{k,51}$
1	0–1	30	0.60
2	2–3	15	0.30
3	4–5	5	0.10
		50	

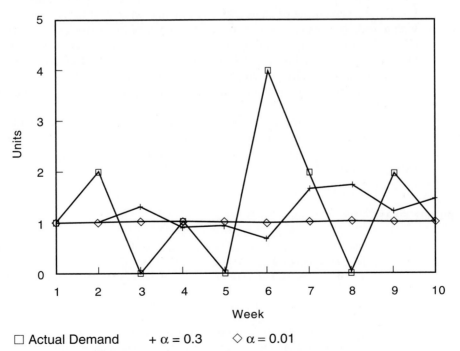

FIGURE 3.3 Exponential smoothing forecasts of a slow-moving item.

This table of empirical data shows that the probability of the demand in the fifty-first week has a 60% probability of falling in the first interval of 0 or 1. If the actual demand in the fifty-first week is 1, the $P_{k,52}$ value would be found in the manner shown earlier in this chapter. The value of α should be low as in the previous section. The q value is 1 for the first interval and 0 for all other intervals.

$$P_{k,52} = \alpha q_{k,51} + (1-\alpha)P_{k,51} = (0.01)\begin{bmatrix}1\\0\\0\end{bmatrix} + (1-0.01)\begin{bmatrix}0.60\\0.30\\0.10\end{bmatrix} = \begin{bmatrix}0.604\\0.297\\0.099\end{bmatrix}$$

SUMMARY

In this chapter we discussed several models for forecasting demand for multiple items belonging to a family or for a single item stocked at several locations. The vector smoothing method provides the updated probabilities of items for every period. These probabilities are then used to prorate individual item forecasts. The blending method gives the weights for every item in the family. Using these weights in an exponential smoothing model, we can generate individual item forecasts. The percentage done method deals specifically with a family of items whose demand is restricted to a season. Finally, the percentage of aggregate demand method is used for seasonal demand items with a peak. It is also possible to forecast a family of items stocked in several locations. Cohen [3] extends the blending method to solve what are known as multiclassified problems. Commercial packages incorporating some of these techniques are available in the marketplace. Finally, we emphasized that better forecasts help to consolidate items being transported to a warehouse and hence reduce the total cost of production and logistics.

Slow-moving items present forecasting difficulty. The standard exponential smoothing method may work if the smoothing constant is very low. The vector smoothing method for a single item can also be used to forecast the probability of demand within intervals.

PROBLEMS

1. The following are the actual demand data for the Knit Picker; group the data into approximately six class intervals: 127, 139, 135, 148, 162, 137, 171, 150, 120, 149, 140, 155, 157, 156, 91, 142, 120, 173, 142, 149, 128, 149, 120, 169, 145, 156, 185, 151, 184, 130.
2. Calculate the probability distribution of demand for the Knit Picker. Given that the actual demand during the thirty-first week was 165, find the updated probabilities for the distribution of demand using the following values for α: 0.1, 0.2, and 0.3.
3. The following are the class intervals and the corresponding frequency distributions of demand for Kailees:

K	Interval	Frequency
1	0–19	15
2	20–39	20
3	40–59	15
4	60–79	10
5	80–100	5

Find the probability distribution for the demand for Kailees. Given that sixty-five Kailees were sold during the most recent period, provide the updated probabilities for various intervals of demand. Use $\alpha = 0.2$.

4. The following is the distribution of the cookie boxes sold by the scoutmaster for the past seventy days:

Number of Cookie Boxes	Frequency
0–14	4
15–29	20
30–44	24
45–59	11
60–74	6
75–89	5

Find the probability distribution for the sales of cookie boxes. Given that fifty-one cookie boxes were sold during the most recent period, find the updated probabilities for various sales ranges. Use $\alpha = 0.2$.

5. A radio manufacturer who has decided to enter the market with a new product has sought the opinion of the new product manager. Her answers for the past thirty days are as follows: 3, 4, 5, 2, 3, 4, 3, 5, 4, 3, 4, 3, 3, 2, 5, 4, 2, 3, 3, 4, 3, 5, 4, 3, 3, 2, 4, 3, 4, 5. Find the probability distribution for the number of models contemplated. Based on the given information, what are the updated probabilities for the next period? Use $\alpha = 0.2$.

6. The following table exhibits the actual demand data for the past five months:

Month	Item 1	Item 2	Item 3
1	100	70	80
2	110	80	75
3	125	60	95
4	105	70	90
5	95	65	98

a. Using the vector smoothing method, generate values of probabilities of demand for individual items.

b. If the actual sales for the sixth month were 100, 70, and 95 for items 1, 2, and 3, respectively, generate current estimates of item forecasts for the seventh month using the vector smoothing method, given the group forecast for the seventh month as 300.

7. Using the sales data given in problem 6, provide an item forecast for the seventh month using the blending method. The variance associated with the group forecast is given as 350.

8. The following table gives sales data for the Piedmont Fertilizer Company for a line of products during the past year:

Item	March	April	May	June	July	August
PX1	10	15	13	16	21	8
PX2	8	10	13	16	19	10
PX3	17	22	18	26	30	20

The demand for these products during March and April and the inventories at the start of the season are as follows:

Item	March	April	Inventory
PX1	11	17	50
PX2	10	13	50
PX3	20	23	70

Based on first week's sales data, provide item forecasts for the remainder of the season, using the percentage done estimating method.

9. In problem 8, how many more of each product in the line should Piedmont expect to produce this season?
   a. Based on March sales data only.
   b. Based on March and April sales data.

10. Donald King caters hamburgers at the beach. The total forecast for this season is 5800. To date, the demand for single, double, and triple hamburgers has been 300, 800, and 400. Using the percentage of aggregate demand method, find the individual hamburger forecast for the season. For the remainder of this season, how much should Donald King expect to sell?

11. The following are the sales data for four different items belonging to a group. The firm forecasts demand only as a group. August forecast is 760 units.

Month	Item 1	Item 2	Item 3	Item 4
January	260	94	125	197
February	251	90	129	188
March	235	100	126	186
April	287	99	137	185
May	304	98	140	173
June	290	96	135	188
July	315	97	142	193

Based on past sales data, generate a forecast for individual items for the month of August.

12. If the actual sales in August were 310, 105, 125, and 205, respectively, for the four items listed in problem 11, and if the September group forecast is 800, generate individual item forecasts for September.

13. Using the sales data given in problem 11, generate an item forecast for August by the blending method. The variance associated with the group forecast is 525.

14. The following are the weekly sales for a family of items during a season last year:

Item	Week 1	Week 2	Week 3	Week 4	Week 5	Week 6	Week 7
1	400	500	600	600	500	400	300
2	100	100	200	200	200	100	50
3	250	300	300	300	200	100	100

If the demands for these items during the first week of this season were 450, 100, and 275, respectively, and the inventories at the start of the season were 2000, 500, and 1100, based on the first week's sales data generate item forecasts for the remainder of the season. How many more of each item should the firm expect to produce this season?

15. If the demands for the items in problem 14 during the second week of the season are 450, 100, and 250, respectively, how would your estimates vary?

16. The demands for the items in problem 14 during the third week of the season are 600, 250, and 350, respectively. What are your new estimates of sales for the remainder of this season?

17. The demands for three items belonging to a group are 260, 94, and 125 up to this period. Total group forecast for the season is 2580. Using the percentage of aggregate demand method, find the individual item forecast for the season. Also find the forecasts for the remainder of the season.

18. The demands for the three items in problem 17 to date were found to be 700, 250, and 390. The total group forecast for the season remains the same. Using the percentage of aggregate demand method, find the individual item forecast for the season.

19. With the information given in problem 18, find the forecast for the remainder of the season.

20. The following table exhibits the sales data for XYZ manufacturing company for a line of products during the past year:

Periods

ITEM	1	2	3	4
1	11	14	18	10
2	26	31	32	30
3	5	6	8	7

If the demand for these products during the first period at the start of the season was 20, 30, and 6, respectively, provide item forecasts for the remainder of the season using the percentage done estimating method.

21. The demand for a slow-moving item over the last twelve weeks has been 1, 0, 0, 0, 3, 0, 2, 0, 1, 0, 2, and 1. Using $\alpha = 0.02$ and $F_0 = 1.3$, find the forecast for each week and for the thirteenth week using the exponential smoothing method.

22. The following is the actual demand history over the last twenty-six weeks for a slow-moving item:

K	Interval	Frequency	$P_{k,27}$
1	0–1	10	
2	2–3	13	
3	4–5	3	
		26	

    a. Find $P_{k,27}$ for each interval.

    b. The demand for week 27 was 3. Find the new $P_{k,28}$ using $\alpha = 0.02$.

    c. The demand for week 28 was 0. Find the new $P_{k,29}$ using $\alpha = 0.02$.

REFERENCES AND BIBLIOGRAPHY

1. R. G. Brown, *Smoothing, Forecasting and Prediction of Discrete Time Series* (Englewood Cliffs, N.J.: Prentice Hall, 1962).

2. R. G. Brown, *Materials Management Systems* (New York: John Wiley & Sons, 1977).

3. G. D. Cohen, "Bayesian Adjustment of Sales Forecasts in Multi Item Inventory Control Systems," *Journal of Industrial Engineering,* Vol. 17, No. 9 (1966), pp. 474–479.

4. P. Hartung, "A Simple Style Goods Inventory model," *Management Science,* Vol. 19, No. 2 (August 1973), pp. 1452–1458.

5. W. P. Hausman and R. S. G. Sides, "Mail Order Demands for Style Goods: Theory and Data Analysis," *Management Science,* Vol. 20, No. 2 (October 1973), pp. 191–202.

6. A. C. Hax and D. Candea, *Production and Inventory Management* (Englewood Cliffs, N.J.: Prentice Hall, 1984).

7. D. B. Hertz and K. H. Schaffer, "A Forecasting Model for Management of Seasonal Style Goods Inventories," *Operations Research,* Vol. 8, No. 2 (1960), pp. 45–52.

8. R. C. Link, "Richard C. Link Report," Winston-Salem, N.C., January 1983.

9. R. Peterson and E. A. Silver, *Decision Systems for Inventory Management and Production Planning* (New York: John Wiley & Sons, 1979).

10. N. T. Thomopoulos, *Applied Forecasting Methods.* (Englewood Cliffs, N.J.: Prentice Hall, 1980).

PART TWO

MATERIALS MANAGEMENT

Part I of this book dealt with forecasting, a component of demand management. Once forecasts are derived, raw materials, components, and assemblies must be available to manufacture necessary end items. Therefore, we need to procure those components parts and manage those inventories (cost) effectively so that items demanded by customers can be manufactured and delivered on time. Hence, material management is an important dimension in the study of manufacturing planning and control.

Arthur Young's framework for competitive advantage is repeated in Figure II.1. Level 3 of Arthur Young's framework encompasses supplier coordination, distribution, and customer service and support as well as necessary manufacturing functions. This level shows the importance of coordination for linking suppliers and customers with the manufacturing planning and control systems of a firm. When dealing with suppliers, our objective is to make sure that items are delivered to us on time and in a cost-effective manner. Similarly, the completed items should reach our customer's destinations in a timely manner while satisfying their needs (customer service levels).

Part II addresses many of these inventory management issues. Chapter 4 deals with the basics of single-item inventory management. We explain why firms hold inventories, we describe economical order quantity concepts, and we illustrate the usefulness of ABC classification in inventory analysis. We also address inventory system implementation issues. Chapter 5 extends these techniques for multiple-item inventory management. In Chapter 6 we introduce lead time, the concept of risk, and safety stock levels. We analyze the effect of lead time on safety stock levels and the trade-off between safety stock levels and associated customer service levels. In addition, Chapter 6 shows how we can jointly determine the order quantity and the reorder point so that the total cost of maintaining inventories are minimized.

Chapter 7 deals with aggregate inventory management where we analyze the trade-offs between inventory holding costs and customer service levels for a number of items

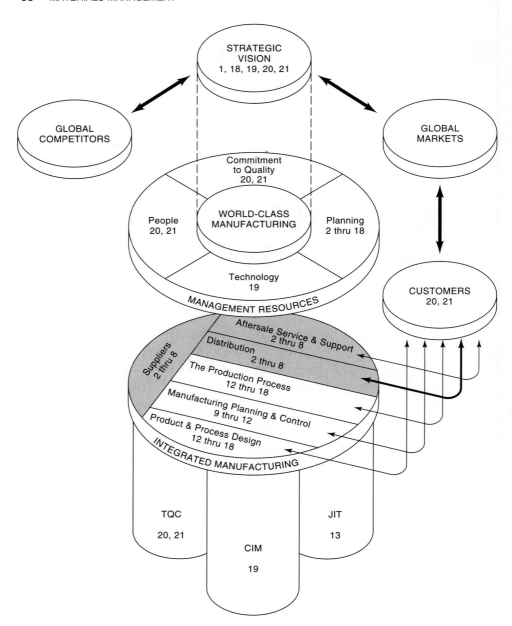

FIGURE II.1   Arthur Young's Manufacturing for competitive advantage framework.[SM] Copyright ® 1987 by Ernst Young (formerly Arthur Young, a member of Arthur Young International). Reprinted by permission. Note: numbers on this figure indicates chapter numbers in the book covering these subjects.

in different situations. We use the LIMIT technique as well as the LaGrange multiplier technique. These models can be applied for managing inventories from the suppliers as well as distribution inventory management at various warehouse locations. Chapter 8 integrates procurement, production, and delivery of goods to the distribution warehouse (or the customer) and thus provides an integrated treatment of manufacturing planning and control. In that chapter we analyze the trade-off between the number of warehouses and corresponding inventory levels required to maintain the same customer service level. We also provide some simple techniques for determining appropriate shipping quantities from single or multiple sources to single or multiple destinations. In Chapter 8 we also describe the concept of distribution requirements planning (DRP).

These chapters can be sequenced in several ways. We chose to deal with the deterministic aspects first and then with risk analysis. In that process we dealt with deterministic inventory management for single and multiple items first in chapters 4 and 5, and then follow with the safety stock level calculations in Chapter 6. These chapters are followed by multiple-item inventory management and risk analysis in Chapter 7. The distribution management addressed in Chapter 8 requires the essentials of material requirements planning (MRP); thus we could have moved this chapter to Part III. We did not, however, want to delay the chapter until MRP is discussed because we wanted to deal with the integration of manufacturing planning and control as well as inventory management aspects as soon as we could. Similarly, MRP can also be considered as a component of material management and the MRP chapter could have been shifted to Part II. We concluded, however, that MRP is more relevant in the context of capacity planning. Hence, the MRP and master production scheduling chapters were included in the manufacturing planning activities part of this book.

The following chapters are in Part II:

Chapter 4, Basic Inventory Systems

Chapter 5, Multi-Item Joint Replenishment

Chapter 6, Inventory Systems under Risk

Chapter 7, Aggregate Inventory Management

Chapter 8, Distribution Inventory Management

Chapter Four

Basic
Inventory
Systems

INTRODUCTION

Inventory control is a critical aspect of successful management. With high carrying costs, companies cannot afford to have any money tied up in excess inventories. The objectives of good customer service and efficient production must be met at minimum inventory levels. This is true even though inflation causes finished goods inventories to increase in value. Putting inventory on the shelf ties up money, and to minimize the amount tied up, a company must match the timing of demand and supply so that the inventory goes on the shelf just in time for the customer to require it.

In this chapter, we develop several systems for handling inventory trade-offs under varying conditions. All the models presented have been implemented in various companies. We discuss how the models work and when they are applicable. It is important to realize that there is no all-purpose, automatic inventory control system; all systems need the intervention and monitoring of intelligent users.

FUNCTIONS AND TYPES
OF INVENTORIES

Inventory is a stock of physical goods held at a specific location at a specific time. Each distinct item in the inventory at a location is termed a stock keeping unit (SKU), and each SKU has a number of units in stock. Each location is a stock point. The local supermarket, for example, is a stock point with a huge inventory of food. Dairy Farms 2% milk in half-gallon containers is an SKU with a specific number of units in stock.

Why do companies keep inventories? Inventories exist because demand and supply cannot be matched for physical and economic reasons. We go to the supermarket to buy a half-gallon container of milk. How could the store supply it without inventorying milk? Our demand obviously cannot be matched to the cow's supply in time, place, or form.

Transaction Stocks

Transaction stocks are those necessary to support the transformation, movement, and sales operations of the firm. Active *work-in-process stocks*—materials currently being worked on or moving between work centers—constitute a large part of transaction stocks, as do *pipeline inventories.* Pipeline or *transportation inventories* are inventories in transit. The size of the pipeline inventory is as much a function of the length of the pipeline as of the rate of sales at the retail stock point. Figure 4.1 compares two pipeline systems to show that a longer pipeline requires a larger inventory to match the same sales rate.

Transaction stocks cannot be reduced, because they support the sales rate directly. There are no frills in transaction stocks.

Organization Stocks

Organization stocks represent investment opportunities to achieve operating efficiencies. *Fluctuation* or *safety stock* is an organization stock designed to buffer against uncertainty. Average daily sales of twenty containers of milk, for example, can be met by a transaction stock of twenty units. Sales above twenty would have to be supported by a buffer stock held to avoid stockouts when sales are higher than expected.

Anticipation inventory or *leveling inventory* may be an attractive investment if it is cheaper to hold stock than to alter short-term production capacity. Seasonal peaks in demand may be met by building inventories earlier during periods of slack demand and excess capacity.

Lot size or *cycle inventories* are held to achieve some payoff from setting up equipment. Having set up equipment, manufacturing people invariably want a long production run to avoid repeating the setup for the same item in the near future. Going to the bank to cash a check, for example, involves travel time and down time from other activities. For

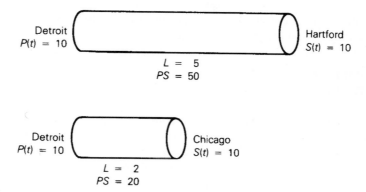

FIGURE 4.1   Pipeline inventories. Key: $P(t)$ = daily production rate, $S(t)$ = daily sales rate, L = lead time in days, PS = pipeline stock.

that reason, most of us carry a lot size or cycle stock in our wallets to avoid going to the bank every time we want to make a purchase.

The last two types of organization stocks are more specialized investment opportunities. *Scheduling stocks* are work-in-process stocks held between operations to allow schedulers a choice of jobs to place on the productive resource. In this way, high resource utilization can be achieved. *Speculative stocks* are those held in anticipation of price increases.

Excess Stock

Excess stock has no purpose. Unlike transaction and organization stock, it owes its existence to oversight rather than to necessity or to operating efficiency.

Levels of Inventory

Within the framework of transaction, organization, and excess stock, inventory may occur at various levels or echelons within the company. An *echelon, level,* or *stage* is a stock point that is under control of the company. Raw materials, work in process, high-level components, and finished products belong to different echelons. *Raw materials* are raw in the sense that the company has not done any work on them. *Work-in-process* inventories are manufacturing inventories that are undergoing processing or are in line at *work centers,* centers with similar personnel/machine capabilities. *High-level components* are parts and assemblies that are ready to be assembled into the finished product. These are often stored ready to be assembled when needed. *Finished goods* are products that are ready to be shipped to the customer.

MEASURES OF INVENTORY SYSTEM PERFORMANCE

Return on investment (ROI) is very important to top managers who are accountable for company profitability. Where do inventories fit in the company scheme? Consider the following simplified ROI analysis:

$$ROI = \frac{sales - cost\ of\ good\ sold}{physical + receivables + inventory}$$

Inventories represent 25% of the assets of many companies. Of all the elements in the ROI formula, inventory offers the most promise for most managers and consultants. A decrease in the inventory investment can lead to a fast improvement in ROI.

Considering the inventory system on its own, however, one finds that the performance measures reflect the interests of the inventory system participants. Marketing creates a customer service measure: A certain number of orders should be shipped complete from stock or a certain percentage of units demanded should be shipped without backorder. Stockouts mean poor service, and an unacceptable stockout record probably means termination for the responsible inventory manager.

Financial managers think in terms of cost: The less inventory, the better. Probably because they do not like borrowing money at high interest rates, financial people fear the holding costs of inventory. Since these costs are relatively easy to quantify, many financial people have undue influence on inventory systems. They erroneously assume that all inventory is excess stock or that the only inventory should be transaction stock. Of course, financial managers who are enlightened by a course in production control understand that organization stocks must be analyzed as an investment.

In addition to holding costs, financial people also think in terms of inventory turns. Inventory turnover is the ratio of the cost of sales in a period to the cost of average inventory on hand. For example, suppose that the cost of sales last month was $100,000, with a beginning inventory of $40,000 and an ending inventory of $60,000 for an average of $50,000. This equates to two turns per period. Although it is interesting, the inventory turns measure varies so greatly by type of business that we recommend it be viewed judiciously if not hostilely.

Manufacturing people are also involved in inventory decisions. Hurt by excessive setups and down time, is it any wonder that they become frustrated by short production runs when they are asked to produce the same item again this week after a very short run two weeks ago? Unfortunately, most companies keep poor records of setup costs and do not factor them into inventory investment targets as well as they might. This is partly because of the difficulty in providing reasonable setup cost estimates. The lower status of the manufacturing people relative to the financial and marketing people in some companies also contributes to a relative lack of attention to setup costs.

Inventory planning and control requires trade-offs between the three major system objectives: customer service, inventory investment, and production efficiency. Explicit or implicit costs associated with these objectives always exist, regardless of whether or not they can be measured accurately.

Holding or *carrying costs* are costs relevant to the inventory decision of when and how much to order. They represent future cash flows that will be changed by decisions to hold more or less inventory. Insurance, obsolescence, deterioration, property taxes, and the cost of capital can add up to 40% of the cost of the item in today's economy. Clearly, holding costs are relevant to the decision to order one more unit. (In this text, we use the terms *holding* and *carrying costs* interchangeably.) If a company faces out-of-pocket costs of $10 per unit to put an item on the shelf and if it must borrow money at 25%, then the cost to carry that extra item is $2.50 plus the cost of insurance, obsolescence, deterioration, and tax. The company would be willing to pay up to $2.50 to retrieve $10. If there were an opportunity to earn 30% on an investment elsewhere in the company, management would be willing to pay up to $3 to retrieve $10. The key concept is one of out-of-pocket costs. If stock already exists at an earlier stage in manufacture, the holding cost at a later stage is the incremental investment of ordering the items through to the next stage, often called the *echelon holding cost.*

Setup costs or *ordering costs* that vary with the frequency of ordering must be considered in inventory decisions. Clerical costs of order preparation and order receiving must be carefully examined to ensure that only marginal clerical costs are counted as ordering costs. Labor costs involved in setting up equipment make up setup costs. *Profits foregone* because of down time, which are often neglected, must be included when equipment is being operated at capacity. In the retail and distribution arena, the term *ordering cost* is used; the term *setup cost* generally refers to the sum of setup and ordering cost in a manufacturing environment.

Stockout costs are almost impossible to measure. A stockout occurs when a customer demands an item for which there is no inventory to meet the demand. The stockout may become a lost sale or a backorder. If it becomes a lost sale, revenue is lost. If it becomes a backorder, extra clerical costs and expediting costs occur. Both cases cause loss of customer goodwill.

Of the various measures of inventory system performance, holding costs have received the greatest attention. For example, a popular inventory control system categorizes inventory by dollar volume with the implicit objective of isolating those SKUs that tie up the most dollars. Since it appears in so many companies, we begin our study of inventory control methods with the holding cost–oriented ABC system.

INVENTORY DISTRIBUTION BY VALUE: THE ABC SYSTEM

Not all customers and not all SKUs are equally important. The company president's golfing buddy should receive top priority, as should SKUs typing up an exorbitant proportion of dollars. The ABC system of inventory planning recognizes that 20% of the SKUs will account for 80% of the dollar value of the inventory. Consider the following case of five SKUs:

SKU	Annual Demand	Cost	Dollar Volume
1	5,000	$2	$10,000
2	1,000	$2	$ 2,000
3	10,000	$8	$80,000
4	5,000	$1	$ 5,000
5	1,500	$2	$ 3,000
			$100,000

Ranking these by dollar volume, we have

Label	SKU	Dollar Volume	% SKUs	% Total Dollar Volume
A	3	$80,000	20%	80%
B	1	$10,000	20%	10%
B	4	$ 5,000	20%	5%
C	5	$ 3,000	20%	3%
C	2	$ 2,000	20%	2%

Usually, the ABC system picks 15% to 20% of the items, representing 80% of dollar value, to be A items. Here SKU 3 would be an A item. Next, about 30% to 40% of the items form a B category, accounting for 15% of the total. Here, SKU 1 and SKU 4 would become B items. The rest are C items. This pattern has been replicated over and over in many companies: A = 20% SKUs/80% value, B = 40% SKUs/15% value, and C = 40% SKUs/5% value.

For purposes of forecasting, inventory control, and scheduling, smart managers keep their eyes on A items personally. No automatic forecasting system or automatic inventory control system will handle these items without managers' continuing intervention.

What kind of control should we exercise in implementing an inventory control system? What are the trade-offs between the cost of controlling the system and the potential benefits derived from the control system? When companies are dealing with hundreds of items in the inventory system, not all items need the same amount of attention. ABC analysis provides the inventory control manager with some useful guidelines for identifying the type of control those items require for effective inventory management.

Because A items are expensive and constitutes a major proportion of the annual revenue, an optimal policy that minimizes the investment in A items should be pursued. For example, these items should be monitored continually while more sophisticated forecasting procedures are adopted. C items should be overstocked, if necessary, so that little or no control is exercised on these items. Large lot sizes of C items could also be used to minimize the frequency of ordering while exercising a minimal degree of control. Judgment is exercised in handling B items. Usually, some of these items are treated like A items while rest are treated like C items. This amounts to classifying all items into an AB type of classification system instead of an ABC classification. Instead, B items could be reviewed periodically and these items could be ordered in groups rather than individually.

As computerized inventory management systems are implemented, accurate and timely records can be obtained more economically on all items. In situations where computerized inventory control systems are installed for all items, the ABC classification system takes a back seat.

INVENTORY SYSTEMS

With a knowledge of the inventory costs involved and the selective perception suggested by the ABC system, we are now ready to study inventory systems designed to handle cost trade-offs. As we might expect, different systems are appropriate for different item categories. The fundamental decisions of inventory management are (1) when to order and (2) how much to order. To answer these questions, we must trade off costs, we must know what is likely to be sold, and we must know how much we have now. Sales forecasting, inventory record keeping, and inventory decision rules form the basis of most inventory control systems.

In this chapter, we confine our attention to the when and how much questions with very simple systems. Although the most elementary models are not encountered in practice, an understanding of them allows us to work with the more complex models. Inventory systems under certainty take demand as fixed at a specified rate. Once we understand these systems, we can modify them to account for uncertainty.

The Basic Order Point/Order Quantity System

Suppose that annual demand for an item is $D = 10{,}000$ units. Throughout this chapter, we will assume a five-day work week with two weeks' vacation in July, giving us a 250-day year. The demand rate would then be $d = 10{,}000/250 = 40$ per day. (We will forgo the use of Greek symbols for rates.) Further, annual per unit holding costs (h) are 40% of the cost of the item. The item's cost, $c = \$10$, gives $h = 0.4 * \$10 = \$4$. Setup costs are $S = \$500$. When and how much should the company produce? Table 4.1 displays three *order quantities* or *lot sizes*. (We use the terms *order quantity* and *production quantity* somewhat interchangeably.) Immediately, we rule out ordering one unit at a time, since the $5,000,000 annual setup cost is prohibitive. With annual demand $D = 10{,}000$ and an order quantity $Q = 1000$, we see that there would be $10{,}000/1000 = 10$ orders in a year; that is, $D/Q = 10$. The annual setup cost is simply the number of orders times the setup cost per order, or $(D/Q) * S$. For $Q = 1000$, we have annual setup cost of $(10{,}000/1000) * \$500$. Annual setup costs increase as Q decreases, as can be seen in Figure 4.2. Marginal analysis reveals that annual setup costs A are decreasing with Q at the rate of $-DS/Q^2$, since $dA/dQ = -DS/Q^2$. Raising Q from 1000 to 2000 saves $2500, but raising it from 4000 to 5000 saves only $250.

At the same time that annual setup costs decrease with increases in Q, annual holding costs increase. With an order size of $Q = 1000$, average inventory would be $Q/2 = 500$. This can be seen in Figure 4.3. Over the first 100 days of the year, we move through four cycles, from 1000 units in inventory to zero units, losing $d = 40$ units per day from inventory. Inventory declines linearly between $A = 1000$ and $B = 0$. For such a linear function, the average falls at the midpoint or geometric balance point of $(A + B)/2$. Since $A = Q$ and $B = 0$, average inventory is $Q/2$. This model thus depends heavily on the assumption of a constant sales rate producing a linear decline in the inventory position. In such a case, the annual holding costs are $(Q/2)h$. In our case, $h = 0.40 * \$10 = \$4$, giving an annual holding cost line of $\$4Q/2 = \$2Q$. This is pictured in Figure 4.4.

Using marginal analysis, the annual holding costs H are increasing with Q at the rate of $2, since

$$\frac{dH}{dQ} = \frac{d\left(\frac{Q}{2} * h\right)}{dQ} = \frac{h}{2} = \frac{4}{2} = 2$$

TABLE 4.1   Setup Costs for Trial Lot Sizes

Order Quantity (Q)	Number of Orders	Annual Setup Cost ($)
10,000	1	500
1,000	10	5,000
1	10,000	5,000,000

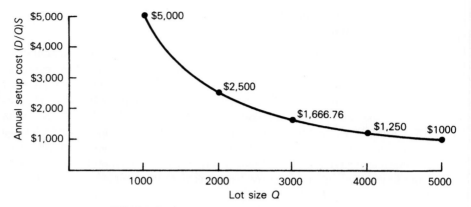

FIGURE 4.2 Annual setup cost as a function of lot size.

Raising Q from 1000 to 2000 units increases annual holding costs by $2000, exactly the same cost increase as would occur in moving from $Q = 4000$ units to $Q = 5000$ units.

We should be willing to raise Q as long as the incremental savings in setup cost overcome the incremental increase in holding cost. Annual setup cost changes by $-DS/Q^2$, which is equivalent to a savings of DS/Q^2. Equating the savings with the incremental holding cost, we have $DS/Q^2 = h/2$. Solving for Q, we have

$$Q* = \sqrt{\frac{2DS}{h}}$$

where $Q*$ is the famous economic order quantity (EOQ) formula of 1915 vintage. Applied to our example, the optimal order quantity is

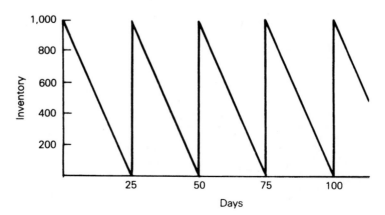

FIGURE 4.3 Inventory cycles.

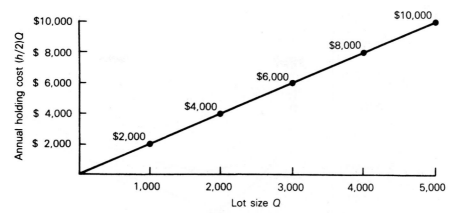

FIGURE 4.4 Annual holding cost as a function of lot size.

$$Q^* = \sqrt{\frac{2 * 10,000 * 500}{4}} = 1581$$

At the EOQ of 1581, the incremental setup cost of $2 per unit is exactly balanced with the incremental holding cost of $2 per unit; that is, $DS/Q^2 = h/2$. Peculiar to the cost functions in this model, annual setup costs happen to equal annual holding costs at the optimal Q; that is, $DS/Q = (Q/2)h$. Equating incremental costs equates annual costs in this case, but you should not accept this as a general principle for other cost functions.

Although the EOQ model has assumed a constant demand rate, the model itself is rather robust. Changes in annual demand or in the ratio of setup to holding costs do not cause severe changes in Q. Similarly, small changes in Q do not cause large changes in total relevant costs (TRC). To see this, consider the inventory cost equation TRC(Q) = $(Q/2)h + (D/Q)S$. We consider only costs that are relevant to the inventory decision, using the symbol TRC (Q) to mean total relevant costs dependent on Q. Figure 4.5 combines the holding costs of Figure 4.4 and the setup costs of Figure 4.2 into a total relevant cost figure.

At the optimal $Q = 1581$, total relevant costs are

$$TRC = \frac{1581}{2} * 4 + \frac{10,000}{1581} * 500 = \$3162 + \$3162 = \$6324$$

Doubling demand from 10,000 to 20,000 will not double Q but will raise it to 2236 or $\sqrt{2}$ times the old EOQ. Inserting doubled demand in the EOQ formula, we obtain the new EOQ:

$$EOQ = \sqrt{\frac{2(2D)S}{h}} = \sqrt{2}\sqrt{\frac{2DS}{h}}$$

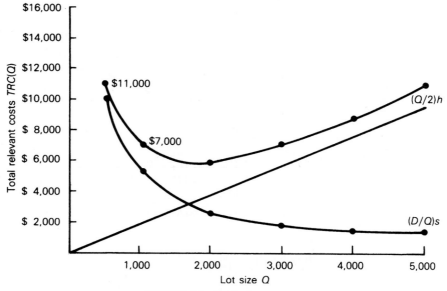

FIGURE 4.5 Total relevant costs.

Similarly, total relevant costs are not very sensitive to small deviations from EOQ. At $Q = 1000$, TRC $(Q) = \$7000$; at $Q = 2000$, TRC$(Q) = \$6500$. Neither figure is far from TRC$(Q) = \$6324$ for the EOQ.

In the scenario so far, we have been showing the next order arriving at precisely the time the first lot is used up. EOQ has answered the question of *how much* to order. Now we need to figure out *when* to order. We shall define *lead time* as the time between placing or releasing the order and receiving it. Assume that the lead time in our example is five days ($L = 5$). Because demand is certain, five days of supply will cover lead time demand. If we set a reorder point (ROP) at $Ld = 5 * 40 = 200$ units to cover lead time demand, then we will order $Q^* = 1581$ units when inventory on hand reaches 200 units.

Production Rate Model

The basic EOQ/ROP model has assumed that the entire lot quantity is delivered at the end of the fixed lead time. In some manufacturing situations, however, the end of the lead time may signal the receipt of the first items of the production run. At a production rate of 100 units per day, it would take $Q/p = 1000/100 = 10$ days to produce a lot size of 1000 units. During that time, $(Q/p)d = 10 * 40 = 400$ units would be lost to sales. Effectively, this reduces the annual holding costs, because we never hold 1000 units in stock but only a maximum of 600 units, $Q - (Q/p)d$. Consequently, we may be able to save money by increasing the lot size to reflect that we are not holding every unit we receive but are passing some along directly to the customer. For example, if I set out to stack a cord of wood at the rate of one cord per week, and if I burn one-tenth of a cord per week, my maximum inventory will be nine-tenths of a cord unless I try to stack more than a cord per week.

Using logic similar to that in the basic EOQ model, we see that annual holding costs $1/2[Q - (Q/p)d]h$ are calculated on half the maximum inventory level. Marginal holding costs would then be

$$\frac{dH}{dQ} = \frac{h}{2}\left(1 - \frac{d}{p}\right)$$

Equating incremental holdings costs and setup savings,

$$\frac{DS}{Q^2} = \frac{h}{2}\left(1 - \frac{d}{p}\right)$$

we can derive an economic production quantity (EPQ):

$$\text{EPQ} = \sqrt{\frac{2DS}{h(1 - d/p)}}$$

With the costs in our example,

$$\text{EPQ} = \sqrt{\frac{2*10,000*500}{4*(1 - 40/100)}} = 2041$$

Figure 4.6 displays the solution, and Table 4.2 gives a comparison of EOQ and EPQ. As we suspected, the EPQ gives a larger Q than the EOQ because the entire EPQ is never held in inventory. The maximum inventory is $Q(1 - d/p) = 2041 * 0.6 = 1225$.

EPQ and Just-in-Time

A closer look at the economic production quantity formula indicates that the setup cost significantly influences the economic production quantity. Tool engineers, in many in-

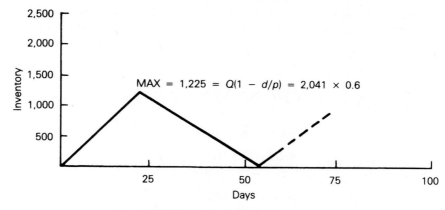

FIGURE 4.6   Production rate model.

TABLE 4.2 Comparison of EOQ and EPQ

	EOQ	*EPQ*
Q	1581	2041
D/Q	6.325	4.9
Annual setup cost	$3162.5	$2450
Annual holding cost	$3162.5	$2450
TRC	$6324	$4900

stances, can simplify the setup process and thus reduce the setup time. Even if other variables remain constant, the reduction in setup time reduces the economical production quantity. In fact, if the setup time can be reduced to a very small quantity, the lot size can be reduced to just one. As we see in Chapter 13, the just-in-time concept attempts to reduce the lot size by reducing the setup time and in-process inventories while increasing the quality of items manufactured. These actions can reduce the number of Kanbans, which dictates the number of containers in the system, and hence reduce the in-process inventory. A reduction in the lot size can also result in a smaller container size, and hence materials are transferred in small quantities from one workstation to another.

Quantity Discounts

In the simple EOQ model, we counted only setup and ordering costs as relevant. If there are volume or quantity discounts for purchasing larger quantities, the purchase price is also a relevant cost. The total relevant cost formula in this case should be

TRC (Q) = annual holding + annual ordering + annual purchase

$$\text{TRC}(Q) = \left(\frac{Q}{2}\right)h + \left(\frac{D}{Q}\right)S + Dp$$

For example, quantities under 1000 may cost $12, those over 1000 but under 4000 may cost $10, and those from 4000 up may cost $8. What, then, should be our order quantity? From our earlier work, we know that the EOQ at a price of $10 is 1581. Should we take the quantity or should we go for the price break? Let's compare TRCs, noting that 1581 falls in the $10 quantity range:

$$\text{TRC}(1581) = \frac{1581}{2} * 4 + \frac{10,000}{1581} * 500 + 10,000 * 10$$

$$= 3162 + 3162 + 100,000 = \$106,324$$

$$\text{TRC}(4000) = \frac{4000}{2} * 3.2 + \frac{10,000}{4000} * 500 + 10,000 * 8$$

$$= 6400 + 1250 + 80,000 = \$87,650$$

There is no comparison! We should definitely go for the price break, but should we go for a quantity higher than the price break? What is the EOQ at a price of $8?

$$Q* = \sqrt{\frac{2*10,000*500}{0.4*8}} = 1767.76$$

At $Q* = 1768$, holding costs would be $2829 and setup costs would be about the same, *but the purchase discount would not be granted.* We would be stuck with a purchase price of $10 per unit for annual total relevant costs of $105,658. If we cannot achieve the EOQ point at the $8 price, our best bet is to go up to the price break quantity of 4000 units. We have no desire to go any higher, since that would only increase our holding costs, which are already too high compared with setup costs. Going to the price break reduces annual purchase price and annual setup cost at the expense of an increase in holding costs. If this results in a net cost reduction, then we will move to the price break quantity.

This suggests a procedure for handling quantity discount problems:

Step 1:   Solve for EOQ at each price:

$$Q*(p) = \sqrt{\frac{2DS}{h(p)}}$$

where $h(p)$ is a function of purchase price, such as $h(p) = 0.4p$, and $Q*(p)$ shows Q as a function of p.

Step 2:   If $Q*(p)$ falls outside the quantity range for which the price can be obtained, throw it away.

Step 3:   Select the price break quantities—those quantities that give us the next lowest price. In this example, there is a price break quantity at 1000 units and another one at 4000.

Step 4:   Compare all remaining EOQs and all price break quantities in TRC(Q) = $(Q/2)h + (D/Q)S + Dp$ and choose the Q giving the smallest TRC(Q).

This procedure can be visualized in Figure 4.7, where the TRC(Q) curves include purchase price. Table 4.3 gives intermediate calculations for Figure 4.7. The three EOQs at the different prices are $Q^*_{12} = 1443$, $Q^*_{10} = 1581$, and $Q^*_8 = 1768$. Q^*_{12} and Q^*_8 do not match with their price ranges. We cannot order Q^*_{12} and select the price break quantities of 1000 and 4000 units. $Q = 1000$ at a price of $10 is uninteresting because we have already seen that the optimal Q at a price of $10 is $Q^*_{10} = 1581$. At the upper end, $Q = 4000$ is interesting because that quantity can get us a price break. Comparing the candidates $Q^*_{10} = 1581$ and $Q = 4000$, we choose $Q = 4000$ with TRC(4000) = $87,650, considerably lower than TRC(1581) = $106,324.

In the quantity discount case, lower annual purchase costs can offset increased an-

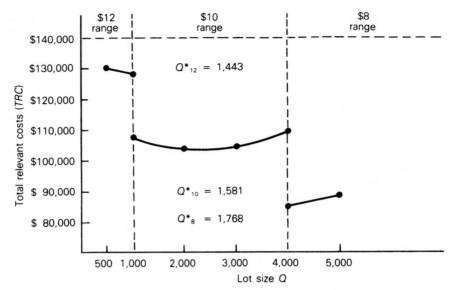

FIGURE 4.7   Total relevant costs for quantity discount model.

nual holding costs. Is it better to be at the optimal point (EOQ) on a high per unit cost curve or to be at a nonoptimal point on a lower per unit cost curve?

EOQ With Shortages

Earlier in this chapter, we mentioned that backorders come at the cost of extra clerical work, additional expediting, and customer dissatisfaction. If this are priced at $\$\pi$ per unit per year, how many backorders should be allowed per cycle (see Figure 4.8)?

In one sense, backorders are simply negative inventory. If we order in lot sizes of Q and allow B backorders per cycle, then the average backorders would be $B/2$, just as the average inventory would be $M/2 = (Q - B)/2$. On each cycle, the maximum inventory

TABLE 4.3   Total Relevant Costs

Q	Per Unit ($12)	Purchase Price ($10)	($8)
500	$131,200	OR[a]	OR
1000	OR	$107,000	OR
2000	OR	$106,500	OR
3000	OR	$107,665	OR
4000	OR	OR	$87,650
5000	OR	OR	$89,000

[a]OR = outside range (i.e., price and quantity do not match).

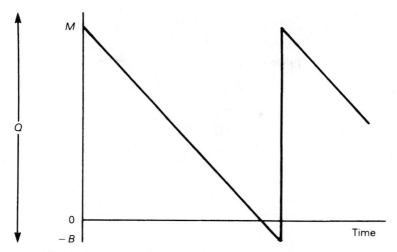

FIGURE 4.8   EOQ cycles with backorders. For example, $M = 90$. $Q = 100$, $B = 10$; $Q = M + B$.

is $M = Q - B$, because the new lot is immediately depleted by the backorder from the previous cycle.

The total relevant cost curve must now reflect backorder costs as well as setup and holding costs. In addition, we now have two decisions: how much to order and how many backorders to allow.

$$\mathrm{TRC}(Q, B) = \frac{(Q - B)}{2} * h * \frac{(Q - B)}{Q} \text{ (annual holding cost)}$$

$$+ \frac{B}{2} * \pi * \frac{B}{Q} \text{ (annual backorder cost)}$$

$$+ \frac{D}{Q} * S \text{ (annual setup cost)}$$

Here $(Q - B)/Q$ represents the time on each cycle when positive inventory exists, and B/Q gives the time when negative inventory exists. For example, a maximum in inventory of $M = Q - B = 100 - 10 = 90$ and a demand rate of one unit per day would have each cycle lasting 100 days, 90 of which have a positive inventory situation. Since the holding cost is on a per unit per year basis, we must apply that cost only to times of positive inventory.

Similarly, B/Q is the fraction of the year with a backorder position. For example, an annual demand of 1000, a Q of 100, and B of 10 would give ten cycles in a year, with only 10% of each cycle spent in a backorder position. Restating the total relevant cost curve, differentiating, and setting the partial derivatives to zero, we can solve the optimal values of Q and B:

$$\text{TRC}(Q, B) = \frac{(Q - B)^2 h}{2Q} + \frac{B^2 \pi}{2Q} + \frac{DS}{Q}$$

$$= \frac{(Q^2 - 2QB + B^2)h}{2Q} + \frac{B^2 \pi}{2Q} + \frac{DS}{Q}$$

$$= \frac{Qh}{2} - Bh + \frac{B^2 h}{2Q} + \frac{B^2 \pi}{2Q} + \frac{DS}{Q}$$

$$\frac{\partial \text{TRC}(Q, B)}{\partial B} = -h + \frac{2Bh}{2Q} + \frac{2B\pi}{2Q} = 0$$

$$\frac{B}{Q}(h + \pi) = h$$

$$B = Q\left(\frac{h}{h + \pi}\right)$$

$$\frac{\partial \text{TRC}(Q, B)}{\partial Q} = \frac{h}{2} - \frac{B^2 h}{2Q^2} - \frac{B^2 \pi}{2Q^2} - \frac{DS}{Q^2} = 0$$

Multiplying through by $2Q^2$, we have

$$Q^2 h - B^2(h + \pi) - 2DS = 0$$

and substituting for B, we have

$$Q^2 h - Q^2 \frac{h^2}{(h + \pi)^2}(h + \pi) - 2DS = 0.$$

Simplifying further, we obtain

$$Q^2 h - Q^2 h\left(\frac{h}{h + \pi}\right) - 2DS = 0$$

$$Q^2 h\left(1 - \frac{h}{h + \pi}\right) = 2DS$$

$$Q^2 = \frac{2DS}{h}\left(\frac{h + \pi}{\pi}\right)$$

$$Q = \sqrt{\frac{2DS}{h}}\sqrt{\frac{h + \pi}{\pi}}$$

Example 4.1

Suppose that $D = 10,000$, $S = 500$, $h = 10$, and $\pi = 40$. Solve for Q, B, and M.

Solution: We obtain

$$Q^* = \sqrt{\frac{2(10,000)500}{10}} \sqrt{\frac{10+40}{40}}$$

$$= 1000 * 1.118 = 1118$$

$$B^* = 1118 \left(\frac{10}{10+40}\right) = 223.6$$

$$M = 1118 - 224 = 894$$

Periodic Inventory Systems

In the EOQ model with shortages, we began to think more about time: how much time was spent in a positive inventory position versus a backorder position. Nevertheless, our model was based on a continuous monitoring of inventory. Implicitly, we assumed that we would place an order to arrive just as our backorder position reached its optimal value.

In reality, however, many firms order periodically without a continuous tracking of inventory positions. With the EOQ-based models developed so far, it is relatively easy to convert to such a periodic system.

Ordering Q units at a time, there will be D/Q cycles in a year, but the time between orders will be Q/D. For example, $Q/D = 1000/12,000$ would have us order every month. Letting T be the time between orders and using our EOQ model with shortages,

$$T = \frac{Q}{D} = \frac{\sqrt{\frac{2DS}{h}} \sqrt{\frac{h+\pi}{\pi}}}{D}$$

$$T^* = \sqrt{\frac{2S}{hD}} \sqrt{\frac{h+\pi}{\pi}}$$

Example 4.2

Suppose that $D = 10,000$, $S = 500$, $h = 10$, and $\pi = 40$. Solve for the optimal time between orders.

Solution: We obtain

$$T^* = \sqrt{\frac{2*500}{10*10,000}} \sqrt{\frac{10+40}{40}} = 0.1*1.118 = 0.1118$$

Then $Q^* = T * D = 0.1118 * 10,000 = 1118$. On a 250-day working year, $0.1118 * 250 \approx 28$, so we should order 1118 units every twenty-eight days.

Hybrid Models

We have described the fixed-order quantity system as well as the fixed time period (periodic) inventory system. In practice, hybrid inventory control systems, which incorporate features from both systems are also used. Some popular systems include (*S,s*) policy, base stock systems, and two-bin systems. These systems are described at the end of Chapter 8.

IMPLEMENTATION ISSUES

Inventory control is a function of material management, and the objective is to keep the total cost associated with the system to a minimum. This requires familiarity with supply sources, price negotiations including bulk quantity discounts, modes of transportation, budgeting, physical handling, record keeping, and monitoring the incoming quality of items. The ABC classification described earlier aids in selecting specific models for managing the item inventory. Some other important issues regarding the implementation of inventory systems are discussed next.

Inventory Transactions

When items are drawn for use or replenished, every transaction should be identified and the system records should be modified to reflect the actual quantities of items on hand. This is true in both manual and a sophisticated computerized inventory management systems. An accurate inventory count helps us to know the individual item inventory status. Then we can order the specific quantities of required items when needed.

Inventory Accuracy

Maintaining a sophisticated computerized inventory system has no value if actual inventory on hand and stock on record differs. This situation could occur if errors in counting and recording accumulate over time. Misplacement or misidentification of items, breakages, theft, and other unauthorized or unidentified actions also cause the document to become inaccurate. Poor design of the system and lack of training of personnel may further aggrevate the situation. Therefore, a method of checking the actual count is necessary to set the records straight. This can be done on an annual basis or on a random basis.

Cycle Counting

In cycle counting, items are counted throughout the year. This minimizes the disruptions to the facility and smooths the workload. The advantage is that a special team can be established and the reconciliation can be done continuously. Cycle counting also allows scheduling of physical counts to ensure that all items are counted over the year to correct errors in the system. A variation of this method is known as event-based cycle counting. Counting is scheduled when a replenishment order for the item is placed. As the quantity on hand is typically low during this period, less effort is required for accomplishing the task.

Inventory Valuations

In many instances, tax laws require annual physical counting for asset valuation purposes. Either cycle counting or annual actual physical counting procedures will satisfy the needs.

SUMMARY

Holding cost, setup cost, and stockout cost trade-offs have been the focus of this chapter. Because these trade-offs arise in the management of organization stocks rather than transaction stocks, they are essentially investment decisions: Should we invest in inventories to avoid excessive setups and downtime or to buffer against uncertain demand?

The amount of money to be invested in inventories depends on the dollar volume of sales. Hence, the ABC classification by dollar volume underlies the rest of the work in the chapter. If we can cheaply obtain reduced setup costs and increased protection against the uncertainties of demand for a particular product, we do not need to be overly concerned about inventory investment control.

Inventory control questions for items requiring significant inventory investment led to the question of how much to order and how often. Starting with the basic economic order quantity (EOQ) model, we traded off holding versus setup costs. We assumed instantaneous resupply, no quantity discounts, no shortages, and steady, deterministic demand.

Because the basic EOQ model was so restrictive, we relaxed the assumptions one by one. Examination of resupply time led to the model of usage during production. Quantity discounts created discontinuities in the relevant cost function and required an algorithm rather than a simple equation. Finally, by allowing shortages or backorders, we created a mirror image problem: How large should the maximum inventory be and how large should the maximum backorder position be?

Throughout our work with deterministic models, we have maintained an interest in the sensitivity of our decisions to changes in demand, setup costs, and holding costs. In addition, implicit or imputed costs arise. For example, specifying the number of cycles in a year provides information about the ratio of setup to holding costs.

In Chapter 6, the deterministic demand assumption will be relaxed, and demand uncertainty will be handled by safety stocks.

PROBLEMS

1. A manufacturer carries stock of an item with an annual demand of 30,000 units. Although the inventory manager cannot estimate setup cost or holding cost precisely, she feels that the ratio of the two is somewhere between 100 to 1 and 150 to 1; that is, $S/h = 100$ to $S/h = 150$. Calculate EOQ on both conditions.
2. How sensitive is the optimal Q to the S/h ratio? If S/h doubles or triples, what happens to Q^*?
3. How sensitive is Q to annual demand? If annual demand doubles or triples, what happens to Q^*?

4. Rather than expressing its order quantity in units, the Posifax Company uses dollars: EOQ = \$500 = 250 * \$2, where \$2 is the unit cost of the product. Further, the holding cost is $r * C$, where r is 40%. Recalling that EOQ = $\sqrt{2DS/rC}$, develop an EOQ formula for EOQ in dollars: $EOQ_\$ = Q * C = ?$

5. With annual demand of 30,000 units, an S/h ratio of 100 to 1, and a lead time of ten days, what reorder point should the Marco Company use? Marco is open for business 250 days per year, and sales are assumed to occur at a constant rate. What would happen if the lead time sometimes went up to fifteen days?

6. A machine produces the product at a rate of 2000 units per day. The annual demand of 200,000 occurs at a constant rate over the 250 business days in the year. Inventory carrying costs are 30% annually, and the unit variable production cost is \$25. The setup cost is \$500. What is the economic production quantity?

7. Two students must wash and dry dishes at a summer camp. There are exactly 100 dishes to be washed. The washer works at a rate of four dishes per minute. The dryer follows at a rate of three dishes per minute. What is the maximum number of dishes washed and waiting to be dried?

8. A company faces an annual demand of 10,000 units. Setup costs are \$200 per order, and the company orders in lot sizes of 1000 units. What must be the company's holding cost per unit per year?

9. A company faces an annual demand of 10,000 units and a holding cost per unit per year of \$5. If the company insists that its lot size of 500 units is the correct one, what must be its setup cost per order?

10. A company annually orders one million pounds of a certain raw material for use in its own curing process. With annual holding costs estimated at 35% of the purchase price of \$50 per 100-pound bag, the purchasing manager wants to decide on an order size. Marginal paperwork costs are \$10 per order. For orders of 500 bags or more, the purchase price falls to \$45 per bag; for orders of 1000 bags or more, the price is \$40 per bag. What is the optimal order size?

11. An item has annual usage of 1000 units. The ordering cost is \$5, and the purchase price is \$3 each. With a carrying cost percentage of 25% and quantity discounts of 5% when 150 units or more are bought and 10% when 300 units or more are bought, what is the optimal order quantity?

12. Inventory costs for an important class of items are found to consist of a per unit storage cost. C_s, based on the maximum inventory level, and a regular holding cost, C_h, expressed as a percentage of the average dollar value of inventory. Develop a formula for EOQ under these conditions.

13. A company faces an annual demand of 1000 units for a particular product, setup costs of \$200 per setup, annual per unit holding costs of 25% of the product's value of \$12, and backorder penalties of \$10 per unit per year. What is the optimal order quantity?

14. A company orders in lot sizes of 2000 units. The holding cost per unit per year is \$8, and the backorder penalty per unit per year is \$15. What should be the optimal inventory held, and what should be the maximum backorder position?

15. With an annual demand of 200 units, setup costs of \$250, and holding costs of \$8 per unit per year, what is the optimal time between orders? Use a 250-day working year and specify the time in days.

16. With an annual demand of 2000 units, setup costs of \$250, holding costs of \$8 per unit per year, and backorder penalty costs of \$24 per unit per year, what is the optimal time between orders? Use a 250-day working year and specify the time in days.

17. In the shortage model, we saw that

$$B^* = Q\left(\frac{h}{h+\pi}\right)$$

Now show that

$$B^* = \sqrt{\frac{2DS}{\pi}}\,\sqrt{\frac{h}{h+\pi}}$$

18. In the shortage model, we saw that

$$B^* = Q\left(\frac{h}{h+\pi}\right)$$

Develop a similar type of formula relating M^* to Q and then show that

$$M^* = \sqrt{\frac{2DS\pi}{h(h+\pi)}}$$

19. In the standard EOQ model, we saw that

$$Q^* = \sqrt{\frac{2DS}{h}}$$

Now express $\mathrm{TRC}(Q^*)$ in terms of D, S, and h only.

20. Modify the usage during production model to allow shortages.

APPENDIX 4A

Re-Engineer the Materials and Procurement Function*
John A. Ferreira, CPIM

The concepts of world class manufacturing have become basic tenets of experienced manufacturing practitioners. Few manufacturers exist that haven't already embarked on programs to improve their operations. The focus for many in implementing world class manufacturing techniques has traditionally been the shop floor, yet this singular focus neglects other areas equally important in determining competitiveness.

*Reprinted from *APICS—The Performance Advantage,* Vol. 3, No. 10, October 1993, pp. 48–51.

Many studies have documented that raw material and purchased components represent the largest cost component of a typical manufactured product. The procurement and management of these materials and the relationship a manufacturer has with its suppliers are also key drivers of product quality, lead time, flexibility and the levels of working capital (inventory and payables) required to operate the business. These factors significantly influence key aspects of overall business competitiveness.

In practice, only a few firms have attempted to capitalize on the potential of improving the materials management and procurement function as a means of achieving competitive advantage. Improvements can directly enhance a firm's competitiveness, often resulting in increased profitability and increased market responsiveness. Like improvements being made to the shop floor, changes required to improve materials and procurement operations are counter intuitive when judged against traditional management practices. For those firms willing to challenge conventional wisdom, dramatic results can be achieved.

RECOGNIZING THE SYMPTOMS

Observations of different manufacturers across multiple industries indicates that there are four common symptoms of poor materials/procurement performance that result in a high-cost operation.

- Supplier proliferation
- Limited emphasis on service level considerations
- Reactive orientation of buyers
- High levels of raw Material inventory

Collectively, these four symptoms can be used as a simple test to determine the competitiveness of operating practices; the impact of each on cost and competitiveness is examined below.

SUPPLIER PROLIFERATION

Most materials/purchasing departments have traditionally viewed their primary responsibility as buying component materials at the lowest possible cost within acceptable quality standards. The traditional approach has been to minimize the price paid for each line item. This practice has, however, led to a significant proliferation of suppliers. Even small manufacturers with revenues of $20 million to $100 million have been observed to have as many as 600 suppliers.

Playing one supplier against another has been the primary method of minimizing line item pricing. Many suppliers are chosen to supply only those items for which they quoted the lowest delivered price. A fallout of this practice is the use of multiple, alternate suppliers.

A second major driver of a proliferating supplier base is an inadequate supplier management process. Often engineering will specify a component that can only be supplied by a very limited number of suppliers. Purchasing finds it difficult to source these unique components at competitive prices. Perceived as "dictating to purchasing," this practice is the direct result of an inadequate requirements review process which hasn't anticipated future product needs. Since selection of one component item often drives other design parameters, and because no preferred supplier exists for an entire group or class of goods, product development by default makes a choice during the design process. After a design roll-out it becomes difficult to go back and simply substitute for the vended part.

The lack of design standardization and part commonality contributes to the proliferation of suppliers. To reduce total product cost, many designers, marketers and accountants have applied standard cost accounting techniques to justify selection of less expensive parts, which although similar to those used in other designs, are slightly different. Applying standard cost accounting techniques to the product design process can lead to savings, but doing so inherently ignores other costs that are not as easily quantified by accounting systems. Part proliferation is one of the largest hidden costs in product design; whenever very similar items with the same function lack commonality, it increases the need to be more accurate in sales forecasting and inventory planning, and compounds material identification, storage, handling and control problems on the shop floor.

Another major hidden cost is the reduced order size with a supplier and increased uncertainty of demand. The end result is higher cost, longer lead times and reduced quality.

As suppliers proliferate, the dollar concentration among them decreases, causing the manufacturer to have a declining level of importance to any single supplier. This is significant, since it is counterproductive to what the buyer is attempting to achieve (low cost and high service levels). Additionally, few manufacturers have contracts with their vendors and, as a result, suppliers know they can be dropped at any time by the buyer. Contrary to the buyer's hope that the threat of lost business will spur the supplier into providing good service, it simply reinforces the supplier's practice of maintaining a good margin on non-core business.

SERVICE LEVEL CONSIDERATIONS

All too often supplier selection is made primarily on the basis of the quoted line item purchase price. Item pricing is, without doubt, the most visible driver of cost. However, there are ten other determinants of cost; each represents a service level provided by the supplier. The limited ongoing monitoring, follow-up and attention paid to supplier service levels by many materials and procurement groups constrains the firm's overall competitiveness.

1. Quality: Poor component/material quality, if not detected up-front by the supplier, leads to incoming inspection cost incurred by the manufacturer. Inspection, however, will only catch the obvious supplier errors; poor incoming material can lead to rework on the shop floor, customer returns, allowances and even a poor market reputation resulting in decreased sales volume.

2. Lead time: The shorter the supplier lead time, the greater flexibility the manufacturer has to respond to uncertain demand patterns, which lessens the dependency on accurate production forecasts. This flexibility allows for a reduced raw material inventory requirement and, thus, lower working capital cost.

3. Fill rates: What percent of an order actually is delivered in the quoted lead time? Uncertainty in fill rate performance results in increased raw material inventory, as the buyer will compensate for uncertainty by ordering larger quantities than needed, resulting in increased working capital cost.

4. On-time delivery performance: How often is an order received on-time, neither early nor late? Uncertainty in the actual material arrival date forces the buyer to "out-guess" the system and bring material in earlier or later than planned. This practice causes releases to fall out of synchronization with established schedules, resulting in increased cost either from early delivery or production shut-downs (late delivery).

5. Responsiveness to demand: To truly have responsive suppliers requires that manufacturers provide reliable and timely demand forecasts and maintain a process to quickly communicate production changes. Failure to provide suppliers with schedule visibility results in exposing them to expedited orders, which jeopardizes quality and increases delivered cost.

6. Technical support: Is there sufficient supplier technical knowledge which can be leveraged? Integrating suppliers as a critical resource and using them as a support team to add or enhance expertise not only helps to avoid making costly material/component decisions but, more importantly, assists the company to become aggressive in product design, ease of assembly, part commonality and cost control.

7. Product warranty and service parts: How readily can replacement components be sourced? Under what circumstances will the supplier absorb the entire cost of replacement/repair once the product has been sold and is in the customer's possession?

8. Freight enhancements: In lieu of multiple suppliers making independent decisions about carrier specification and modes of transportation, manufacturers should seize the opportunity to reduce overall freight costs by concentrating shipments among as few transportation carriers as possible and by specifying these carriers to their suppliers. Indirect benefits of consolidated carriers include lowered cost of partial shipments, more frequent delivery and shortened lead times.

9. Payment terms: Even though orders always specify payment terms, all suppliers are familiar with stretched receivables well beyond the contracted period. Since it is common for a supplier to expect delayed payment, this carrying cost of the receivable is added to the overall cost of the product being purchased. Few manufacturers have worked out agreements with suppliers that incorporate a lower base cost for a pledge of consistent on-time payment. Although early payment discounts are available, they don't always lead to lower cost since they require the manufacturer to typically pay very early with only a minor, one-time cost impact.

10. Ordering practices: Is the supplier capable of receiving orders or releases in such a way as to reduce the administrative cost burden of placing orders? Even a simple technique such as utilizing blanket purchase orders with a release can eliminate effort and cost. Technology solutions such as electronic data interchange (EDI) can be another means of reducing administrative cost.

Suppliers need to be viewed as critical strategic alliances, not as replaceable component vendors. Viewed as a resource, suppliers can help solve problems with product design and manufacturing performance. Strong cross-functional communication with suppliers develops true partnerships that stimulate knowledge transfer and build a more competitive manufacturer.

The less emphasis given to these 10 service level areas, the more difficult it will be for the organization to establish the type of supplier relationships necessary to drive total cost down. Without a real working relationship with suppliers, manufacturers perpetuate buying on line item price alone.

REACTIVE ORIENTATION OF BUYERS

A characteristic of many purchasing departments is one of buyers habitually on the phone chasing down needed parts from some "shortage" list. Many organizations find it necessary to establish full-time expediters to handle material delivery problems. Short-ages may be a fact of manufacturing; but when they regularly consume the department's time, it generally indicates that the procedures and systems in place are inefficient and ineffective.

Managing to a stable production schedule with limited surprises can significantly reduce the "reactive" mode. In many organizations, buyers are given uncoordinated production schedules from multiple, independent demand sources, such as sales, customer service and spare parts. With these independent demands arriving at different times, buyers are left to consolidate requirements and play the "can you squeeze it in" game.

A well coordinated master production schedule (MPS), developed with input from all affected parties (sales, finance, manufacturing and materials) is designed to consolidate independent demands, reduce schedule changes and limit surprises. However, few manufacturing organizations have implemented effective MPS processes and even fewer have focused on the process of translating sales forecasts into production schedules. As a result, few purchasing agents have effective systems and tools in place to assist them in removing the guess work from supplier releases. Suppliers cannot be expected to anticipate a build schedule that the firm itself doesn't know about until the last minute. Cost increases as uncertainty increases, suppliers want higher prices for rush orders, freight costs are at a premium and service levels decline causing further uncertainty and reactive behavior.

Lack of supplier visibility into the manufacturer's forecasted build schedule creates demand variability and uncertainty for the supplier. If suppliers have no advanced knowledge of demand trends and approximate size, they are unable to effectively adjust their operations and cannot pass their expected material/component demand onto their suppliers. The less visibility a manufacturer gives its supplier, the lower the resulting service levels will be. As supplier service levels fluctuate, buyers are forced inevitably into a reactive mode.

Taking the master production schedule through the next step in production scheduling, material requirements planning (MRP), for time-phased material buys only creates false security unless ongoing maintenance of both supplier lead times and internal processing lead times are managed. If schedule changes are the rule, as service levels fluctuate, and if lead times are not updated, it becomes very difficult or impossible to maintain an effective MRP system and resulting material control.

Other negative effects of reactive behavior are increased administrative burdens created by a for-the-order review to determine how many can be shipped, how soon and by whom. Such a costly administrative burden increases the net price paid. In addition, the more reactive the organization is, the more they lose sight of the true market price of the component.

RAW MATERIAL INVENTORY LEVELS

A visible impact of poor materials and procurement practices is increased raw materials inventory. Well over half of all manufacturers have sufficient total levels of raw material on hand to satisfy more than a month's production. The assumption by many firms is that its raw material inventory will be used to build production for the next month or two. Upon close examination of what is actually in inventory, one realizes that the inventory is not matched to future production requirements and there are too many of some items while shortages of other items abound.

While it is true that the cheapest place to hold inventory is in raw materials, where the least value has been added, this practice is still costly. Because material is an overwhelming percentage of the cost of goods sold, large raw material levels require the firm's commitment of expensive working capital earlier than necessary. The cost of funding working capital to carry excess raw material can be significant. For example, with an annual cost of capital of 14 percent, there is a resulting 1 percent inventory carrying cost for each month material is held. The cost of carrying excess or mismatched inventory essentially increases the firm's net material cost by an equal amount.

The following example, using a 1 percent per month carrying cost, illustrates the cost implication of excess inventory. Assume company XYZ operates with an average of three months raw material. If material accounts for 65 percent of cost of goods sold, and the existing gross margin is 30 percent, then profitability will be reduced by nearly 1 percent from the level the company would have had they operated with only one month of raw material inventory. Given that inventory financing charges are generally not captured in cost of goods sold but rather in interest charges, this impact is largely ignored by many firms. Early purchases, obsolete material, slow moving and excess inventory all consume working capital by incurring carrying cost.

In addition to inventory carrying costs, there is also the risk of increased material obsolescence. As raw materials go unmatched to production requirements and obsolescence sets in, the value of the material declines rapidly. Other problems resulting from large excess inventories include increased space for material storage and utility cost.

REFERENCES

CARR, L. P. and C. D. ITTNER, "Measuring the Cost of Ownership," *Journal of Cost Management*, Fall 1992, pp. 42–51.

SADRIAN, A. A. and Y. S. YOON, "Business Volume Discount: A New Perspective on Discount Pricing Strategy," *National Association of Purchasing Management, Inc.*, April 1992, pp. 43–46.

HARRISON, A. and C. VOSS, "Issues in Setting up JIT Supply", *International Journal of Operations & Production Management,* Vol. 10, No. 2 (1990), pp. 84–93.

LANDEROS, R. and R. M. MONCZKA, "Cooperative Buyer/Seller Relationships and a Firm's Competitive Posture," *Journal of Purchasing and Materials Management,* Fall 1989, pp. 9–18.

SHERIDAN, J. H. "Are You A Bad Customer?" *Industry Week,* August 19, 1991, pp. 24–34.

VOLLMANN, T. and W. L. BERRY, D. C. WHYBARK, *Manufacturing Planning and Control Systems,* (Illinois: Irwin, 1988).

Chapter Five

Multi-Item Joint Replenishment

INTRODUCTION

Chapter 4 dealt with the management of basic inventory systems for single-item situations. The demand was assumed to be deterministic in nature. We discussed the economic order quantity model, economic production quantity model, and quantity discount model as well as trade-offs between carrying inventories and incurring shortages. We also discussed some models involving periodic and hybrid inventory systems. Rarely, however, does a real-life situation exist where only one item is managed by a firm. This chapter extends some of the concepts of Chapter 4 to a multi-item environment generally known as the joint replenishment inventory model.

ECONOMIES OF JOINT REPLENISHMENT

Joint replenishment can occur when a firm is either purchasing a number of items from an outside vendor or producing internally. A group of items belonging to the same family may need a common major setup and a unique minor setup for individual items. Hence, in many instances it may be possible for a number of items to share the fixed cost associated with a major setup or replenishment. For example, when a product is packaged soon after manufacture into more than one size, a savings can be realized if these items are manufactured jointly and then packaged individually. In many cases, by combining order quantities of several items a firm can reduce either shipping costs, obtain a discount based on total dollar volume of purchase, or both. In general, the fixed cost associated with the purchase of several items from a single vendor is independent of the number of items purchased at any given order. The fixed cost is analogous to the major setup cost incurred in manufacturing several items with a common setup. In many instances the capacity of the firm may be limited. By grouping families of these items, valuable capacity that would otherwise be spent on several unnecessary setups might be saved. Therefore, it is necessary to decide how much of each item should be produced (purchased) during any given setup (order). Integrated models where the total of order cost, inventory carrying cost, and transportation cost can be minimized are covered in Chapter 8. Regardless of whether one is addressing a purchasing or a manufacturing situation, the decision variables that should be addressed are similar:

1. The dollar value or quantity of individual items produced or ordered during each cycle
2. The total dollar value or quantities of all items produced or ordered during each cycle
3. The frequency at which these items are ordered or produced

ASSUMPTIONS

Similar to the economic order quantity model discussed in Chapter 4, we assume that the demands, lead times, costs, and inventory carrying percentage for all items are given and are deterministic. First, we deal with the joint replenishment model such that all items are ordered each time an order in placed (every cycle). Next, we attempt to improve the solution by varying the replenishment cycles of individual items in the family. In the latter case, every item may or may not be ordered during each order cycle. Finally, we present the joint production quantity model for determining the quantities of individual items within a group of manufactured items.

JOINT REPLENISHMENT ORDER QUANTITY MODEL

The objective of all models presented in this chapter is to minimize the total relevant costs for a group of items jointly purchased or produced. The relevant costs generally include the setup cost and inventory carrying costs. Therefore, models presented in this chapter determine the economical order (production) quantities for a group of items by minimizing the total cost of inventories and setups per period. Once the optimal dollar value for a group of items is determined, the total is prorated to individual item dollar value or quantities. Now, let us define the following:

S = fixed cost of placing an order for a group of items

s_i = item-dependent marginal cost of placing an order associated with an additional item i

$A\$$ = annual dollar value of all items in the group ordered

$a\$_i$ = annual dollar value of item i in the group ordered

C_i = unit cost of item i

D_i = annual demand for item i in number of units

I = inventory carrying charge expressed as a decimal

$Q\$$ = total dollar value of all items ordered during a cycle

$Q\$_i$ = dollar value of item i ordered during a cycle

Q_i = quantity of item i ordered during a cycle

N = number of cycles per year

T = time between orders in years

Given these definitions, we can establish the following relationships.

1. The total dollar value of all items ordered per year is the sum of the dollar value of individual items ordered per year.

$$A\$ = \Sigma\, a\$_i$$

2. The total dollar value of all items ordered during a cycle is the sum of the dollar value of individual items ordered during each cycle.

$$Q\$ = \Sigma\, Q\$_i$$

3. The annual dollar value of item i ordered is also equal to the dollar value of items ordered per cycle multiplied by the number of cycles per year.

$$a\$_i = N * Q\$_i$$

4. The units of demand per year for item i is also equal to the annual dollar value usage of the item divided by the unit cost of item i.

$$D_i = \frac{a\$_i}{C_i}$$

5. Similarly, the units of item i ordered during the cycle is equal to the dollar value of item i ordered during the cycle divided by the unit cost of the item i.

$$Q_i = \frac{Q\$_i}{C_i}$$

6. The time between orders can be obtained from one of the following ratios:

$$T = \frac{1}{N} = \frac{Q\$}{A\$} = \frac{Q\$_i}{a\$_i}$$

Because placing an order involves a fixed cost S and a variable costs s_i for each item i, the total cost of placing an order for a group of items amounts to $\{S + \Sigma s_i\}$. Therefore,

Total cost of placing N orders per year $= N\{S + \Sigma s_i\}$

Annual average cost of inventories carried $= I\{Q\$/2\}$

Total annual relevant cost $= N\{S + \Sigma s_i\} + I\{Q\$/2\}$

To minimize the total annual relevant cost of ordering and carrying inventories, we take the first derivative of the total cost expression with respect to $Q\$$ and solve for the value of $Q\$$. It can easily be proven that the optimal dollar value of all items ordered during a cycle is

$$Q\$ = \sqrt{\frac{2[S + \Sigma s_i]A\$}{I}} \tag{5.1}$$

The economical dollar value formula for a group of items is strikingly similar to that of an individual item formula. Once the dollar value of all items contained in the group is determined, the dollar value of an individual item i can be calculated by multiplying this value and the proportion of the annual usage value of the item i to the total usage value of items in the group:

$$Q\$_i = Q\${a\$_i/A\$} \tag{5.2}$$

The quantity of item i ordered during each cycle can be found as

$$Q_i = Q\$_i/C_i \tag{5.3}$$

The following problem illustrates the computations involved in this model.

Example 5.1

Iowa Abe sporting goods company order five different tennis rackets from a major distributor. The annual demand, cost, and other data are exhibited in Table 5.1. The major order cost associated with the group of items is $75 per order, and the annual inventory carrying percent is 15% of the cost of items. Find the economical order quantity in dollar value and units for all items.

TABLE 5.1 Data for Example 5.1

Brand Name	Annual Demand ($)a$_i$	Unit Cost C_i	Annual Demand Quantity	Item Preparation s_i	Order Size~ Dollars Q$_i$	Order Size~ Size Quantity Q_i
1	5,000	5	1000	5	986.75	197.35
2	4,000	8	500	5	789.40	98.68
3	10,000	10	1000	8	1973.51	197.35
4	18,000	12	1500	8	3552.32	296.03
5	1,000	20	50	10	197.35	9.86
Total	38,000			36	7499.33	

Using the joint replenishment formula, we find the economical order size to be $7499.33 per order cycle.

$$Q\$ = \sqrt{\frac{2\,[75+36]38,000}{0.25}}$$

Once the total dollar value per order cycle is obtained, the proportional dollar value and quantity of individual items to be ordered per order can be calculated using equations (5.2) and (5.3) as

$Q\$_1 = 7499.33 * 5000/38,000 = \$ \ 986.75; Q_1 = \ 986.75/5 \ = 197.35$

$Q\$_2 = 7499.33 * 4000/38,000 = \$ \ 789.40; Q_2 = \ 789.40/8 \ = \ 98.68$

$Q\$_3 = 7499.33 * 10,000/38,000 = \$1973.51; Q_3 = 1973.51/10 = 197.35$

$Q\$_4 = 7499.33 * 18,000/38,000 = \$3552.32; Q_4 = 3552.32/12 = 296.03$

$Q\$_5 = 7499.33 * 1000/38,000 = \$ \ 197.35; Q_5 = \ 197.35/20 = \ 9.86$

Number of order per year $N = A\$/Q\$ = 38,000/7499.33 = 5.067$

Time between orders $= 1/N = 0.1974$ year

Annual cost of ordering $= 5.067 * \{75 + 36\} = \$562.45$

Average annual inventory costs $= 0.15 * \{7499.33/2\} = \$562.45$

We find the solution to this problem consists of fractional values, which is common when working with these problems. The quantities could be rounded to the nearest integer without sacrificing the optimality conditions of the solution because we know that the total cost curve for the economic order quantity is almost flat for any point near the optimal quantity, as shown in Figure 4.5.

Although we calculated the annual order costs based on the number of cycles per year, we can also arrive at the same value by prorating the major setup cost, to individual items according to the annual dollar value $a\$_i$ of item i to the total annual dollar demand $A\$$.

The following steps show how the calculations are performed. The reason for these tedious calculations will become apparent in the next section.

Step 1: Prorate the major setup cost for item i based on the ratio $a\$_i/A\$$. For item 1, we can determine this as $(5000/38{,}000) * 75 = \$9.87$

Step 2: Add the minor setup cost of item i to the prorated major setup cost. For item 1, we can determine this as $\$9.87 + 5 = \$14.87$

Step 3: Find the annual setup cost for item i by multiplying the cost obtained in step 2 and N, the number of cycles per year. Again, for item 1, we can determine the annual order cost as $\$14.87 * 5.067 = \$75.35$.

Similarly, we can calculate the annual order costs for all other items. They are summarized in Table 5.2a.

We can also calculate the inventory carrying costs of individual items on the basis of the average inventory $(Q\$_i/2)$ carried by each item i instead of computing the total using the average cycle inventory $Q\$/2$. The detailed calculations are exhibited in Table 5.2b.

Of course, the total annual inventory carrying costs as well as the sum of the annual setup costs are the same regardless of the method used in determining the total costs. This is obvious because these quantities were determined using the economical order quantity formula.

TABLE 5.2a   Annual Setup Cost Summary

Item	Annual Setup Cost
1	[[5000/38,000] * 75 + 5] * 5.067 = $ 75.35
2	[[4000/38,000] * 75 + 5] * 5.067 = $ 65.31
3	[[10,000/38,000] * 75 + 8] * 5.067 = $140.58
4	[[18,000/38,000] * 75 + 8] * 5.067 = $220.56
5	[[1000/38,000] * 75 + 10] * 5.067 = $ 60.65
Total	$562.45

Table 5.2b   Annual Inventory Cost Summary

Item	Annual Inventory Costs
1	[(986.75/2) * 0.15] = $ 74.02
2	[(789.40/2) * 0.15] = $ 59.14
3	[(1,973.51/2) * 0.15] = $148.04
4	[(3,552.82/2) * 0.15] = $266.44
5	[(197.35/2) * 0.15] = $ 14.81
Total	$562.45

JOINT REPLENISHMENT ORDER MODEL WITH VARYING ITEM CYCLE

Although the total annual ordering costs and the inventory carrying costs are equal in Example 5.1, this does not hold true when ordering and inventory carrying cost of individual items are compared (see Tables 5.2). We know from the single-item economic order quantity formula that the annual order costs and inventory carrying costs should be equal when the solution to the given problem is optimal. Therefore, it may be possible to improve the solution. Brown [2], Silver [10], Goyal [4], Goyal and Stair [5], and Kaspi and Rosenblatt [7,8] among others provide algorithms to improve the results obtained earlier by modifying the cycle time of individual items (that is, some items may be ordered, for example, every second or third cycle). Next we present these algorithms.

Brown's Algorithm

Brown's algorithm follows an iterative procedure. The procedure starts with $n_i = 1$ for all items and then calculates the values of multiples n_i for each item i using a formula. When the values of multiples during two consecutive iterations do not change significantly the computations are terminated. Although the calculations appear to be complicated, it is straightforward. The following steps illustrate the algorithm using Example 5.1.

Step 1:   Calculate the order interval T from the aggregate order solution.

$$T = Q\$/A\$ = 7499.33/38,000 = 0.1974$$

Step 2a: Calculate the interval multiple n_i using the following relationship:

$$n_i = \frac{1}{T} \sqrt{\frac{2 * s_i}{I * a\$_i}} \tag{5.4}$$

Step 2b: Round off the values of n_i to the integer values of k shown in the following table:

Range of n obtained using the formula	Integer value of n to be used
0 to 1.414	1
1.414 to 2.449	2
2.449 to 3.464	3
3.464 to 4.472	4
4.472 to 5.477	5
5.477 to 6.480	6
6.480 to 7.4583	7
For $n > 6$, use the integer part of $n + 0.52$	

The values of n should satisfy the following equation as indicated by Brown:

$$k(k - 1) \leq n^2_i \leq k(k + 1) \tag{5.5}$$

where k represents the integer value of n shown in the table. For our problem,

$$n_1 = \frac{1}{0.1974}\sqrt{\frac{2*5}{0.15*5000}} = 0.585 => n_1 = 1$$

$$n_2 = \frac{1}{0.1974}\sqrt{\frac{2*5}{0.15*4000}} = 0.654 => n_2 = 1$$

$$n_3 = \frac{1}{0.1974}\sqrt{\frac{2*8}{0.15*10,000}} = 0.523 => n_3 = 1$$

$$n_4 = \frac{1}{0.1974}\sqrt{\frac{2*8}{0.15*18,000}} = 0.390 => n_4 = 1$$

$$n_5 = \frac{1}{0.1974}\sqrt{\frac{2*10}{0.15*1000}} = 1.85 \;=> n_5 = 2$$

Step 3:   Determine the revised value of T using the following relationship that accounts for the possibility of different-order intervals for individual items:

Step 4:   Go to step 2 and recalculate the values of n_i's are same. If the order multiples change, repeat Steps 2 through 4 until two consecutive values of n_i's are same.

$$T = \sqrt{\frac{2*[S+\Sigma(s_i/n_i)]}{I\Sigma n_i * a\$_i}} = \sqrt{\frac{2(75+31)}{0.15*3900}} = 0.19$$

Using the formula again we find that the n_i's did not change significantly during the second iteration. Given these multiples, we calculate the aggregate dollar value, the quantity, and the annual inventory carrying cost as well as the order cost for each item. We find that the new multiples reduce the annual total cost by $1124.9 - 1114.37 = $10.53 as shown in Table 5.3. We also notice that the annual order costs and inventory carrying costs are more closer now compared with the solution obtained earlier.

TABLE 5.3   Order Size Using Brown's Algorithm

Item	a_i	$Q\$_i$	Q_i	Inventory Cost	Order Cost	Total Cost
1	5,000	986.75	197.35	74.00	75.35	$ 149.35
2	4,000	789.40	98.675	59.21	65.31	$ 124.52
3	10,000	1,973.51	197.35	148.02	140.59	$ 288.61
4	18,000	3,552.32	296.03	266.42	220.56	$ 486.98
5	1,000	394.70	19.74	29.60	35.31	$ 64.91
Total						$1114.37

Silver's Algorithm

In many instances, Brown's algorithm may require several iterations before reaching the final values of these multiples. To reduce the computational burden, Silver presented a simplified algorithm. Silver's algorithm for finding the multiples is slightly less tedious because it does not require an iterative process. As a first step, the algorithm compares the ratios of minor setup costs with annual dollar demand for each item. For the item with the smallest ration of $a\$_i/s_i$, the interval multiple is set equal to 1. Then the multiples for other items are calculated using Steps 2 and 3 as shown below.

Step 1: Find $\min[s_i/a\$_i]$ and let the item be j. (5.6)

Step 2: Find

$$\frac{a\$_j}{S+s_j}$$ (5.7)

Step 3: Find the multiple n_i's for all other items using the following formula. If the multiples were nonintegers, round them to the nearest integer greater than zero as shown earlier in Brown's method.

$$n_i = \sqrt{\frac{s_i}{a_i}\frac{a\$_j}{S+s_j}}$$ (5.8)

The calculations are exhibited in Table 5.4. The total cost obtained using the Silver's algorithm is the same as that of Brown's algorithm because the multiples are the same. Although the iterations can be repeated, similarly to Brown's algorithm, Kaspi and Rosenblatt [7] recommend that we terminate the calculations after one iteration because any further improvements are negligible.

Kapsi and Rosenblatt [7] indicate that Goyal and Belton improved the algorithm by choosing the smallest $[(S + s_i)/a\$_i]$ instead of minimum of $[s_i/a\$i]$. Counterexamples proved that Goyal and Belton's algorithm do not always improve the results, however. Kaspi and Rosenblatt [7] further showed that the accuracy of these results can be improved by following a combined approach, described below.

TABLE 5.4   Computation of Multiples

Item	$a\$_i$	s_i	$s_i/a\$_i$	$a\$_i/[S + s_j]$	n_i
1	5,000	5	0.001	216.87	$0.4656 => n_1 = 1$
2	4,000	5	0.00125	216.87	$0.52 \quad => n_2 = 1$
3	10,000	8	0.0008	216.87	$0.42 \quad => n_3 = 1$
4	18,000	8	0.00044		$n_4 = 1$
5	1,000	10	0.01	216.87	$1.5 \quad => n_5 = 2$

Kaspi and Rosenblatt's Algorithm

In earlier sections of this chapter we dealt with algorithms developed by Brown as well as by Silver. Although Silver's algorithm reduced the necessary computations for finding the values of these multiples, the amount of computations required could still be prohibitive when we deal with hundreds of items in a firm. Extensive simulation studies comparing these algorithms were conducted by Kaspi and Rosenblatt [8]. The results indicated that these algorithms can further be improved by first choosing the initial values suggested by Silver as a first step and then implementing Goyal's heuristic [4] as a second step. The combined procedure is fully illustrated by the following example.

Step 1:   Find

$$\min_{j} \quad \frac{S + s_j}{a\$_j} \tag{5.9}$$

Step 2:   Find

$$k_i^2 = \frac{\left(\dfrac{2s_i}{I * a\$_i} \right)}{T^2} \tag{5.10}$$

using the existing value of T initially.

Step 3:   Find the value of n_i that satisfies the condition as explained earlier.

$$k_i \, (k_i - 1) \le n_i^2 \le k_i \, (k_i + 1)$$

Step 4:   Find T from Silver's algorithm by using the values of n_i obtained in step 3.

$$T = \sqrt{\frac{2 \, [S + \Sigma(s_i / n_i)]}{I * \Sigma(n_i * a\$_i)}} \tag{5.11}$$

Now, using the value of T generated at step 4, continue to repeat steps 2 through 4 until two consecutive values of n_i remain the same. Kaspi and Rosenblatt also explain that the first iteration gives about 90% improvement. They recommend the discontinuation of computations because any improvement is not due to global nature but to local conditions. Table 5.5 exhibits the necessary calculations for Example 5.1 using Kaspi and Rosenblatt's algorithm.

We find, again, that item 4 has the smallest ratio for this problem. The remaining steps of this algorithm yield the same results obtained through the Silver's algorithm here, although it could have been different in another problem.

TABLE 5.5   Summary of Results from Kaspi and Rosenblatt's Algorithm

Item	$a\$_i$	s_i	$S + s_i$	$(S + s_i)/a\$_i$
1	5,000	5	80	0.016
2	4,000	5	80	0.020
3	10,000	8	83	0.0083
4	18,000	8	83	0.0044
5	1,000	10	85	0.085

JOINT REPLENISHMENT PRODUCTION QUANTITY MODEL

The previous section dealt with joint replenishment of items. Here we address the joint replenishment situation when a family of items can be produced using the same facility. Similar to that of the joint replenishment model, we assume that a major setup for the family is necessary and that a minor setup for each item in the family is also required. Therefore, the cost structure is same as that of the joint replenishment formula; the major difference is that the inventory is being consumed as production resumes. We now define the following additional variables:

$d\$_i$ =   dollar value of demand per period (daily, weekly)

$p\$_i$ =   dollar value of production per period as expressed by the demand

EPQ\$ =   economical production in dollar value of a group of items to be determined

Given these variables, we can express the annual setup costs as well as the annual inventory carrying costs by using a formula similar to that used for the economical production quantity in Chapter 4.

Number of setups per year $N = A\$/EPQ\$$

Annual setup costs $= N * [S + \Sigma s_i]$

Average dollar value of inventories $= EPQ\$[1 - (\Sigma d\$_i/\Sigma p\$_i)]/2$

Annual inventory carrying costs $= I * EPQ\$ [1 - (\Sigma d\$_i/\Sigma p\$_i)]/2$

Total Cost $= (A\$/EPQ\$)*[S + \Sigma s_i] + I*EPQ\$[1 - (\Sigma d\$_i/\Sigma p\$_i)]/2$

By taking the first derivative of the total cost equation with respect to EPQ\$ and setting it to equal to zero, we can find the following equation to determine the total dollar value of an economical production run for a group of items:

$$EPQ\$ = \sqrt{\frac{2[S + \Sigma s_i]A\$}{I * [1 - \{\Sigma d\$i / \Sigma p\$_i\}]}} \tag{5.12}$$

The following example illustrates the mechanics of the procedure.

Example 5.2

MultiMedia produces video promotion tapes. The tapes are different lengths and require different minor setups for each item in addition to a major setup cost of $50. As MultiMedia produces, it is also meeting the demands. The company operates 250 days per year. The annual carrying percent amounts to 25% of the cost of the item. Given the following information on the operational capability of MultiMedia,

(a) What is the dollar value of the aggregate lot size?
(b) What should be the dollar value of economic production run for each item?
(c) What are the values of economic production quantity for each item?

Item	Annual Usage $a\$_i$	Cost per Unit C_i	Annual Usage Quantity D_i	Weekly Usage $d\$_i$	Weekly Production $p\$_i$	Minor Setup s_i
1	8,000	4	2000	160	480	4
2	6,000	6	1000	120	320	3
3	4,900	7	700	98	150	5
4	3,200	8	800	64	100	6
Total	22,100			442	1050	

The economical value of the production run can be found by using equation 5.12.

$$EPQ\$ = \sqrt{\frac{2[S + \Sigma\, s_i]A\$}{I * [1 - \{\Sigma\, d\$i / \Sigma p\$_i\}]}}$$

$$= \sqrt{\frac{2 * [50 + 18] * 22,100}{0.25 * [1 - 442/1050]}} = \$4556.57$$

Once the economical production dollar value of the aggregate quantity is known, the dollar value of individual items can be found by using the following relationship:

$$EPQ\$_i = EPQ\$ * (a\$_i/A\$) \tag{5.13}$$

Thus

$$EPQ\$_1 = 4556.57 * (8000/22,100) = \$1649.43 = 412.38 \text{ units}$$

$$EPQ\$_2 = 4556.57 * (6000/22,100) = \$1237.08 = 206.19 \text{ units}$$

$$EPQ\$_3 = 4556.57 * (4900/22,100) = \$1010.28 = 144.33 \text{ units}$$

$$\text{EPQ\$}_4 = 4556.57 * (3200/22{,}100) = \ \$659.78 = 82.48 \text{ units}$$

$$\text{Annual inventory cost} = Q\$ * I/2$$

$$= 4556.57 * 0.25/2 = \$329.80$$

$$\text{Annual setup cost} = (A\$/Q\$) * (S + \Sigma s_i)$$

$$= (22{,}100/4556.57) *68 = \$329.80$$

SUMMARY

In this chapter we dealt with analytical models for joint replenishment of inventory and production problems. The basic premises behind these models have been to determine the multiples or cycles at which individual items need to be ordered (produced) and the quantities of these items to be ordered during each cycle. In many instances, the solution provided by these models may not be integer values of items that may need to be rounded to the nearest integer. The number of days specified by the cycle may not be an integer. Again, for practical purposes, it may be necessary to round the values to the nearest integer. Jackson, Maxwell, and Muckstadt [6] have attempted to improve the solutions to these problems where the multiples are restricted to the power of two instead of to any number. They have also extended their model to include multiple warehouses in the system and thus include the total cost of inventory and setup as well as transportation costs. Stochastic models dealing with joint replenishment problems have also been addressed in the literature. An excellent survey of these models can be found in Goyal and Stair [5]. As with many analytical models in inventory management, it is important to realize that the accuracy of results are dependent on the quality of costs used in solving these problems.

PROBLEMS

1. Colorado Ski Company needs to order the following five equipments from a major distributor. The annual demand, cost, and other data are exhibited below. The major order cost associated with the group of items is $100 per order, and the annual inventory carrying percent is 25% of the cost of items. Find the economical order quantity in dollars and units for all items.

Item	Annual Demand ($)a$_i$	Unit Cost C_i	Annual Demand Quantity	Item Preparation s_i
1	3,000	10	300	10
2	30,000	30	1,000	10
3	72,000	60	1,200	15
4	60,000	75	800	20
5	10,000	100	100	20

2. Given problem 1, is it possible to improve the solution by adopting multiple cycles using Brown' algorithm?
3. Given problem 1, is it possible to improve the solution by adopting multiple-cycle criteria using Silver's algorithm?
4. For the results obtained in problems 2 and 3 using problem 1 data, find the revised values of annual inventory carrying costs and the annual cost of orders. Compare the total cost with the total cost obtained in the original problem.
5. The annual demand and associated data for four different types of films at BMS Images are provided below. The major order costs are $60 per order and inventory carrying charges are 25% of the cost of items. Based on the information, compute the economical dollar value and quantities of individual items per cycle. Also calculate the annual inventory carrying costs as well as the annual order costs based on your solution.

Item	ASA Speed	Annual Demand ($)	Unit Cost ($)	Annual Demand Quantity	Minor Setup Cost
1	100	18,500	1.85	10,000	$5
2	200	16,800	2.10	8,000	$5
3	300	14,750	2.95	5,000	$5
4	400	7,100	3.55	2,000	$5
Total		57,150			$20

6. It is possible to improve the solution obtained in Problem 5 by adopting Brown's multiple-cycle method?
7. Is it possible to improve the solution by adopting Silver's multiple-cycle approach?
8. The annual demand for the items produced by Zelex Company Inc., the major and minor setup costs, and other pertinent information are given below. Based on the data, calculate the economical production quantity as well as the dollar value of these individual items per cycle. Also given are $S = \$180$, $I = 25\%$, and the number of operating days per year = 250.

Item	Annual Demand ($)	Unit Cost ($)	Annual Demand Quantity	Minor Setup Cost	Daily Production Rate
1	20,000	2.0	10,000	$10	$200
2	18,000	2.5	7,200	$10	$150
3	16,000	4.0	4,000	$20	$160
4	10,000	5.0	2,000	$20	$ 60
Total	64,000			60	$570

REFERENCES AND BIBLIOGRAPHY

1. J. Banks and W. J. Fabrycky, *Procurement and Inventory Systems Analysis* (Englewood Cliffs, N.J.: Prentice Hall, 1987).
2. R. G. Brown, *Decision Rules for Inventory Management,* (New York: Holt, Reinhart and Winston, 1967).

3. S. E. Elmaghraby, "The Economic Lot Scheduling Problem (ESLP): Review and Extensions," *Management Science,* Vol. 24, NO. 6 (February 1978), pp. 587–598.

4. S. K. Goyal, "Optimal Order Policy for a Multi-Item Single Supplier System," *Operational Research Quarterly,* Vol. 25, 1974, pp. 293–298.

5. S. K. Goyal and A. T. Stair, "Joint Replenishment Inventory Control: Deterministic and Stochastic Models," *European Journal of Operational Research,* Vol. 38, No. 38 (1989), pp. 2–13.

6. P. Jackson, W. Maxwell, and J. Muckstadt, "The Joint Replenishment Problem with a Powers-of-Two Restriction," *IIE Transactions,* Vol. 15 (1983), pp. 25–32.

7. M. Kaspi and M. J. Rosenblatt, "An Improvement of Silver's Algorithm for the Joint Replenishment Problem," *IIE Transactions,* Vol. 15 (1983), pp. 264–267.

8. M. Kaspi and M. J. Rosenblatt, "The Effectiveness of Heuristic Algorithms for Multi-Item Inventory Systems with Joint Replenishment Costs," *International Journal of Production Research,* Vol. 23, No. 1 (1985), pp. 109–116.

9. S. Love, *Inventory Control* (New York: McGraw-Hill, 1979).

10. E. A. Silver, "A Simple Method of Determining Order Quantities for Joint Replacement for Deterministic Demand," *Management Science,* Vol. 22 (1976), pp. 1351–1361.

11. E. A. Silver and R. Peterson, *Decision Systems for Inventory Management and Production Planning,* second edition, (New York: John Wiley & Sons, 1985).

12. M. K. Starr and D. W. Miller, *Inventory Control Theory and Practice* (Englewood Cliffs, N.J.: Prentice Hall, 1962).

APPENDIX 5A

Designing an Integrated Distribution System at DowBrands, Inc.*
*E. Powell Robinson, Jr., Li-Lian Gao, and Stanley D. Muggenborg

In 1985 Dow Consumer Products, Inc., a manufacturer of food-care products (for example, Saran Wrap and Ziploc storage bags), acquired the Texize home-care product lines of Morton Thiokol, Inc. and formed DowBrands, Inc., subsidiary of Dow Chemical Company. The combination gave Dow much needed marketing muscle and the potential for increased distribution efficiency if it could successfully merge the two complementary product lines.

By early 1986 the sales organizations were integrated, but the distribution organizations remained, for the most part, separate. Each organization employed an echelon of full-service central distribution centers (CDCs) the received products from their respective manufacturing plants, performed product mixing operations, stored inventories, and provided truckload (TL) and less-than-truckload (LTL) delivery to customers. In addition, the food-care distribution network maintained a second echelon of satellite or regional distribution centers (RDCs) that received product from CDCs and provided only LTL delivery to customers.

Each organization felt that its distribution system was best for its products. But to achieve the anticipated economic benefits of the merger, they needed an integrated distri-

Reprinted from *Interfaces,* Vol. 23, No. 3 (May–June 1993), pp. 107–117.

bution system. This propelled management to explore alternative consolidation strategies for the product lines.

Management was particularly interested in finding out whether a two-echelon system consisting of CDCs and RDCs or a single-echelon of CDCs was preferable. In addition, it wanted to know (1) the best number and location of facilities at each distribution echelon, (2) the best service assignments by product and shipment size for each facility, (3) the best assignment of customer demand to facilities, and (4) the best shipment routings by product and shipment size through the distribution system.

DowBrands, Inc., asked us to discuss these issues and to suggest a DSS that could analyze its problem. In reviewing the literature, we uncovered several applications of optimization procedures for locating a single-echelon of facilities [Erlenkotter 1978; Fitzsimmons and Allen 1983; and Geoffrion and Graves 1974]. In addition, Kaufman, van den Eede, and Hansen [1977], Ro and Tcha [1984], and Tcha and Lee [1984] discuss mathematical models and optimization procedures for two-echelon distribution system design. However, we did not find any procedure capable of solving a two-echelon design problem of the size and complexity of DowBrands'.

During the next two years, we developed an efficient optimization-based DSS for solving two-echelon distribution system design problems. Concurrently, DowBrands used scenario evaluation to guide integration of the home-care product line into the two-echelon food-care distribution system.

THE OPERATIONS AND COMPETITIVE ENVIRONMENT

DowBrands, Inc., manufactures and distributes over 80 products nationwide with additional sales in international markets. Based on consumer application, and manufacturing and distribution characteristics, it classifies the products into food-care and home-care product lines. Both product lines are sold to retailers whose order size ranges from 100 pounds for a mom-and-pop operation to multiple truckloads for national supermarket chains. The sales price includes the product price plus transportation costs to the customer's dock.

DowBrands operates multiple manufacturing plants for both product lines. Each plant focuses on a single product line. The distribution system consists of two-echelons of facilities, CDCs and RDCs, which are hierarchically linked. CDCs receive TL shipments from the manufacturing plants; perform product mixing operations; and maintain seasonal, order cycle, and safety-stock inventories. The CDCs ship TL quantities to replenish RDC inventories and service large-volume customers and LTL quantities to supply small-volume customers.

RDCs operate as CDC satellites. Each RDC is supplied from a single CDC and maintains regional cycle and safety-stock inventories but only minimal season inventories. RDCs provide only LTL delivery.

Public warehousing is used for product storage and order processing. Costs vary by location and consist of a handling cost per unit throughput and a storage cost per unit per unit time. Leases are flexible to permit the addition or deletion of capacity on a three-month notice.

All of DowBrands' freight movements are by common carrier. Freight rates are negotiated based on a Class 80 rating for food-care products and a Class 55 rating for home-care products. Most shipments to customers are in TL quantities. LTL shipment sizes range from 100 to 20,000 lbs. Several customers maintain private fleets and provide their own transportation services. These customers are credited at TL rates for picking up their own orders. Consequently, DowBrands bears all transportation costs from the manufacturing plants to the customer's door.

Competition in the consumer products industry is intense; several regional and national manufacturers offer rival products and vie for limited retailer shelf space. Although product features, product quality, promotions, and price are important factors in the customer's selection process, logistics performance is key to a firm's success.

Retailers often employ just-in-time inventory concepts requiring distributors to bear the bulk of inventory maintenance costs and provide quick and reliable deliveries. Poor logistics performance leads to lost sales due to stockouts and can result in lower allocations of retail shelf space. In addition, TL deliveries are frequently assigned delivery time windows of two hours at the customer's dock; missing a time window means the delivery must be rescheduled for the following day. A missed time window yields not only poor customer service and a potential loss of sales, but ties up a truck and driver for an extra day, adding to the cost of delivery.

Industry norms require that LTL deliveries in each customer zone originate from a single facility. TL shipments may originate at different facilities for each product line. An average order cycle lead time of seven to 10 days is standard for the industry as long as delivery occurs when promised. As a rule of thumb, the maximum delivery distance for LTL shipments is set at 500 miles to avoid consolidation or breakbulk terminals, which add to the mean and variance of delivery lead time. TL shipment distances are not constrained since they bypass consolidation centers, and their transit lead times are more predictable.

FIXED-CHARGE NETWORK PROGRAMMING MODEL

We modeled the problem as a fixed-charge network programming model in which nodes represent customer zones and candidate facility locations, and arcs represent potential shipment paths. The graphical aspects of the network model provide a convenient, comfortable, and effective instrument for communicating and structuring the problem.

A unique feature of the network model is the representation of RDCs by both their location and the supplying CDC. This maintains the structural and cost relationships of DowBrands's problem and is a critical component of the model. Other two-echelon facility location problem formulations do not have this characteristic.

Figure 1 shows the network model's structure and draws on the notation in Jensen and Barnes [1980]. The first two echelons of arcs are fixed-charge arcs which correspond to facility open and close decisions at potential CDC and RDC locations, respectively. If any product is shipped into a facility node, then the facility must be opened and the supplying arc's fixed-charge (that is, annual facility overhead costs) is incurred. Otherwise, the facility is closed and the fixed-charge is not applied.

We aggregated customers into three demand classes (LTL demand for both product lines, TL demand for food-care products, and TL demand for home-care products), and identified these with a unique node for each demand class in each customer zone.

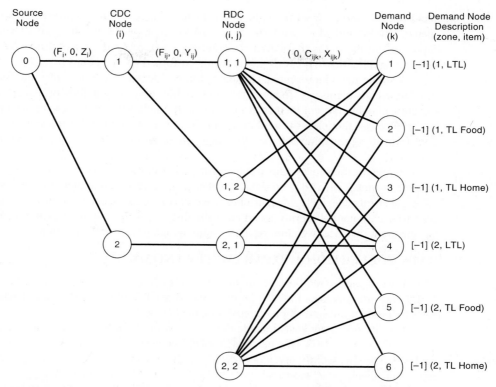

FIGURE 1:   The network model with two CDCs, two RDCs, and two customer zones. The first two echelons of fixed-charge arcs correspond to facility open and close decisions. The third echelon of arcs represents potential shipment paths to customers. The parameters above the arcs are the fixed-charge, the unit flow cost, and the decision variable associated with the arc. The objective is to find the least-cost routing of the six units of supply at the source node such that each demand node receives exactly one unit.

The third echelon of arcs connects the potential facilities to the demand nodes. Each arc represents a specific shipment path through the two echelons of facilities. The unit arc cost is the total variable transportation and facility throughput costs for using the replenishment path.

A dummy RDC j is defined for each CDC i, where $F_{ij} = 0$ when when $i = j$. These facilities maintain a strict hierarchical linkage of nodes and arcs for algorithm implementation and provide a mechanism for serving TL and LTL demand from CDCs. CDCs can serve (through their dummy RDCs) both TL and LTL demand nodes, while satellite RDCs (that is, $i \neq j$) are restricted to serving only LTL demand nodes.

We describe a mathematically equivalent mixed-integer programming formulation of the network model in the appendix. Gao and Robinson [1992] describe the solution algorithm.

Because public warehousing is abundant at all potential locations, we assumed facilities are uncapacitated and allowed the model to determine what the facility capacities should be at each location. This assumption also guarantees that each facility and demand node is served from a single shipment path in the optimal solution. Hence, each opened RDC is linked to one CDC, such that the costs for establishing an RDC are not duplicated in the optimal solution. In addition, each customer demand node is served from a single facility as set forth in the firm's customer-service policy.

A single node represents LTL demand in each customer zone. Since the food-care and home-care plants are focused by product line and geographically dispersed, it is sometimes more efficient to supply TL demand through different CDCs for each product line. Hence, we modeled TL food-care and TL home-care demand using separate nodes. This level of aggregation satisfies the single-sourcing customer-service policy at minimum cost.

A prespecified maximum distance for LTL shipments serves as a surrogate measure for in-transit delivery lead time. In calculating the cost parameters for the replenishment paths into the LTL demand nodes, we checked the distance from each customer zone to each potential facility location. If the distance exceeded the maximum permissible shipment distance, we eliminated the shipment path from consideration.

DATA COLLECTION AND PARAMETER ESTIMATION

The network model requires parametric estimates for facility operating costs, TL and LTL transportation costs for each product line over all shipment paths, and demand forecasts by product line and shipment size for each customer zone.

Facility operating costs include the cost of storing inventory and processing orders at the CDCs and RDCs. We calculated RDC inventory storage costs using regression analysis to predict facility inventory level I_{RDC} given facility throughout V_{RDC}. The RDC regression equation in thousand units is $I_{RDC} = 78.544 + 0.09636 (V_{RDC})$ with an R-squared of 0.775. Given management's minimum acceptable facility throughput level V_{MIN} and the per-unit holding cost S, the fixed cost for opening the minimum-size facility is $F_{ij} = S ((78.544 + 0.09636 (V_{MIN}))$.

The slope of a second regression equation with the Y-intercept set equal to V_{MIN} provides an estimate of the incremental inventory required to service each unit of throughput at the RDC (R-squared = 0.705). This slope, multiplied by the storage rate, yields the unit throughput cost for RDC inventory storage. We added unit processing and handling costs to unit storage costs to derive the total unit variable cost for RDC throughput.

Due to the confounding effects of seasonal inventories, regression analysis of CDC inventory level and CDC throughput failed to provide a strong causal relationship between the two variables. Using available data and management's experience, we estimated the minimum inventory level excluding seasonal inventory for opening a CDC to be five times that of the minimum-sized RDC. Given this estimate, we calculated the CDC operating costs as we did those of the RDCs.

For the study, we aggregated demand into 93 customer zones and stated it in LTL and TL shipment sizes by product line. A single transportation rate structure applies for all TL movements. Separate LTL rate structures apply to food-care and home-care products, reflecting differences in the freight classifications of the products. At this level of aggregation, the model requires approximately 6,700 freight charges.

Although actual transportation charges are based on specific origin-destination pairs, regression analysis indicates that over 80 percent of the variation in transportation rates for each shipment size and commodity are distance-related. Consequently, we used individual regression equations for each shipment size and product line to calculate mileage-based freight rates per hundred weight (cwt). We multiplied these rates by the appropriate shipment weight and distance to determine the freight charge. We used transportation distances from the *Standard Highway Mileage Guide* [1985] in the study.

COMPUTER IMPLEMENTATION AND SOFTWARE PERFORMANCE

The computer code was written in FORTRAN and implemented on an IBM 4381 mainframe computer at Indiana University. We set the problem dimensions at 13 CDC locations, 23 RDC locations, and 93 market zones with three demand classes in each zone. This equated to 13 fixed-charge arcs for CDCs, 299 fixed-charge arcs for RDCs, and 83,421 shipment arcs. The total number of feasible distribution configurations exceeded 68.7 billion, from which we determined the optimal system design. The central processing unit (CPU) times for problem solution and report generation ranged from 0.33 to 28.55 seconds with an average time of 7.7 seconds. During the study, we evaluated over 60 different cost and customer-service scenarios.

PERFORMING THE SYSTEM DESIGN STUDY

The system design study consisted of four parts: verification of the model and data set, sensitivity analysis, analysis of cost-service trade-offs, and what-if evaluation.

VERIFICATION

We solved several problems to verify the accuracy of the data set and computer code. We checked the internal validity of the algorithm by solving identical small test problems with both the specialized computer code and a general purpose mixed-integer programming computer code. Both computer codes found identical solutions, which gave us confidence in the numerical accuracy of the solution procedure.

Next, we solved the problem using the complete data set and studied the output reports to verify the external validity of the network model and the data set. We focused on establishing whether the model accurately reflected the assumptions of the problem and produced reasonable solutions. We identified and corrected several minor inconsistencies in the data set during this process.

SENSITIVITY ANALYSIS

Once we had accepted the validity of the model and data set, we investigated the sensitivity of the optimal solution to changes in cost parameters. We were particularly interested in the potential impact of errors in the estimation of facility fixed-cost structures. Using the estimates from the regression analysis as the base case, we evaluated eight different cost scenarios with the fixed costs ranging from 50 to 200 percent of the base case. Since the purpose of this evaluation was to study cost interactions, we imposed no limits on the maximum distance for LTL product shipments.

This analysis provided three major observations. First, the optimal system design was relatively insensitive to errors in fixed-cost estimation. The number of opened CDCs and RDCs ranged from four to five and from zero to two, respectively. All of the eight optimal solutions included combinations of the same five CDCs indicating their strong dominance over the other eight candidate CDCs that did not appear in any of the optimal system designs. Second, the solutions did not utilize RDCs until the RDC fixed costs were 45 percent of the base case. This suggested that, considering only cost, a single-echelon of CDCs would be moire efficient than the two-echelon system of CDCs and RDCs. Finally, the analysis provided insight into how the distribution system would evolve given changes in facility cost structures.

Sensitivity analysis for different CDC/RDC fixed-cost ratios, transportation cost structures, and demand rates confirmed the relative insensitivity of the optimal system design to changes in these problem parameters.

CUSTOMER-SERVICE TRADE-OFFS

In the next phase of the study, we investigated the relationship between the least-cost system design and customer service, as defined by the maximum shipment distance for LTL deliveries. We evaluated eight different levels of customer service ranging from a distance of 500 miles to a distance of 2,000 miles. The maximum LTL distance of 500 miles reflected management's existing service policy. The service constraints were not binding in the 2,000-mile scenario.

Figure 2 shows the results of our analysis. The cost-service curve follows the elbow shape typically associated with these curves [Rosenfield, Shapiro, and Bohn 1985].

The graph helps management understand the cost-service trade-offs and provides evidence to support decisions affecting customer service. For example, total costs de-

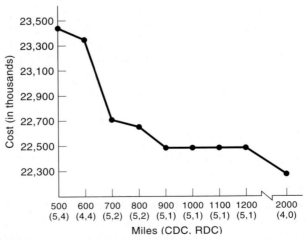

FIGURE 2: The cost-service curve indicates the relationship between total system costs and maximum LTL service distances. The two numbers in parentheses below the X-axis are the optimal open number of CDCs and RDCs for the specified maximum LTL shipment distance.

crease by only $100,000 when the maximum LTL transit distance is increased from 500 to 600 miles. This may support management's preference for a 500-mile service level. However, costs are reduced approximately $700,000 if the maximum service distance is increased from 500 to 700 miles. Management must decide whether the 500-mile service policy is appropriate. Do customers perceive a sufficient difference between a 500-mile shipment distance and a 700-mile distance to warrant the $700,000 cost increase?

The cost-service analysis also clarifies the roles of the two facility types. Without customer-service restrictions, the company handles all demand from CDCs. It adds RDCs as it decreases the maximum LTL shipment distance. RDCs thus provide the least-cost alternative for providing higher LTL customer-service levels. Consequently, for DowBrand's problem, the justification for a two-echelon distribution system rests on a differentiated distribution strategy for TL and LTL movements.

The customer-service analysis provided additional insights into what constitutes a good distribution system design for DowBrands. Of the possible 13 CDC sites, only six entered into the optimal solutions. Of these five were at plant locations. Of the 23 possible RDCs, the model suggested opening only seven. These RDCs were located primarily in the Midwest and Northwest. The specific RDCs the model suggested varied considerably depending upon the customer-service policy. For example, service policies of 700 and 800 miles were most economically served by five CDCs and two RDCs. However, RDCs in Kansas City and Seattle were associated with the 700-mile service policy, whereas the model opened facilities in Minneapolis and Portland to meet an 800-mile service policy.

WHAT-IF EVALUATION

In the final phase of the study, we merged the most attractive distribution system configurations we identified in the sensitivity and cost-service analyses with management's experience to generate several what-if configurations for evaluation. These what-if configurations typically involved forcing an existing facility that was not recommended in the optimal solutions into the set of opened facilities. We then solved the model to optimality and evaluated the what-if scenario.

An example illustrates the benefits of the what-if analysis. An existing CDC in Chicago was not recommended in any of the computer-generated solutions. This concerned the management team since Chicago was a high demand market area. Furthermore, Denver, which was recommended as a satellite RDC, could be served from Chicago with frequent and reliable LTL service, eliminating the need for an RDC in Denver. Based on these observations, we evaluated a what-if scenario.

In the scenario, we forced Chicago open and solved the resulting problem. In this solution, Chicago replaced an existing CDC in Ohio, for an annual cost increase of $36,000 over the previous solution. Given the relatively small cost difference, management could make the choice between keeping the Chicago CDC open or keeping the Ohio CDC open primarily on the basis of customer service.

However, management was uncomfortable closing either one of the existing CDCs. We conducted a second what-if analysis, forcing both CDCs open. This solution increased costs by approximately $400,000 annually over the scenarios in which either the Chicago or the Ohio CDC was open and increased service only moderately. These what-if analyses allowed management to study the cost and service trade-offs associated with a nonoptimal system design.

IMPLICATIONS, BENEFITS, AND CONCLUSIONS

DowBrands, Inc., began consolidating its distribution system in 1986 right after it acquired Texize, Inc. It reduced the initial configuration of five CDCs and 19 RDCs to six CDCs and seven RDCs by 1988. However, management felt uncertain as it moved into the final stages of the design process.

It needed a better understanding of the roles of CDCs and RDCs, whether it should use a single-echelon or a two-echelon system, and what type, number, and location of facilities would correspond to the least-cost system design for current (or forecast) cost and demand parameters. In addition, management did not understand the impact of alternative customer-service policies on the optimal distribution system configuration and on cost.

The optimization procedures gave management the analytical support it needed to eliminate these uncertainties and develop guidelines for change. The sensitivity analysis, the cost-service trade-off analysis, and the optimization-based what-if analysis clarified all the major cost and service trade-offs, and gave managers confidence that the distribution strategy they proposed was economically sound.

The savings projected for the recommended system configuration with a maximum 500-mile LTL distance constraint over the one in operation in 1988 were $762,400 per year. When this constraint was relaxed to 700 miles, as recommended by the management team, the projected savings were $1,466,800 per year. These savings were primarily due to a reduction in the number of RDCs.

The academics involved in the project gained considerable benefits as well. Our preliminary discussions about the problem provided a meaningful research topic that culminated in a PhD dissertation [Gao 1988] and two research papers [Robinson and Gao forthcoming; and Gao and Robinson 1992]. The solution algorithms performed well using hypothetical data, but the actual data and research support DowBrands provided helped us to document the true applicability of the new solution procedures. In essence, we had a living laboratory in which to test our research findings. The lessons we learned will serve well both in the classroom and in research.

ACKNOWLEDGMENTS

Special thanks go to Robert Larson, vice-president of material flow and Sharon Gillie, manager of analysis, at DowBrands, Inc. The study benefited immensely from their encouragement, insight, and suggestions.

APPENDIX

We use the following notation in the mathematical problem statement:

m = the number of candidate central distribution centers (CDCs),

n = the number of candidate regional distribution centers (RDCs), and

q = the number of customer zones. Each customer zone represents the demand

in a specific geographic area, for a particular product and shipment size.
For each $i \, \varepsilon \, \{1, 2, \ldots, m\}, j \, \varepsilon \, \{1, 2, \ldots, n\}$, and $k \, \varepsilon \, \{1, 2, \ldots, q\}$

Z_i = the binary decision variable for opening CDC i,

Y_{ij} = the binary decision variable for opening RDC j and supplying it from CDC i,

X_{ijk} = the continuous decision variable for the fraction of zone k's demand that is served through CDC i and RDC j,

F_i = the annualized fixed cost for opening CDC i,

F_{ij} = the annualized fixed cost for opening RDC j and supplying it from CDC i,

C_{ijk} = the cost of serving zone k's demand through CDC i and RDC j where $C_{ijk} = t_{ijk}d_k$,

ti_{ijk} = the unit throughput cost at CDC i, and RDC j, and transportation costs from the plant to CDC i and from CDC i through RDC j to zone k, and

d_k = the annual demand at zone k.

The mixed-integer programming model of the two-echelon uncapacitated facility location problem is

$$\min Z = \sum_{i=1}^{m} F_i Z_i + \sum_{i=1}^{m} \sum_{j=1}^{n} F_{ij} Y_{ij}$$
$$+ \sum_{i=1}^{m} \sum_{j=1}^{n} \sum_{k=1}^{q} C_{ijk} X_{ijk},$$

$$(1)$$

subject to

$$\sum_{i=1}^{m} \sum_{j=1}^{n} X_{ijk} = 1, \quad k = 1, \ 2, \ldots, \ q, \tag{2}$$

$$-Z_i + Y_{ij} \leq 0, \, i = 1, 2, \ldots, m$$
$$j = 1, 2, \ldots, n, \tag{3}$$

$$-Y_{ij} + X_{ijk} \leq 0, \, i = 1, 2, \ldots m$$
$$j = 1, 2, \ldots n \ \ k = 1, 2 \ldots, q, \tag{4}$$

$$0 \leq Z_i \leq 1 \text{ and integer } i$$
$$= 1, 2, \ldots, \text{m}, \tag{5}$$

$$0 \leq Y_i \leq 1 \text{ and integer } i = 1, 2, \ldots, \text{m}$$
$$j = 1, 2, \ldots, n, \tag{6}$$

$$0 \leq X_{ijk} \leq 1, \, i = 1, 2, \ldots, m$$
$$j = 1, 2, \ldots, \text{n} \ \ k = 1, 2, \ldots, q. \tag{7}$$

The objective function represents the fixed costs for establishing CDCs and RDCs and the variable costs for serving customers. Constraint set (2) requires that all demand be served. Constraint set (3) prevents a CDC from supplying an RDC unless the CDC is opened. Constraint set (4) prevents an RDC from supplying a customer zone unless the RDC is opened.

REFERENCES

ERLENKOTTER, D. 1978, "A dual-based procedure for uncapacitated facility location," *Operations Research,* Vol. 26, No. 6, pp. 992–1009.

FITZSIMMONS, J. A. and ALLEN, L. A. 1983, "A warehouse location model helps Texas comptroller select out-of-state audit offices," *Interfaces,* Vol. 13, No. 5, pp. 40–46.

GAO, L. 1988, "A revised formulation and two complementary optimization procedures for the two-echelon uncapacitated facility location problem," PhD diss., Indiana University, Bloomington, Indiana.

GAO, L. and ROBINSON, E. P. 1992, "A dual-based optimization procedure for the two-echelon uncapacitated facility location problem," *Naval Research Logistics,* Vol. 39, No. 2, pp. 191–212.

GEOFFRION, A. M. and GRAVES, G. W. 1974, "Multicommodity distribution system design by benders decomposition," *Management Science,* Vol. 20, No. 5, pp. 822–844.

JENSEN, P. A. and BARNES, J. W., eds. 1980, *Network Programming,* John Wiley and Sons, New York, New York.

KAUFMAN, L; VAN DEN EEDE, M.; and HANSEN, P. 1977, "A plant and warehouse location problem," *Operational Research Quarterly,* Vol, 28, No. 3, pp. 547–554.

RO, H. and TCHA, D. 1984. "A branch-and-bound algorithm for the two-level uncapacitated facility location problem with some side constraints," *European Journal of Operational Research,* Vol. 18, pp. 349–358.

ROBINSON, E. P. and GAO, L. forthcoming, "A new formulation and linear programming based optimization procedure for the two-echelon uncapacitated facility location problem," *The Annals of the Society of Logistics Engineers,* Vol. 2, No.

ROSENFIELD, D. B.; SHAPIRO, R. D.; and BOHN, R. E. 1985, "Implications of cost-service trade-offs on industry logistics structures," *Interfaces,* Vol. 15, No. 6, pp. 47–59.

Standard Highway Mileage Guide 1985, Rand McNally and Col., Chicago, Illinois.

TCHA, D. and LEE, B. 1984, "A branch-and-bound algorithm for the multi-level uncapacitated facility location problem," *European Journals of Operational Research,* Vol. 18, pp. 34–43.

Chapter Six

Inventory Systems under Risk

INTRODUCTION

Inventory models discussed in Chapters 4 and 5 assumed that demands are constant and known. Once we begin to move closer to reality, we must recognize that demand is never certain but that it occurs with some probability. Inventory models that consider risk or probability attempt to manage the chance of stockout by trading off holding costs, setup costs, and stockout costs.

In this chapter, we begin with a single-period trade-off between foregone profit, if not enough inventory is held, and loss on excess inventory sold at salvage. This simple trade-off focuses attention on stockout risk (the probability of stocking out), a concept underlying the entire chapter. Specifically, profit and cost if available lead us to an implied optimal stockout risk. If foregone profit and cost figures are unavailable, then at least their ratio can be imputed from the stockout risk the decision maker is willing to accept.

Before embarking on our journey into more difficult EOQ terrain, we need to recognize that many practitioners disparage EOQ models, sometimes with justification and sometimes without. Nevertheless, EOQ keeps coming back. Why?

There are cases in which EOQ should not be applied. [An entire later chapter is devoted to material requirements planning (MRP) for such cases.] Nevertheless, EOQ's staying power results from some simple characteristics of forecasting systems and inventory trade-offs. First, forecasting systems are assumed to yield normally distributed error terms. If the error terms are not normally distributed, a good forecasting expert will be able to produce more accurate forecasts by incorporating another term to remove more explained variance from the forecasts. Second, end item or finished goods inventories embody cost trade-offs, whether or not they are explicitly recognized. The very act of specifying a lot size—any lot size—says something about managers' risk aversion and their concept of the relative magnitude of setup versus holding costs.

Throughout this chapter, we constantly examine acceptable stockout risks. We can impute a stockout penalty for a stockout risk, just as we can derive an acceptable stockout risk if we are given a stockout penalty cost. This, then, will be a chapter on trade-offs, beginning with the most ubiquitous of all inventory problems, the trade-off of holding too little or too much inventory for a specified season or time.

SINGLE-PERIOD MODEL

In New York City a number of years ago, an enterprising graduate student made $10,000 one December selling Christmas trees. He purchased the trees at $4 per tree and sold them at $9. Although $10,000 was enough to pay for a year's expenses at the university then, the student was greedy. The next year, he expanded his business and managed to lose $20,000. (After Christmas, a tree may be worth $1, for firewood.)

Such single-period problems may be the most common inventory problem faced by businesses. Newspapers, milk, clothing, and many other consumer goods have a specified shelf life, after which they lose much of their value.

Assume that the graduate student's location in the city will generate for him an average demand of 2000 trees ($\bar{x} = 2000$). The standard deviation of quantity demanded is 100 trees. Assuming that demand is normally distributed, there is a 0.50 probability that demand could be larger than 2000 trees and 0.16 probability that demand could be larger than 2000 plus one standard deviation—that is, $2000 + 100 = 2100$ trees. For a z value of 1.0, the area in the upper tail of the unit normal curve is 0.1587 (see appendix 6A).

k or z value	SOR
1.0	0.1587

What stockout risk should the student be willing to take? Good business majors would surely reply, "That depends on the rewards." The larger the per unit profit on a tree, the less willing the student should be to accept stockout risk; conversely, the larger the per unit loss on trees remaining unsold after Christmas, the more willing he should be to accept stockout risk so as to avoid being stuck with too many trees. Marketing majors might be tempted to create a Christmas tree tradition for Easter, but we prefer a more conservative solution to the problem.

We shall try to answer the question of how much risk the student should take. Let S = sales price, C = cost, and V = salvage value, so that the example here yields $S = \$9$, $C = \$4$, and $V = \$1$. Relying again on marginal analysis, the marginal profit (MP) on a tree is $S - C = \$5$ and the marginal loss (ML) is $C - V = \$3$. To order Q rather than $Q - 1$ trees, the expected profit on the marginal tree must be greater than or equal to the expected loss. Let p be the probability of selling the marginal tree. Then our profit criterion can be manipulated into a decision rule:

expected profit $\geqslant$ expected loss

$$p(\text{MP}) \geqslant (1 - p)\text{ML}$$

$$p(\text{MP}) \geqslant \text{ML} - p(\text{ML})$$

$$p(\text{MP} + \text{ML}) \geqslant \text{ML}$$

Applying this rule to our example, we derive a minimum acceptable probability of selling the Qth tree:

$$p \geqslant \frac{3}{5 + 3} = \frac{3}{8} = 0.375$$

Therefore, the graduate student should buy the Qth tree provided that he has a 0.375 or better chance of selling it. Equivalently, the probability that quantify demanded will equal or exceed Q must be 0.375. Stockout risk (SOR), however, is precisely the probability that quantity demanded will exceed supply. Hence, our enterprising student is willing to accept SOR = 0.375. If we were to raise the marginal profit on a tree to \$6, he should be less willing to accept a higher stockout risk and, using $p \geq \text{ML}/(\text{MP} + \text{ML})$, he would lower p to 0.33.

The rest is mechanical. The graduate student had enough foresight to save his production control text. Checking out Appendix 6A for an upper tail area of 0.375, he finds a z value of 0.32. He should then order the mean plus 0.32 standard deviations worth of trees—that is $\bar{x} + zs_x = 2000 + 0.32(100) = 2032$. (Note that we are using the normal distribution rater than the t because $n \geqslant 30$.) In capsule form, the single-period model has

$$\text{SOR} \geqslant \text{ML}/(\text{MP} + \text{ML})$$

$$Q = \bar{x} + z_{\text{SOR}} \, s_x$$

MP $\uparrow$ implies SOR $\downarrow$

ML $\uparrow$ implies SOR $\uparrow$

In summary, we can say that if the graduate student carries thirty-two Christmas trees as a buffer or safety stock, he can reduce the stockout risk from 50% to 37.5%. The complement of stockout risk, that is 62.5%, is known as the service level.

SERVICE LEVEL AND SAFETY STOCK

The Christmas tree problem provides an insight into the concept of service level and safety stocks. Although we used marginal analysis to provide an estimate of the stockout risk, in practice it is often difficult to determine the cost of stockouts (or marginal loss), which includes intangibles such as the loss of customer goodwill or the cost of potential delays to other part of the system. Also, in some instances the customers may be willing to wait if the response time is reasonable. The purpose of the safety stock is to cover the random variations in demand or lead time. The safety stock is not intended to cover 100% of the variations during that period. The amount of variation covered by the safety stock depends on the desired stockout risk or the customer service level. When stockout costs are not available, a common surrogate is the customer service level, which generally refers to the probability that a demand or a collection of demands are met. For example, in a very competitive environment, a firm may choose to keep stockouts as small as possible for fear of losing customers to competitors. On the other hand, when dealing with perishables or dated items such as Christmas trees, the same firm may choose to lower its customers service level because there may not be any demand for the trees left after the season. In any event, it is important for a firm to specify the service level so that the appropriate level of safety stock can be calculated as a starting point to determine the amount of investment required to carry those safety stocks. A firm may decide to change those safety stock levels as it learns more about the demand patterns of individual items. If adequate investments are not available a firm could use a trade-off analysis between the safety stock investment and associated service levels, as discussed later in this chapter. In fact, the service level can be classified in two different ways: (1) order or cycle service level and (2) unit service level.

Order Service Level (OSL)

The order service level (OSL) can be interpreted as the proportion of cycles that the customer demand was satisfied. In the Christmas tree problem, we assumed an order was placed only once during the year. But many consumer items are sold year round. Items are ordered several times (cycles) during the year, as illustrated in the EOQ/EPQ problems of Chapter 4. Therefore, the order service level represents the probability of not having a stockout during the placement of an order. If several cycles or orders exist during the year, then the stockout probability times the number of cycles provides the probability of not having stockouts during the year. Suppose that OSL = 0.90 and the number of cycles per year = 20. Then customer demand will be satisfied during eighteen (0.90 * 20 = 18) cycles on average, whereas during the two (20 − 2 = 18 or 0.10 * 20 = 2) cycles on average we can expect a stockout during the year.

The order stockout risk does not tell us exactly how many units were short or not filled during any cycle. In many production situations, if shortages occur during any operation, the succeeding operation cannot be performed. Regardless of the quantity of items short, the remaining operations have to be curtailed during the cycle. Therefore, the order service level or the order stockout risk is appropriate for manufacturing environment.

Unit Service Level (USL)

The unit service level (USL) indicates the percentage of units of demand filled during any period of time, whereas the unit stockout risk (USOR) specifies the quantities of units unfilled or short during that period. The unit service level can provide us with the exact quantities of units of customer demand filled during each cycle or during any time period, and hence it is a more appropriate measure for many consumer-good applications. Suppose that USL = 0.95 and that the annual demand D = 5000 for an item. Then 0.95 * 5000 = 4750 units of customer demand will be filled on average during the year. We can also expect 250 units of stockout during the year. Further, if there were ten cycles during the year, then on average we may not be able to fill twenty-five units of demand during each cycle.

PERCENT ORDER SERVICE
SAFETY STOCK
AND THE MULTIPERIOD CASE

Returning to our EOQ/ROP multiperiod model, we need to take account of stockout risk in a similar fashion. Usually, an item forecast is fed to the inventory system as an estimate of $\bar{x}_L$, the mean demand during lead time. Most forecasting systems also provide an estimate of the mean absolute deviation of lead time demand (MAD_L). Although the standard deviation is the preferred measure, habit has enthroned MAD, even though its ease of calculation is no longer important thanks to modern computers.

For the normal distribution, MAD is approximately 0.8σ. Because the sample mean absolute deviation is an unbiased estimate of the mean absolute deviation, just as the sample standard deviation is an unbiased estimate of the standard deviation, we can equivalently take MAD = $0.8s$ or s = 1.25MAD, where we adopt the sloppy but conventional notation that MAD stands for either the population or the sample mean absolute deviation.

Our multiperiod EOQ/ROP system can now be captured by Figure 6.1. Here we assume that $\bar{x}_L$ = 200, MAD_L = 32, L = 5 days, D = 10,000, S = 1500, and h = 30, using $Q^* = \sqrt{2DS/h}$, Q^* = 1000 for this example. Also, there are ten cycles, because D/Q^* = 10,000/1000 = 10. Figure 6.1 shows only one cycle (see also Figure 6.2).

For the moment, suppose that the inventory manager is willing to accept a 0.375 probability of stockout on any cycle. What, then, should be the reorder point? Defining *percent order service* as the probability of satisfying all demand on a cycle, we have

$$OSL = 1 - OSOR$$

where OSOR, *order stockout risk,* is the probability that lead time demand will exceed lead time supply and OSL, the *percent order servive level,* is the probability that lead time demand can be satisfied by the reorder quantity—the lead time supply. In our example, OSL = 1 − 0.375 = 0.625. In a manner identical to the order quantity calculation in the Christmas tree model, we calculate the reorder quantity.

With $\bar{x}_L$ = 200, s_L = 1.25MAD_L = 1.25(32) = 40, and $z_{0.375}$ = 0.32, the desired lead time supply or reorder quantity is

$$R = \bar{x}_L + z_{SOR}\, s_L$$

$$R = 200 + 0.32(40) = 213 \text{ units}$$

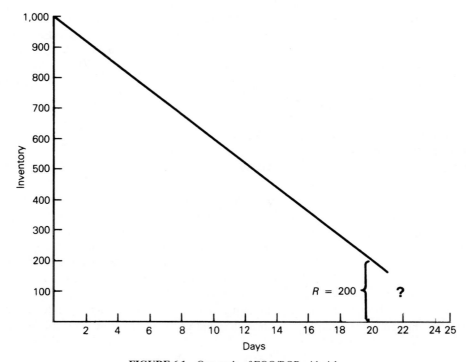

FIGURE 6.1   One cycle of EOQ/ROP with risk.

Although thirteen units of safety stock and a percent order service level of only 62.5% may appear extremely low at first glance, these thirteen units actually buy considerable protection against stockout. We are expecting a stockout position in fewer than eight of twenty cycles, or about four times per year with the present order quantity. With this percent order service level, what percentage of units demanded do we expect to supply from stock?

PERCENT UNIT SERVICE SAFETY STOCK

The *percent unit service level* answers the question of what percentage of units demanded can be supplied from stock. Sometimes known as the *fill rate,* it is the opposite of the *stockout rate* that specifies the percentage of units demanded that cannot be supplied from stock and that become either lost sales or backorders.

With a percent order service level of 0.625 and a lead time MAD of thirty-two units, what would be the percent unit service level? To answer this question, we shall work through the stockout rate first and then simply convert from the stockout rate to the percent unit service level.

Figure 6.3 shows R as the reorder point on the normal distribution of lead time de-

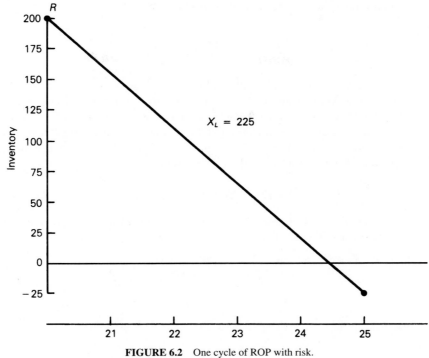

FIGURE 6.2   One cycle of ROP with risk.

mand and k as the reorder point on the unit normal distribution of lead time demand. The familiar standardization gives $R = (d_L - \bar{d}_L)/\sigma_L$. For the special case of R, $k = (R - \bar{d}_L)/\sigma_L$.

Fortunately, tables exist to give us the expected number of stockouts on the unit normal distribution. These tables are calculated from

$$E(Z-k)^+ = \int_0^k 0 f(z)\, dz + \int_k^\infty (z-k)f(z)\, dz$$

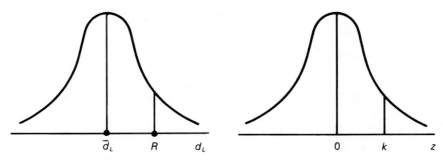

FIGURE 6.3   Stockouts on the normal versus unit normal.

which is the standardized equivalent of

$$E(X - R)^+ = \int_0^R 0 f(x)\, dx + \int_R^\infty (x - R) f(x)\, dx$$

For demand less than the reorder point, the stockout quantity is zero; for demand greater than the reorder point, the stockout quantity is demand minus the reorder point. Appendix 6A provides a table for the partial expectation, $E(Z - k)^+$ or $g(k)$.

In our example we determined the reorder point using 0.32 standard deviations of safety stock to provide a 62.5% order service level. The expected number of stockouts per cycle on the unit normal curve would be

$$E(Z - k)^+ = g(k) = g(0.32) = 0.26$$

Conversion back to the normal distribution of lead time demand stems from observing the standardization relationships:

$$z - k = \left(\frac{x - \overline{x}_L}{\sigma_L} \right) - \left(\frac{R - \overline{x}_L}{\sigma_L} \right) = \frac{x - R}{\sigma_L}$$

so that

$$x - R = \sigma_L (z - k)$$

and

$$E(X - R)^+ = \sigma_L * E(Z - k)^+ = \sigma g(k)$$

In our case, $\sigma_L = 1.25\ \text{MAD} = 1.25(32) = 40$, and $g(k) = 0.26$. Hence, $E(X - R) = 40(0.26) = 10.4$. On every cycle, the expected number of stockouts is 10.4. On average, each cycle faces $Q = 1000$ units of demand, so that the unit stockout rate (USOR) would be USOR $= 10.4/1000 = 0.0104$. The associated percent unit service level (USL) is 0.9896.

Suppose that the inventory manager indicates that she really had meant a 99% unit service level. How would we calculate the appropriate safety stock directly?

$$USOR = \frac{E(X - R)^+}{Q}$$

(e.g., $10/1000 = 0.01$), but $E(X - R)^+ = \sigma_L g(k)$, where σ_L is the standard deviation of lead time demand, giving

$$USOR = \frac{\sigma_L g(k)}{Q}$$

or

$$g(k) = \frac{Q * \text{USOR}}{\sigma_L} = \frac{Q(1 - \text{USL})}{\sigma_L}$$

and the unit service level can be expressed as $1 - \text{USOR}$, that is,

$$\text{USL} = 1 - \frac{g(k)\sigma_L}{Q}$$

In our example, $g(k) = (1000 * 0.01)/40 = 0.25$. Finding the k in the table corresponding to $g(k) = 0.25$, we get $k = 0.35$. Hence, 0.35 standard deviations of safety stock (SS = $0.35 * 40 = 14$ units) gives a 99% unit service level.

The following steps are necessary for calculating σ_L-based safety stock from a unit service objective:

1. Specify USL; USL = 0.99.
2. Calculate USOR = $1 - \text{USL}$; USOR = 0.01.
3. Calculate $g(k) = (Q/\sigma_L) \text{USOR}$; $g(k) = (1000/40)0.01 = 0.25$.
4. Find the safety factor k giving $g(k) = 0.26$; $k = 0.35$.
5. Set safety stock at k standard deviations; SS = $k\sigma_L = 0.35(40) = 14$.
6. Set the reorder point to cover lead time demand plus safety stock;
   $R = \bar{x}_L + \text{SS} = 200 + 14 = 214$.

BACKORDER OR LOST SALE COSTS

As most inventory managers will do, ours has specified a 99% service level without explicitly considering stockout costs. Because we already have holding cost estimates, we can now derive the *implicit* or *imputed* stockout costs that the inventory manager used intuitively. Our analysis will distinguish between the backorder and the lost sales case. The analysis is important in its own right, because it will allow us to specify safety stocks when we are provided with stockout cost estimates but no desired service levels.

At this point, it is worth noting that increasing the safety stock level by one more unit has the same effect as adding one more unit to the reorder point. Therefore, adding one unit to R is analogous to increasing SS by one additional unit.

Backorder Case

In the *backorder* case, we note that raising the reorder point by one unit will cost us $h(Q/D) = \$3$ per cycle. This is because we will hold the unit for almost the entire cycle, regardless of whether or not it is used to satisfy demand. If we do not add the unit to the reorder point, we suffer a backorder penalty of $\$\pi$ per unit (expediting costs, extra paperwork, ill will) with probability SOR. Balancing these marginal costs gives

Per cycle marginal cost of adding 1 unit to R
= per cycle marginal cost of *not* adding 1 unit to R

$$h(Q/D) = \text{OSOR} \ (\pi)$$

$$\text{OSOR} = \frac{hQ}{\pi D} \quad \text{or} \quad \pi = \frac{hQ}{\text{OSOR} * D}$$

In our case, we must determine π, the *imputed* backorder cost given the inventory manager's 99% unit service level. Recall that we calculated a safety stock of fourteen units for a 99% USL. Such a safety stock yields a 36.3% order stockout risk (OSOR). This is because $z\sigma_L = z(40) = 14$ gives a z value of 0.35. With $h = \$30$ per unit per year, $Q = 1000$, $D = 10,000$, and OSOR = 0.363, we estimate the backorder cost to be \$8.26 per unit:

$$\pi = \frac{30(1000)}{0.363(10,000)} = \$8.26$$

Our inventory manager has second thoughts, however; \$8.26 sounds too low! She really thinks the backorder penalty is more in the area of \$15 per unit.

Undaunted, the inventory analyst checks this out in terms of implied OSOR, USOR, and safety stock:

$$\text{OSOR} = \frac{hQ}{\pi D} = \frac{30(1000)}{15(10,000)} = 0.2$$

$$\text{SS} = z_{0.2}\sigma_L = 0.84(40) \approx 34$$

$$\text{USOR} = \frac{\sigma_L g(k)}{Q} = \frac{40g(0.84)}{1000} = \frac{40(0.112)}{1000} = 0.004$$

$$\text{USL} = 1 - 0.004 = 0.996$$

In the backorder case, stockout costs of \$15 per unit lead to a 99.6% unit service level. But what about the lost sales case? Why would that be any different?

Lost Sales Case

In the lost sales case, the stockout cost π^1 includes foregone revenue in addition to customer ill will. Marginal analysis also yields slightly different results because of the impact of lost sales on inventory levels. Adding one unit to R costs $h(Q/D)$ minus OSOR * $h(Q/D)$, the unused inventory holding cost in the event that one unit of safety stock is used (ignoring the holding costs during the lead time) in holding. If we do not add the unit, the penalty in the event of stockout includes π^1. To see this, suppose that $D = 20$ and $Q = 10$. A stockout on the first cycle implies that eleven units of demand have occurred, while ten units have been supplied. To start the next cycle, the full supply of $Q = 10$ units is on hand. (Compare this with the backorder case, where the backorder would deplete the beginning inventory down to nine units.) With $D = 20$, only nine units of demand remain for the second cycle. Hence, one *extra* unit of inventory will be held through the second cycle.

Marginal cost of adding one unit to reorder point = Marginal cost of not adding or expected savings

$$h \ (Q/D) - \text{OSOR} * h(Q/D) = \text{OSOR} * \pi^1$$

$$h(Q/D) = \text{OSOR} * [\pi^1 + h(Q/D)]$$

$$\text{OSOR} = \frac{h(Q/D)}{\pi^1 + h(Q/D)}$$

$$\text{Therefore,} \ \ \text{OSOR} = \frac{hQ}{hQ + \pi^1 D}$$

Assuming a lost sale penalty of $15 per unit, we would have

$$\text{OSOR} = \frac{30(1000)}{30(1000) + 15(10,000)} = 0.167$$

An OSOR of 0.167 calls for z value of about 0.97 and, consequently, about thirty-nine units of safety stock (SS = 0.97 * 40). Converting to percent unit service level can be done readily:

$$\text{USL} = 1 - \frac{g(k)\sigma_L}{Q}$$

$$= 1 - \frac{g(0.97)(40)}{1000}$$

$$= 1 - \frac{0.088(40)}{1000} = 0.996$$

Comparing the backorder and lost sales case for the same stockout penalty, the lost sales case uses slightly more safety stock.

We have learned to calculate the safety stock level for any given OSL or USL, but for cases in which service levels are unknown but the stockout costs are given, or the stockout costs are unknown but service level is known. Table 6.1 summarizes the necessary steps required for calculating the safety stock level in all cases. It is worth noting that if the service level is given, we can find the imputed cost of stockouts and vice versa.

TABLE 6.1 Reorder Points for Various Safety Stock Situations

	Order Service Level (OSL)	Unit Service Level (USL)
Stockout cost known Service level unknown	Backorder penalty specified as $8.26 $$\pi = 8.26$$ $$Q = 1000$$ $$h = 30$$ $$D = 10,000$$ $$OSOR = \frac{hQ}{\pi D} = \frac{30(1000)}{8.25(10,000)}$$ $$OSOR = 0.363$$ $$z_{0.363} = 0.35$$ $$\bar{x}_L = 200,\ \sigma_L = 40$$ $$R = 200 + 0.35(40) = 214$$	Backorder penalty specified as $8.26 $$\pi = 8.26$$ $$Q = 1000$$ $$h = 30$$ $$D = 10,000$$ $$OSOR = 0.363$$ $$k = 0.35$$ and $$g(k) = 0.248$$ $$R = 200 + 0.35(40) = 214$$ $$USL = 1 - \frac{g(k)\sigma_L}{Q}$$ $$USL = 1 - \left[\frac{0.248(40)}{10,000}\right]$$ $$USL = 0.99008$$
Stockout cost unknown Service level known	OSL specified as 0.625 $$OSOR = 0.375$$ $$z_{0.375} = 0.32$$ $$\bar{x}_L = 200,\ \sigma_L = 40$$ $$R = 200 + 0.32(40) = 213$$	USL specified as 0.99 $$USOR = 0.01$$ $$Q = 1000$$ $$\sigma_L = 40$$ $$g(k) = \frac{Q * USOR}{\sigma_L}$$ $$g(k) = \frac{1000 * 0.01}{40}$$ $$g(k) = 0.25$$ $$k = 0.35$$ $$R = 200 + 0.35(40)$$ $$R = 214$$

JOINT DETERMINATION OF Q AND R

Had Q been 500 rather than 1000 units, more safety stock would be needed to cover the increased number of cycles in the year. Returning to the backorder case, recall that the safety stock was fourteen units, with ten cycles per year. Now, with $D/Q = 10,000/500 = 20$ cycles per year, we must search for a k to yield $g(k) = 0.125$, because, by formula,

$$g(k) = \frac{Q(USOR)}{\sigma_L} = \frac{500(0.01)}{40} = 0.125$$

$$SS = k\sigma_L$$
$$SS = 0.78 * 40 = 31.2$$

Looking for k in Appendix 6A, we find $k = 0.78$. The corresponding safety stock is about thirty-one units. Since $\bar{x}_L = 200$, the reorder point for $Q = 500$ is 232. The interrelationships between Q and R can be visualized by this simple comparison between two order sizes. The larger order size,

Q	SS	R	D/Q	$Q/2$
500	32	232	20	250
1000	14	214	10	500

yields lower safety stock and fewer setups but higher cycle inventory. The smaller order size yields just the reverse. Thus, to determine Q and R, we must examine them together.

Because the stockout penalty is assessed on the number of units stocked out, the expected stockout cost per cycle is $\sigma_L g(k) * \pi$. In the previous example, the expected number of units stocked out per cycle was $\sigma_L g(k) = 40 * 0.125 = 5$ units. With a penalty cost of $8 per unit stocked out, the expected stockout cost per cycle was $5 * \$8 = \$40$.

By analogy to setup cost, we can modify the EOQ formula to include the expected stockout cost per cycle. Adding what we already know about the determination of the reorder point for the backorder case, we have a set of formulas to determine Q and R jointly:

$$Q^* = \sqrt{\frac{2D[S + \pi\sigma_L g(k)]}{h}}$$

$$g(k) = \frac{Q * \text{USOR}}{\sigma_L}$$

$$\text{OSOR} = \frac{hQ}{\pi D}$$

$$R = \bar{x}_L + k * \sigma_L$$

We shall try it and improvise as we go, again using the same example with $S = 1500$ and USL = 99% versus USL = 95%. First, for USL = 99%:

Step 1: Calculate Q^*, ignoring stockouts.

$$Q^* = \sqrt{\frac{2*10,000*1500}{30}} = 1000$$

Step 2: Calculate $g(k)$ for $Q^* = 1000$.

$$g(k) = \frac{1000*0.01}{40} = 0.25$$

$$k = 0.35$$

Step 3: From Appendix 6A, determine OSOR = 36.3%, with $z = k = 0.35$.
Step 4: Calculate $\pi = (hQ/\text{OSOR} * D) = (30 * 1000)/(0.363 * 10,000) = 8.26$.

$$Q^* = \sqrt{\frac{2*10,000*(1500 + 8.26*40*0.28)}{30}}$$

$$Q^* = 1027$$

Step 5:   Calculate $g(k)$ for the new $Q* = 1027$.

$$g(k) = \frac{1027 * 0.01}{40} = 0.257$$

$$k \simeq 0.33$$

Step 6:   Repeat steps 4 and 5 until two successive values of Q (or k) are the same. We must settle, then on $Q = 1028$, $k = 0.33$, and $R = 200 + 0.33 * 40 = 213$.

Now, for USL = 95%:

Step 1:   Calculate $Q*$, ignoring stockouts.

$$Q* = \sqrt{\frac{2 * 10,000 * 1500}{30}} = 1000$$

Step 2:   Calculate $g(k)$ for $Q* = 1000$.

$$g(k) = \frac{1,000 * 0.05}{40} = 1.25$$

$$k = -1.19$$

(Read down the far right-hand column of Appendix 6A until you reach 1.25.)
Step 3:   Calculate OSOR = $1 - 0.117 = 0.883$, with $z = k = -1.19$.
Step 4:   Calculate $Q* = 1055$, with $\pi = 3.40$ [from $\pi = h * Q/(OSOR * D)$]
Step 5:   Calculate $g(k)$ for $Q* = 1055$:

$$g(k) = \frac{Q * USOR}{\sigma_L}$$

$$g(k) = \frac{1055 * 0.05}{40} = 1.32$$

$$k \simeq -1.27$$

Step 6:   We must settle here on $Q = 1057$, $k = 1.32$, and $R = 200 - 1.32 * 40 = 147.2$.

To follow the sequence further, consider the case where the desired service level cannot be specified but the backorder cost is estimated at $15 per unit. In that case, we must do some extra calculations to find the USOR.

Step 1:   Calculate $Q*$, ignoring stockouts:

$$Q* = 1000$$

Step 2:   Calculate $g(k)$ for $Q^* = 1000$:

$$\text{OSOR} = \frac{hQ}{\pi D} = \frac{30 * 1000}{15 * 10,000} = 0.2$$

$$z_{0.2} = 0.84$$

$$k = z_{0.2} = 0.84$$

$$g(k) = g(0.84) = 0.112$$

$$\text{USOR} = \frac{g(k) * \sigma_L}{Q^*} = \frac{0.112 * 40}{1000} = 0.00448$$

$$\text{USL} = 0.995$$

Step 3:   Calculate Q^* for $g(k) = 0.112$

$$Q^* = \sqrt{\frac{2 * 10,000 * (1500 + 15 * 40 * 0.112)}{30}}$$

$$Q^* = 1022$$

Step 4:   Calculate $g(k)$ for $Q^* = 1022$:

$$g(k) = \frac{Q * \text{USOR}}{\sigma_L} = \frac{1022 * 0.0048}{40} = 0.114$$

Step 5:   Repeat steps 3 and 4 until Q or k repeats:

$$Q^* \simeq 1023$$

$$k \simeq 0.14$$

Remember that the expected stockout cost per cycle acts analogously to the setup cost: The higher the expected stockout cost per cycle, the larger the EOQ; and the higher the setup cost, the larger the EOQ.

LEAD TIME ADJUSTMENTS

To this point, our entire analysis has been based on $\bar{x}_L$ and s_L, the mean and standard deviation of lead time demand. Most forecasting systems in use today yield monthly forecasts. Since the lead time may not be in months, some conversion is necessary. When lead time is variable, this conversion can be cumbersome. Many suggest simply forecasting demand during lead time as a data series, rather than combining separate estimates of lead time and demand per unit time. In many companies, lead time demand is not recorded. For that reason, we should take a close look at lead time adjustments.

Suppose that our lead time is two weeks and our forecast is for a four-week month.

Obviously, the forecast for lead time demand would be simply half the four-week forecast, but what about MAD for lead time demand? To answer this question, we must resort to some elementary statistics. If two random variables are identically and independently distributed, then the variance of the sum is the sum of the variances:

$$\text{var}(X_1 + X_2) = var(X_1) + var(X_2)$$
$$\text{S.D.}(X_1 + X_2) = \sqrt{var\,(X_1) + var(X_2)}$$

where S.D. is the standard deviation.

Intuitively, this bit of statistics reveals that the standard deviation of a sum is less than the sum of the standard deviations. Suppose that the standard deviation of X_1 is 100 and the standard deviation of X_2 is 100. The standard deviation of $X_1 + X_2$ would be $\sqrt{100^2 + 100^2} \approx 141$, only 40% more than the variance of one of the variables taken alone. If we have several variables, all with the same standard deviation, we can simplify the situation:

$$\text{var}(X_1 + X_2 + \ldots + X_n) = n\,\text{var}\,(X_1)$$
$$\text{S.D.}(X_1 + X_2 + \ldots + X_n) = \sqrt{n} * \text{S.D.}\,(X_1)$$

In our example, S.D. $(X_1 + X_2) = \sqrt{2} * \text{S.D.}\,(X_1)$.

Applying these concepts to lead time demand, we first need to calculate the ratio of lead time to forecast interval:

$$r = \frac{\text{lead time}}{\text{forecast interval}} = \frac{2\text{ weeks}}{4\text{ weeks}} = \frac{1}{2}$$

Then calculate $\bar{x}_L = r * \bar{x}_f$, where $\bar{x}_f$ is the four-month forecast. If $s_f = 1.25\text{MAD}$ is the standard deviation of forecast error, then we have $s_L = \sqrt{r}\,s_f$, where s_L is the standard deviation of lead time forecast error. We can summarize lead time conversion in four formulas:

$$r = \frac{\text{lead time}}{\text{forecast interval}} \qquad \left(\text{e.g., } r = \frac{1}{2}\right)$$
$$\bar{x}_L = r * \bar{x}_f \qquad \left(\text{e.g., } 500 = \frac{1}{2}(1000)\right)$$
$$s_L = \sqrt{r} * s_f \qquad \left(\text{e.g., } 70.7 = \sqrt{\frac{1}{2}(100)}\right)$$
$$\text{MAD}_L = \sqrt{r} * \text{MAD}_f \qquad \left(\text{e.g., } 56.6 = \sqrt{\frac{1}{2}(80)}\right)$$

Continuing our example with $r = 1/2$, suppose that $\bar{x}_f = 1000$ and $\text{MAD}_f = 80$. Then $s_f = 1.25 * 80 = 100$, $s_L = 70.7$, and $\bar{x}_L = 1/2 * 1000 = 500$.

LEAD TIME VARIABILITY

Lead time variability must also be accounted for. Although the preferred approach is to measure actual demand during each lead time and then to forecast lead time demand with its associated standard deviation, most forecasting and inventory control systems are not designed to do this. Another approach is to *manage* the lead time so that lead time can be ignored in safety stock calculations. An approach of last resort is to measure lead time variability and to merge it with demand variability.

If we assume that lead times are normally distributed with mean $\overline{LT}$ and standard deviation σ_{LT}, it is possible to derive an estimate of the standard deviation of demand during uncertain lead time. Brown [1, p. 376] showed that variance of lead time demand equals variance of lead time demand given average lead time plus variance of lead time demand given average demand:

$$\sigma_{LTD}^2 = \overline{LT}\,\sigma_f^2 + \overline{x}_f^2\,\sigma_{LT}^2$$

where f refers to the forecast interval.

The variance of lead time demand given average lead time is the variance stemming from demand variability. To derive $\overline{LT}\sigma_f^2$, we proceed using the variance rule: variance (sum) = sum (variances), where sum = $d_1 + d_2 + \ldots + d_L$.

The variance of lead time demand given average demand is the variance stemming from lead time variability. To derive $\overline{x}_f^2\sigma_{LT}^2$, we must use the relation var$(kX) = k^2$ var(X); that is, the variance of the product of a constant times a variable is a product of the square of the constant times the variance of the variable.

Brown [3, p. 148–151] claims that MAD is no longer appropriate to the real world of computers. He also advises steering clear of mathematical models of the standard deviation over randomly varying lead times. Because Brown originally developed both these ideas, we must assume that he has seen the error of his youthful ways. Nevertheless, MAD persists and, short of simulation, an estimate of the variance of lead time demand during uncertain lead times cannot easily be avoided given current forecasting and data collection practices.

TIME-VARYING DEMAND

In a later chapter we recommend the use of a minimum cost per period algorithm to determine order quantities for highly fluctuating demand rates. For the typical seasonal sales pattern, however, the EOQ approach developed in this chapter should be satisfactory. At this point, we might merely consider the reorder point under time-varying demand:

$$R = \overline{x}_L + \text{SS}$$

When the forecasts indicate a trough in demand, the reorder point will decline, only to increase when the forecasts pick up again. The reorder point constantly adjusts to the projected activity level.

ANTICIPATED PRICE INCREASES

As we examine real-world variations on the models in this chapter, we must face the reality of inflation. For items that we sell, inflation has little effect on our models, since it is always better to satisfy customer demand by holding inventory for as little time as possible. For items that we purchase, anticipated price increases can entice us to purchase early.

Again, marginal analysis yields a simple decision rule, suggested by McClain and Thomas [4]. If the price of an item is about to increase by $X permanently, what should the order quantity be? Ordering Q units, we will hold the last unit ordered for Q/D years for a marginal holding cost of $h(Q/D)$. Other savings will result from postponing one setup, but we will ignore these as they are small. Equating the marginal benefits and costs $X = [h(p)Q]/D$ and solving for Q^* (the optimal Q), we find

$$Q^* = X\frac{D}{h(p)}$$

For example, a $5 anticipated increase from $25 to $30, with 30% holding costs and an annual demand of 1000 units, would give an optimal order quantity

$$Q^* = \frac{5 * 1000}{0.3 * 25} = 667$$

INVENTORY CONTROL SYSTEMS
IN PRACTICE

Various inventory control systems may be encountered in practice. Here we present some of these systems as variations of the models given in this chapter.

1. *Periodic system:* Every T years (e.g., 0.2 years), we order an amount to bring us up to a target level S. Hence, Q is variable, as shown in Figure 6.4. If we

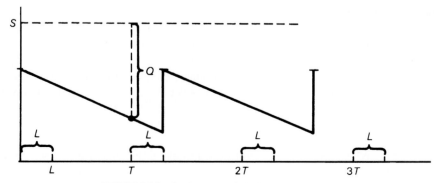

FIGURE 6.4   Cycles in a periodic inventory system.

add one unit to S, we will hold an item for a fraction T of the year at a cost of h per year. The holding cost for that unit is then hT. This will be worth the money, provided that hT is not greater than the expected stockout cost SOR(π). The decision rule then becomes: Determine S so that $hT = $ SOR(π), or

$$SOR = \frac{hT}{\pi}$$

$$S = \bar{x}_p + z_{SOR} * \sigma_p$$

where σ_p is the standard deviation of demand over the planning interval $T + L$.

Because $T = Q/D$, we can easily determine the optimal T from the EOQ formula:

$$Q^* = \sqrt{\frac{2DS}{h}}$$

$$T^* = Q^*/D = \sqrt{\frac{2S}{hD}}$$

We can then calculate S and Q:

$$S = \bar{x}_p + z_s \sigma_p$$

where x_p is the mean usage during the planning interval $T + L$. Similarly, σ_p is the standard deviation of planning interval demand. $Q = S - I$, where I is the current on hand plus on order.

The periodic system had more appeal in precomputer days, because it does not require continuous inventory records. The long planning interval, with its associated demand variation, causes more safety stock to be held than in the similar (Q, R) system. Of course, joint determination of S and T can also be considered.

2. (Q, Q) is a two-bin policy. When inventory falls to the reorder point Q, an order is placed for Q units. This system makes stores clerks happy, because no fancy record keeping is needed. They merely open the second bin and place an order.

3. (S, s) is the most robust policy and is optimal under the greatest number of conditions. When inventory falls to s, an order is placed up to the level S; that is, the order is the difference between current inventory and S. If demand occurs one unit at a time, then a (Q, R) policy would be identical. Otherwise, current inventory may be less than R when the order is placed, and the (Q, R) system would order Q, whereas the (S, s) system would order S minus current on hand and on order (I).

4. $(S, S - 1)$ is a base stock policy. When inventory falls below S, an order is

placed up to S. Known in industry as the sell-one/buy-one policy, this procedure is used for low-demand items. We see it again in Chapter 8 on multilocation inventories.

SUMMARY

In this chapter, we dealt with a company facing demand uncertainty. The models of the preceding chapter were extended and modified to cope with this uncertainty. Using marginal analysis, we found that we could derive safety stock rules. Just as the preceding chapter examined the setup versus holding cost trade-off, this chapter balanced stockout costs against the costs of holding safety stock. In addition to demand uncertainty, a company can also face lead time uncertainty. Safety stock should cover lead time demand. Hence, forecasts must be converted to a lead time base, and the variance of lead time forecasts then becomes the basis for safety stock calculations.

Imputed or implicit costs appeared throughout this chapter. The act of specifying a desired or acceptable service level reveals information about the relative magnitude of the costs involved.

Finally, the uncertain real world offers the challenge of time-varying demand and price increases. Generally, a good forecasting system will project seasonal demand rates so that we can adjust inventory levels over the year. With inflation, anticipated price increases can also cause us to change our order quantity.

Since the end-of-chapter problems will require a fair amount of student effort, we present here a summary approach for tackling the problems. The main equations are

$$Q = \sqrt{\frac{2D[S + \pi\sigma_L g(k)]}{h}}$$

$$R = \mu_L + k\sigma_L$$

$$\text{OSOR} = \frac{hQ}{\pi D}$$

$$\text{USOR} = \frac{\sigma_L g(k)}{Q}$$

The following points should be kept in mind:

1. k depends on the desired demand coverage.
2. Lead time demand is the basis of the system.
3. Lead time may vary over the year.
4. Price increases would automatically change the holding costs and thus must be considered.

Classifying the problem as shown in Table 6.2 will aid in developing a solution.

TABLE 6.2   Classification of Safety Stock Calculation Methods

	Order Service Level	Unit Service Level
Stockout cost known; optimal service level derived	$OSOR = \dfrac{hQ}{\pi D}$	$OSOR = \dfrac{hQ}{\pi D}$ implies a k value $USOR = \dfrac{g(k) * \sigma_L}{Q}$
Stockout cost imputed; desired service level specified	$\pi = \dfrac{hQ}{OSOR * D}$	$g(k) = \dfrac{Q * USOR}{\sigma_L}$ implies a k value and consequent OSOR

We encourage the student to embellish the scheme as necessary to help in the solution of the chapter problems. For any further work of this type, Brown [2] can serve as a comprehensive reference to the appropriate use of order points and order quantities.

PROBLEMS

For all problems, assume that the company is open for business five days a week for fifty weeks a year.

1. Seasonal demand for air conditioners at the Koolair store is normally distributed, with mean equal to 100 units and standard deviation equal to 20. The retail price per unit is $275, and the cost to the store is $170. For ordering once to cover seasonal demand, the store manager needs an optimal number of air conditioners to order. In the past, the manager has sold off units at a 50% discount once the peak heat spell is over. How many units should the manager order?

2. A perishable item is ordered only once each demand period. Acquisition cost is $3, selling price is $5, and salvage value is $1.50. What is the optimal order quantity? Given:

Demand	Probability
100	0.1
110	0.2
120	0.2
130	0.3
140	0.1
150	0.1

3. Find the optimal reorder point for the following item, using percent order safety stock. Mean weekly demand is 400 units, with a standard deviation of 100 units. Setup costs are $50 per order, and inventory carrying costs are $1.30 per unit per year. The desired order service level is 95%, and the lead time is one week.

4. The monthly demand for an item was recorded as follows:

1	2	3	4	5	6	7	8	9	10	11	12
335	295	275	305	304	338	290	305	285	275	295	311

Calculate the mean absolute deviation and the standard deviation of forecast error if the forecast is 300 per month. What safety stock would be required if orders are received monthly and if two stockout occurrences in twelve months are acceptable.

5. A company orders in lot sizes of 100 units while facing annual demand of 2000. Holding costs are $10 per unit per year, and the desired order service level is 95%. What is the implicit stockout cost, assuming the backorder case?

6. A company orders in lot sizes of 100 units while facing annual demand of 3000. Holding costs are $8 per unit per year, and the desired order service level is 98%. What is the implicit stockout cost assuming the lost sales case?

7. A forecasting system yields weekly forecasts. The current forecast is 100 units for the next week, and the weekly MAD is 20. Lead time is one week. Annual demand is 8000 units, and the company uses an order quantity of 500 units. To provide 99% unit service, what safety stock should be kept?

8. A forecasting system yields forecasts. The current forecast is 100 units per week, and the weekly MAD is 40. Lead time is one week. To provide 99% unit service, the company holds a safety stock of about 45 units. Annual demand is 9000 units, and the company uses an order quantity of 500 units. To preserve the same service level, what order size would allow the company to carry no safety stock? How many inventory cycles would occur in a year?

9. A company decides to order in larger lot sizes to avoid carrying safety stock. Specifically, the company increases its order quantity from 500 units to 600 units. At the same time, a safety stock of 30 units is completely eliminated. Which order quantity yields lower annual holding costs?

10. Annual demand for an item is 1000 units. Unit price is $1.50, and the annual holding cost is 35% per year. Backorder costs are $3 per unit, and setup costs are $15 per lot. Lead time is eight days, and the standard deviation of lead time demand is eight units. What are the joint optimal Q and R levels?

11. Annual demand for an item is 1000 units. Unit price is $5, and the annual holding cost is 35% per year. Backorder costs are impossible to estimate, although management desires a 98% unit service level. Setup costs are $10 per lot. Lead time is eight days, and the standard deviation of lead time demand is eight units. What are the joint optimal Q and R levels? What are the *implicit* backorder costs?

12. For a particular item, management desires s 98% unit service level. Weekly demand averages 1000 units, with a standard deviation of 100 units. Setup costs are $200, and holding costs are $10 per unit per year. Lead time is three weeks on average, with a standard deviation of three days. Determine the joint optimal Q add R levels. Point out any possible weakness in this application of inventory control models. Assume a five-day work week.

13. Orders for an item are received one week after they are placed, and the ordering cost is $10. Demand during this lead time is normally distributed, with a mean of nine units and a standard deviation of two units. Holding costs are $1.50 per unit per week, and the stockout cost is $10 per unit of unsatisfied demand. All unsatisfied demand is lost. What are the joint optimal Q and R levels?

14. Find the optimal reorder point for the following item, using percent order safety stock.

Mean yearly demand is 20,000 units, with a standard deviation of 1000 units. Setup costs are $50 per order, and inventory carrying costs are $1.30 per unit per year. The desired order service level is 95%, and the lead time is two weeks.

15. A forecasting system yields weekly forecasts. The current forecast is 100 units per week, and the weekly MAD is 20. Lead time is approximately normally distributed, with a mean of sixteen working days and a standard deviation of three working days. Annual demand is 9000 units, and the company uses an order quantity of 1000 units. To provide 98% unit service, what safety stock should be kept? Assume the backorder case. Assume a five-day work week.

16. Reliable rumor has it that our petroleum-based raw material will undergo a 20% price increase next month. We are currently placing an order for 1000 units at $20 per unit. Given an annual demand of 10,000 units and a holding cost percentage of 35%, should we revise our order quantity? If so, what should it be?

17. A forecasting system yields weekly forecasts. The current forecast is 100 units for the next week, and the weekly MAD is 20. Lead time is approximately normally distributed, with a mean of five working days and a standard deviation of three working days. Annual demand is 8000 units, and the company uses an order quantity of 1000 units. To provide 98% unit service, what safety stock should be kept?

18. A forecasting system yields weekly forecasts. The current forecast is 100 units for the next week, and the weekly MAD is 20. Lead time is approximately normally distributed, with a mean of sixteen working days and a standard deviation of three working days. Annual demand is 8000 units, and the company uses an order quantity of 1000 units. To provide 98% unit service, what safety stock should be kept?

19. For a 5% order stockout risk, what is the implied cost of a backorder for the following item? Using a combined (Q, R) model, the company faces annual sales of 500 units. The inventory holding charge is $25 per unit per year. Setup cost is $20, and the standard deviation of lead time demand is 25 units.

20. For a certain item, weekly demand is forecast at 120 units, with a standard deviation of 20 units. The inventory carrying charge is 25% of the unit value of $20. Setup costs are $25. With a lead time of eight days, determine the appropriate reorder point to maintain a 90% order service level. Now assume that the per unit backorder cost is $3. Determine the appropriate order quantity and order point using a joint optimal (Q, R) policy.

REFERENCES AND BIBLIOGRAPHY

1. R. G. Brown, *Smoothing, Forecasting and Prediction of Discrete Time Series,* (Englewood Cliffs, N.J.: Prentice Hall, 1963).
2. R. G. Brown, *Decision Rules for Inventory Management* (New York: Holt, Rinehart and Winston, 1967).
3. R. G. Brown, *Materials Management Systems* (New York: John Wiley & Sons, 1977).
4. J. O. McClain and L. J. Thomas, *Operations Management* (Englewood Cliffs, N.J.: Prentice Hall, 1980).

APPENDIX 6A

Table of the Unit Normal Distribution

Number of Standard Deviations of Safety Stock (k or z)	Stockout Risk (SOR)	Expected Stockouts [g(k) or g(z)]a	Expected Stockouts [g(−k) or g(−z)]a
0.00	0.5000000	0.3989423	0.3989243
0.01	0.4960106	0.3939622	0.4039622
0.02	0.4920216	0.3890221	0.4090221
0.03	0.4880335	0.3841218	0.4141218
0.04	0.4840465	0.3792614	0.4192614
0.05	0.4800611	0.3744409	0.4244409
0.06	0.4760777	0.3696602	0.4296602
0.07	0.4720968	0.3649193	0.4349193
0.08	0.4681186	0.3602182	0.4402182
0.09	0.4641435	0.3555569	0.4455569
0.10	0.4601721	0.3509353	0.4509353
0.11	0.4562046	0.3463535	0.4563535
0.12	0.4522415	0.3418112	0.4618112
0.13	0.4482832	0.3373086	0.4673086
0.14	0.4443300	0.3328455	0.4728455
0.15	0.4403823	0.3284220	0.4784220
0.16	0.4364405	0.3240379	0.4840379
0.17	0.4325051	0.3196931	0.4896931
0.18	0.4285763	0.3153877	0.4953877
0.19	0.4246546	0.3111216	0.5011216
0.20	0.4207403	0.3068946	0.5068946
0.21	0.4168339	0.3027068	0.5127068
0.22	0.4129356	0.2985579	0.5185579
0.23	0.4090459	0.2944480	0.5244480
0.24	0.4051652	0.2903770	0.5303770
0.25	0.4012937	0.2863447	0.5363447
0.26	0.3974319	0.2823511	0.5423511
0.27	0.3935802	0.2783960	0.5483960
0.28	0.3897388	0.2744794	0.5544794
0.29	0.3859082	0.2706012	0.5606012
0.30	0.3820886	0.2667612	0.5667612
0.31	0.3782805	0.2629594	0.5729594
0.32	0.3744842	0.2591956	0.5791956
0.33	0.3707000	0.2554697	0.5854697
0.34	0.3669283	0.2517815	0.5917815
0.35	0.3631694	0.2481310	0.5981310
0.36	0.3594236	0.2445181	0.6045181
0.37	0.3556913	0.2409425	0.6109425

Number of Standard Deviations of Safety Stock (k or z)	Stockout Risk (SOR)	Expected Stockouts $[g(k)$ or $g(z)]^{a}$	Expected Stockouts $[g(-k)$ or $g(-z)]^{a}$
0.38	0.3519728	0.2374042	0.6174042
0.39	0.3482683	0.2339030	0.6239030
0.40	0.3445783	0.2304388	0.6304388
0.41	0.3409030	0.2270114	0.6370114
0.42	0.3372428	0.2236207	0.6436207
0.43	0.3335979	0.2202665	0.6502665
0.44	0.3299686	0.2169487	0.6569487
0.45	0.3263552	0.2136671	0.6636671
0.46	0.3227581	0.2104215	0.6704215
0.47	0.3191775	0.2072119	0.6772119
0.48	0.3156137	0.2040379	0.6840379
0.49	0.3120670	0.2008996	0.6908996
0.50	0.3085375	0.1977966	0.6977966
0.51	0.3050257	0.1947288	0.7047288
0.52	0.3015318	0.1916960	0.7116960
0.53	0.2980559	0.1886981	0.7186981
0.54	0.2945985	0.1857348	0.7257348
0.55	0.2911597	0.1828060	0.7328060
0.56	0.2877397	0.1799116	0.7399116
0.57	0.2843388	0.1770512	0.7470512
0.58	0.2809573	0.1742247	0.7542247
0.59	0.2775953	0.1714320	0.7614320
0.60	0.2742531	0.1686728	0.7686728
0.61	0.2709309	0.1659469	0.7759469
0.62	0.2676288	0.1632541	0.7832541
0.63	0.2643472	0.1605942	0.7905942
0.64	0.2610862	0.1579671	0.7979671
0.65	0.2578460	0.1553724	0.8053724
0.66	0.2546269	0.1528101	0.8128101
0.67	0.2514288	0.1502798	0.8202798
0.68	0.2482522	0.1477814	0.8277814
0.69	0.2450970	0.1453147	0.8353147
0.70	0.2419636	0.1428794	0.8428794
0.71	0.2388520	0.1404754	0.8504754
0.72	0.2357624	0.1381023	0.8581023
0.73	0.2326950	0.1357600	0.8657600
0.74	0.2296499	0.1334483	0.8734483
0.75	0.2266273	0.1311670	0.8811670
0.76	0.2236272	0.1289157	0.8889157
0.77	0.2206499	0.1266943	0.8966943
0.78	0.2176954	0.1245026	0.9045026
0.79	0.2147638	0.1223404	0.9123404
0.80	0.2118553	0.1202073	0.9202073
0.81	0.2089700	0.1181032	0.9281032
0.82	0.2061080	0.1160278	0.9360278
0.83	0.2032693	0.1139809	0.9439809

Number of Standard Deviations of Safety Stock (*k or z*)	Stockout Risk (*SOR*)	Expected Stockouts [*g(k) or g(z)*][a]	Expected Stockouts [*g(−k) or g(−z)*][a]
0.84	0.2004541	0.1119623	0.9519623
0.85	0.1976625	0.1099718	0.9599718
0.86	0.1948945	0.1080090	0.9680090
0.87	0.1921502	0.1060738	0.9760738
0.88	0.1894296	0.1041659	0.9841659
0.89	0.1867329	0.1022851	0.9922851
0.90	0.1840601	0.1004312	1.0004312
0.91	0.1814112	0.0986038	1.0086038
0.92	0.1787864	0.0968028	1.0168028
0.93	0.1761855	0.0950280	1.0250280
0.94	0.1736088	0.0932791	1.0332791
0.95	0.1710561	0.0915557	1.0415557
0.96	0.1685276	0.0898578	1.0498578
0.97	0.1660232	0.0881851	1.0581851
0.98	0.1635431	0.0865373	1.0665373
0.99	0.1610871	0.0849142	1.0749142
1.00	0.1586553	0.0833155	1.0833155
1.01	0.1562477	0.0817410	1.0917410
1.02	0.1538642	0.0801904	1.1001904
1.03	0.1515050	0.0786636	1.1086636
1.04	0.1491700	0.0771601	1.1171602
1.05	0.1468591	0.0756801	1.1256801
1.06	0.1445723	0.0742230	1.1342230
1.07	0.1423097	0.0727886	1.1427886
1.08	0.1400711	0.0713767	1.1513767
1.09	0.1378566	0.0699871	1.1599871
1.10	0.1356661	0.0686195	1.1686195
1.11	0.1334996	0.0672736	1.1772736
1.12	0.1313569	0.0659494	1.1859494
1.13	0.1292382	0.0646464	1.1946464
1.14	0.1271432	0.0633645	1.2033645
1.15	0.1250720	0.0621035	1.2121035
1.16	0.1230245	0.0608630	1.2208630
1.17	0.1210005	0.0596429	1.2296429
1.18	0.1190002	0.0584429	1.2384429
1.19	0.1170233	0.0572628	1.2472628
1.20	0.1150697	0.0561024	1.2561024
1.21	0.1131395	0.0549613	1.2649613
1.22	0.1112325	0.0538395	1.2738395
1.23	0.1093486	0.0527366	1.2827366
1.24	0.1074878	0.0516525	1.2916525
1.25	0.1056498	0.0505868	1.3005868
1.26	0.1038347	0.0495394	1.3095394
1.27	0.1020424	0.0485100	1.3185100
1.28	0.1002726	0.0474985	1.3274985
1.29	0.0985254	0.0465045	1.3365045

Number of Standard Deviations of Safety Stock (k or z)	Stockout Risk (SOR)	Expected Stockouts [g(k) or g(z)][a]	Expected Stockouts [g(−k) or g(−z)][a]
1.30	0.0968006	0.0455279	1.3455279
1.31	0.0950980	0.0445684	1.3545684
1.32	0.0934176	0.0436258	1.3636258
1.33	0.0917592	0.0427000	1.3727000
1.34	0.0901227	0.0417906	1.3817906
1.35	0.0885081	0.0408975	1.3908975
1.36	0.0869150	0.0400204	1.4000204
1.37	0.0853435	0.0391591	1.4091591
1.38	0.0837934	0.0383134	1.4183134
1.39	0.0822645	0.0374832	1.4274832
1.40	0.0807567	0.0366681	1.4366681
1.41	0.0792699	0.0358680	1.4458680
1.42	0.0778039	0.0350826	1.4550826
1.43	0.0763586	0.0343118	1.4643118
1.44	0.0749337	0.0335554	1.4735554
1.45	0.0735293	0.0328131	1.4828131
1.46	0.0721451	0.0320847	1.4920847
1.47	0.0707809	0.0313701	1.5013701
1.48	0.0694367	0.0306690	1.5106690
1.49	0.0681122	0.0299813	1.5199813
1.50	0.0668072	0.0293067	1.5293067
1.51	0.0655217	0.0286451	1.5386451
1.52	0.0642555	0.0279963	1.5479963
1.53	0.0630084	0.0273600	1.5573600
1.54	0.0617802	0.0267360	1.5667360
1.55	0.0605708	0.0261243	1.5761243
1.56	0.0593800	0.0255246	1.5855246
1.57	0.0582076	0.0249367	1.5949367
1.58	0.0570534	0.0243604	1.6043604
1.59	0.0559174	0.0237955	1.6137955
1.60	0.0547993	0.0232420	1.6232420
1.61	0.0536989	0.0226995	1.6326995
1.62	0.0526161	0.0221679	1.6421679
1.63	0.0515507	0.0216471	1.6516471
1.64	0.0505026	0.0211369	1.6611369
1.65	0.0494715	0.0206370	1.6706370
1.66	0.0484572	0.0201474	1.6801474
1.67	0.0474597	0.0196678	1.6896678
1.68	0.0464786	0.0191982	1.6991982
1.69	0.0455140	0.0187382	1.7087382
1.70	0.0445654	0.0182878	1.7182878
1.71	0.0436329	0.0178469	1.7278469
1.72	0.0427162	0.0174151	1.7374151
1.73	0.0418151	0.0169925	1.,7469925
1.74	0.0409295	0.0165788	1.7565788
1.75	0.0400591	0.0161739	1.7661739

Number of Standard Deviations of Safety Stock (k or z)	Stockout Risk (SOR)	Expected Stockouts [g(k) or g(z)][a]	Expected Stockouts [g(−k) or g(−z)][a]
1.76	0.0392039	0.0157776	1.7757776
1.77	0.0383635	0.0153897	1.7853897
1.78	0.0375379	0.0150103	1.7950103
1.79	0.0367269	0.0146389	1.8046389
1.80	0.0359303	0.0142757	1.8142757
1.81	0.0351478	0.0139203	1.8239203
1.82	0.0343794	0.0135727	1.8335727
1.83	0.0336249	0.0132327	1.8432327
1.84	0.0328841	0.0129001	1.8529001
1.85	0.0321567	0.0125750	1.8625750
1.86	0.0314427	0.0122570	1.8722570
1.87	0.0307218	0.0119461	1.8819461
1.88	0.0300540	0.0116421	1.8916421
1.89	0.0293789	0.0113449	1.9013449
1.90	0.0287165	0.0110545	1.9110545
1.91	0.0280655	0.0107706	1.9207706
1.92	0.0274289	0.0104931	1.9304931
1.93	0.0268034	0.0102220	1.9402220
1.94	0.0261898	0.0099570	1.9499570
1.95	0.0255880	0.0096981	1.9596981
1.96	0.0249978	0.0094452	1.9694452
1.97	0.0244191	0.0091981	1.9791981
1.98	0.0238517	0.0089568	1.9889568
1.99	0.0232954	0.0087211	1.9987211
2.00	0.0227501	0.0084908	2.0084908
2.01	0.0222155	0.0082660	2.0182660
2.02	0.216916	0.0080465	2.0280465
2.03	0.0211782	0.0078322	2.0378322
2.04	0.0206751	0.0076229	2.0476229
2.05	0.0201821	0.0074186	2.0574186
2.06	0.0196992	0.0072192	2.0672192
2.07	0.0192261	0.0070246	2.0770246
2.08	0.0187627	0.0068347	2.0868347
2.09	0.0183088	0.0066493	2.0966493
2.10	0.0178644	0.0064684	2.1064684
2.11	0.0174291	0.0062920	2.1162920
2.12	0.0170030	0.0061198	2.1261198
2.13	0.0165857	0.0059519	2.1359519
2.14	0.0161773	0.0057881	2.1457881
2.15	0.0157776	0.0056283	2.1556283
2.16	0.0153863	0.0054725	2.1654725
2.17	0.0150034	0.0053205	2.1753205
2.18	0.0146287	0.0051724	2.1851724
2.19	0.0142621	0.0050279	2.1950279
2.20	0.0139034	0.0048871	2.2048871
2.21	0.0135525	0.0047498	2.2147498
2.22	0.0132093	0.0046160	2.2246160

Number of Standard Deviations of Safety Stock (k or z)	Stockout Risk (SOR)	Expected Stockouts [g(k) or g(z)]ᵃ	Expected Stockouts [g(−k) or g(−z)]ᵃ
2.23	0.0128737	0.0044856	2.2344856
2.24	0.0125454	0.0043585	2.2443585
2.25	0.0122244	0.0042347	2.2542347
2.26	0.0119106	0.0041140	2.2641140
2.27	0.0116038	0.0039964	2.2739964
2.28	0.0113038	0.0038819	2.2838819
2.29	0.0110106	0.0037703	2.2933703
2.30	0.0107241	0.0036617	2.3036617
2.31	0.0104441	0.0035558	2.3135558
2.32	0.0101704	0.0034527	2.3234527
2.33	0.0099031	0.0033524	2.3333524
2.34	0.0096419	0.0032546	2.3432546
2.35	0.0093867	0.0031595	2.3531595
2.36	0.0091375	0.0030669	2.3630669
2.37	0.0088940	0.0029767	2.3729767
2.38	0.0086563	0.0028890	2.3828890
2.39	0.0084242	0.0028036	2.3928036
2.40	0.0081975	0.0027205	2.4027205
2.41	0.0079763	0.0026396	2.4126396
2.42	0.0077603	0.0025609	2.4225609
2.43	0.0075494	0.0024844	2.4324844
2.44	0.0073436	0.0024099	2.4424099
2.45	0.0071428	0.0023375	2.4523375
2.46	0.0069469	0.0022670	2.4622670
2.47	0.0067557	0.0021985	2.4721985
2.48	0.0065691	0.0021319	2.4821319
2.49	0.0063872	0.0020671	2.4920671
2.50	0.0062097	0.0020041	2.5020041
2.51	0.0060366	0.0019429	2.5119429
2.52	0.0058678	0.0018833	2.5218833
2.53	0.0057031	0.0018255	2.5318255
2.54	0.0055426	0.0017693	2.5417693
2.55	0.0053862	0.0017146	2.5517146
2.56	0.0052336	0.0016615	2.5616615
2.57	0.0050850	0.0016099	2.5716099
2.58	0.0049400	0.0015598	2.5815598
2.59	0.0047988	0.0015111	2.5915111
2.60	0.0046612	0.0014638	2.6014638
2.61	0.0045271	0.0014178	2.6114178
2.62	0.0043965	0.0013732	2.6213732
2.63	0.0042693	0.0013299	2.6313299
2.64	0.0041453	0.0012878	2.6412878
2.65	0.0040246	0.0012470	2.6512470
2.66	0.0039701	0.0012073	2.6612073
2.67	0.0037926	0.0011688	2.6711688
2.68	0.0036812	0.0011314	2.6811314
2.69	0.0035726	0.0010952	2.6910952

Number of Standard Deviations of Safety Stock (k or z)	Stockout Risk (SOR)	Expected Stockouts [g(k) or g(z)][a]	Expected Stockouts [g(−k) or g(−z)][a]
2.70	0.0034670	0.0010600	2.7010600
2.71	0.0033642	0.0010258	2.7110258
2.72	0.0032641	0.0009927	2.7209927
2.73	0.0031668	0.0009605	2.7309605
2.74	0.0030720	0.0009293	2.7409293
2.75	0.0029798	0.0008991	2.7508991
2.76	0.0028901	0.0008697	2.7608697
2.77	0.0028029	0.0008412	2.7708412
2.78	0.0027180	0.0008136	2.7808136
2.79	0.0026355	0.0007869	2.7907869
2.80	0.0025552	0.0007609	2.8007609
2.81	0.0024771	0.0007358	2.8107358
2.82	0.0024012	0.0007114	2.8207114
2.83	0.0023275	0.0006877	2.8306887
2.84	0.0022557	0.0006648	2.8406648
2.85	0.0021860	0.0006426	2.8506426
2.86	0.0021183	0.0006211	2.8606211
2.87	0.0020524	0.0006002	2.8706002
2.88	0.0019884	0.0005800	2.8805800
2.89	0.0019263	0.0005604	2.8905604
2.90	0.0018659	0.0005415	2.9005415
2.91	0.0018072	0.0005231	2.9105231
2.92	0.0017502	0.0005053	2.9205053
2.93	0.0016949	0.0004881	2.9304881
2.94	0.0016411	0.0004714	2.9404714
2.95	0.0015889	0.0004553	2.9504553
2.96	0.0015383	0.0004396	2.9604396
2.97	0.0014891	0.0004245	2.9704245
2.98	0.0014413	0.0004099	2.9804099
2.99	0.0013950	0.0003957	2.9903957
3.00	0.0013500	0.0003819	3.0003819
3.01	0.0013063	0.0003687	3.0103687
3.02	0.0012639	0.0003558	3.0203558
3.03	0.0012228	0.0003434	3.0303434
3.04	0.0011830	0.0003314	3.0403314
3.05	0.0011443	0.0003197	3.0503197
3.06	0.0011068	0.0003085	3.0603085
3.07	0.0010704	0.0002976	3.0702976
3.08	0.0010351	0.0002871	3.0802871
3.09	0.0010009	0.0002769	3.0902769
3.10	0.0009677	0.0002670	3.1002670
3.11	0.0009355	0.0002575	3.1102575
3.12	0.0009043	0.0002483	3.1202483
3.13	0.0008741	0.0002394	3.1302394
3.14	0.0008448	0.0002308	3.1402308
3.15	0.0008164	0.0002225	3.1502225

Number of Standard Deviations of Safety Stock (k or z)	Stockout Risk (SOR)	Expected Stockouts [g(k) or g(z)][a]	Expected Stockouts [g(−k) or g(−z)][a]
3.16	0.0007889	0.0002145	3.1602145
3.17	0.0007623	0.0002068	3.1702068
3.18	0.0007364	0.0001993	3.1801993
3.19	0.0007114	0.0001920	3.1901920
3.20	0.0006872	0.0001850	3.2001850
3.21	0.0006637	0.0001783	3.2101783
3.22	0.0006410	0.0001718	3.2201718
3.23	0.0006190	0.0001655	3.2301655
3.24	0.0005977	0.0001594	3.2401594
3.25	0.0005771	0.0001535	3.2501535
3.26	0.0005571	0.0001478	3.2601478
3.27	0.0005378	0.0001424	3.2701424
3.28	0.0005191	0.0001371	3.2801371
3.29	0.0005010	0.0001320	3.2901320
3.30	0.0004835	0.0001271	3.3001271
3.31	0.0004665	0.0001223	3.3101223
3.32	0.0004501	0.0001177	3.3201177
3.33	0.0004343	0.0001133	3.3301133
3.34	0.0004189	0.0001091	3.3401091
3.35	0.0004041	0.0001050	3.3501050
3.36	0.0003898	0.0001010	3.3601010
3.37	0.0003759	0.0000972	3.3700972
3.38	0.0003625	0.0000935	3.3800935
3.39	0.0003495	0.0000899	3.3900899
3.40	0.0003370	0.0000865	3.4000865
3.41	0.0003249	0.0000832	3.4100832
3.42	0.0003132	0.0000800	3.4200800
3.43	0.0003018	0.0000769	3.4300769
3.44	0.0002909	0.0000740	3.4400740
3.45	0.0002803	0.0000711	3.4500711
3.46	0.0002701	0.0000684	3.4600684
3.47	0.0002603	0.0000657	3.4700657
3.48	0.0002508	0.0000632	3.4800632
3.49	0.0002416	0.0000607	3.4900607
3.50	0.0002327	0.0000583	3.5000583
3.51	0.0002241	0.0000560	3.5100560
3.52	0.0002158	0.0000538	3.5200538
3.53	0.0002078	0.0000517	3.5300517
3.54	0.0002001	0.0000497	3.5400497
3.55	0.0001927	0.0000477	3.5500477
3.56	0.0001855	0.0000458	3.5600458
3.57	0.0001785	0.0000440	3.5700440
3.58	0.0001718	0.0000423	3.5800428
3.59	0.0001654	0.0000406	3.5900406
3.60	0.0001591	0.0000390	3.6000390
3.61	0.0001531	0.0000374	3.6100374

Number of Standard Deviations of Safety Stock (k or z)	Stockout Risk (SOR)	Expected Stockouts [g(k) or g(z)][a]	Expected Stockouts [g(−k) or g(−z)][a]
3.62	0.0001473	0.0000359	3.6200359
3.63	0.0001417	0.0000345	3.6300345
3.64	0.0001364	0.0000331	3.6400331
3.65	0.0001312	0.0000318	3.6500318
3.66	0.0001261	0.0000305	3.6600305
3.67	0.0001213	0.0000292	3.6700292
3.68	0.0001166	0.0000280	3.6800280
3.69	0.0001122	0.0000269	3.6900269
3.70	0.0001078	0.0000258	3.7000258
3.71	0.0001037	0.0000248	3.7100248
3.72	0.0000996	0.0000237	3.7200237
3.73	0.0000958	0.0000228	3.7300228
3.74	0.0000920	0.0000218	3.7400218
3.75	0.0000884	0.0000209	3.7500209
3.76	0.0000850	0.0000201	3.7600201
3.77	0.0000816	0.0000192	3.7700192
3.78	0.0000784	0.0000184	3.7800184
3.79	0.0000753	0.0000177	3.7900177
3.80	0.0000724	0.0000169	3.8000169
3.81	0.0000695	0.0000162	3.8100162
3.82	0.0000667	0.0000155	3.8200155
3.83	0.0000641	0.0000149	3.8300149
3.84	0.0000615	0.0000143	3.8400143
3.85	0.0000591	0.0000137	3.8500137
3.86	0.0000567	0.0000131	3.8600131
3.87	0.0000544	0.0000125	3.8700125
3.88	0.0000522	0.0000120	3.8800120
3.89	0.0000501	0.0000115	3.8900115
3.90	0.0000481	0.0000110	3.9000110
3.91	0.0000462	0.0000105	3.9100105
3.92	0.0000443	0.0000101	3.9200101
3.39	0.0000425	0.0000097	3.9300097
3.94	0.0000408	0.0000092	3.9400092
3.95	0.0000391	0.0000088	3.9500088
3.96	0.0000375	0.0000085	3.9600085
3.97	0.0000360	0.0000081	3.9700081
3.98	0.0000345	0.0000077	3.9800077
3.99	0.0000331	0.0000074	3.9900074

[a]If k is negative, the stockout risk is one minus tabled number.

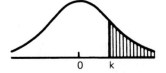 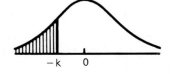

Shaded area is the tabled number.

Chapter Seven

Aggregate Inventory Management

INTRODUCTION

In this chapter, we develop techniques for managing all inventory items together rather than each inventory item separately. Although we could have named this chapter multi-item inventory management, the type of problems addressed here are popularly known in the literature as aggregate inventory management. Chapter 5 dealt with the joint replenishment of multiple items as well as production lot sizing of multiple items. The solutions given by these models assume that adequate capacity is available; however, it may not always be feasible. To conserve the available capacity in addition to combining the production of a group of items requiring a common setup, we might have to increase the lot sizes of individual items. Such aggregation into larger EOQs means heavier inventory investment but reduced setup expense. In Chapter 6 we calculated the safety stock requirements necessary to meet uncertainty in demand and lead times under several conditions. Adequate resources may not always be available to carry the safety stock requirements, however. Increased safety stock means better customer service but, again, a heavier total inventory investment. For an overall investment target, we must determine the optimal order quantities.

Even with only two inventory items, aggregate inventory management problems appear. Suppose, for example, that the first item has an EOQ of 1000 units but the second item has an EOQ of 500 units. Further, suppose that the firm has an inventory investment target of $500. If each item were valued at $1 per unit, the average inventory investment would be $750, missing the investment target by $250.

This chapter explores such issues and problems and will present techniques for dealing with them. Specifically, this chapter presents a simple and yet a powerful technique known as LIMIT to analyze the trade-offs between the suggested lot sizes and the total available setup capacity. Several models to analyze the trade-offs between safety stock requirements and associated service levels for handling multi-item situations are also presented. In addition, the application of a powerful mathematical model known as the LaGrange multiplier technique is explored to solve these types of problems.

LIMIT

The lot-size inventory management interpolation technique (LIMIT) was developed to aid practitioners in handling aggregate inventory trade-offs between holding and setup costs. Because the notation of Plossl and Wight [3] has become standard in seminars on LIMIT, we shall use it here. Our derivation of the appropriate formulas, however, is much simpler than theirs.

We shall take D as annual demand, S as setup cost, I as the inventory carrying percentage, and c as the per unit value of the item. The subscript i, when used, refers to the ith inventory item.

Suppose that we have a trial lot size Q_T, calculated from our best estimates of holding and setup costs: $D = 10,000$, $S = \$125$, $I = 0.25$, and $c = \$10$:

$$Q_T = \sqrt{\frac{2DS}{Ic}} = \sqrt{\frac{2 * 10,000 * 125}{0.25 * 10}} = 1000$$

That lot size depends specifically on I and yields $D/Q = 10$ setups in the year. If we were limited to fewer setups in the year—such as five—then our *limit order quantity* would have to be increased to 2000 units. The very action of limiting the number of setups to five implies a lack of adequate capacity or a disbelief in our original ratio of holding to setup costs. Given a Q of 2000 units and the same setup cost, the implied holding percentage would be 6.25%:

$$Q_L = \sqrt{\frac{2 * 10,000 * 125}{0.0625 * 10}} = 2000$$

Letting I_L be the limit carrying percentage and I_T the trial percentage, we can show the relationship between the trial and the limit lot size. The relationship is a function of the trial and limit carrying percentages:

$$Q_L = \sqrt{\frac{2DS}{I_L c}}$$

$$Q_T = \sqrt{\frac{2DS}{I_T c}}$$

$$\frac{Q_L}{Q_T} = \frac{\sqrt{2DS}}{\sqrt{I_L c}} * \frac{\sqrt{I_T c}}{\sqrt{2DS}} = \frac{\sqrt{I_T}}{\sqrt{I_L}}$$

$$Q_L = Q_T \sqrt{I_T / I_L} = M * Q_T$$

with $M = \sqrt{I_T / I_L}$

In our example,

$$Q_L = 1000 * \sqrt{0.25/0.0625} = 2000$$

and $M = 2$.

If we could always know the limit carrying percentage, we could stop now and use the foregoing formula to adjust lot sizes. Generally, however, we know only the trial carrying cost and the limit on the available annual setup hours. From the annual setup hours limit, we must adjust our trial lot sizes. We shall proceed by defining H_L as the limit on total setup hours across all SKUs. Also, H_T represents total setup hours across all SKUs, based on the trial lot sizes. Letting h_i be the hours required per setup for item i and recognizing that $Q_L = M * Q_T$, we have

$$\frac{H_T}{H_L} = \frac{\sum\left(\frac{D_i h_i}{Q_{Ti}}\right)}{\sum\left(\frac{D_i h_i}{Q_{Li}}\right)} = \frac{\sum\left(\frac{D_i h_i}{Q_{Ti}}\right)}{\sum\left(\frac{D_i h_i}{Q_{Ti} * M}\right)} = \frac{\sum\left(\frac{D_i h_i}{Q_{Ti}}\right)}{\frac{1}{M}\sum\left(\frac{D_i h_i}{Q_{Ti}}\right)} = M$$

That is,

$$\frac{H_T}{H_L} = M$$

But $M = \sqrt{I_T/I_L}$ and so $(H_T/H_L)^2 = I_T/I_L$.

This gives up two fundamental formulas for LIMIT applications:

$$I_L = I_T\left(\frac{H_L}{H_T}\right)^2$$

$$M = \frac{H_T}{H_L} = \sqrt{\frac{I_T}{I_L}}$$

Suppose that our present setup hours total 49 and our limit is 35. How can we adjust our order quantities to accommodate this limit? We have $H_T = 49$ and $H_L = 35$, which gives us $M = H_T/H_L = 49/35 = 1.4$. In Example 7.1, we use the same numbers and also show how they might arise. If we raise our order quantities by 1.4, we should reduce our annual setup hours by 1.4. Is this plausible? Remember that annual setup hours on an individual item would be D/Q times hours per setup. Increasing our order size to $1.4Q$ would given annual setup hours of $D/(1.4Q)$ times hours per setup.

Suppose that $D = 10,000$, $Q = 845$, and there are two hours per setup. Then $D/Q *h = (10,000/845) * 2 = 23.67$ total trial setup hours. Now, raising the order quantity by 1.4

would give $[10,000/(1.4 * 845)] * 2 = 23.67/1.4 = 16.9$ hours. The trial setup hours are 1.4 times the revised setup hours $(23.67 = 1.4 * 16.9)$.

Example 7.1

The two items in our inventory have been managed in a "seat of the pants" fashion for several years. The following table shows the current situation:

Item	Annual Usage	Setup Hours per Order	Unit Cost of Item	Present Order Quantity	Yearly Setup Hours
A	10,000	2	10	769	13 * 2 = 26
B	5,000	3	15	1667	3 * 3 = 9
Total					35

Current setup costs are $62.50 per hour. Now a bright, young inventory analyst has criticized our present order quantities. The inventory manager insists that we cannot afford more than 35 setup hours per year for these items (and has indicated that the inventory analyst is wrong). How can this situation be handled using LIMIT? First, we shall specify $H_L = 35$ hours. Further, our best estimate of the carrying cost percentage is 35%. Calculating trial lot sizes as a starting point, we obtain the following:

Item	Cost per Setup	Trial Q	Approximate Yearly Setup Hours
A	$125.00	845	24
B	$187.50	598	25
Total			49

$$M = \frac{H_T}{H_L} = \frac{49}{35} = 1.4$$

$$Q_L = M * Q_T = 1.4 * Q_T$$

This leads us directly to the LIMIT order quantities:

Item	LIMIT Quantity	Approximately Yearly Setup Hours
A	1.4 * 845 = 1183	10,000/1183 * 2 = 17
B	1.4 * 598 = 837	5000/837 * 3 = 18
Total		35

Now we are in a position to examine the merits of our gyrations. Recall that the present order quantities had no particular merit. The trial order quantities gave a good setup versus carrying cost trade-off but a faulty setup to carrying cost ratio. The LIMIT quantity should give a good trade-off, based on the carrying to setup ratio implied by a

TABLE 7.1   LIMIT Lot Size Trade-offs at a 17.86% Holding Rate

	Lot Size	Annual Holding Cost ($)	Annual Setup Cost ($)	Total Relevant Costs ($)
Q_P—A	769	687	1625	2312
Q_P—B	1667	2233	563	2796
Total		2920	2188	5108
Q_T—A	845	755	1479	2234
Q_T—B	598	801	1568	2369
Total		1556	3047	4603
Q_L—A	1183	1057	1057	2114
Q_L—B	837	1120	1120	2240
Total		2177	2177	4354

limit on the setup hours. Table 7.1 summarizes these trade-offs. To develop this table, we note that the true carrying percentage is 17.86%; based on limit order quantity that is,

$$I_L = I_T \left(\frac{H_L}{H_T}\right)^2 = 0.35\left(\frac{35}{49}\right)^2 = 0.1786$$

Based on this implied carrying cost percentage of 17.86%, we next compare the total costs implied by the present, the trial, and the LIMIT order quantities.

Table 7.1 reveals that our LIMIT lot sizes yield the lowest total relevant costs possible within a limit of thirty-five hours annual setup. Approximately seventeen setup hours on item A and eighteen hours on item B make more sense than twenty-six on A and only nine on B. It is not hard to see that we could have obtained the LIMIT order quantities if we had used a carrying cost percentage of 17.86% rather than 35% in the EOQ formula. We did not know, however, that the correct percentage was 17.86 until we calculated M.

EXCHANGE CURVES

The LIMIT example demonstrated that the optimal lot size depends on the holding percentage. At a holding rate of 35%, the trial lot sizes were optimal. Then, limiting annual setups to thirty-five hours, we had a holding rate of 17.86%, and the LIMIT lot sizes became optimal. Figure 7.1 provides a rough sketch of the situation. The exchange curve sketched in the figure shows the optimal exchange between setup and holding costs. The present situation, represented by Q_P, involves higher holding costs than necessary for the given level of aggregate setup cost.

Out of curiosity, you should wonder what the *optimal* annual setup costs would be for the annual holding costs represented by Q_P. That point is shown in Figure 7.1 as Q_R. To rephrase the question: Given a willingness to invest $2920 in holding costs, what order quantities would give the lowest setup costs? To solve this problem, we shall move to a more powerful form of analysis that will allow us to handle general exchange curves.

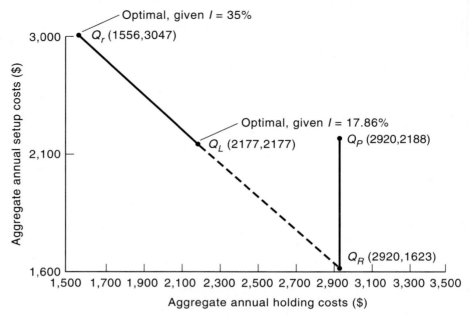

FIGURE 7.1 Holding/setup exchange curve *17.86% holding percentage)

LAGRANGE MULTIPLIERS

For a modest investment of time and effort, you can learn one of the most useful analytical techniques in businesss and economics. The use of Lagrange multipliers allows us to maximize profits or minimize costs subject to constraints. We shall use only basic calculus to establish the plausibility of the technique.

Suppose that we have a profit function we are trying to maximize subject to some constraints: Maximize $f(x, y) = x + 3y$ subject to $g(x, y) = x^2 + y^2 - 10 = 0$. We normally would see the constraint in the form $x^2 + y^2 = 10$, the equation of a circle. As in linear programming, we can draw the objective function in two dimensions by specifying arbitrary profits and then solving for x and y. For example, the two isoprofit lines shown in Figure 7.2 come from $x + 3y = 9$ and $x + 3y = 3$. From the graph in Figure 7.2, it is apparent that maximum profit occurs at the point of tangency of the profit line and the constraint line. But how do we solve for that point?

First, examine the profit line: $\pi = x + 3y$. Rewriting this in terms of y as a function of x would give us $y = (\pi/3) - (1/3)x$. The slope of the profit line is thus $-1/3$. By calculus, $dy/dx = y_x = -1/3$.

What about the slope of the constraint circle? Given $g(x, y) = x^2 + y^2 - 10 = 0$, we can solve for y in terms of x:

$$y^2 = 10 - x^2$$
$$y = \sqrt{10 - x^2}$$

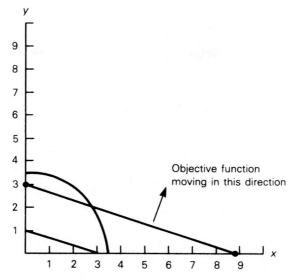

FIGURE 7.2   Lagrangian geometry in two dimensions

and differentiating, we obtain

$$y_x = \frac{1}{2}(10 = x^2)^{-1/2} * (-2x)$$

$$= \frac{-x}{\sqrt{10 - x^2}}$$

$$= \frac{-x}{y}$$

Equating the slope of the profit line and the constraint line gives

$$-\frac{1}{3} = -\frac{x}{y}$$

or $y = 3x$. But $y^2 = 10 - x^2$, and so

$$(3x)^2 = 10 - x^2$$

$$9x^2 = 10 - x^2$$

$$x = 1$$

With $x = 1$, we have $y^2 = 10 - 1 = 9$ and $y = 3$. Our conclusion, then, is that profits are maximized at $(x, y) = (1, 3)$ with profits of $x + 3y = 10$.

As a shortcut, we might note that the profit function and the constraint equation can be differentiated easily using a rule from calculus: Where

$$f(x, y) = 0, \qquad y_x = -\frac{f_x}{f_y}$$

In our case, we have

$$\pi = x + 3y$$

or

$$f(x, y) = x + 3y - \pi = 0$$

and

$$y_x = \frac{f_x}{f_y} = -\frac{1}{3}$$

Similarly,

$$g(x, y) = x^2 + y^2 - 10 = 0$$

gives

$$y_x = -\frac{g_x}{g_y} = -\frac{2x}{2y} = -\frac{x}{y}$$

As before, we now equate the slopes, $-1/3 = -x/y$.

The Lagrange multiplier technique utilizes the principle of equating the slopes of the profit and constraint equations. Given a profit function $f(x, y)$ and a constraint function $g(x, y)$, the technique calls for the creation of a Lagrangian function $f(x, y) + \lambda g(x, y)$. When we equate the derivatives with respect to x, y, and λ to zero, we reproduce the slope condition at the same time that we satisfy the constraint equation:

$$L(x, y, \lambda) = f(x, y) + \lambda g(x, y) \text{ (the Lagrangian function)}$$

$$L_x = f_x + \lambda g_x = 0$$

implies

$$\lambda = -\frac{f_x}{g_x} \tag{7.1}$$

$$L_y = f_y + \lambda g_f = 0$$

implies

$$\lambda = \frac{f_y}{g_y} \tag{7.2}$$

$$L_\lambda = g(x, y) = 0 \tag{7.3}$$

Equating 7.1 and 7.2 gives the slope condition:

$$-\frac{f_x}{g_x} = -\frac{f_y}{g_y} \quad \text{or} \quad -\frac{f_x}{f_y} = -\frac{g_x}{g_y} = y_x$$

and equation 7.3 provides satisfaction of the constraint. We shall apply the technique to our example: Maximize $f(x, y) = x + 3y$ subject to $g(x, y) = x^2 + y^2 - 10$. Form the Lagrangian function

$$L(x, y, \lambda) = x + 3y - \lambda(x^2 + y^2 - 10)$$

and differentiate with respect to x, y, and λ:

$$L_x = 1 - 2\lambda x = 0 \tag{7.4}$$

$$L_y = 3 - 2\lambda y = 0 \tag{7.5}$$

$$L_\lambda = x^2 + y^2 - 10 = 0 \tag{7.6}$$

Solve for λ:

$$\lambda = \frac{1}{2x} \tag{7.7}$$

$$\lambda = \frac{3}{2y} \tag{7.8}$$

Equations 7.7 and 7.8 gives

$$\frac{1}{2x} = \frac{3}{2y} \tag{7.9}$$

$$y = 3x$$

Solving for y in equation 7.6 gives

$$y = \sqrt{10 - x^2} \tag{7.10}$$

and equating this with 7.9 allows us to solve for x:

$$y = 3x = \sqrt{10 - x^2}$$

$$9x^2 = 10 - x^2$$

$$x = 1$$

Hence, $y = 3$. The solution to our maximization problem is then $x = 1$ and $y = 3$, giving the profit $f(1, 3) = 1 + 3 = 4$.

Applying this method to our original pursuit of Q_R, we can specify the minimization problem and then form the Lagrangian function. The problem is to minimize the total setup costs, that is,

$$\frac{D_1}{Q_1}(S_1) + \frac{D_2}{Q_2}(S_2) = \frac{10,000}{Q_1}(125) + \frac{5000}{Q_2}(187.50)$$

subject to an inventory investment constraint

$$\frac{Q_1}{2}(Ic_1) + \frac{Q_2}{2}(Ic_2) = 2920 = \frac{Q_1}{2} * 0.1786 * 10 + \frac{Q_2}{2} * 0.1786 * 15$$

Forming the Lagrangian function, we obtain

$$L = \frac{D_1 S_1}{Q_1} + \frac{D_2 S_2}{Q_2} + \lambda\left[\frac{Q_1}{2}(Ic_1) + \frac{Q_2}{2}(Ic_2) - 2920\right]$$

$$\frac{\partial L}{\partial Q_1} = \frac{D_1 S_1}{Q_1^2} + \frac{\lambda Ic_1}{2} = 0 \tag{7.11}$$

$$\frac{\partial L}{\partial Q_2} = \frac{D_2 S_2}{Q_2^2} + \frac{\lambda Ic_2}{2} = 0 \tag{7.12}$$

$$\frac{\partial L}{\partial \lambda} = \frac{Q_1}{2}(Ic_1) + \frac{Q_2}{2}(Ic_2) = 2920 \tag{7.13}$$

From equation (7.11) and (7.12), we obtain

$$Q_1 = \sqrt{\frac{2D_1 S_1}{\lambda Ic_1}} = \frac{1}{\sqrt{\lambda}} * EOQ_1 \tag{7.14}$$

$$Q_2 = \sqrt{\frac{2D_2 S_2}{\lambda Ic_2}} = \frac{1}{\sqrt{\lambda}} * EOQ_2 \tag{7.15}$$

where

$$EOQ_1 = \sqrt{\frac{2 * 10,000 * 125}{0.1786 * 10}} = 1183$$

$$EOQ_2 = \sqrt{\frac{2 * 5000 * 187.50}{0.1786 * 15}} = 837$$

Substituting Q_1 and Q_2 into the constraint equation, we can solve for λ:

$$\frac{1}{\sqrt{\lambda}} \left[\frac{EOQ_1}{2}(Ic_1) + \frac{EOQ_2}{2}(Ic_2) \right] = 2920$$

$$\frac{1}{\sqrt{\lambda}} = (2177) = 2920$$

$$\frac{1}{\sqrt{\lambda}} = \frac{2920}{2177} = 1.342$$

$$\lambda = 0.555$$

Finally, substituting the value of λ in equation (7.14) and (7.15), we obtain

$$Q_1 = \frac{1}{\sqrt{\lambda}} * EOQ_1$$

$$= 1.342 * 1183 = 1587$$

$$Q_2 = \frac{1}{\sqrt{\lambda}} * EOQ_2$$

$$= 1.342 * 837 = 1122$$

For the given values of Q_1 and Q_2, we obtain the minimum value of inventory holding costs as (see Figure 7.1)

$$(Q_1/2)h_1 + (Q_2/2)h_2 = (1587/2) * 10 * 0.1786 + (1122/2) * 15 * 0.1786$$

$$= 2920$$

With holding costs of \$2920, the setup costs are (see Figure 7.1)

$$\frac{D_1}{Q_1}(S_1) + \frac{D_2}{Q_2}(S_2) = \frac{10,000}{1587} * 125 + \frac{5000}{1122} * 187.50 = 1624$$

These costs should be compared with the present annual setup costs of $2.188. The results are plotted in Figure 7.1.

UNKNOWN COSTS

If the per unit inventory holding percentage is unknown, the Lagrange multiplier takes on an interesting interpretation. In such a case, we can minimize the annual setup costs subject to a constraint based on the average value of the inventory; that is, minimize $(D_1/Q_1) * S_1 + (D_2/Q_2) * S_2$ subject to $(Q_1/2)c_1 + (Q_2/2)c_2 = Y$. Lagrangian analysis would give

$$Q_1 = \sqrt{\frac{2D_1 S_1}{\lambda c_1}}$$

$$Q_2 = \sqrt{\frac{2D_2 S_2}{\lambda c_2}}$$

but then the Lagrange multiplier is simply the unknown holding cost percentage.

Table 7.2 shows a comparison of the company's present order quantities with those generated by constraining the inventory holding investment. For the same investment, the setup costs are $564 lower when the lot sizes are adjusted to get away from the relatively high number of setups for item A.

HOLDING/SERVICE EXCHANGE

So far we have analyzed the inventory holding/setup cost exchange. Let us now look at the inventory holding/service level exchange. Safety stocks are based on a number of standard deviations of lead time demand. The higher the number of standard deviations, the higher the service level and the corresponding holding cost. Correcting and expand-

TABLE 7.2 Comparison of Present Quantities and Investment Constrained Quantities at a 17.86% Holding Rate

	Lot Size	Annual Holding Cost ($)	Annual Setup Cost ($)	Total Relevant Costs ($)	Annual Demand	Cost per Setup ($)	Unit Cost ($)
Q_P—A	769	687	1625	2312	10000	125.0	10
Q_P—B	1667	2233	563	2796	5000	187.5	15
Total		2920	2188	5108			
Q_R—A	1587	1417	788	2205	10000	125.0	10
Q_R—B	1122	1503	836	2339	5000	187.5	15
Total		2920	1624	4544			

ing an example from McClain and Thomas [2], we will analyze the service level versus holding cost trade-off.

We begin by reviewing some of our findings from single-item inventory control before moving to the aggregate level. Consider item X from the following two-item table:

Item	Annual Demand	Cost per Unit	Holding Cost per Unit per Year	Cost per Setup	Backorder Cost	Standard Deviation of Lead Time Demand
X	10,000	$5	$1	$50	$10	50
Y	1,000	$5	$1	$50	$10	5

$$Q^* = \sqrt{\frac{2 * 10,000 * 50}{1}} = 1000$$

Recall now that marginal analysis in the backorder case calls for adding another unit to safety stock, as long as the number of cycles times the probability of a backorder without the marginal unit times the per unit backorder penalty is greater than or equal to the cost of holding the marginal unit for a year:

$$\frac{D}{Q} * OSOR * \pi \geq h$$

Solving for the optimal OSOR, we have

$$OSOR^* = \frac{hQ}{\pi D}$$

In this example,

$$OSOR^* = \frac{1 * 1000}{10 * 10,000} = 0.01$$

For simplicity, we will not jointly determine Q and R. For OSOR = 0.01, k would be 2.32. Hence, the appropriate safety stock would be $2.32\sigma_L = 2.32 * 50 = 116$ units of safety stock.

Similar calculations for item Y would yield OSOR = 0.1 and $1.28 * 5 = 6.4$ units of safety stock. Now, adding items X and Y together, we would have 122.4 units of safety stock, or holding costs of $122.40 at $1 per unit.

UNIT STOCKOUT OBJECTIVE

Suppose that we are now asked to limit ourselves to $100 in holding costs on safety stock. What should the safety stocks be? If we use Lagrangian analysis in the unit stockout objective case, we find

$$OSOR_i = \lambda \left(\frac{h_i Q_i}{\pi_i D_i} \right)$$

and

$$k_1 \sigma_1 h_1 + k_2 \sigma_2 h_2 = 100$$

For our example, we shall try different λ values. By trial and error, we can adjust the safety stock investment until it approximates our target. With $\lambda = 1.5$:

$$OSOR_1 = 1.5(0.01) = 0.015$$
$$OSOR_2 = 1.5(0.1) = 0.15$$
$$k_1 = 2.17 \text{ (from Appendix 6A)}$$
$$k_2 = 1.04$$
$$2.17(50) + 1.04(5) = 113.7$$

With $\lambda = 2$:

$$OSOR_1 = 2(0.01) = 0.02$$
$$OSOR_2 = 2(0.1) = 0.2$$
$$k_1 = 2.06$$
$$k_2 = 0.84$$
$$2.06(50) + 0.84(5) = 107.2$$

With $\lambda = 2.5$:

$$OSOR_1 = 2.5(0.01) = 0.025$$
$$OSOR_2 = 2.5(0.1) = 0.25$$
$$k_1 = 1.96$$
$$k_2 = 0.68$$
$$1.96(50) + 0.68(5) = 101.4$$

Since this gives us a value close to \$100 worth of safety stock, we will stop with the suggestion that further search could give us an exact value. Nevertheless, the procedure we have used gives us a new breakdown of safety stock. Previously, we employed 116 units of item X safety stock and 6.4 units of item Y safety stock. With our investment constraint at \$100, we employ 98 units of X safety stock and 3.4 units of Y safety stock.

The key to the unit stockout situation, the Lagrange multiplier adjusts the ratio of holding cost to stockout penalty. Rewriting the stockout probability relationship as a function of Q_i/D_i gives

$$\text{OSOR}_i = \frac{\lambda h_i}{\pi_i} * \left(\frac{Q_i}{D_i} \right) \quad \text{rather than} \quad \lambda \left(\frac{h_i Q_i}{\pi_i D_i} \right)$$

We started with a ratio $h_i/\pi_i = 1/10 = 0.1$, but that ratio gave us total holding costs of \$122.40, which are too high. To reduce the holding costs, we need a larger ratio of holding cost to stockout penalty. Our revised ratio of $\lambda(h_i/\pi_i) = 2.5(0.1) = 0.25$ forces us to hold less inventory.

The formal statement of this problem with a unit stockout objective would be as follows: Minimize $\Sigma(D_i/Q_i)\sigma_i g(k_i)\pi_i$ subject to $\Sigma\ k_i\sigma_i h_i = I = 100$; that is, minimize the number of cycles times the expected number of stockouts per cycle times the per unit backorder penalty, subject to a constraint on the aggregate safety stock holding.

We do not present the Lagrangian calculations here, because they involve differentiating an integral. The interested reader is referred to Silver and Peterson [4] for a discussion of special results for differentiating the normal integral or the normal loss integral. The result of the Langrangian analysis is the relationship:

$$\text{OSOR}_i = \lambda \left(\frac{h_i Q_i}{\pi_i D_i} \right)$$

Unknown Costs

As an appealing benefit of the Lagrangian approach, cost specification loses its importance. Suppose that management is willing to specify a constraint on total safety stock investment, $\Sigma\ k_i\sigma_i c_i = \$500$. Suppose also that the order quantities, Q_i, have been previously determined. We can find the optimal safety stock without knowing either the holding costs or the stockout penalties, because management's specification of safety stock investment implies a ratio of holding cost to stockout penalty. In our example, the ratio is 0.25, as we shall see:

$$\text{OSOR}_i = \lambda \left(\frac{Q_i}{D_i} \right) = 0.25 \left(\frac{Q_i}{D_i} \right)$$

Here, rather than having $\lambda(h_i/\pi_i)$ as the holding cost to stockout penalty ratio with known h_i and π_i, we simply have λ as the ratio. We then try values of λ until we find one that gives a total safety stock holding investment of $500. A value of $\lambda = 0.25$ comes close:

$$\text{OSOR}_1 = 0.25\left(\frac{Q_1}{D_1}\right) = 0.25\left(\frac{1,000}{100}\right) = 0.025$$

$$\text{OSOR}_2 = 0.25\left(\frac{Q_2}{D_2}\right) = 0.25\left(\frac{100}{100}\right) = 0.25$$

$$k_1 = 1.96$$

$$k_2 = 0.68$$

$$k_1\sigma_1 h_1 + k_2\sigma_2 h_2 = 1.96(50)(5) + 0.68(5)(5) = 507$$

Note that the formulation with unknown costs gives one ratio of holding to stockout costs for all items in the inventory.

STOCKOUT SITUATION OBJECTIVE

Stockout situations and units stocked out are not the same. Each stockout situation can lead to several units being stocked out. In the order service level, we look at the probability of having a stockout position or situation on a cycle. The expected number of units stocked out per cycle goes into our calculation of unit service level.

Recall that our original unconstrained solution involved $2.32\sigma_1$ units of item X safety stock and $1.28\sigma_2$ units of item Y safety stock. This gave us $2.32(50) + 1.28(5) = 122.4$ units of safety stock, for a holding cost of $122.40. The question now is whether we can have fewer stockout situations for the same holding cost.

With the data from our problem,

Item	D	Q	h	π	σ_L	K	OSOR
X	10,000	1000	1	10	50	2.32	0.01
Y	100	100	1	10	5	1.28	0.10

the original solution would give

$$\frac{D_1}{Q_1}\,\text{OSOR}_1 + \frac{D_2}{Q_2}\,\text{OSOR}_2 = \frac{10,000}{1000}(0.01) + \frac{100}{100}(0.1)$$

$$= 0.10 + 0.10 = 0.2 \text{ stockout situation}$$

The safety stock holding cost would be

$$k_1\sigma_1 h_1 + k_2\sigma_2 h_2 = 2.32(50) + 1.28(5) = 122.4$$

It would not be very surprising to get fewer stockout situations with a situation-oriented objective function: Minimize $(D_1/Q_1) * OSOR(k_1) + (D_2/Q_2) * OSOR(k_2)$ subject to $k_1\sigma_1c_1 + k_2\sigma_2c_2 = Y$, where c_i is the item cost per unit.

Forming the Lagrangian in our example gives

$$L(k_1, k_2, \lambda) = \frac{D_1}{Q_1} * OSOR(k_1) + \frac{D_2}{Q_2} * OSOR(k_2)$$
$$+ \lambda(k_1\sigma_1c_1 + k_2\sigma_2c_2 - Y)$$

Setting the first partial derivatives to zero gives

$$\frac{\partial L}{\partial k_1} = -\frac{D_1}{Q_1} f(k_1) + \lambda\sigma_1c_1 = 0$$

$$\frac{\partial L}{\partial k_2} = -\frac{D_2}{Q_2} f(k_2) + \lambda\sigma_2c_2 = 0$$

$$\frac{\partial L}{\partial \lambda} = k_1\sigma_1c_1 + k_2\sigma_2c_2 - Y = 0$$

Notice that $f(k)$ is the density function for the unit normal distribution, as in Figure 7.3. Intuitively, as k increases, the area OSOR (k) decreases at the rate $f(k)$. For a proof, see Silver and Peterson [4]. Solving for k, we find

$$f(k_i) = \lambda \frac{Q_i}{D_i} \sigma_i c_i$$

and $k_1\sigma_1c_1 + k_2\sigma_2c_2 = Y$.

Noting that

$$f(k) = \frac{1}{\sqrt{2\pi}} \exp(-k^2/2)$$

FIGURE 7.3   Density function for unit normal distribution.

we can solve to find

$$k = \sqrt{2 \ln \left(\frac{D}{\lambda \sqrt{2\pi Q c \sigma}} \right)}$$

In our example, $\lambda = 1/745$ gives $k_1 = 2.23$ and $k_2 = 2.23$. Check it out.

$$k_1 = \sqrt{2 \ln \left(\frac{745(10,000)}{\sqrt{2(3.1416)(1000)(5)(50)}} \right)}$$

Substituting these values of k_1 and k_2 into the objective function, we can determine the expected number of stockout situations in a year:

$$\frac{D_1}{Q_1}(\text{OSOR}_1) + \frac{D_2}{Q_2}(\text{OSOR}_2) = \frac{10,000}{1000}(0.0129) + \frac{100}{100}(0.0129) = 0.14$$

By comparison, the previous solution gave 0.2 expected stockout situations per year.

What does λ mean? At the optimal, we know now that

$$L(k_1, k_2, \lambda) = \frac{D_1}{Q_1}\text{OSOR}_1 + \frac{D_2}{Q_2}\text{OSOR}_2 + \lambda(k_1\sigma_1 c_1 + k_2\sigma_2 c_2 - Y)$$

$$= 10(0.0129) + 1(0.0129) + \lambda(612 - Y)$$

Differentiating with respect to Y, we have $dL/dY = -\lambda$. Note that $L(k_1 k_2 \lambda)$ is the expected number of stockout situations if Y is \$612. Then λ represents the reduction in expected stockouts per \$1 increase in Y.

If the decision were whether or not to invest \$1 more in safety stock, marginal analysis would require that the cost to carry an extra dollar of safety stock equal the reduction in expected stockout situations times the cost per stockout situation:

$$I = \lambda B \quad \text{or} \quad \lambda = I/B$$

where I is the carrying percentage and B is the cost per stockout situation.

To obtain a value for λ, we use the method of successive approximations:

1. Guess a value of λ.
2. Solve for k_1, k_2. If the value under the square root is negative, return to 1. Let

$$X_i = \frac{D_i}{\sqrt{2\pi Q_i c_i \sigma_i}}$$

Then calculate

$$k_i = \sqrt{2 \ln (X_i/\lambda)}$$

3. Solve for implied inventory investment.
4. Calculate the ratio of desired inventory investment to the investment generated in step 3 and call that ratio INRAT. Clearly, a ratio greater than unity means that we need to revise the k values upward or, correspondingly, the λ value. Multiply the k values by INRAT.
5. Solve for λ. We know that

$$k_1 = \sqrt{2 \ln (x_1/\lambda)}$$

Solving now for λ, we would find

$$\lambda = \frac{X_1}{\exp(k_1^2/2)}$$

6. Return to step 2 and repeat the process until two successive iterations yield approximately the same λ value.

Let us go through the steps with our example.

1. Guess $\lambda = 1/100$.
2. $X_1 = 10,000/[2.51(1000)(5)(50)] = 0.0159$; $k_1 = \sqrt{2 \ln[100(0.0159)]} = 0.963$; $X_2 = 100/[2.51(100)(5)(5)] = 0.0159$; $k_2 = \sqrt{2 \ln[100(0.0159)]} = 0.963$.
3. Inventory investment $= k_1\sigma_1c_1 + k_2\sigma_2c_2 = 0.963(50)(5) + 0.963(5)(5) = 264.83$.
4. Desired inventory investment is \$612. Then new k values are (612/264.55) * 0.963 = 2.23.
5. Because $X_1 = X_2 = 0.0159$, no further adjustments need be made and

$$\lambda = \frac{0.0159}{\exp(2.23^2/2)} = 0.00132$$

When $X = X_1 = X_2$, then $k = k_1 = k_2$ and $k = \sqrt{2 \ln (x/\lambda)}$. Problem 12 at the end of this chapter provides an example where $X_1 \neq X_2$.

COMPARISON OF THE SAFETY STOCK POLICIES

Retracing our steps, we have now worked with three different sets of safety stock values for three different objectives. To gain some perspective, we now compare these policies on the measures of dollar investment, expected stockout situations, and expected stock-

TABLE 7.3 Comparison of Safety Stock Policies

	Unit Objective		Situation Objective	
	Unlimited Holding Cost Budget	$100 Budget on Holding Cost	$122.40 Budget on Holding Cost	$100 Budget on Holding Cost
k_1 value	2.32	1.96	2.23	1.82
k_2 value	1.28	0.68	2.23	1.82
Expected number of stockout situations	0.2	0.5	0.14	0.15
Holding costs	$122.40	$101.40	$122.40	$100.10
Expected stockout costs	$ 19.63	$ 54.64	$ 22.67	$ 68.53
Inventory investment	$612.00	$507.00	$612.00	$500.50

out costs. Table 7.3 presents only solutions; we ask that you make the intermediate calculations to assure yourself that you know what is going on.

So far, we have considered only the trade-off of holding cost or investment versus service level. In Chapter 6, we stressed the interaction between safety stock and order quantities in the (Q, R) model. Recall that larger Qs mean fewer exposures to stockout and hence allow lower safety stock in the reorder point calculation (R = expected lead time demand plus safety stock).

What we need is a formulation to cover both trade-offs: investment versus service and investment versus setup. With a unit stockout objective, we would have the following problem: Minimize

$$\sum \frac{D_i}{Q_i} \sigma_i g(k_i)$$

subject to

$$\sum \frac{Q_i}{2} c_i + \sum k_i \sigma_i c_i = Y$$

We could solve this problem by the Lagrangian method coupled with the method of successive approximations. Gardner and Dannenbring [1] do this, and you should be able to understand their article after studying this chapter.

SUMMARY

This chapter examined aggregate inventory trade-offs. The LIMIT technique provides an algebraic way of adjusting lot sizes to meet an overall constraint on setup hours. Using the LIMIT technique, we soon discovered that an overall constraint on setup hours implies a carrying cost percentage, and so we saw yet another example of imputed or implicit costs.

In analyzing the holding cost to setup cost trade-off subject to overall budget constraints, we found that the Lagrange multiplier technique can handle the same problems as LIMIT. Not only can this powerful technique handle other holding/setup questions, but it also allows us to analyze holding cost versus inventory investment constraints. Returning to our study of unit- and situation-oriented service objectives, we employed Lagrange multipliers to develop optimal safety stocks within budget constraints.

In the various trade-offs made, the Lagrange multiplier always seemed to have a meaning. In a model minimizing annual setup hours, the Lagrange multiplier was the holding cost percentage. In meeting a unit stockout objective, the multiplier appeared as the ratio of per unit holding cost to per unit stockout penalty. Finally, the multiplier showed up as the ratio of the holding cost percentage to the cost per stockout situation in a model designed to meet a stockout situation objective.

From basic inventory models to aggregate inventory models, we have constantly discovered implicit or imputed costs. In this chapter on aggregate inventories, budget statements about setup hours or inventory investment were viewed as statements about cost ratios.

The next chapter changes the pace. Not only will we be concerned about aggregate inventories, but we also consider where to locate inventories—whether it is better to hold inventory centrally or to duplicate inventories across the country. Because the problems of multilocation inventories are so complex, we no longer find conclusive results or formulas. Chapter 8 discusses the controversy about the best approach.

PROBLEMS

1. Suppose that $D = 5000$, $S = \$150$, $c = \$15$, and we require five setups in the year. What is the implied annual holding cost percentage per unit per year?

2. Suppose that $D = 5000$, $S = \$150$, $c = \$15$, and $I = 10\%$. Now we are asked to revise I to 35%. What multiplier M should be used in revising the original Q? What will the new Q be? Check this against the EOQ formula, with $I = 0.35$. What is the ratio of the number of setups under the original to the number of setups under the revised carrying percentage?

3. Two inventory items have been managed in a "seat of the pants" fashion for several years. The following table shows the current situation:

Item	Annual Usage	Setup Hours per Order	Unit Cost	Present Quantity	Setup Cost
A	8000	2	$16	1000	$75/hour
B	2000	4	$8	1000	$75/hour

For trial purposes, assume that $1 = 0.35$ and calculate lot sizes. Within the constraint of twenty-four hours of setup per year, revise the lot sizes. What is the implied carrying percentage? Does the initial carrying percentage guess make any difference?

4. Solve the following problem using the Lagrange multiplier method: Maximize $f(x, y) = 3x + 2y$ subject to $x^2 + y^2 = 20$.

5. Solve the following problem using the Lagrange multiplier method: Minimize $f(x, y) = 1/x + 4/y$ subject to $x + 1.4y = 10$.

6. Using the Lagrange multiplier technique, we go through the same steps for finding a maximum or a minimum. How do we know which we have? An analogy to the second derivative test could be used, but a pragmatic approach suggests that we simply graph the objective function. In problem 4, did we get a maximum? In problem 5, did we get a minimum?

7. Solve the following problem using the Lagrange multiplier method: Minimize $f(x, y) = 1,250,000/x + 937,500/y$ subject to $0.895x + 1.345y = 2926$.

Problems 8 through 13 use the following data:

Item	Annual Demand	Cost per Unit	Holding Cost per Unit per Year	Cost per Setup	Backorder Cost	Standard Deviation of Lead Time Demand
A	8,000	10	$1	$500	$20	40
B	8,000	20	$2	$300	$20	30

8. What are the optimal stockout probabilities for the two items? Assume the backorder case. What safety stocks should be set for the two items based on single-item inventory control methods?

9. With a unit stockout objective and a constraint of $200 worth of holding costs on safety stock, calculate the revised safety stocks.

10. With the same two items, A and B, determine whether a different allocation of safety stocks will minimize stockout situations subject to an inventory holding constraint of $200. What would the allocation be?

11. Develop a table similar to Table 7.3 to specify expected stockout situations and expected stockout costs, with k values of $k_A = 2.12$ and $k_B = 2.07$.

12. Develop a table similar to Table 7.3 to compare expected stockout situations and expected stockout costs for the unit and situation objectives. Use the $200 holding cost budget constraint.

13. Write a computer program to find optimal λ values in the unit stockout objective case. Use the method of successive approximations.

14. For items X and Y we allow backorders. Further, we require that safety stock investment costs per year be limited to 200 units for these two items. The MAD on X is 100 and on Y is 50. Holding and stockout costs are unknown. Item X has annual demand of 1000 and an order quantity of 100. Item Y has annual demand of 2000 and an order quantity of 500. What safety stock should be held? What is the implied ratio of holding to stockout cost?

15. Two inventory items have the following data:

Item	Annual Usage	Cost per Setup	Value per Unit
A	10,000	$125.00	$10
B	5,000	$187.50	$15

What carrying cost percentage makes the order quantities $Q_A = 1586$ and $Q_B = 1122$ optimal?

16. Two inventory items have the following data:

Item	Annual Usage	Order Quantity	Value per Unit	Standard Deviation of Lead Time Demand
A	10,000	10,000	5	40
B	100	100	5	5

The inventory investment budget is $200. Find the optimal amounts of safety stock for each item, beginning with a Lagrange multiplier of 1/100.

17. Examine Table 7.3. for a $100 budget on holding costs, why does the unit objective employ k values of 1.96 and 0.68 while the situation objective employs equal values of 1.82? What are the order stockout risks for these k values?

18. Suppose that we decide to minimize average inventory investment subject to a constraint on total setup hours. Is that sensible? What would be the appropriate order quantities? How would you interpret the Lagrange multipler?

19. Two items in inventory have the following data:

Item	Annual Usage	Cost per Setup	Value per Unit
A	10,000	$125.00	$10
B	5,000	$187.50	$15

At a holding cost percentage of 17.9%, we found the optimal Q's to be $Q_A = 1586$ and $Q_B = 1122$. Our budget constraint of $2920 was exactly satisfied, and the Lagrange multiplier was 0.555. What would be the decrease in annual setup costs for $1 increase in the holding cost budget?

20. Repeat problem 3 with present quantities for A and B as 2000 and 500, respectively.

21. Repeat problem 9 with a constraint of $180 of holding costs on safety stock.

REFERENCES AND BIBLIOGRAPHY

1. E. S. Gardner and D. D. Dannenbring, "Using Optimal Policy Surfaces to Analyze Aggregate Inventory Trade-offs," *Management Science,* Vol. 25, No. 8 (August 1979), pp. 709–720.
2. J. O. McClain and L. J. Thomas, *Operations Management* (Englewood Cliffs, N.J.: Prentice Hall, 1980), pp. 389–392.
3. G. W. Plossl and O. W. Wight, *Production and Inventory Control* (Englewood Cliffs, N.J.: Prentice Hall, 1967).
4. G. A. Silver and R. Peterson, Decision systems for Inventory Management and Production Planning, second edition, (New York: John Wiley & Sons, 1985).

APPENDIX 7A

Hewlett-Packard Gains Control of Inventory and Service through Design for Localization
Hau L. Lee, Corey Billington, and Brent Carter

At Hewlett Packard (HP) Company, design for manufacturability has recently been adopted as a principle for product design and development. Frequently overlooked is the relationship between design and the eventual customization, distribution, and delivery of the product to multiple markets. Different markets may have different requirements for the product due to differences in taste, language, geographical environment, or government regulations. We use design for localization or design for customization for design processes that take into account the operational and delivery service considerations for the multiple market segments. We developed an inventory model that the HP's Deskjet-Plus Printer Division used to evaluate alternative product and process designs for localization. Significant benefits can be obtained by properly exploring the opportunities in this design for localization concept.

In recent years, interest has increased in bridging the gap between product design and manufacturing. Product design evaluations should include consideration of the impact of the designs on manufacturability, cost, and quality. Consequently, such concepts as "design for manufacturability," "design for assembly," "design for testability," "design for producibility," and "design for quality" have received widespread attention and support [Whitney 1988, Dean and Susman 1989, and Taguchi and Clausing 1990]. One critical element in product design has been overlooked: the relationship between design and the eventual customization, distribution, and delivery of the product to multiple market segments. These segments may have different requirements of the product due to differences in taste, language, environment, or government regulations. The design of the product and of the manufacturing process can affect the company's operational costs and its delivery service to its customers. We use the terms *design for localization* or *design for customization* to represent design processes that take into account the operational and delivery service considerations for multiple market segments.

Usually, in manufacturing products suitable for different market segments, manufacturers produce basic products that contain most of the features and components of the finished products and assemble the final products with some additional components that differentiate the products for different market segments. For example, a computer manufacturer might consider different countries as the market segments. The computer products would differ in the power supply module to accommodate local voltage, frequency, and plug conventions. Keyboards and manuals would be produced to suit the local language. Telecommunication products may also be differentiated by the communication protocols supported. In some cases, the need for localized versions of a product results from government-imposed local content requirements.

Reprinted from *Interfaces,* Vol. 23, No. 4 (July–August 1993), pp 1–11.

Depending on the design of the product and the production process, the localization or customization might be performed differently; for some product designs, the main factory assembles differentiating modules. For some, distributors assemble the differentiating modules just before shipping them to customers; and for some, the customer performs the final customization.

Where customization is performed can greatly affect the company's inventory and service trade-off. Delaying customization increases the company's flexibility to respond to changes in the mix of demands from different market segments. The company can improve its responsiveness to orders to reduce its investment in inventory.

Hence, the designs of the product and of the production process can affect the degree of localization, the site at which localization can be done, and the cost of localization. To design for localization, design engineers should consider the cost and service implications in designing product for market segments.

We developed an inventory model to address the consequences in inventory costs and delivery service that result from different design alternatives. The model has been used to support the manufacture of Deskjet-Plus printers at the Vancouver, Washington Division of Hewlett-Packard Company (HP).

THE DESKJET-PLUS SUPPLY CHAIN

The Deskjet-Plus is one of several printers manufactured by the Vancouver Division of HP. The manufacturing process has two stages: (1) printed circuit board assembly and test (PCAT); and (2) final assembly and test (FAT). In PCAT, electronic components such as ASICs (application-specific integrated circuits), ROM (read-only memory), and raw printed circuit boards used to make logic boards and print head driver boards for the printers are assembled and tested. In FAT, such subassemblies as motors, cables, key pads, plastic chassis and "skins," gears, and the printed circuit boards from PCAT are assembled to produce working printers, which are then tested. The sources for the components are other HP divisions and external suppliers worldwide.

To localize the Deskjet-Plus for different countries, HP packages the appropriate power supply module (with the correct voltage and plugs) and the appropriate manual with the printer. In the past, the factory performed this step. Hence, the factory produced finished printers destined for all other countries. It then sorted them into three groups for the distribution centers (DCs) in North America, Europe, and Asia and the Pacific. We refer to this process as *factory-localization*.

TheVancouver plant ships to the three distribution centers by sea. We use the term *supply chain* to describe the network of manufacturing and distribution sites [Cohen and Lee 1988].

The printer industry is highly competitive. HP's computer dealers like to carry as little inventory as possible, but must supply products to end-users quickly. Consequently, HP is under increasing pressure as a manufacturer to provide high levels of availability at the DCs for the dealers. In response, HP management has decided to operate the DCs in a make-to-stock mode to provide very high levels of availability to the dealers. Hence, the DCs operate as inventory stocking points with large safety stocks to meet a target off-the-shelf fill rate, where the replenishment of products comes from manufacturing.

Manufacturing, on the other hand, operates in a pull mode. It sets production plans to replenish the DCs just in time to maintain the target safety stocks. To ensure material availability, HP also sets up safety stocks for incoming materials at the factory.

Three major sources of uncertainty can affect the supply chain: (1) delivery of incoming materials (late shipments, wrong parts, and so forth); (2) internal process (process yields and machine downtimes); and (3) demand. The first two can cause delays in replenishing stocks at the DCs. Demand uncertainties can lead to inventory buildup or to backorders at the DCs. Under "factory-localization," HP ships the different versions of the Deskjet-Plus to the two non-US DCs by sea, with a transit time of about a month. Because of this long lead time, the DC has limited ability to respond to fluctuations in the demand for the different versions of the product. To ensure prompt service for the customers, the European and Far East DCs have to maintain high levels of safety stocks. For the North American DC, the situation is simpler; most of the demand is for the US version, and there is little localization product-mix fluctuation.

Limiting the inventories in the Deskjet-Plus supply chain and at the same time providing the high level of service needed has been quite a challenge to the Vancouver Division's management. The manufacturing group in Vancouver has worked hard on supplier management to reduce the uncertainty caused by delivery variability, to improve process yields, and to reduce machine downtimes at the plant. The progress made has been admirable. However, improving the forecast accuracy of product-mix demands remains a formidable task. This is why it considered product/process designs as a way to improve the effectiveness of inventory in the supply chain. Design changes can ameliorate the impact of poor forecasts.

The linkage between manufacturing and distribution is one opportunity. If the shipment lead times between manufacturing and distribution were drastically reduced, for example, by shipping by air, then the manufacturing plant's ability to respond to product-mix fluctuations at the DCs would be greatly enhanced. However, air shipment is very costly, and HP found it to be an ineffective alternative.

Localization at the DCs provides a more attractive alternative. The factory would manufacture and ship a generic Deskjet-Plus printer without the power supply module and manual. The DCs would then localize the generic product to the different specific options as needed. From standard inventory theory, such a change would have the benefit of the risk pooling effect, and consequently, HP would need less safety stock in finished goods. We will term such a strategy "DC-localization." However, because Vancouver is not far from the US DC, and because the whole US demands essentially one option of the product, it is more efficient for Vancouver to localize the US option at the factory. Hence, the more realistic alternative HP considered is for Vancouver to manufacture two types of deskjet printers: (1) a fully localized US option; and (2) a generic product without the power supply module and manual, to be shipped to Europe and the Far East for localization there (Figure 1).

To implement DC-localization, HP had to make some design changes to the product. It has to redesign the product so that the power supply module would be the last component added on and so that the DC could add it on easily. To minimize the time that it takes to localize the product, HP wanted this assembly to be a simple operation, for example, a simple plug-in. Moreover, even if the addition of the power supply module and

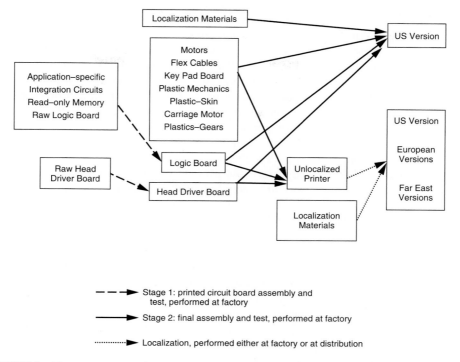

Stage 1: printed circuit board assembly and
test, performed at factory

Stage 2: final assembly and test, performed at factory

Localization, performed either at factory or at distribution

FIGURE 1: The two-stage manufacturing process, the bill of materials, and the two localization strategies for the Deskjet-Plus. The arrows represent the two stages in manufacturing, as well as the localization step. The boxes represent inputs and outputs.

the manuals to the product is a simple kitting process, HP had to make some investment to equip the international DCs with such a capability.

In terms of inventory control, as long as all complete printers were assembled at the factory, it kept all safety stocks of the power supply modules and manuals; when the DCs localize finished products, they also keep safety stocks. The Vancouver factory keeps safety stocks of the US power supply modules and manuals to produce fully localized US printers, whereas the remote DCs keep stocks of the international power supply modules and manuals. Semi-finished goods, that is, unlocalized versions of the product, are stocked at the European and the Far East DCs. When they receive actual orders, the DCs quickly localize the generic version into the specific products required.

Since the DC-localization alternative involves some product redesign and investment at the DCs, HP needed to evaluate the cost-effectiveness of such an alternative by quantifying the inventory benefits. Such an analysis was a critical input to management's decision.

AN INVENTORY MODEL FOR THE DESKJET-PLUS SUPPLY CHAIN

Although the planners use many local rules and adjustments in managing the inventory levels at each of the DCs, the basic principles they use follow: at each DC, management sets a target inventory level for each product so as to achieve the desired fill rate. The target inventory level for a particular product is a function of the length and variability of the lead time to replenish the stock from the factory and the level and variability of demand for the product. At HP, such a target inventory level is usually expressed in terms of weeks of supply. The planners review the actual inventory position (inventory level plus inventory in the pipeline) each week. This weekly review period corresponds to the frequency with which products are shipped from the factory to the DCs in Europe and Asia. Studies on transportation economics have shown that this frequency allows Vancouver to maximize its use of containers for shipment. The quantity needed to bring the inventory position back to the target level constitutes the production requirement (to satisfy that DC) at the factory. Since the factory holds no finished goods inventory, the lead time for the factory to replenish the DC's stockpile is the sum of the transportation time from the factory to the DC, the manufacturing flow time at the factory, and any possible delay due to material shortages, congestion (queuing) effects, process downtimes, and other unexpected disruptions (power outages, special celebrations at the plant, and so forth).

With ample plant capacity, the factory can assume that manufacturing lead time is unaffected by the replenishment quantities and their variabilities. If this were not the case, then the factory should explicitly model manufacturing lead time as a function of the replenishment quantity. An example is a linear function with the coefficient given by the manufacturing cycle time. In that case, the manufacturing lead time would be a random variable whose mean and variance are functions of the mean and variance of the replenishment quantity.

To estimate the delays due to these various causes, we used empirical data based on the production time-log in Vancouver. The time-log documents the frequency and duration of process disruptions. We used these data to estimate the man and variance of the delays at the factory, and consequently, of the replenishment lead time from the factory to the DCs.

Hence, we can model the inventory system at each DC as a standard periodic review, order-up-to system [Nahmias 1989] where the period is a week. We have also assumed that the demand per week for a product is stationary and normally distributed. The stationarity assumption is reasonable for a mature product such as the Deskjet printer. The uncertainties such an inventory system faces include demand and replenishment lead time. Such a simple inventory model is easy to use, but we should recognize its limitations. For example, the basic model assumes that demands in different time periods are independent and that successive replenishment lead times are also statistically independent. We also make the assumption that the demands for the different product versions of the printer are statistically independent. By treating each DC as a single inventory system, we also ignore the multi-echelon nature of the network, and consequently we do not consider such issues as lateral resupply among the DCs, allocation of stock by the factory to the DCs in times of limited supply, and correlation of demands across the DCs. Nevertheless, we found the resulting model to be adequate for the following reasons. First, delays due to material shortages and process downtimes are usually short (relative to the period of a week), and they tend to occur independently. Second, since the DCs are located on three different continents, lateral transshipments are possible but uneconomical and inefficient, so that they are seldom used. Hence, the demands across the DCs are quite independent of one another. Third, initial validation of the model (described later) indicated that the model is of sufficient accuracy for policy evaluation.

In the appendix, we outline the basic inventory model we used. For the factory-localization alternative, each of the product versions at a DC has its own replenishment lead time, and independent target inventory levels can be set. The replenishment lead times at the DCs differ by the transit time from the factory to the DCs. For the DC-localization alternative, each DC stocks the generic product, whose mean and variance of demand equals the sums of the means and variances of the amount of different product versions stocked at the DC.

We validated the inventory model by considering one of the Vancouver products. We selected two months (November 1989 and April 1990) for tracking inventory levels, material shortage, process downtime profiles, and service levels. Using the target inventory levels set by management for those two months, we used the inventory model to predict fill rates at the DCs. We compared these fill rates to those observed in the field. For the two months chosen, the actual fill rates and the predicted fill rates were very close (Table 1). We made similar observations at the individual DC level.

We then used the model to evaluate the two localization alternatives. Under different target fill rates that range from 80 to 99.9 percent, the model computed the respective target inventory levels for the finished goods (one localized and one unlocalized) at the DCs, plus the material costs at the factory and the DCs, under the two localization alternatives.

ANALYSIS OF LOCALIZATION ALTERNATIVES

Table 2 shows the average inventory levels (in weeks of supply) of finished goods (localized or unlocalized Deskjet-Plus printers) and localization materials at the DCs and the factory under the two localization alternatives with a fill rate target of 98 percent. For the DC-localization alternative, localization materials are stocked at the Far East DC and at the European DC.

For this alternative, localization materials are also stocked at the factory for localizing the product for the US market. Given the close proximity of the factory to the US DC, the inventory levels of the current partial DC-localization alternative would not be

TABLE 1: The Model was validated by comparing predicted fill rates with actual fill rates in two months

	Downtime Frequency in Vancouver	Finished Goods Inventory	Predicted Fill Rate	Actual Fill Rate
Nov 89	High	DC1: 8.4 wks	8%	86%
		DC2: 2.8 wks		
		DC3: 5.1 wks		
Apr 90	Low	DC1: 9 wks	99%	99.9%
		DC2: 2.6 wks		
		DC3: 5.8 wks		

TABLE 2: Inventory levels at all sites of the supply chain to achieve a service target of 98 percent were compared under the two localization strategies. In factory localization, Vancouver localizes printers for all distribution centers which stock localized printers. For DC localization, the distribution centers stock unlocalized printers, which are then localized at the distribution centers on demand.

		Factory-Localization (Weeks of Supply)	DC-Localization (Weeks of Supply)
Far East Distribution Center	Printers	13.4	9.8
	Localization Materials	0	11.0
European Distribution Center	Printers	5.2	3.5
	Localization Materials	0	5.2
US Distribution Center	Printers	3.2	3.2
	Localization Materials	0	0
Vancouver Factory	US Localization Materials	2.8	2.8
	Other Localization Materials	3.9	0
Worldwide Travel (Weighted Average)	Printers	4.4	3.5
	Localization Materials	3.3	4.0

too different from a full-scale DC-localization alternative where the US DC localizes the US printers.

We consider the finished goods inventory (FGI) levels for a given service target to be a measure of performance. FGI is the average inventory levels at the DCs. Under the DC-localization alternative, FGI levels are much lower, while localization materials inventory levels are higher at the Far East and European DCs (Table 2). Under such an alternative, there is less total printer inventory, although there are higher stocks of localization materials at the DCs. Since the values of printers and localization materials differ, we evaluated the alternatives in terms of dollar investment.

The reduction in the total dollar value of FGI investment from factory-localization to DC-localization is 21 percent, whereas the corresponding increase in localization materials worldwide is 24 percent. The overall dollar impact is that DC-localization would lead to a reduction of 18 percent in total inventory investment in the supply chain, with no change in the DCs' service to the customers. The dollar value of that 18 percent is in millions. Figure 2 shows the FGI levels under the two alternatives at various levels of fill rate target.

We considered a range of material shortage and downtime profiles, and sensitivity analysis showed similar results.

Our analysis indicates that the inventory savings to be gained from changing from factory-localization to DC-localization is substantial. By continuing the current investment in inventory, HP could greatly improve customer service (off-the-shelf fill rate) by changing from factory-localization to DC-localization. Our analysis quantifies the inventory or service benefits to support management in its evaluation of the design and process changes needed to introduce DC-localization.

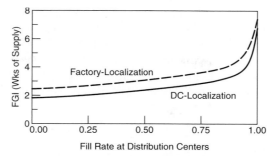

FIGURE 2:   DC-localization achieves lower inventory levels for different service targets than factory-localization.

ADDITIONAL CONSIDERATIONS

Many other factors should be considered in a comprehensive evaluation of the localization alternatives: We based our analysis on the demands for the printers at the different DCs. As the Asian-Pacific and European markets for printers grow, the dollar value of the inventory savings for DC-localization will increase. Because a localized printer contains localization materials, it has a higher value than an unlocalized printer. Hence, the capital tied up in inventory in transit is also lower when localization is performed at the end of the chain. Although the magnitude of such a difference is much smaller than the difference in inventories held at the DCs and the factory, this factor should be included in a thorough evaluation.

Of further interest is how to value the savings associated with reducing inventory. This requires a careful analysis of the real holding and opportunity costs incurred by committing capital to inventory. The quantification of the difference in inventory investment from the model enables management to perform sensitivity analysis based on alternative rates of holding and opportunity costs of inventory.

Shipping unlocalized printers to the Far East and European DCs under the DC-localization alternative provides HP with a major opportunity to reduce transportation cost, another point to consider. An unlocalized printer is much less bulky than a localized one, because the localization materials and many of the final packaging materials that are needed for the customers do not have to be bundled with the printer. The factory can ship the unlocalized printers in bulk pallets and cut the cost of transportation.

To implement DC-localization, HP needs to change the product design. This requires valuable engineering resources. Moreover, it would have to add final configuration and packaging capability to the Far East and European DCs—at a cost. DC-localization of subsequent Vancouver products and products from other divisions will also provide substantial savings to offset this one-time investment.

Other nonquantifiable factors should be considered as well. For example, increasing "local content" and local "manufacturing" presence can make a product more marketable, a factor that supports localization of the products at the non-US DCs [Cohen and Lee 1989]. There is also a need to develop a local supply base of the localization materials for the DCs. The pros and cons of decentralizing or regionalizing supply of materials should be considered [Cohen and Lee 1989]. Finally, since DC-localization requires the DCs to perform some operations that are traditionally viewed as manufacturing activities, they may have cultural and organizational barriers to overcome.

After considering both the quantifiable and the nonquantifiable factors, the Vancouver Division estimated that the combined net savings of adopting the DC-localization alternative for the Deskjet-Plus printers would be well over several million dollars per year. The division is now beginning to design or redesign all current and future products to support the DC-localization strategy.

CONCLUSION

The Vancouver, Washington Division of Hewlett-Packard is redesigning all its current products to support DC-localization. This approach has become an integral part of its product development process.

Most companies have not fully explored the concept of design for localization. It offers tremendous opportunities for increased flexibility, reduced inventory investment, and improve service. Design for localization should be an important strategy consideration for companies that wish to compete successfully in a global market.

ACKNOWLEDGMENT

We gratefully acknowledge the programming support provided by Paul Gibson, and helpful comments by Tom Davis. We also thank David Archambault and Allan Gross of the Vancouver Division for their support and encouragement.

APPENDIX

Inventory Control Model for a DC
For a product at a DC, let

S = order-up-to level (target stock level);

L = supplier lead time (in weeks);

D = demand per week; and

X = demand during lead time plus one week.

Let $X(x)$, $s(x)$ and $Var(x)$ denote the mean, standard deviation and variance of a random variable, respectively. We can compute $Var(X)$, and consequently, $s(X)$, as

$$Var\,(X) = E[Var\,(X|L)] + Var\,[E(X|L)],$$

$$= E[(L + 1)Var(D)] + Var[(L + 1)E(D)],\ \text{and}$$

$$= [E(L)+1]Var\,(D)]^2 Var(L).$$

The order-up-to level is then given by $S = E(D)[E(L) + 1] + ks(X)$, where k is the safety-stock factor. To determine the safety-stock factor, we use the approximation formula in Nahmias (1988, p. 653) for the estimation of fill rate:

$$\text{Fill rate} = 1 - [\exp(-0.92 - 1.19k \tag{1}$$
$$- 0.37k^2)]s(X)/E(D).$$

For a given fill rate target, the safety-stock factor is found by solving equation (1).

The average inventory level is the sum of safety and cycle stock, and is given by $ks(X) + E(D)/2$.

REFERENCES

COHEN, M. A. and LEE, H. L. 1988, "Strategic analysis of integrated production-distribution systems: Models and methods," *Operations Research,* Vol. 36, No. 2, pp. 216–228.

COHEN, M. A. and LEE H. L. 1989, "Resource deployment analysis of global manufacturing and distribution network," *Journal of Manufacturing and Operations Management,* Vol. 2, No. 2, pp. 81–104.

DEAN, J. W., Jr. and SUSMAN, G. I. 1989, "Organizing for manufacturable design," *Harvard Business Review,* Vol. 67, No. 1 (January-February), pp. 28–36.

NAHMIAS, S. 1989, *Production and Operations Analysis,* Richard Irwin, Homewood, Illinois.

TAGUCHI, G. and CLAUSING, D. 1990, "Robust quality," *Harvard Business Review,* Vol. 68, No. 1 (January-February), pp. 65–75.

WHITNEY, D. E. 1988, "Manufacturing by design," *Harvard Business Review,* Vol. 66, No. 4 (July-August), pp. 83–91.

Chapter Eight

Distribution Inventory Management

INTRODUCTION

This chapter deals with inventory control issues arising from the reality that customers are not conveniently located next to the factory. Often, inventory must be stored in several locations, as shown in Figure 8.1.

We are concerned here with place utility. A customer wants to buy a portable television, for example, from a local store. Because shipment is not instantaneous, the local store will hold an inventory of televisions. When the local store needs resupply, the store wants immediate delivery from a company in its vicinity. Otherwise, delivery might take too long for the customer to wait. How often have you been told that your order will take two weeks because the store is out of stock and the factory is 3000 miles away on the opposite coast, or in Taiwan or Stuttgart?

Immediate supply can be guaranteed from 3000 miles away if the company is willing to pay expediting costs. If a company supplies high-dollar-value, small items, such as medical sutures, orchids, and printed circuit boards, expediting may be appropriate. Normally, a company that is able to supply items quickly has a marketing advantage. Such rapid response can be achieved either through information and transportation or by holding inventories close to the customers at the time the customers want the items.

Referring to Figure 8.1, the factory warehouse provides inventory storage at the production site. The regional distribution centers may be located in the Northeast, the Southeast, and so on, to serve customers in the various regions of the country. About five or six can cover the United States within twelve hours of 90% of the population. Finally, the local service center is closest to the customer. A service center in each metropolitan area might distribute supplies to retail outlets. In this scheme, the company owns the factory, the factory warehouse, the regional distribution center, and the local service center.

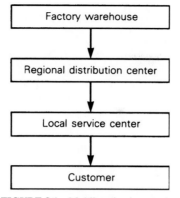

FIGURE 8.1   Multilocation inventories.

Our objective in this chapter is to explain the trade-offs involved in holding inventories at various locations so that the reader can appraise the advantages and disadvantages of such systems as distribution requirements planning (DRP). The weaknesses of single-item, single-location inventory control methods should be apparent when they are applied in the multilocation setting. The main issues are (1) where to have warehouses and what to stock and (2) how to replace stocks, given the answer to the first issue.

MULTILOCATION SYSTEM DEFINITIONS

Various multilocation systems can be classified using electronics and forestry analogies (see Figure 8.2). *Arborescent systems* have branches spreading apart, with the products flowing to different branches. *Coalescent systems* have materials coming together into one end item. *Series systems* have locations feeding each other in a direct path.

Several cycles or lead times make up the lead time separating the customer from the original raw materials. A firm with short lead times can respond more quickly to customer desires. The *procurement* cycle is the lead time necessary for the plant to obtain raw materials from its suppliers. The replenishment lead time is the time it takes to replenish stock at the distribution center, from the time necessary to place an order with the factory through the time of receiving the order. Scheduling, production, and shipment time make up the factory's portion of replenishment lead time, unless the factory is buffered by finished goods stock. Finally, the *order* cycle is the time taken between the distribution center or the service center and the retail outlet.

Lead Time	Starting Event	Ending Event
(a) Procurement	Factory order or vendor	Receivie raw materials
(b) Production	Generate work order	Receive finished goods at factory

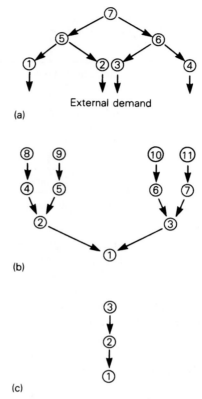

(a)

External demand

(b)

(c)

FIGURE 8.2 Multilocation systems: (a) arborescent distribution system (arrows point in direction of flow to customer); (b) coalescent assembly system; and (c) series system.

| (c) Distribution or replenishment | Generate distribution order | Receive finished goods at warehouse |
| (d) Order | Generate order | Receive finished goods at next echelon |

Measures of multilocation inventory system performance parallel those we found in single-location inventory problems:

1. *Fill rate,* or percent unit service, gives the average fraction of unit demand satisfied from stock on hand (eg, 90%).
2. *Fills* are the number of units demanded and satisfied per unit time; that is, fills = (fills rate)(demand rate). For example, given a 90% fill rate and a demand rate of twenty units per week, we have an average of eighteen fills per week (0.9 × 20 = 18).
3. *Expected number of backorders* is the time-weighed average number of backorders outstanding at a stocking location. Including times of zero backorders,

this measure depends on the fill rate—that is, the probability of *no* backorders at a random point in time. Expected backorders per year = (% time when no backorders possible) (0) + (% time when backorders possible)(average back-orders).

Example 8.1

Suppose that annual demand is 1000 units (D = 1000), the order quantity is 100 units (Q = 100), and the maximum backorder position is 10 units (B = 10). What is the expected number of backorders?

Solution: On any cycle, the average backorder position is $B/2$ = 5. The percentage of time when backorders are possible is B/Q = 10/100 = 0.1. Hence, the expected number of backorders per year is

$$\frac{B}{Q} * \frac{B}{2} = \frac{10}{100} * \frac{10}{2} = 0.5$$

4. *Expected delay* is the average time necessary to satisfy a unit of demand. The unit may be supplied instantaneously, or there may be some wait for expediting. Expected number of units backordered = (expected delay)(demand rate). This relationship comes from simple pipeline flow theory, in which the expected number of customers in the system equals the arrival rate times the expected waiting time. For example, if two customers per minute arrive at a car wash and if each customer's car takes six minutes to move through the waiting line and be washed, then there will be an average of 2 * 6 = 12 customers in the system either waiting or being washed. Similarly, if the expected number of backorders in an inventory system is twelve and two units are demanded per day, the average time to supply each unit must be six days.

Example 8.2

Suppose that annual demand is 1000 units (D = 1000), the order quantity is 100 units (Q = 100), and the maximum backorder position is 50 units (B = 50). What is the expected delay or the average time to satisfy a unit of demand?

Solution: On a 250 working day year, the demand rate is d = 1000/250 = 4 units per day.

$$\frac{\text{Expected number of backorders}}{\text{Demand rate}} = \frac{(B/Q)(B/2)}{d} = \frac{B^2}{2dQ}$$

$$= \frac{(50/100)(50/2)}{4} = \frac{2500}{800}$$

$$= 3.125 \text{ days}$$

5. *Inventory holding cost* is the cost of holding inventory, including insurance, obsolescence, deterioration, property taxes, and the cost of capital. Inventory holding cost may also be a management policy variable.
6. *Setup costs* and *ordering costs* are the costs of preparing or receiving an order. In the factory, these are the costs of setting up the equipment. Elsewhere, they are the clerical costs of order processing and receiving (including inspection).
7. *Stockout costs* are the costs of lost sales or backorders. In the lost sales case, revenues are foregone. With backorders, *expediting, transshipment* from an alternative supplier, or *substitution* of a similar item in stock all create costs in addition to the costs of *delay*. Stockout costs may not be observed in any operational way.
8. *System stability* costs are costs associated with overreaction to changes in demand rates. These costs are also usually nebulous.
9. Several of the measures can be found in Brown [4].

Echelon Inventory and Echelon Holding Costs

For analytical purposes, *echelon inventory* at a stocking point includes all inventory that either is at that stocking point or has passed through that stocking point. Defined by Clark and Scarf [5] in 1960, this concept was picked up by Orlicky [11, p. 54], who argued that "for a low-level inventory item, the quantity that exists under its *own identity,* as well as any quantities existing as *(consumed)* components of parent items, parents of parent items, etc. must be accounted for."

The *echelon holding cost rate* at a given inventory stocking point is the incremental cost of holding a unit of system inventory at that stocking point rather than at an earlier or predecessor point. For example, the echelon holding cost at a service center would be the incremental cost of holding inventory at the service center rather than at the regional warehouse.

Now we have arrived at a key point. From an accounting standpoint, the same total dollar holding cost will be computed whether the inventory analyst uses conventional holding rates applied to conventional inventory or echelon holding cost rates applied to echelon inventory. Economic order quantity calculations will be quite different, though, and may not apply at all to field stocks.

Example 8.3

In a serial system, suppose that there is an average of 100 units of inventory at Stage 2 and 80 units at Stage 1. Under both the conventional and the echelon system, total holding costs would be $2200. In Table 8.1 the total inventory holding costs are calculated by both methods. Stage 1 is the stage closest to the customer.

Example 8.4

In a serial system, suppose that Stage 2 has an annual demand of 2000 units, a setup cost of $500, and conventional holding costs of $10 per unit per year. Standard EOQs and simple echelon EOQs are shown in Table 8.2. The simple echelon EOQ uses the echelon holding rate.

TABLE 8.1 Equivalence of Echelon and Conventional System Costs

Stage	Conventional Holding Rate	Echelon Holding Rate	Average Inventory	Average Echelon Inventory	Conventional Holding Cost	Echelon Holding Cost
2	$10	$10	100	180	$10(100) = \$1,000$	$10(180) = \$1800$
1	$15	$ 5	80	80	$15(80) = \$1,200$	$5(80) = \$400$
Total					$2,200	$2200

TABLE 8.2 Simple Echelon EOQ versus Conventional EOQ

Stage	Conventional Holding Rate	Echelon Holding Rate	Conventional EOQ	Simple Echelon EOQ	Demand and Setup Data
2	$10	$10	$\sqrt{\dfrac{2(2000)500}{10}} = 447$	$\sqrt{\dfrac{2(2000)500}{10}} = 447$	$D_2 = 2000$ $S_2 = \$500$
1	$15	$ 5	$\sqrt{\dfrac{2(2000)100}{15}} = 164$	$\sqrt{\dfrac{2(2000)100}{5}} = 283$	$D_1 = 2000$ $S_1 = \$100$

We have found, then, that echelon rates wok well from an accounting point of view while they isolate the setup cost versus holding cost trade-off at each stage. The EOQs should not be blindly calculated at Stage 1, since the conventional holding rate would cause us to hold too little inventory.

More complicated EOQ formulas have been derived [6] to handle the problem of interaction between lot sizes at various stages in the system. System-myopic heuristics treat two stages at a time. For our purposes, however, the main lesson is the impact of echelon holding rates on the geographic distribution of inventory.

Several features of an echelon holding rate system center on a general shift of inventories toward the customer. Comparing echelon systems with conventional systems, we find the following:

1. The same size and frequency of orders will be placed by the stocking point farthest from the customer, because the conventional and the echelon rates will be the same at this level.
2. The same total system inventory will be held. The larger lot sizes at the levels closer to the customer indicate that lots will not stay long at predecessor levels. This is illustrated in Figure 8.3. If the lot size at Stage 1 is increased, the average inventory at Stage 2 will decrease.
3. Inventories will be shifted toward the retail level.

INDUSTRIAL DYNAMICS

When production is separated from demand by several echelons, each with its own ordering rules and cycles, oscillations become amplified through the system. On an upswing in sales, for example, all service centers are likely to place their orders quickly in batch sizes large enough to contend with the upward trend. During a downswing, their inventories will carry them for a longer time.

As an illustration of the amplification phenomenon, we consider the simple accelerator effect from retail sales to service center inventories.

Example 8.5

A service center seeks to maintain inventories at double the level of current sales; that is,

$$I_t^* = 2S_t$$

Lead time at the service center is one month.

Consider sales as shown in Table 8.3. Using an order decision rule that attempts to restore inventory to the desired level of twice current sales, the increase in sales is magnified in the service center's ending inventories. Table 8.3 show this amplification.

Two root causes for amplification of fluctuations can be distinguished: (1) the ordering policy and (2) lead times. Table 8.4 shows the same sales leading to smaller fluctuations in service center inventories because there is no time lag between ordering and receiving the inventory.

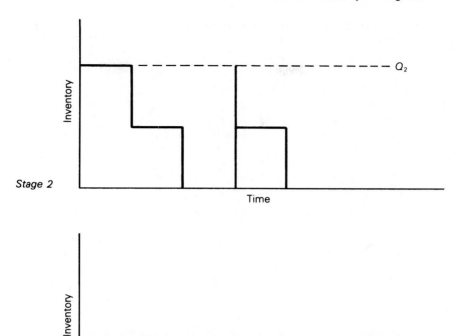

FIGURE 8.3 Inventory in a series system.

CENTRALIZATION OF INVENTORIES

So far, we have been concerned mainly with order decision rules in the multilocation setting. There is also a question, however, of where to locate safety stocks. This section examines order decision rules and safety stock rules together to demonstrate their combined pressure to centralize inventories. A later section will isolate safety stock considerations in an attempt to allocate stocks through the system.

Consider a single stocking site with a *(Q, R)* system. Using notation similar to that in earlier chapters, D = annual demand, S = setup cost, h = holding cost per unit per year, σ = standard deviation of lead time demand, k = number of standard deviations of lead time demand used to determine safety stock, SS = safety stock, and TRC = total relevant cost. The relevant annual cost equation would be

$$\text{TRC}(Q, R) = \frac{D(S)}{Q} + \frac{Q(h)}{2} + h(\text{SS}) \tag{8.1}$$

TABLE 8.3 Inventory Amplification with Order Lead Time

Month	Beginning Inventory (B_t)	Retail Sales (S_t)	Desired Inventory (I_t^*)	Ending Inventory (I_t)	Order Size (O_t)
0					200
1	400	200	400	200	200
2	400[a]	200	400	200	200
3	400	200	400	200	200
4	400	210	420	190	230
5	390	220	440	170	270
6	400	230	460	170	290
7	440	230	460	170	290
8	460	230	460	230	230
9	520	230	460	290	170
10	520	220	440	300	140
11	470	210	420	260	160
12	400	200	400	200	200
13	360				

[a]Open order for 200 units arrives

Noting that SS = $k\sigma$ and letting $Q = \sqrt{2DS/h}$, we can rewrite the relevant cost equation as

$$TRC(Q, R) = \sqrt{2ShD} + hk\sigma \qquad (8.2)$$

Compare this with Problem 19 in Chapter 4.

If we now had N such stocking points, each with annual demand D_i and standard deviation σ_i, the total *decentralized* relevant cost equation would be

TABLE 8.4 Inventory Amplification without Order Lead Time

Month	Beginning Inventory (B_t)	Retail Sales (S_t)	Desired Inventory (I_t^*)	Ending Inventory (I_t)	Order Size (O_t)
1	400	200	400	200	200
2	400	200	400	200	200
3	400	200	400	200	200
4	400	210	420	190	230
5	420	220	440	200	240
6	440	230	460	210	250
7	460	230	460	230	230
8	460	230	460	230	230
9	460	230	460	230	230
10	460	220	440	240	200
11	440	210	420	230	190
12	420	200	400	220	180

$$TRC = \left(\sqrt{2Sh} \sum_{i=1}^{N} \sqrt{D_i} \right) + \left(hk \sum_{i=1}^{N} \sigma_i \right) \tag{8.3}$$

assuming the same h and k for all stocking points.

By comparison, we could centralize these inventories at one location. Ignoring transportation costs, the total relevant costs would be

$$TRC = \sqrt{2Sh} \sqrt{D} + hk\sigma \tag{8.4}$$

where

$$\sqrt{D} = \sqrt{\sum_{i=1}^{N} D_i} \leqslant \sum_{i=1}^{N} \sqrt{D_i}$$

and

$$\sigma = \sqrt{\sum_{i=1}^{N} \sigma_i^2} \leqslant \sum_{i=1}^{N} \sigma_i$$

assuming independence between locations.

Example 8.6

Two inventory locations have annual demands and costs as shown.

S	h	D	k	σ	$\sqrt{D}$
100	10	1000	1.64	50	31.62
100	10	2000	1.64	50	44.72
		3000		70.7	54.77

The *decentralized* system is given by equation (8.3):

$$TRC = \sqrt{2(100)10} \ (\sqrt{1000} + \sqrt{2000}) + 10(1.64)(50 + 50)$$

$$= 5054$$

For the *centralized* system, equation (7.4) gives

$$TRC = \sqrt{2(100)10} \ \sqrt{3000} + 10(1.64)70.7$$

$$= 3609$$

The advantages of centralization hinge on two separate effects:

1. The square root of total demand is less than the sum of the square roots of individual location demand.
2. The standard deviation of total lead time demand is less than the sum of the standard deviation of individual location demand.

Although the quantitative treatment of distribution inventory planning points out the advantage of centralization of inventories, there could be some valid reasons for stocking items at multiple regional distribution centers instead of a central warehouse. It is evident from Example 8.5 as well as from Table 8.3 and 8.4 that amplification in inventory is inevitable as the lead time increases. We already know that as the number of levels in the distribution system increases, the total lead time also increases. In many instances it is difficult to provide quick response to customer orders unless orders are shipped right away. To avoid exorbitant shipping and handling costs or to provide satisfactory customer service levels, a firm may decide to carry inventories at more than one location. The additional inventory costs may also be offset by the economies of consolidating shipments to the distribution center.

SAFETY STOCKS

Centralization pressure on both lot size and safety stocks tells only part of the story. The remainder of the story centers on the measure of stockouts. If a stockout incidents measure is used, there is no incentive to maintain backup stock at the distribution center to speedily handle stockout situations at the service center. Only under a time-weighted stockout measure does backup stock make sense—that is, only when it matters how long we are out of stock.

Even with a time-weighted stockout measure, backup stock has only an indirect effect on expected delay at stage 1 or on the fill rate at stage 1. Consider a two-stage system, with stage 1 as the retail outlet and stage 2 as the service center. Recall that the fill rate (F_1) is the fraction of retail unit demand satisfied from stock on hand—that is, without delay. Customer expected delay (T_1) is the average time to fill a unit of demand at the retail level.

Service Center	*Retailer*
Stage 2 $\longrightarrow$	Stage 1 $\longrightarrow$
$D_2 = 1000$	$D_1 = 1000$
$B_2 = 20$	$S_1 = 20$
$h_2 = 5$	$h_1 = 5$
$Q_2^* = 200$	$Q_1 = 200$
$d = 4$ per day (250-day year)	$d = 4$ per day
$L_2 = 10$ days	$L_1 = 5$ days

For the moment, consider the time delay at stage 2 only. If there is safety stock at stage 2, the expected time delay would be zero. All orders placed on stage 2 by stage 1 should be met from stock. If stage 2 has negative safety stock, however, such as a maximum backorder position of $B_2 = 20$, there will certainly be a delay in filling such orders. The average backlog would be $B_2/2 = 20/2 = 10$.

We have already shown in this chapter that the expected delay can be calculated as expected backorders/demand rate. Letting T be the expected delay and using the concept of Example 8.2, we have

$$T_2 = \frac{B_2^2}{2dQ_2}$$

In our example, the expected delay would be given as

$$\frac{B_2^2}{2dQ_2} = \frac{20^2}{2(4)200} = 0.25 \quad \text{or one-quarter day}$$

At stage 1, the effective stage 1 safety stock (ESS_1) is stage 1 safety stock minus the expected demand during expected stage 2 delay: $\text{ESS}_1 = S_1 - dT_2$; for example, $\text{ESS}_1 = 20 - 4(0.25) = 19$.

If trade-offs are to be made between reducing the expected backorders at stage 2 and increasing safety stocks at stage 1, what should be done? A unit of stock added to stage 2 would reduce the maximum backorder position to nineteen units and the expected stage 2 delay would be

$$T_2 = \frac{19^2}{2 * 4 * 200} = 0.23$$

Hence, the effective safety stock would be

$$\text{ESS}_1 = 20 - 4(0.23) = 19.08 \text{ units}$$

On the other hand, increasing stage 1 safety stock by one unit would increase stage 1 effective safety stock by one unit:

$$\text{ESS}_1 = 21 - 4(0.25) = 20$$

These findings are consistent with the material requirements planning (MRP) philosophy that safety stock should be kept at end item levels only, not at intermediate levels. In distribution systems, however, these findings must be balanced against centralization pressures.

DISTRIBUTION INVENTORY SYSTEMS

Now that we have uncovered some of the principles of the multilocation inventory problem, we are ready to examine some systems in use. Each of these systems solves the problem by ignoring some of its aspects and concentrating on others.

Level Decomposition Systems

Level decomposition systems set aggregate service level objectives for all items at an echelon. For example, the objective at the main distribution center might be 95% service, interpreted as a 95% fill rate. With n items in the inventory, the problem can be stated as follows:

$$\text{Minimize} \sum_{i=1}^{n} \left(\begin{array}{c} \text{unit value} \\ \text{on item } i \end{array} \right) \left(\begin{array}{c} \text{safety stock} \\ \text{on item } i \end{array} \right)$$

$$\text{subject to} \sum_{i=1}^{n} \left(\frac{\text{item demand rate}}{\text{aggregate demand rate}} \right) \left(\begin{array}{c} \text{item} \\ \text{fill rate} \end{array} \right) \geqslant 0.95$$

Example 8.7

Two items (A and B) are valued at \$10 and \$15, respectively, with demand rates of three per day and seven per day. With an aggregate service level objective of 0.95, the problem would be: Minimize $10s_1 + 5s_2$ subject to 0.3(fill rate on A) + 0.7(fill rate on B) $\geqslant 0.95$. If the fill rate on item A were 0.90, the fill rate on item B would have to be 0.97. With unit service objectives on A and B, we could calculate the safety stocks required for these item fill rates and then evaluate the costs in the objective function.

Although this problem can be solved using the Lagrange multiplier technique, our objective here is to describe the basics of the system, not the mathematics. Level decomposition systems will have some items with high fill rates and some with low, depending on the costs involved. Items with high fill rates at one echelon will wind up with high fill rates at another echelon. Safety stocks will be duplicated. In level decomposition systems, no mechanism allows safety stocks at one echelon to be related to safety stocks at another echelon.

Multiechelon Systems

Multiechelon systems, sometimes called differentiated distribution or item decomposition systems, focus on effective safety stock. Muckstadt and Thomas [10] examine multiechelon methods applied to low-demand-rate items. Because the mathematics of such systems is complex, we present only simple differentiated distribution concepts and encourage the interested reader to pursue the Muckstadt and Thomas [10] article.

Heskett, Glaskowsky, and Ivie [7] mention that dual (differentiated) distribution systems are the outgrowth of ABC inventory policies. The ABC system differentiates

TABLE 8.5   Differentiated Distribution System

Local Service Level	Type of Item	Where Located	Central Service Level
High	High volume/high price	Locally	Low
High	High volume/low price	Locally[a]	Low
Low	Low volume/high price	Centrally[a]	High
Low	Low volume/low price	Centrally	High

[a]Ideal candidates for these locations.

items on the basis of such variables as sales volume, unit value, or customer importance. Rather than the standard ABC breakdown by dollar volume, a breakdown by unit volume makes sense for distribution systems. The effect of such a breakdown is shown in Table 8.5. With such a breakdown, single-location methods can be used to determine appropriate ordering policies.

Distribution problems are very common in everyday life. Simple versions, of course, are cases where each source and destination is handled as an individual set. In those cases, the total cost of setups, inventory, and shipping can be minimized. The next level of complexity occurs when a number of sources send different shipments to a single destination or when a single source is sending shipments to several destinations. This is illustrated in Figure 8.4.

The logical way to solve this situation is to look at each link containing the source and a destination and to solve individual links for optimality. When multiple sources are sending shipments to several destinations, however, solving these problems optimally becomes more complex. The situation is illustrated in Figures 8.5 and 8.6. As the number of levels in the system increases, obviously the problem becomes hopelessly difficult to solve optimally. Blumenfeld et al. [2] have approached this problem from a practical point of view with General Motors automotive components shipments. Although not optimal, their research presents a simple method of solving these cases. Their solution

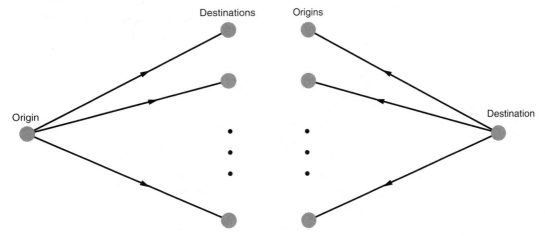

FIGURE 8.4   (a) One origin with several destinations. (b) Several origins with one destination.

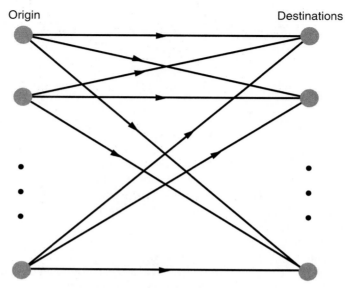

Origin   Destinations

FIGURE 8.5   Several origins each with several destinations.

methodology is similar to the formulas we derived in the multi-item inventory situation in Chapter 5. In this section we explore their models with some simple problems. We also point out the similarity of results between Blumenfeld et al. [2] and multi-item formulas considered in Chapter 5.

One Origin Shipping to Many Destinations

This problem addresses a case where one origin ships items to many destinations. In those cases where each destination requires a different item, it is necessary to change production regularly so as to satisfy the demand of each destination. If we assume that

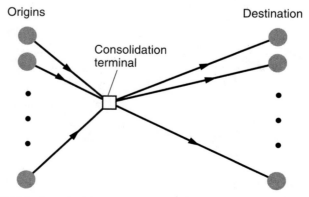

Origins   Destination

Consolidation
terminal

FIGURE 8.6   Several origins with a consolidation terminal to several destinations.

the items are consumed or are sold at a constant rate (the same assumptions made in the economical order quantity formulas), then the problem faced here is same as the joint replenishment economical production quantity problem, considered in Chapter 5, with the assumption that each item is dispatched to a different destination. Therefore, the same formula can be adopted here.

Several Origins Shipping to One Destination

This problem addresses a case where different items are made at different locations and are shipped to a single location. In this case the economical production quantity for each origin can be computed independently because each origin is producing a different item. Therefore, the single item EOQ and EPQ (economical production quantity) formula discussed in Chapter 4 can be adopted here.

Several Origins Shipping to Several Destinations

Problems involving multiple origins and multiple destinations are difficult to solve for optimality. They can be complicated further when the same origin manufactures more than one product. Muckstadt and Roundy [9] and Anily and Federgruen [1] address these problems using mathemantical programming and network theory techniques. Such techniques are beyond the realm of this text; the interested reader is referred to their work.

This type of problem becomes manageable if we assume that all origins ship their products to a single consolidation terminal and that all items are distributed to destinations as demanded from the consolidation terminal (see Figure 8.6). The formulas developed by Blumenfeld et al. [2] are by no means optimal; they provide, however, a practical solution to this problem by simplifying many underlying assumptions. Next we present their algorithm with an example. Let us define the following:

d_{ijk} = quantity of demand from origin or source i for destination j for product k

p_k = cost or price per unit of item k

$D_{i,k}$ = demand at source i for item k from all destinations

$$= \sum^{j} d_{ijk}$$

S_{ic} = freight cost of a load from source i to consolidation terminal c

S_{ck} = freight cost of a load from consolidation terminal c to destination k

W_{ic} = capacity of vehicle from source i to consolidation terminal c

W_{ck} = capacity of vehicle from consolidation termination c to destination k

T_{ic} = lead time/travel time from source i to consolidation terminal c

T_{ck} = lead time/travel time from consolidation terminal c to destination k

F_{ic} = total quantity of items flowing per period from source i to consolidation terminal c

$$= \sum^{j} \sum^{k} d_{ijk}$$

F_{cj} = total quantity of items flowing per period from consolidation terminal c to destination k

$$= \sum_{i} \sum_{k} d_{ijk}$$

I = inventory carrying percentage

Given these definitions, assuming fifty periods per year, the economical shipment quantity Q_{ic} from source i to consolidation terminal c and the economical shipment quantity Q_{cj} from the consolidation terminal c to destination j can be derived [2].

The shipping quantity from source i to consolidation terminal c is given by the following:

$$\min[Q_{ic}, W_{ic}]$$

and the shipping quantity from the consolidation terminal c to destinations is given by the following:

$$\min[Q_{cj}, W_{cj}]$$

where

$$Q_{ic} = \left[\frac{S_{ic}\left[\sum_{j} \sum_{k} d_{ijk}\right]50}{I\left(\sum_{j} \sum_{k} p_k d_{ijk} \Big/ \sum_{j} \sum_{k} d_{ijk}\right)} \right]^{1/2}$$

and

$$Q_{cj} = \left[\frac{S_{ck}\left[\sum_{i} \sum_{k} d_{ijk}\right]50}{I\left(\sum_{i} \sum_{k} p_k d_{ijk} \Big/ \sum_{i} \sum_{k} d_{ijk}\right)} \right]^{1/2}$$

Although the formulas appear to be complicated, they are actually simple in concept and merely mirror the formula derived in Chapter 5 for multi-item situations. The numerator contains the setup cost and the total demand summarized from all destinations. The denominator consists of the average cost per part and the inventory carrying percentage. In summary, the total demands from all sources are added and the economical order quantity is calculated using a slightly different formula. It can be proven easily that the results are identical, except that the formulas presented here deal in quantities instead of dollar values and a missing 2 in the denominator. The order quantity is determined as the smaller of the vehicle capacity or the economical order quantity provided by the formula. Then the individual quantities are prorated according to their demand pattern.

TABLE 8.6 Destination Demand and Origin Capacity

Product	Cost per Unit	Demand per Period at: Destination 1	Destination 2	Source Location
1	20	8	4	1
2	25	6	10	1
3	25	5	8	2
4	30	6	8	2

Example 8.8

The demand for products at destinations 1 and 2 and the source of these are presented in the table. The capacity of vehicles, relevant setup costs, lead time between locations, and other pertinent data are given in Tables 8.6 and 8.7.

Assume that the inventory carrying charges amount to 20% and that the firm operates fifty periods (weeks) per year. Find the economical quantity to ship from each source to the terminal and from the terminal to each destination.

Solution:

Step 1: Calculate the total annual demand for items 1 and 2 in units for source 1.

$$[8 + 6 + 4 + 10] * 50 = 1400$$

Step 2: Calculate the average cost per part at source 1.

$$\frac{12 * 20 + 16 * 25}{28} = 22.86$$

Step 3: Calculate the economical shipment quantity flowing from source 1 to the consolidation terminal.

$$Q_{1c} = \left(\frac{45 * 1400}{0.20 * 22.86} \right)^{1/2} = 118$$

TABLE 8.7 Setup Cost, Vehicle Capacity, and Lead Time Data

From To	Source 1 Terminal	Source 2 Terminal	Terminal Destination 1	Terminal Destination 2
Setup cost	45	25	30	35
Vehicle capacity	150	200	150	100
Lead time (days)	4	2	3	4

Step 4: Find the minimum of Q_{1c} and W_{1c}.

$$s_{1c} = \min[118, 150] = 118$$

Step 5: Calculate the quantities of individual items 1 and 2 flowing from origin 1 to the consolidation terminal.

$$q_{1c1} = \frac{s_{1c} \overset{j}{\underset{}{*}} \overset{j}{\sum} d_{1j1}}{\overset{j}{\underset{}{\sum}} \overset{k}{\underset{}{\sum}} d_{ljk}}$$

$$= \frac{118 \ * \ 12}{28} = 51$$

$$q_{1c2} = \frac{118 \ * \ 16}{28} = 67$$

Step 6: Repeat the calculations for source 2. Following the same procedure, we obtain the quantities of individual items 3 and 4 flowing from origin 2 to the consolidation terminal.

$$Q_{2c} = 78$$

$$s_{2c} = \min[78, \ 200] = 78$$

$$q_{2c3} = \frac{78 \ * \ 13}{27} = 38$$

$$q_{2c4} = \frac{78 \ * \ 14}{27} = 40$$

The same procedure can be followed to obtain the shipment quantities from the consolidation terminal to destinations 1 and 2, respectively.

Step 1: Calculate the total annual demand for all items at destination 1.

$$[8 + 6 + 5 + 6] * 50 = 1250$$

Step 2: Calculate the average cost per part at destination 1.

$$\frac{[20 \ * \ 8 + 25 \ * \ 6 + 25 \ * \ 5 + 30 \ * \ 6]}{25} = 24.6$$

Step 3: Calculate the economical shipment quantity for the total flow from the consolidation terminal to destination 1.

$$Q_{c1} = \left(\frac{30 * 1250}{0.2 * 24.6} \right)^{1/2} = 87.3$$

Step 4:   Find the minimum of Q_{c1} and W_{c1}.

$s_{c1} = \min[87.3, 150] = 87.3$

Step 5:   Calculate the quantities of individual items 1 through 4 flowing from the consolidation terminal to terminal 1. We rounded the digits below.

$$q_{c11} = \frac{s_{c1} * \overset{i}{\underset{k}{\Sigma}} d_{i11}}{\overset{i}{\underset{k}{\Sigma}} \overset{k}{\Sigma} d_{i1k}}$$

$$= \frac{88 * 8}{25} = 28$$

$$q_{c12} = \frac{88 * 6}{25} = 21$$

$$q_{c13} = \frac{88 * 5}{25} = 18$$

$$q_{c14} = \frac{88 * 6}{25} = 21$$

Step 6:   Repeat the calculations for destination 2. Following the same procedure, we obtain the quantities of items 1 through 4 flowing from the consolidation terminal to terminal 2.

$$Q_{c2} = 102$$

$$s_{c2} = \min [102, 100] = 100$$

$$q_{c21} = \frac{100 * 4}{30} = 13$$

$$q_{c22} = \frac{100 * 10}{30} = 33$$

$$q_{c23} = \frac{100 * 8}{30} = 27$$

$$q_{c24} = \frac{100 * 8}{30} = 27$$

Calculation of Safety Stock
at Various Locations

Once we calculate the flow of individual item quantities for every shipment, it is possible to write a simulation program either in a spreadsheet format or by using a BASIC/FOR-TRAN program for calculating the quantity of individual item safety stocks for every location.

Distribution Requirements Planning
(DRP) Systems

A unique problem exists for those firms producing items in a manufacturing facility and shipping them to several regional distribution centers. In this environment each regional distribution center makes its own forecast. In traditional independent inventory management systems, each location receives a fixed quantity of items at certain intervals that may be based on the past average demands. It is possible that at any given time some distribution centers are experiencing higher-than-average demand while others are facing lower-than-average demand. Then certain centers are left with excess inventories while others have stockouts. If the requirements for each center are known, DRP offers a set of tools for avoiding inventory buildup in a location when it is not really necessary. The distribution requirements planning system enables a firm to better anticipate future needs in various distribution centers and to coordinate the material supply more closely to meet the actual forecasts of demand and to make adjustments to the production schedule more rapidly when changes in the marketplace occur.

Distribution requirements planning (DRP) simply translates material requirements planning (MRP) logic to a distribution system. In what follows, we assume that the reader has at least a rudimentary knowledge of MRP. If you have not studied MRP, please skip ahead to Chapter 11 and read the sections on the parts requirement problem, the bill of materials, and the mechanics of MRP.

In essence, the planned shipment of orders are netted to obtain the projected on-hand inventories for each distribution center and then summed into a total warehouse demand at the plant. The production planning system at the plant uses this data to plan shipments of orders that, in turn, are also netted to obtain a projected on-hand inventory in the plant. Thus the DRP system plays a central coordinating role by providing a linkage between the marketplace demand and the production schedule with a time-phased information on field inventories as demand patterns change.

Consider a system in which a distribution center in North Carolina supplies service centers in Connecticut and California. Lead time from North Carolina to California is three weeks; to Connecticut, one week. The North Carolina distribution has a one-week replenishment lead time.

Even in this simple example, we will see both the attraction and the oversimplification of the DRP approach. Clearly, the attraction lies in treating the North Carolina demand as dependent or derived demand. Rather than the conventional approach of using past demand at North Carolina to forecast future demand, we use only the derived demand as gross requirements. At the distribution center, then, no safety stock needs to be held against demand variance, only against replenishment lead time variance from the factory. All safety stock against demand uncertainty will be held at the service centers, as in the MRP approach.

Just as practitioners have criticized operations researchers for over-sophistication, practitioners have fallen into an oversimplification trap with DRP. Notice in Table 8.8 that EOQs have been used at the service centers. Technically, these EOQs should be calculated on echelon holding costs, not on standard holding costs. Except for transportation expense, there is then little difference between holding stock at the service center rather than at the distribution center. Hence, on-hand inventory at the distribution center ought to be shipped out as soon as possible to provide extra protection at the service centers.

In Table 8.9, we show some improvement by lot sizing at the factory only. At least we have pruned away meaningless lot sizes. Indeed, this version of DRP comes closest to what we think is the best approach to the problem—the "fair shares" approach invented by R. G. Brown in about 1960 in Akron, Ohio (well before DRP ever came on the scene).

TABLE 8.8 DRP Example with Lot Sizes

CAL—Service Center	0	1	2	3	4	5	6	7	8	9
Forecast		25	25	25	25	25	25	25	25	25
On hand	125	100	75	50	25	0	175	150	125	100
Planned order release				(200)						

$L = 3, Q = 200$

CONN—Service Center	0	1	2	3	4	5	6	7	8	9
Forecast		50	50	50	50	50	50	50	50	50
On hand	110	60	10	160	110	60	10	160	110	60
Planned order release			(200)				(200)			

$L = 1, Q = 200$

NC—Distribution Center	0	1	2	3	4	5	6	7	8	9
Gross requirements			200	200			200			
On hand	265	265	65	65	65	65	65	65	65	65
Planned order release			(200)			(200)				

$L = 1, Q = 200$

NC—Factory	0	1	2	3	4	5	6	7	8	9
Gross requirements			200			200				
On hand	300	300	100	100	100	400	400	400	400	400
Planned order release			(500)							

$L = 3, Q = 500$

Note: Example assumes that service center requirements continue at the forecast level indefinitely.

TABLE 8.9   DRP Example with Lot Sizes at Factory Only

CAL—Service Center	0	1	2	3	4	5	6	7	8	9	10	11	12
Forecast		25	25	25	25	25	25	25	25	25	25	25	25
On hand	125	100	75	50	25	0	0	0	0	0	0	0	0
Planned order release				25	25	25	25	25	25	25	25ᵃ	25	25

$L = 3$

CONN—Service Center	0	1	2	3	4	5	6	7	8	9	10	11	12
Forecast		50	50	50	50	50	50	50	50	50	50	50	50
On hand	110	60	10	0	0	0	0	0	0	0	0	0	0
Planned order release			40	50	50	50	50	50	50	50	50	50	

$L = 1$

NC—Distribution Center	0	1	2	3	4	5	6	7	8	9	10	11	12
Gross requirements			40	75	75	75	75	75	75	75	75	75	75
On hand	265	265	225	150	75	0	0	0	0	0	0	0	0
Planned order release						75	75	75	75	75	75	75	

$L = 1$

NC—Factory	0	1	2	3	4	5	6	7	8	9	10	11	12
Gross requirements						75	75	75	75	75	75	75	75
On hand	300	300	300	300	300	225	150	75	0	425	350	275	200
Planned order release						(500)							

$L = 3$. $Q = 500$

ᵃto cover period 13.

Fair Shares Allocation Systems

Examining the same situation with Brown's [3] fair shares allocation procedure, we see the difference between a push system and a pull system. DRP is a pull system; it depends on service center or retail EOQs to "pull" inventory through the system. Fair shares allocation is a push system, pushing inventory from the distribution center to the service center according to factory or distribution center stock levels. Both systems project requirements on the manufacturing location, giving it visibility in specifying when to produce.

Using the same example, we show the fair shares allocation approach in Table 8.10. To follow the table, consider the following fair shares allocation steps:

1. Perform lot-for-lot (new requirements by period) explosions from the service centers.
2. The gross requirements resulting at the distribution center or factory are called net shipping requirements. Any desired safety stocks should be added to give a final figure for net shipping requirements.

TABLE 8.10 Fair Share Example

CAL—Service Center	0	1	2	3	4	5	6	7	8	9	10	11	12
Forecast		25	25	25	25	25	25	25	25	25	25	25	25
On hand	125	100	75	50	100	175	150	125	100	75	50	25	0
Time-phased net (cumulative) requirement (monthly)				25	50	100	25	25	25	25	25	25	25
Planned receipt						(75)	100						

$L = 3$ | 75 covers through week 5

CONN—Service Center	0	1	2	3	4	5	6	7	8	9	10	11	12
Forecast		50	50	50	50	50	50	50	50	50	50	50	50
On hand	110	60	200	350	300	250	200	150	100	50	0	284	234
Time-phased net (cumulative) requirement (monthly)			40	90	140	190	50	50	50	50	50	50	50
Planned receipt			(190)	200								334	

$L = 1$ | 190 covers through week 5

fair share $= 334 = \dfrac{50}{75}(500)$

NC—Distribution Center	0	1	2	3	4	5	6	7	8	9	10	11	12
Net shipping (cumulative) requirements (monthly)			40	115	190	265	75	75	75	75	75	75	75
On hand	265	0	0	0	0	0	0	0	0	0	0	0	0
Planned shipment		(265)	300[a]									500	
Time-phased net requirement						75	75	75	75	75	75	75	75
Planned receipt			300								500		

$L = 1$ 265 covers through week 5

NC—Factory	0	1	2	3	4	5	6	7	8	9	10	11	12
Gross requirements						75	75	75	75	75	75	75	75
On hand	300	0	0	0	0	0	0	0	0	0	0	0	0
Planned shipment		300[b]								(500)			
Planned order release							(500)						

$L = 3$. $Q = 500$

[a]This planned shment includes a 200-unit fair share to CONN and a 100 fair share shipment to CAL.
[b]This planned shipment represents four weeks of shipping requirements (weeks 5 through 8).

3. Economic order quantity calculations or other trade-off calculations are made with regard to setups and scheduling at the factory. In our example, we show the factory producing in lot sizes of 500 units.
4. When inventory is available at the distribution center, fair shares are calculated. How many weeks of national (total) net shipping requirements can the

distribution center inventory support? A fair share for a warehouse is the net shipping requirements to that warehouse through the time that the source's available stock will cover national net shipping requirements. In Table 8.10, the 265 units on hand at the distribution center will support net shipping requirements through week 5. Time-phased net requirements through week 5 for California calls for a fair share of 75 units.

In distinguishing between DRP with lot-for-lot explosions and fair shares allocation, it is important to recognize that fair shares allocation separates shipping decisions from requirements explosions. Although a DRP system could be pathworked to mimic fair shares, a typical DRP system would plan on many small shipments to match the lot-for-lot explosions. Rather than parceling out inventory according to planned orders, fair shares allocates on-hand inventory according to up-to-date net requirements.

Underlying this difference between lot-for-lot DRP and fair shares allocation is a different approach to the use of information. A fair shares system waits until the last moment before determining shipment sizes. All available information is utilized. Inventory status at the source and *current* net shipping requirements at each warehouse are considered. DRP logic thus mirrors the real-world demand and converts the physical distribution system into a detailed production planning system.

Base Stock Systems

Whether fair shares or DRP techniques are used, low order rates at the service center level are quite common. DRP allows the service center to establish its own ordering rules. In a fair shares system, the service center simply passes along demand rate information, facilitating centralized production and distribution decisions. Because low order rates are so pervasive, we need to examine one final system based on decentralized ordering: the base stock system.

The base stock approach, known as the sell-one/but-one system, can be characterized as an (S,S) order up to policy. At one Sears location, for example, three sofas of a particular style might be kept in stock as a target level S. Should the store sell one, one more is ordered, because the inventory has dropped below the reorder point. With a two-week lead time, a recorder point (S) of three, and an average order rate of one unit every two weeks, the system might behave as shown in Table 8.11.

In about 1957, Kimball distinguished replenishment orders from base stock orders. Base stock orders modify net demand on the source but preserve information about end consumption. If we sell 100 but reduce stock by 10, the ordinary system orders 90. Base stock system orders would show a replenishment order of 100 and a base stock order of − 10, giving a net order of 90. The source can then see that the pipeline is changing, not end consumption.

For such a system, the decision variable S must be determined. With constant lead times, marginal analysis yields a familiar relationship. Let OSOR(S) be the probability of stockout as a function of the target level/reorder point S. Let π be the per unit stockout cost and h be the holding cost per unit per year. Then we have the condition for an optimal solution:

TABLE 8.11   Base Stock System

Time (Weeks)	On Hand (End of Week)	Sell	Order	Order Receipt
1	3	0		
2	2	1	1	
3	2	0		
4	3	0		1
5	2	1	1	
6	1	1	1	
7	2	0	0	1
8	3	0	0	1

Expected holding cost = expected stockout cost

$$h[1 - \text{OSOR}(S)] = \pi[\text{OSOR}(S)]$$

$$\text{OSOR}(S)\,(\pi + h) = h$$

$$\text{OSOR}(S) = \frac{h}{\pi + h}$$

But we cannot end this chapter with such a simple, single-location statement. First, note that constant lead times do not make much sense in our multiechelon structure, because a great deal depends on whether or not the supply echelon has the item in stock. We are again faced with the question of expected delay. Second, we should use echelon holding cost; but use of echelon holding cost will most probably result in an extremely low optimal stockout probability.

To date, the best work in this area has been done by Muckstadt and Thomas [10] and Brown [4]. At this time, no systematic comparison of the three main systems has been made, but this chapter's objective has been met if the reader now understands the main ideas behind DRP, fair shares, and base stock systems.

SUMMARY

This chapter surveyed multilocation inventory principles and systems. Recognizing that the larger problem encompasses the question of facilities location, we nevertheless have been content to examine questions of how best to allocate stock within a given system.

Can any conclusions be drawn from our study? As mentioned earlier, this field still has no uniformly accepted methods, but this chapter has examined some dimensions of the problem:

1. If we use EOQ concepts, they should be based on echelon holding costs. This approach questions simplistic use of DRP.
2. Because low order rates are so common, base stock concepts coupled with multiechelon methods ought to yield good answers. The mathematics involved

in this approach will probably preclude any practical applications unless someone can invent a simple DRP/fair shares–type system employing base stock concepts.

3. Safety stock generally should be kept at the levels closest to the customer, even though pressures for centralization arise from the smaller variance of central demand.

For further research in the field, the reader is referred to Schwarz [12] and Stenger and Cavinato [13].

PROBLEMS

1. Consider two products, A and B, supplied by a factory (stage 2) to a distribution center (stage 1). Costs and demand rates are as follows:

| | Stage 2 | | Stage 1 | | |
| | Conventional | | Conventional | | Demand |
Item	Holding Cost	Setup	Holding Cost	Setup	Rate
A	$10	$500	$15	$25	1000
B	$12	$500	$18	$25	2000

   a. What are the appropriate EOQs, using echelon holding rates?
   b. What are the average echelon holding costs at the two stages?
   c. What assumption about stage 2 demand rates makes your answer to part b questionable?

2. A service center seeks to maintain inventories at three times current sales. Lead time at the service center is one week. Sales are 100 units for the first four weeks and 200 units per week thereafter. With a beginning inventory of 300 units and an open order for 100 units to arrive at the beginning of week 2, create a table showing beginning inventory, sales, desired ending inventory, actual ending inventory, and orders for twelve months.

3. Two inventory locations have annual demands and costs as shown:

S	h	D	k	σ
$250	$10	2000	2.08	60
$250	$10	3000	2.08	60

   where S = setup cost, h = holding cost per unit per year, D = annual demand, k = safety factor, and σ = standard deviation of lead time demand. Assume that the centralized location would exhibit the same s, h, and σ.

   a. What dollar advantage comes from the EOQ effect on centralized inventory?
   b. What dollar advantage comes from the standard deviation effect on centralized inventory?
   c. What is the total dollar advantage of centralization versus decentralization?

4. In a two-stage system with a distribution center supplying a service center, the expected delay at the distribution center is one-half day. The demand rate at the service center is 20 units per day, the service center safety stock is 30 units, and the service center EOQ is 150 units.

  a. What is the service center effective safety stock?
  b. What is the expected delay at the service center?
5. Suppose that we have a factory supplying a distribution center. At the factory, annual demand for an item is 10,000 units, holding costs are $10 per unit per year, and the cost per setup is $750. How would you allocate 80 units of safety stock between the two locations?

Problems 6 through 11 use the following data: A factory in Chicago serves warehouses in Nashville and St. Louis. Normal lead time from the factory to either warehouse is one week. The Nashville warehouse faces demand of 40 units per week, and the St. Luis warehouse faces demand of 30 units per week, with order quantities of 200 units at each warehouse and on-hand quantities of 60 units and 40 units, respectively. The EOQ at the factory is 500 units, the replenishment lead time is one week, and there are 100 units on hand. For each problem, develop tables for all three locations in the format shown in Table 8.12.

6. Development the DRP factory shipment schedule for the next nine weeks.
7. Develop the fair shares factory shipment schedule for the next nine weeks.
8. Suppose that the warehouses use an order point system with safety stocks of ten units. The reorder point (ROP) for each warehouse would be

$$ROP_{Nashville} = 40 + 10 = 50$$

$$ROP_{St. Louis} = 30 + 10 = 40$$

Develop the order point factory shipment schedule for the next nine weeks.
9. Revise the DRP factory shipment schedule to include the ten units of safety stock at each warehouse.
10. Revise the fair share system to include the ten units of safety stock at each warehouse.
11. Ignoring safety stocks, develop the DRP factory shipment schedule, permitting lot sizing at the factory only.

TABLE 8.12 Table Format for Problems 6 through 11

Nashville	0	1	2	3	4	5	6	7	8	9
Gross requirements		40	40	40	40	40	40	40	40	40
On hand	60									
Planned order release										

St. Louis	0	1	2	3	4	5	6	7	8	9
Gross requirements		30	30	30	30	30	30	30	30	30
On hand	40									
Planned order release										

Chicago	0	1	2	3	4	5	6	7	8	9
Gross requirements										
On hand	100									
Planned order release										

12. The number of bin trips (N) to put away stock in the field can be represented by annual demand divided by order quantity (D/Q). Show that $N = \sqrt{(Dh)/(2S)}$, where h is the holding cost per unit per year and S is the setup cost.

13. Compare bin trips for the DRP system and a fair shares system. Assume that a factory supplies four warehouses such that annual demand is divided equally among the four. Further, each warehouse faces an ordering or acquisition cost of A, where A is considerably smaller than the factory setup cost S. What is the ratio of annual bin trips under fair shares to annual bin trips under DRP?

14. Suppose that annual demand is 6000 units, the order quantity is 500 units, and the maximum backorder position is 60 units. What is the expected number of backorders?

15. Suppose that annual demand is 600 units, the order quantity is 500 units, and the maximum backorder position is 60 units. What is the expected delay or the average time to satisfy a unit of demand?

16. A service center seeks to maintain inventories at exactly the three-month moving average of sales. Consider sales as shown, a current moving average of 200 units, and a beginning inventory of 400 units.

Monthly	1	2	3	4	5	6	7	8	9	10	11	12
sales	200	200	200	210	220	230	230	230	230	220	210	200

With a service center lead time of one month, develop a table to show beginning inventory, desired inventory, ending inventory, and order sizes. Ordering decisions are made at the end of each month according to the current sales information and that of the previous two months. Compare your results to Table 8.3 in the chapter. Can you draw any conclusions?

17. For two warehouses with equal demand, $D_1 = D_2$, national demand is $D = D_1 + D_2$. Show that the square root of national demand is less than or equal to the sum of the square roots of warehouse demand.

18. We saw that $ESS_1 = 5 - \lambda T_2$ represented effective stage 1 safety stock. Also,

$$T_2 = \frac{B_2^2}{2\lambda Q_2}$$

where B_2 is the maximum backorder position at stage 2, λ is the demand rate, Q_2 is the stage 2 order quantity, and T_2 is expected stage 2 delay. What are the partial derivatives of ESS_1 with respect to S_1, B_2, λ, and Q_2? Provide some justification for your results.

19. Using the simple backorder model from Chapter 4, we found that the expected delay at stage 2 was represented by

TABLE 8.13 Destination Demand and Origin Capacity Data

Item	Cost per Unit	Demand per Period		Origin or Source
		Destination 1	Destination 2	
1	15	80	40	1
2	20	60	100	2
3	18	50	80	2
4	10	60	80	1

TABLE 8.14 Setup Cost, Vehicle Capacity, and Lead Time Data

From To	Origin 1 Terminal	Origin 2 Terminal	Terminal Destination 1	Terminal Destination 2
Setup cost	450	250	300	350
Vehicle capacity	1500	2000	1500	1000
Lead time	4	2	3	4

$$T_2 = \frac{B_2^2}{2\lambda Q_2}$$

where B_2 is the maximum backorder position at stage 2, λ is the demand rate, and Q_2 is the stage 2 order quantity. What are the partial derivatives of T_2 with respect to B_2, λ, and Q_2^2? Provide some justification for your results.

20. The manager of a furniture department at Sofa's scoffs at quantitative approaches to management. Nevertheless, she always seems to maintain about three units of a certain style couch on hand. If one is sold, she replaces it. With three on hand, she is adamant that she has very little chance of stocking out. Because the couches have a wholesale value of $500 each ($1000 retail) and because Sofa's has investment opportunities at 12% annual return, the holding cost per couch appears to be $65 per year (including storage and the like). What, then, is the per unit stockout cost that exists only in the manager's intuition? As far as we can determine, "little chance" means 10% order stockout risk to the manager.

21. The demands for items 1 through 4 at destinations 1 and 2 from origins 1 and 2, pertinent costs, lead times, and capacity of vehicles are exhibited in Tables 8.13 and 8.14.
    The firm operates fifty weeks per year. Assuming the inventory carrying percentage is 25% per year, find the economical order quantities flowing from each origin to consolidation terminals and from the consolidation terminal to each destination.

22. Repeat problem 21 using the concept of the joint replenishment of inventories discussed in Chapter 5.

23. Repeat Example 8.8 in the text using the concept of the joint replenishment of inventories discussed in Chapter 5.

REFERENCES AND BIBLIOGRAPHY

1. S. Anily and A. Federgruen, "One Warehouse Multiple Retailer Systems with Vehicle Routing Costs," *Management Science* Vol. 36, No. 1 (January 1990) pp. 92–114.

2. D. E. Blumenfeld, L. D. Burns, J. D. Diltz, and C. F. Daganzo, "Analyzing Trade-Offs between Transportation, Inventory, and Production Costs on Freight Networks," *Transportation Research*, Vol. 19B, No. 5 (1985), pp. 361–380.

3. R. G. Brown, *Materials Management Systems* (New York: Wiley-Interscience, 1977).

4. R. G. Brown, *Advanced Service Parts Inventory Control,* 2nd ed. (Norwich, Vt.: Materials Management Systems, Inc., 1982).

5. A. J. Clark and H. Scarf, "Optimal Policies for a Multi-echelon Inventory Problem," *Management Science,* Vol. 6, No. 4 (July 1960), pp. 475–490.

6. S. C. Graves and L. B. Schwarz, "Single Cycle Continuous Review Policies for Arborescent

Production/Inventory Systems," *Management Science,* Vol 23, No. 5 (January 1977), pp. 529–540.

7. J. L. Heskett, N. A. Glaskowsky, and R. M. Ivie, *Business Logistics,* 2nd ed. (New York: Ronald Press, 1973).

8. P. Jackson, W. Maxwell, and J. Muckstadt, "The Joint Replenishment Problem with a Powers-of-Two Restriction," *IIE Transactions,* Vol. 17, No. 1 (March 1985), pp. 25–32.

9. J. A. Muckstadt and R. O. Roundy, "Multi-Item, One-Warehouse, Multi-Retailer Distribution Systems," *Management Science,* Vol 33, No. 12 (December 1989), pp. 1643–1621.

10. J. A. Muckstadt and L. J. Thomas, "Are Multi-Echelon Inventory Methods Worth Implementing in Systems with Low-Demand-Rate Hems?" *Management Science,* Vol. 26, No. 5 (May 1980), pp. 483–494.

11. J. Orlicky, *Material Requirements Planning*(New York: McGraw-Hill, 1975).

12. L. B. Schwarz, "Physical Distribution: The Analysis of Inventory and Location," *AIIE Transactions* (June 1981), pp. 138–150.

13. A. Stenger and J. Cavinato, "Adapting MRP to the Outbound Side—Distribution Requirements Planning," *Production and Inventory Management Journal* (4th quarter 1979), pp. 1–14.

APPENDIX 8A

Integrated Production, Distribution, and Inventory Planning at Libbey- Owens-Ford

Clarence H. Martin, Denver C. Dent, and James C. Eckhart

Flagpol, a large-scale linear-programming model of the production, distribution, and inventory operations in the flat glass business of Libbey-Owens-Ford deals with four plants, over 200 products, and over 40 demand centers in a 12-month planning horizon. Annual savings from a variety of sources are estimated at over $2,000,000.

Libby-Owens-Ford Glass Company was formed in 1930 when the Edward Ford Plate Glass Company and the Libbey-Owens Sheet Glass Company merged. LOF's long-standing commitment to technical research has resulted in a long list of product innovations from Thermopane insulating glass in the 1950s to the EZ COOL windshield in 1990. Today LOF operates as an autonomous operating company of the Pilkington Group (which has over 400 subsidiary and related companies in the global glass market). LOF has approximately 9,000 employees and annual sales of $900 million.

LOF comprises three strategic business units: the original equipment (OE) group, automotive glass replacement (AGR) group, and flat glass products (FGP) group. We developed the FLAGPOL model for the FGP group. The FGP group manufactures clear and tinted heat-absorbing float glass, and ECLIPSE and Mirropaine E.P. filmed products for the architectural, mirror and furniture, and residential markets.

The FGP group supplies approximately 300 customers, including international customers in South and Central America and the Caribbean, from four float glass facilities.

Reprinted from *Interfaces,* Vol. 23, No. 3 (May–June 1993), pp. 68–78.

The Laurinburg, North Carolina facility has two float glass tanks that produce glass primarily for the architectural and mirror and furniture markets. The Lathrop, California facility has one tank that produces glass primarily for the architectural market and OE automotive market. The Ottawa, Illinois facility has one tank and produces glass primarily for the residential market. The Rossford, Ohio facility has two tanks and produces glass for the residential and architectural markets.

THE FLOAT PROCESS

The float process for manufacturing glass, a Pilkington development announced in 1959, is recognized as the world standard for flat glass production. Since 1962, 35 glass manufacturers have licensed the float process from Pilkington. Today a new float line costs between $80 and $120 million dollars, and the operating life of a furnace, called a campaign, is from seven to ten years. At the end of the campaign, the float must be rebuilt and relined with new refractory materials. This rebuild costs between $10 and $15 million and takes three to four months to complete.

In the float process, a large gas-fired melding furnace, or "tank," operating at temperatures near 2,700 degrees Fahrenheit, melts a batch of raw materials, including silica sand, limestone, soda ash, dolomite, cullet (which is crushed recycled glass), and additives to give the glass the appropriate color of bronze, gray, or blue-green.

A continuous ribbon of molten glass mixture floats from the tank over a bath of molten tin where its speed and temperature are computer monitored and controlled to give the finished glass its proper thickness and characteristics. There are three separate temperature zones in the bath. The first zone is the heating zone where irregularities in the glass surface are melted out and both surfaces become flat and parallel. The fire polishing zone is where the glass acquires its brilliant surfaces. The final zone is the cooling zone, where the glass cools sufficiently for it to touch the rollers without spoiling the fire-polished surfaces.

When an on-line pyrolitically-coated product, such as ECLIPSE or Mirropane E.P., is being produced, a chemical vapor is released in the float bath over the semimolten surface of the ribbon. The reaction of the vapor with the glass surface forms the reflective coating.

The ribbon of glass then moves from the bath onto the annealing lehr where precise gradual cooling relieves stresses in the glass. Following cooling and a series of quality-control inspections, the continuous ribbon enters the wareroom area where the glass is cut into sizes for storage, distribution, or fabrication into value-added products.

When the ribbon of glass enters the wareroom from the lehr, the ribbon is inspected by automatic inspection equipment for the presence of flaws. There are three quality levels of final product: factory run, glazing, and mirror. Factory run is the lowest-quality, glazing is an intermediate quality, and mirror is the highest quality.

Once the product is separated from the ribbon, it automatically moves down the conveyor system to the destined packing station, where it is packed for storage or shipment. Some products are packed using automated packing equipment, while others are packed manually.

OPERATIONAL ISSUES AT LOF FLAT GLASS PRODUCTS

Planning operations at a flat glass plant is complicated by a number of factors, including the necessity to plan for approximately 205 different product types. The glass is produced in four different colors (clear, gray, bronze, and blue-green), 26 thicknesses, three quality levels, four packaging modes, two cutting classifications (cut-size and uncut), and a number of further fabrication options (tempered, coated, and so forth).

The single most important consideration in planning flat glass operations is the transition schedule, that is, the scheduling of production time for the four colors. Changing color from clear to a tint (bronze, gray, or blue-green) results in up to eight days production of off-color glass that can not be sold as finished product. Changing from one tint to another results in two to four days lost production. Because of these large transition losses, the plants naturally schedule long color runs that take as much as 10 months to complete a full color cycle. The sequence of colors at a plant is generally unalterable because of the inventory levels held from previous color cycles. However, deciding the duration of color runs is critical since poor choices can result in stockouts in a color that may not be produced again for many months.

Another factor that is crucial in operating a flat glass plant is maintaining a high percentage of on-line cutting. Standard sized glass sheets are cut on-line in the wareroom as the glass is being produced. However, glass that is cut to customer order may be cut either on-line or off-line after a period of storage. On-line yields are higher for a number of reasons. However, some level of off-line cutting is mandatory. For example, if a customer orders a special size of, say, bronze glass while some other color is scheduled on the float, that order must be cut off-line from standard bronze sizes in inventory. Since customer orders for cut-sizes are rarely known more than a month in advance, most orders for cut-size products, other than clear, must be cut off-line.

A final important area of concern is inventory management. Because plants take so long to make transitions, their inventories can be very large. To manage inventory successfully, the plants must balance the risks of obsolescence against those of stockouts. Managers specify minimum safety-stock levels in terms of equivalent days of sales. Furthermore, to cover demand for products of a particular color during the interval between successive occasions when the float is producing that color, plants maintain minimum cycle stocks. For a given product, this cycle stock must be sufficient to cover all demand for that product plus all demand for products cut off-line from the given product.

PRODUCTION PLANNING PRIOR TO FLAGPOL

Prior to the development of FLAGPOL, a corporate production planner was responsible for generating production plans for each of the floats based on the marketing forecasts developed for each plant individually. Marketing assigned customers to plants, typically on the basis of past practice. Therefore the production plan could best be described as a best fit plan based on such factors as minimizing product shortfall at each plant rather than maximizing the overall financial return of the entire enterprise.

ORGANIZATION OF MODELING EFFORT

The vice-president of research and development initiated discussions concerning the feasibility and potential value of a formal production planning model in 1985. He envisioned four major benefits: (1) cost savings from planning on a system-wide basis, that is, simultaneous consideration of all manufacturing facilities and distribution requirements; (2) the ability to efficiently incorporate additional plants into a coordinated production and distribution system (in 1985, only the plants at Laurinburg and Lathrop were significantly involved in FGP operations); (3) providing a means through which the occasionally conflicting views of various functional areas (marketing, manufacturing, transportation and so forth) could be examined and resolved; and (4) providing management with an approach to addressing strategic issues of system design, structure, and capacity.

With the enthusiastic support of its vice-president, the flat glass products division organized a task force to discuss and examine the various issues concerning the goals, scope, and design of the model. This task force included corporate staff members from finance, marketing, MIS, materials management, transportation, production planning, and representatives from the plants, such as managers of production scheduling and cost analysts. The task force stated its goals for the modeling effort as follows:

1. To develop a linear-programming model to optimize production, inventory, and distribution in a multi-plant system based on a 12-month planning horizon along with the supporting data base;

2. To validate the model by analyzing and evaluating the benefits from implementing the operating methods and planning procedures suggested by the model;

3. To identify and obtain the resources necessary to incorporate the model as the standard centerpiece of production planning practices; and

4. To develop the capability to analyze scenarios and evaluate options in other areas, such as marketing, transportation, and capacity planning.

The process of determining the proper scope of the model rarely involved compromises of the sort common in such projects. Most attempts at reducing the complexity or level of detail were viewed by one or more potential users as seriously impairing their ability to achieve meaningful insights from the model. As a result, the model became extremely large.

The task force decided that the basic thrust of the model would be to determine optimum monthly levels of production (both on-line and off-line), inventory, and distribution. The model would not address detailed specification of the monthly production schedule (the precise sequence and timing of individual products) or secondary operations like tempering or coating. Managers would specify the actual transition schedule as input to the model; the model would not determine it. However, they correctly anticipated that the model would furnish valuable information (through shadow prices or dual variables) that would help them to develop transition schedules.

A project team (the authors) undertook actual development of the FLAGPOL model. We assumed complete responsibility for all activities through implementation, including about two years of actual on-line use. We named the linear-programming model

FLAGPOL (*FLA*t *G*lass *P*roducts *O*ptimization Mode*L*). We describe its salient features in the appendix.

REPORT GENERATION

In designing a model that involves about 100,000 decision variables, one must take great care to avoid overwhelming potential users with a mountain of data while providing them with access to the wealth of information actually available. In general, one must work closely with the users (1) to insure that, whenever possible, report formats from the model mirror the formats of documents used previously, (2) to design new reports targeted to specific uses by specific groups, (3) to explain and modify reports as requested, and (4) to provide output files that can be used directly by other information systems.

DATA-BASE DEVELOPMENT

From the outset we realized that developing the data base to support FLAGPOL would be a major undertaking. Because of the scope of the model, we had to assemble data produced by a number of different information systems and check it for accuracy and compatibility. Maintaining data integrity would become an ongoing effort. The individual data sources include the following:

1. Market demand and sales prices.
2. Freight rates,
3. Production rates and yields,
4. Manufacturing costs,
5. Inventory levels, and
6. Interplant rail schedules.

THE IMPLEMENTATION PROCESS

We took about nine months to design and develop the initial FLAGPOL code. The corporate production planner, plant personnel, and FGP management then tested it and reviewed the results thoroughly. Several extensions and revisions were required. For example, plant personnel quickly pointed out that the first version completely overlooked off-line cutting. The financial staff of FGP identified a shortcoming that required the development of the piecewise linear treatment of ending inventory value (see the description of final inventory variables in the appendix). They also realized that differences in the cost accounting systems of the plants needed to be addressed. The corporate production planner identified the need for greater precision in the specification of both safety-stock levels and end-of-horizon cycle stock. Input from FGP and plant management resulted in the extensive redesign of reports and the development of new reports.

After nearly two years of effort, we brought FLAGPOL to operational status. However, the process of upgrades continues to this day. Typically the upgrades are to develop new reports, to improve the user interface, or to allow the users control of specific decisions.

THE ROLE OF FLAGPOL IN PLANNING

A number of decisions made at the corporate level restrict the planning options available to FLAGPOL. These can include the production of certain products at specific plants only, the assignment of customers to specific facilities, safety-stock levels, and the general parameters of the transition schedules.

FLAGPOL is used to support the annual budgeting process by specifying the level of production for each plant during the upcoming fiscal year. The plant controller at each facility uses this information in developing the facility budgets.

At a tactical level, the FGP production planner runs FLAGPOL each month to develop production plans for each of the float facilities. These are then transmitted to the production schedulers at the individual facilities who develop detailed daily float schedules that reflect the monthly production plans developed by FLAGPOL.

When unplanned events occur, the planners run FLAGPOL to develop reaction procedures, and when strategic scenarios of interest to FGP managers arise, they use FLAGPOL to evaluate the options under consideration.

To provide the production plans, the FGP planner spends approximately one week per month updating the data base, running the model, analyzing the outputs, and releasing the plan to the production facilities and FGP corporate staff. In addition, various scenarios are often run throughout the month to develop reaction procedures or to evaluate proposed decisions concerning the FGP business unit. The model is run 10 to 20 times each month to determine the production plan and address scenarios of interest. These model runs take about 10 to 20 hours of computer time per month. Originally the model was run on an IBM 3084 but now is run on a stand-alone Sun Sparcstation 4/330.

BENEFITS FROM INTEGRATED SYSTEM PLANNING

A number of the benefits of the FLAGPOL effort come from the company's ability to plan on a system-wide basis rather than by plants in isolation as was previously necessary. Examples of these benefits follow:

1. The model suggested many realignments in the assignment of customer orders to plants. Among the major opportunities from the potential reassignments, a set of about 10 provided savings in excess of $700,000 a year.

2. The model identified opportunities for the greater use of interplant rail shipments that justified investment in rail-car capacity and yielded annual savings in excess of $700,000.

3. The model indicated that the company could save $1,800,000 by reducing the system-wide off-line cutting from current levels. With an increased awareness of the cost of off-line cutting, management is trying to insure that the standard size products are cut on-line as much as possible.

4. The model showed that a contemplated elimination of a color at one plant would be counter-productive. The plant produces this color only a few days a year; its elimination would save several days of transition losses. However, the model showed that the elimination of transition losses would be more than offset by increased transportation expenses resulting in a net loss of $900,000 annually.

5. The model provides centralized modeling of capacity issues, which provides in-

sights valuable in capital resource planning. For instance, management quickly authorized expenditures to increase inventory storage at one of the plants based on modeling recommendations.

6. With centralized management of inventory volume and composition, management is better able to utilize capacity and specialized marketing tactics.

7. Centralized management of capacity shortfalls has enhanced customer service. Because it can quantify the problems and costs, the company routinely employs cross country delivery of products to preserve customer relationships and market share.

BENEFITS FROM DEVELOPING REACTION PROCEDURES

The model is an important tool in developing reaction procedures, that is, short-term adjustments to unforeseen changes in the operating environment of FGP. The model can evaluate the consequences of alternative courses of action on a systemwide basis one year into the future and can expose the shortcomings in proposals based on localized or short-term views. FLAGPOL has been used to develop reaction procedures in the following situations:

1. As market demand changes, should new orders be accepted? How should production plans be changed to accommodate new orders or order cancellations?

2. When capacity changes over the short term because equipment must be modified, repaired, or rebuilt, how should production be rescheduled?

3. How should the company deal with quality problems, periods of lower than anticipated yields, and new fabrication requirements?

BENEFITS IN DEALING WITH STRATEGIC ISSUES

Although we originally conceived of FLAGPOL as a tactical/operational model, it has been used to address a number of strategic issues:

1. Planning the transition from a two-plant system to a four-plant system and deciding how extensive the product mix capabilities of new facilities should be,

2. Introducing new products or eliminating existing products from plants,

3. Developing schedules for major construction and maintenance activities that require significant downtime, and

4. Investigating location and sizing issues for new facilities.

CONCLUSION

The FLAGPOL model is an integrated part of the FGP planning process. FGP uses the model monthly to plan production for the plants so that FGP can operate in a proactive manner rather than reactive. This monthly planning is important for decisions pertaining to inventory management, interplant movement of products, and product mix adjustments. However, its greatest opportunity for increasing profits is at times when the production facilities are at full capacity. At such times, it is especially crucial for FGP to identify the optimal system-wide production and distribution schedules. In addition, the model enhances customer service by managing the demand shortfall.

APPENDIX

The FLAGPOL model is a specialized version of a production, distribution, and inventory model. To turn it into a model that represents the activities of the FGP Group of LOF requires a large set of parameters that specify the structure and technological factors unique to this particular operation:

The structural parameters specify the size and structure of the operation:

- Number of plants,
- Number of floats,
- Number of products,
- Number of colors,
- Number of quality levels,
- Number of float speeds,
- Number of demand centers,
- Number of months, and
- Number of weeks in month.

The product list parameters are alphanumeric descriptions and numeric codes for six product descriptors: color, thickness, quality level, cutting classification, required shipping equipment, and fabrication. *The plant-product parameters* include

- Manufacturing costs,
- Throughput rates,
- Safety stocks in weeks,
- Inventory holding costs, and
- Inventory valuations.

The transition schedule parameters include

- Days available each month by float by color and
- Quality level percentages by float by month.

The sales-demand parameters are

- Sales prices by product by demand center,
- Forecasted demand by product by demand center by month, and
- Demand shortfall penalty by product.

The inventory parameters are

- Warehouse capacity by plant by product class,
- Safety stock by plant by product in weeks of sales, and
- Transition cycle stock by plant by product in days of sales.

The distribution parameters are a

- Freight rates to demand centers from plants by required shipping equipment,
- Interplant rail rates, and
- Interplant shipment limits by shipping plant by month.

The off-line cutting parameters are a

- List of parent-child off-line cutting pairs of products by plant,
- Off-line yields by pair by plant,
- Off-line cutting cost by pair by plant by product, and
- Off-line cutting capacity in input tons by plant by month.

The decision variables are factors involved in the production, distribution, and inventory activities of the FGP Group that are to be determined by FLAGPOL. In FLAGPOL, each decision variable is specified in units of finished tons and is indexed or described by a number of factors, such as product, plant, month, and so forth. The model currently has approximately 100,000 decision variables. Table 1 shows the sets of decision variables and their corresponding indices.

The final inventory variable is used to model a piecewise linear value function of inventory at the end of a 12-month planning horizon. Any product in final inventory is valued at a percentage of its sales realization value based on the number of days of sales as shown in Table 2.

A standard linear treatment of final inventory in the objective function results in solutions in which end-of-horizon inventory levels are unacceptable. This occurs because for each combination of float and color, any available production time in excess of that required to cover demand and minimum inventory requirements is allocated to the production of the single most attractive product as measured by its objective function contribution. This can produce end-of-horizon inventory levels equivalent to several years of sales. The piecewise linear approach shows significant advantages over alternative approaches (such as placing maximum limits on end-of-horizon inventory levels).

TABLE 1:   Each set of decision variables is characterized by index sets.

Decision Variable	*Indices*
On-line production	Plant, float, product, and month
Off-line cutting	Plant, parent product, product, and month
Inventory (beginning)	Plant, product, and month
Distribution	Plant, demand center, product, and month
Demand shortfall	Demand center, product, and month
Interplant shipment	Shipping plant, receiving plant, product, and month
Purchasing	Plant, product, and month
Final inventory	Plant, product, and days of sales interval (see discussion)

TABLE 2: End of horizon inventory value is adjusted by the days of sales held.

Days of Sales	Percentage	Days of Sales	Percentage
1–60	90%	181–240	40%
61–120	80%	241–360	20%
121–180	60%	361 and above	0%

OBJECTIVE FUNCTION

The objective function for FLAGPOL is the maximization of total margin over the 12-month planning horizon. This includes operating revenues and costs and also reflects changes in the value of inventory over the horizon.

CONSTRAINT SET

The constraint set of the FLAGPOL model consists of about 26,000 constraints organized as eight sets of constraints as follows:

1. The first set of constraints places limits on glass production corresponding to the transition schedules. Three constraints (one for each quality level: mirror, glazing, and factory run) are defined for each combination of float, color, and month. Each constraint involves the total production time (in hours) required by a set of products of the given combination of float, color, and month. The constraint for factory run involves products of all quality levels and states that the total time must be equal to the time specified in the transition schedule multiplied by the quality time percentage for factory run (usually 100 percent). The constraint for glazing involves products of quality levels glazing and mirror and states the total production time must not exceed the product of the transition schedule time and the quality time percentage for glazing. Finally, the constraint for mirror quality involves only products of mirror quality and states that the total time must not exceed the product of transition schedule time and the quality time percentage of mirror.

2. The second constraint set is a set of material balance equations. One such constraint is defined for each combination of plant, product, and month. The form of the constraint is (on-line production) + (purchasing) + (beginning inventory) + (inbound interplant shipment) + (off-line cutting-child) = (ending inventory) + (off-line cutting-parent) + (outbound interplant shipment) + (distribution). For each constraint, at most one off-line cutting entry occurs. For child products, only the first occurs; for parent products, only the second occurs; and for products not involved in off-line cutting, neither occurs. The parent entry is the off-line cutting decision variable. The child entry is the product of the off-line cutting decision variable and the off-line cutting yield parameter.

3. The third set of constraints places limits on the inventory space available. One such constraint is defined for each combination of plant, month, and inventory classification.

4. The fourth set of constraints is used to implement the piecewise linear reduction in inventory value based on days of sales.

5. Constraint set 5 places limits on the quantity of interplant shipment. One such constraint is defined for each combination of shipping plant, receiving plant, interplant shipment class (that is, the type of storage equipment used in transporting the glass, such as case, stoce, ministoce, and rack) and month. The total interplant shipment for the particular combination must not exceed the maximum limit parameter.

6. The sixth set of constraints places limits on the total off-line cutting each month at the plants. One such constraint is defined for each plant (that has off-line capability) in each month.

7. Constraint set 7 is defined by plant, by month, and by a number of product subsets for each plant. Each product subset is split into two groups, A and B. The constraint states that the total monthly production time for products in group A must not exceed a multiple (which is a variable input parameter specified separately for each subset) of the total monthly production time for products in group B. This is useful in expressing a number of operational restrictions.

8. Constraint set 8 is based on the stated demand for each combination of demand center, product, and month. For each such combination, the sum of all the decision variables representing distribution from the various plants plus the demand shortfall variable must be equal to the stated demand. They are treated as GUB (generalized upper bound) constraints.

The decision variables are further constrained by simple upper and lower bounds. Usually these are based on problem restrictions, such as minimum and maximum inventory levels. They can also be used to allow an analyst to specify any desired value for a decision (by setting both the lower and upper bounds to that desired value).

TECHNICAL ASPECTS

FLAGPOL consists of a general purpose optimizer called GUBKOD and a number of routines developed specifically for this application. The source code consists of a main program and 36 subroutines (26 in GUBKOD) coded in 8,200 lines of FORTRAN (3,600 in GUBKOD). The general technical features of GUBKOD are the following:

- The revised simplex method,
- Primal and composite phases,
- Super-sparse storage of constraint matrix,
- LU representation of basis inverse,
- Simple lower and upper bounds handled implicitly,
- Generalized upper bounds handled implicitly, and
- Advanced basis start customized for FLAGPOL.

Currently the model dimensions are approximately 99,000 variables and 26,000 constraints, of which 7,000 are handled explicitly. The remaining 19,000 GUB constraints are handled implicitly. Typical execution times are three to four hours on the Sun Sparcstation. However, once a given problem is solved, additional scenarios can be completed in as little as five minutes starting from an optimum basis to the original problem and their estimate of at least $2,000,000 in annual savings.

"The FLAGPOL model has evolved into a management tool of great significance to LOF Flat Glass Products. It has allowed integration of planning efforts that previously would have occurred in isolation and with less precision. It has provided unbiased insights into operational issues frequently yielding novel solutions and approaches that would have otherwise gone undiscovered. The model has resulted in the identification of many significant opportunities and has enhanced the monthly production planning process, with the greatest operational and financial benefits being derived during periods when the manufacturing facilities are at full capacity."

PART THREE

PLANNING ACTIVITIES

To ensure that the resources to accomplish the mission of an organization are available, operations must be planned before they can be undertaken. Without advance planning, a firm may not be able to produce for an increase in demand at some future date. The capacity may not be available at that future date, and sales may be lost. With advance planning, future demand could be produced earlier to meet later demand.

Figure III.1 repeats Arthur Young's framework for competitive advantage. This part of the text provides the details of the section called manufacturing planning and control in the figure. The framework shows that this segment is linked to suppliers and customers. The planning processes discussed in this part illustrate this connection. Customers provide inputs through booked orders and anticipated orders through forecasting. The planning processes then determine the production plan to meet these demands and provide output that can be used by the suppliers to provide the needed inputs at the right time and place.

Meal [2] uses the term *hierarchical production planning* (HPP) to indicate how planning is done hierarchically through the organization. Figure III.2 shows how long-term decisions are made at the corporate level and how the time frame shrinks as the planning process works toward first-level supervision.

Part III discusses three basic planning processes: aggregate planning (Chapter 9), master production scheduling (Chapter 10), and materials requirements planning (Chapter 11). These planning methods work from the aggregate to the plant level to the individual item level. The normal process within planning organizations is to develop aggregate plans at the plant level to balance demand with capacity and inventory levels for the entire operation. As Arthur Young's model shows, customers provide the inputs to this process by providing orders. These plans can then be disaggregated to specific items. The master production schedule will take the forecast demands and determine a manu-

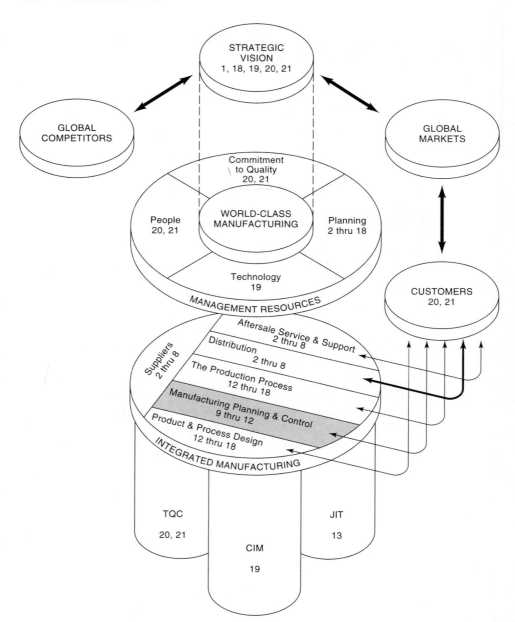

FIGURE III.1 Arthur Young's Manufacturing for competitive advantage framework. From Copyright ® 1987 by Ernst & Young (formerly Arthur Young, a member of Arthur Young International). Reprinted by permission. *Note:* numbers on this figure indicates chapter numbers covering these subjects.

Decision Level	Decision Process	Forecasts
Corporate	Allocate production to plants	Annual demand
Plant	Seasonal production plan by product type	Monthly demand by product type
Shop	Monthly production plan for items	Monthly demand by item

FIGURE III.2   Hierarchical planning process.

facturing plan, which is used as input to the materials requirement planning system. The result is a manufacturing plan that should satisfy the demands and meet capacity requirements. The materials requirement planning system provides the requirements of items that can then be ordered through the suppliers.

Meal's HPP process starts with the decision on plant sizing and location. This imposes constraints on equipment type and amount, which leads to production allocation among plants, which leads to plant capacity planning, which in turn leads to item production scheduling. The planning process is presented in a hierarchical manner in relation to the HPP steps.

The following chapters are in Part III:

Chapter 9, Aggregate Planning

Chapter 10, The Master Production Schedule

Chapter 11, The Planning of Material Requirements

REFERENCES AND BIBLIOGRAPHY

1. Thomas G. Gunn, *Manufacturing for Competitive Advantage: Becoming a World-Class Manufacturer,* (Cambridge, Mass.: Ballinger Publishing, 1987).
2. Harlan C. Meal, "Putting Production Decisions Where They Belong," *Harvard Business Review,* Vol. 2, No. 2 (March–April 1984).

Chapter Nine

Aggregate Planning

INTRODUCTION

In Chapter 2, we explained that demand forecasting can be for the long, medium, or short range. Long-range forecasts help management formulate capacity planning strategies. Managers ask many policy-related questions: Do we need to increase or curtail our operations at various locations? Do we need to build or expand facilities at existing or new locations? Do we need to negotiate or renegotiate our supply of parts from vendors? Once long-range capacity decisions are made, intermediate-range plans should be made that are consistent with long-range policies. Management must work within the resources allocated by long-range decisions. The plans do not necessarily have to be so detailed as to provide specific instructions for daily or weekly operations, such as loading, sequencing, expediting, and dispatching. These specifics will be dealt with in later chapters.

THE NATURE OF THE AGGREGATE PLANNING DECISIONS

Inventories provide a means of storing excess capacity during intermediate slack periods and assist us in smoothing the impact of demand fluctuations on personnel levels. In most productive systems we must be concerned with scheduling equipment and the workforce in addition to managing inventories. Given the sales forecast, the factory capacity, aggregate inventory levels, and the size of the workforce, the manager must decide at what rate of production to operate the plant over the intermediate term. Intermediate-range planning is generally known as *aggregate planning*. Aggregate plans and master schedules provide common points at which capacity and inventories are considered jointly in the light of firm's long-range plans, and they provide inputs to the financial plan, the marketing plan, and requirements planning and detailed scheduling decisions. Several crucial decisions have to be made while generating an aggregate plan. Management may ask many inventory- and workforce-related questions: To what extent should inventories be used for absorbing changes in demand that might occur during the intermediate term? Should we absorb the fluctuations by varying the size of the workforce? Should we keep the workforce constant and absorb fluctuations by overtime, shorter time schedule hours, and part-time workforce? Should we maintain a fairly stable workforce, as well as production rate, and subcontract the fluctuating order rates? Should we vary the prices to counterinfluence the demand pattern? Generally, a mixture of strategies is preferred and is feasible. We briefly discuss the effects of various policies next. In this chapter we first cover various aggregate planning strategies and associated costs. Then we present some qualitative and quantitative methods to solve these problems. Finally, we compare these methods and evaluate their advantages and limitations.

An aggregate plan is a valuable procedure to help in the development of operating budgets. As discussed above, an aggregate plan will determine workforce levels, overtime, and inventory levels with the objective of minimizing cost. These results will be useful to the operating manager when determining an operating budget. Workforce levels will be translated into the labor budget, and inventory levels can be used to determine requirements for storage space.

AGGREGATE PLANNING DEFINED

Aggregate output planning generally consists of planning a desired output over an intermediate range of three months to one year. The aggregate plan needs some logical, common unit of measuring output such as gallons of paint in a paint factory, number of dresses in a garment factory, cases of beer in a brewery, and perhaps equivalent machine hours in manufacturing industries. Product group forecasts are generally more accurate than an individual item forecast. The further the forecast goes into the future, the less likely it is to be accurate. Recognizing this, planning and control of production is done on the basis of group demand over the intermediate and long range. In the short range, however, as better forecasts become available for individual products, disaggregation and detailed scheduling become feasible. Choosing meaningful groups requires a thorough knowledge of the products as well as manufacturing processes. The groups are not necessarily the same as those used by the marketing/sales department or the inventory control system department. The chosen groups must be meaningful in terms of, for example, demand on manufacturing facilities for products that go through similar manufacturing operations or that are processed in common manufacturing facilities. Comparable Ford and Mercury cars, for example, require the same capacity for body construction, although assembly department requirements may vary. The disaggregation of products groups and the master scheduling process is dealt with in another chapter.

Many aggregate planning strategies are available to the manager. These strategies involve the manipulation of inventory, production rate, manpower needs, capacity, and other controllable variables. When we vary any one of the variables at a time to cope with changes in product output rates, we use what are known as *pure strategies. Mixed strategies,* in contrast, involve the use of two or more pure strategies to arrive at a feasible production plan. We briefly discuss, first, the effect of adopting pure strategies—that is, changing the level of individual variables.

Pure Strategies

Changing Inventory Levels.   If we accumulate inventories during slack periods of demand, working capital and costs associated with obsolescence, storage, insurance, and handling will increase. Conversely, during periods of increasing demand, changes in inventory levels or backlogs might lead to poorer customer service, longer lead times, possible lost sales, and potential entry of new competitors in the market.

Changing Workforce Levels.   The manager may change the size of the workforce by hiring or laying off production employees to match the production rate so as to meet the demand exactly. In many instances new employees require training and the average productivity is temporarily lowered. A layoff frequently results in lower worker morale and lower productivity, and the remaining employees may retard output to protect themselves against a similar fate.

In some instances it is possible to maintain a constant workforce and vary the working hours. During upswings of demand, however, there is a limit on how much overtime is practical. Excessive overtime may wear workers out, and their productivity may go down. The incremental costs associated with shift premium, supervision, and overhead may be significant. In period of slack demand, the firm also faces the difficult task of absorbing the workers' idle time.

Subcontracting.   As an alternative to changing workforce or inventory, perhaps the company could subcontract some work during the peak demand periods and increase the capacity to satisfy the demand. Again, a potential danger exists of opening doors to competition.

Influencing Demand.   Because changing demand is a chief source of aggregate planning problems, management may decide to influence the demand pattern itself. For example, telephone companies level their loads by offering evening rates and some utility companies are experimenting with similar strategies. The airline industry offers weekend discounts and winter fares. It is not always possible, however, to balance the demand pattern and the desired production plan.

Mixed Strategies

We see that every pure strategy has a countervailing cost associated with it, and pure strategies are often infeasible. Therefore, a combination of strategies, or a mixed strategy, is often used. Mixed strategies involve the use of two or more controllable variables to arrive at a feasible production plan. Such strategies could include, for example, a combination of subcontracting and overtime or overtime and inventory, as illustrated by some examples later in this chapter. When one considers the possibility of mixing these strategies—the infinite variety of ratios for blending various strategies—one can realize how challenging the problem is. In any case, it is the responsibility of top management to set guidelines for aggregate planning activity, because these planning decisions frequently involve company policies. The production control department, in conjunction with the marketing department, should generate master schedules commensurate with company policies and specific operating procedures. The complexity of the decision process leads us to a discussion of the value of decision rules for solving aggregate planning problems.

THE VALUE OF DECISION RULES

Determining when to change production level is a difficult decision. It involves a great deal of time and money, as millions of strategies are available to the decision maker. By establishing decision rules, the production control manager, in conjunction with the operations manager, is in effect setting the rules of the game once, rather than each time it is played. Once a policy is established regarding what justifies a change, weekly decisions can be concerned with the specific actions to be taken to accomplish the change in the most economical and effective manner. The aim of these rules is to help the manager keep out of trouble, rather than get out of trouble. Industry experience indicates that having some kind of rule, even if it is not optimal, significantly improves operations.

To optimize any production plan, we need to review the behavior of cost structures. After a brief exposition of these costs, we will present some important approaches and techniques for solving aggregate planning problems.

COSTS

Regular Payroll and Overtime Cost

Detailed empirical cost studies by Hold, Modigliani, Muth, and Simon [12] examined regular payroll, overtime, hiring and layoff costs, and inventory and backorder costs. These studies covered several years' operation of a paint factory. The typical regular time production cost versus production rate relationship was found to be a monotonic increasing curve, as shown in Figure 9.1. The sudden jump in cost could be attributed to the cost of additional equipment acquired for achieving higher production rates. A significant portion of regular time production cost is spent as wages to the full-time workforce. Such costs can be approximated to increase linearly with the size of the workforce. Industry surveys [23], however, indicate that firms attempt to maintain a constant workforce size because of social pressures, public opinion, union contracts, and the high cost of training and severance associated with changes in the workforce. Under these circumstances, the cost of the workforce becomes a constant, as illustrated by the dotted line in Figure 9.2.

The general shape of overtime costs for a given workforce size is shown in Figure 9.3. The costs are kept to a minimum when the facilities are operated at optimum level. The cost increases when the plant is operated below the designed capacity. With continued increases in demand, more and more production is scheduled, and the cost curve rises sharply at higher levels of production. The increases can be attributed to shift pre-

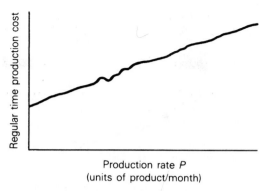

FIGURE 9.1   Costs of regular time production.

mium, supervision, and the decrease in productivity of workers as they toil through long hours.

The Cost of Changing the Production Rate

The costs of changing the production rate can be attributed primarily to changes in the size of the workforce. The typical incremental cost of hiring and layoff is depicted in Figure 9.4. When the size of the workforce is increased, the firm incurs costs of hiring, training, and possible reorganization, resulting in lower productivity in the initial periods. Similarly, when employees are laid off, terminal pay, decreased morale in the remaining employees, and possible decreased productivity from the fear of losing their jobs increases the cost of production. Rarely is a laid-off worker rehired for the same job. In addition, social pressures, company image, and other factors prevent excessive hiring and firing. In many instances, union contracts and supplemental unemployment benefits (SUB) programs make it very costly for a firm to lay off workers. The incremental cost of increasing the production rate could be different from the incremental cost of decreas-

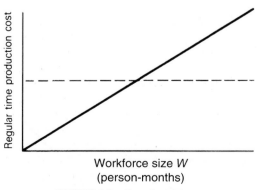

FIGURE 9.2   Cost of workforce.

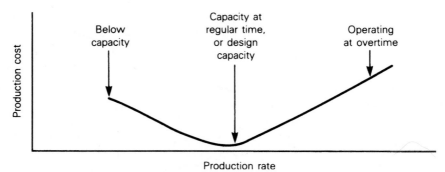

FIGURE 9.3 Cost of overtime and idle time.

ing the production rate, as illustrated by the shape of the curve along the vertical axis in Figure 9.4.

In responding to changes in production levels, management should consider the costs of hiring and training and other associated layoff costs against the costs of overtime and undertime and the possible decrease in productivity caused by prolonged working hours.

Inventory, Backorder,
and Shortage Costs

There is a cost associated with funds used for inventories. The optimal aggregate inventory level may be approximated by the sum of the average safety stock plus half the optimal batch size, as determined for individual items. The aggregate inventory for an item is shown in Figure 9.5. If $\bar{I}_t$ represents the amounts of inventory tied up in inventory at time t and r is the unit cost of inventory for item i, then the total cost of inventory (C_I) for n periods is

$$C_I = \bar{I}_1 r + \bar{I}_2 r + \bar{I}_3 r + \ldots + \bar{I}_n r = r \sum_{t}^{n} \bar{I}_t$$

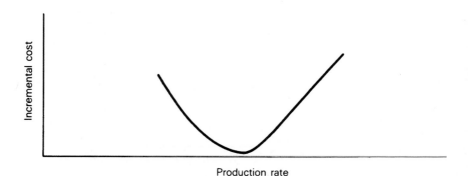

FIGURE 9.4 Cost of changing workforce level.

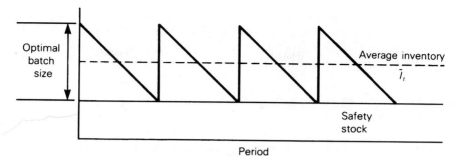

FIGURE 9.5   Aggregate inventory level.

The cost per year of carrying inventory typically ranges from 5% to 50% of the value of items. The total inventory for all items is obtained by summing individual item inventory costs.

The cost of backordering and lost sales could also be treated in the same manner. If lost sales occur too often, an easy path to competition might be opened, and hence the cost could be high. The cost of lost sales is very difficult to estimate. For any given sales forecast (demand), the cost of inventory, backorder, and shortage can be approximated by the curve exhibited in Figure 9.6.

We have discussed the determination of cost curves pertaining to optimal inventory associated with a sales forecast S_t. In fact, a whole set of such curves can be drawn, one for each value of S_t, as discussed by Holt et al. [12].

Subcontracting Costs

As an alternative to changing production levels and carrying inventory, a firm may elect subcontracting to meet peak demands. Subcontracting may not be profitable, however, since the contractor may charge a much higher price than the firm usually pays employees. Subcontracting may also open the doors for competition. It is also hard, in many instances, to find a reliable supplier who delivers on time. Difficulties in forecasting the right quantities could result in excessive inventory or shortage costs.

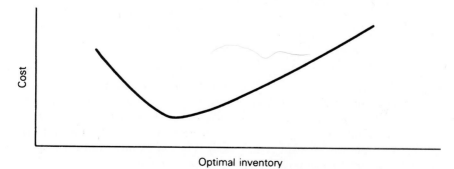

FIGURE 9.6   Cost of inventory and shortage.

So far, we have discussed various costs that could be used in a cost model; the costs are quantified in a later section of this chapter. It is important to keep in mind, however, that cost structures change over time. Hence, these cost models (estimates) must be continually updated to reflect ongoing changes.

AGGREGATE PLANNING STRATEGIES

Aggregate planning is usually the responsibility of the operations manager. He or she must devise a strategy to meet changing needs so that total cost is minimized and the objectives of the firm are met. Aggregate planning methods can be grouped into two major strategy groups. The first is the traditional top-down strategy that uses the concept of an average or composite product in formulating the overall plan. The composite product plan will be disaggregated in due course for detailed planning purposes, which is discussed in Chapter 10. The second method is the bottom-up approach, which is also known as capacity requirements planning. With the availability of computers and material requirements planning systems, it becomes feasible to compile the plans for a total product line in one overall picture. The overall aggregate plan can then be evaluated in light of available resources and can be revised if necessary. This approach is dealt with in detail in the next chapter. As discussed earlier, production planning involves a great deal of time and money, and millions of combinations of resources are possible. In the next section we deal with some decision rules that have significant practical value in industry.

AGGREGATE PLANNING METHODS

Several methods for solving aggregate planning problems exist including both qualitative and quantitative methods. The qualitative methods include consensus among groups and inventory ratios. The quantitative methods consist of heuristic rules, explicit mathematical solutions, simulation, and other sophisticated search procedures. First, we briefly summarize the qualitative methods of solving the problem. Second, we describe the graphical and charting techniques. Finally, a detailed discussion of some viable quantitative approaches to the aggregate planning problem is provided.

Nonquantitative or Intuitive Method

In almost all organizations, there are always conflicting goals and views. The marketing department desires many product varieties and large buffer inventories. The manufacturing department prefers to keep as few products as possible and thus avoid any unnecessary setup costs. The financial controller would like as few inventory items as possible to minimize the investment in inventory and the carrying cost, whereas marketing desires to increase the service level by increasing safety stock inventories. The actual policy is usually dictated by the most persuasive individual, rather than according to economic measures. In general, this method is not desirable. In many industry situations, management takes the previous plan and adjusts it slightly upward or downward to meet the present situation. Such a decision is not dependable if the previous plans were not close to optimal. Unfortunately, in many instances they are not, and hence management is locked into a series of poor plans.

Inventory Ratios

Turnover ratio is a concept that is often used in production planning because the performance of managers is often measured by the turnover ratios their facility achieves. The ratio is defined as follows:

$$\text{Turnover ratio} = \frac{\text{average sales}}{\text{average inventory}}$$

Using turnover ratios for controlling production capacity, however, has a drawback. It leads to large gyrations in the inventory level for a fluctuating demand pattern [25], as illustrated by the following example. For simplicity, we consider the economic order quantity (EOQ) formula.

$$\text{Optimum inventory} = \left(\frac{1}{2}\right)\sqrt{\frac{2SD_t}{h}}$$

$$= K\sqrt{D_t}$$

where K is constant, S is the setup cost, h is the holding cost, and D_t is the demand during period t. Based on the optimum inventory model, the best turnover ratio is

$$\frac{D_t}{K\sqrt{D_t}} = \frac{\sqrt{D_t}}{K}$$

which is a function of demand. Since the demand is fluctuating, the turnover ratio is not a constant, and hence it is fallacious.

Charting and Graphical Methods

Charting and graphical techniques are easy to understand and convenient to use. These techniques basically work with a few variables at a time on a trial-and-error basis. They require only minor computational effort. The essence of an aggregate planning problem is best illustrated by means of production requirements charts and cumulative workload projections.

Example 9.1

ABC Corporation has developed a forecast for a group of items that has the following seasonal demand pattern:

Quarter	Demand	Cumulative Demand
1	220	220
2	170	390
3	400	790
4	600	1390
5	380	1770
6	200	1970
7	130	2100
8	300	2400

1. Plot the demand as a histogram. Determine the production rate required to meet average demand, and plot the average demand forecast on the graph.
2. Plot the actual cumulative forecast requirements over time and compare them with the available average forecast requirement. Indicate the excess inventories and backorders of the graph.
3. Suppose that the firm estimates that it costs $100 per unit to increase the production rate, $150 to decrease the production rate, $50 per quarter to carry the items on inventory, and an incremental cost of $80 per unit if subcontracted. Compare the cost incurred if pure strategies are used.
4. Given these costs, design a mixed strategy solution for this problem.

The histogram and the cumulative requirements graph show how the forecast deviates from the average requirements (see Figures 9.7 and 9.8). Using pure strategies, it is possible to come up with several plans as follows:

Plan 1:   *Varying the Workforce Size.* Demand can be met exactly by varying the workforce size. The plan involves hiring and firing as necessary. The production rate will equal the demand. The cost of this plan is $138,000, as computed in Table 9.1. Notice that inventory and backorders are both zero in each quarter, and the resulting costs for those are zero.

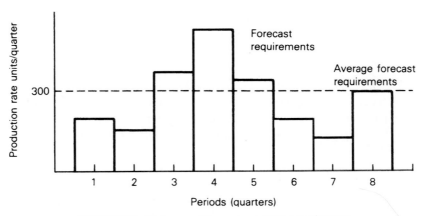

FIGURE 9.7   Histogram of forecast and average requirements.

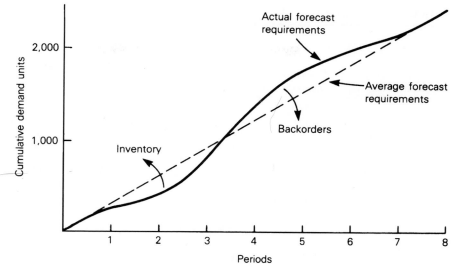

FIGURE 9.8 Cumulative and average forecast graph.

Plan 2: *Changing Inventory Levels.* Suppose that a firm wants to avoid frequent hiring and layoffs. It might choose a production level equal to its average demand and meet the variations in demand by holding inventory. The cost of such a plan is computed in Table 9.2. The plan incurs a maximum shortage of 270 units during period 5. Since a certain amount of uncertainty is involved in any forecast, the firm might decide to carry the inventory from the beginning of period 1. Adjusted inventories and cost of carrying inventories are shown. The total cost of the plan is $96,500. Notice, however, that if the item in question is high-fashion apparel, the firm might not want to carry unnecessary inventory, even though Plan 2 is less costly than Plan 1.

TABLE 9.1 Varying the Workforce Size to Meet the Demand

Quarter	Demand Forecast	Cost of Increasing Production Level: Hiring ($)	Cost of Decreasing Production Level: Layoff ($)	Total Cost of Plan ($)
1	220	—	—	—
2	170	—	7,500	7,500
3	400	23,000	—	23,000
4	600	20,000	—	20,000
5	380	—	33,000	33,000
6	200	—	27,000	27,000
7	130	—	10,500	10,500
8	300	17,000	—	17,000
Total				138,000

TABLE 9.2 Changing Inventory Levels to Meet the Demand

Quarter	Demand Forecast	Cumulative Demand	Production Level	Cumulative Production	Inventory	Adjusted Inventory with 270 at Beginning of Period 1	Cost of Holding Inventories ($1000s)
1	220	220	300	300	80	350	17.5
2	170	390	300	600	210	480	24.0
3	400	790	300	900	110	380	19.0
4	600	1390	300	1200	−190	80	4.0
5	380	1770	300	1500	−270	0	0
6	200	1970	300	1800	−170	100	5.0
7	130	2100	300	2100	0	270	13.5
8	300	2400	300	2400	0	270	13.5
Total							96.5

Plan 3: *Subcontracting.* A firm might prefer to produce an amount equal to its lowest requirements and meet the rest of the demand by subcontracting. The cost of such a plan amounts to $108,000, as computed in Table 9.3. Again, it may not always be feasible or desirable to subcontract. This decision leads us to a fourth plan, involving a mixed strategy.

Plan 4: *Mixed Strategy.* As a compromise, a firm might combine the pure strategies, thus designing a mixed strategy. This mixed strategy varies production capacity slightly up or down as aggregated demand varies. Drastic changes in production capacity are curtailed, and frequent hiring and layoff situations are avoided. For example, based on past experience and available personnel, management may decide to maintain a constant production rate of 200 per quarter and permit 25% overtime when the demand exceeds the production rate. To meet any further demand, the firm chooses to hire and lay off workers. Remember, in a mixed strategy, a host of other alternatives exist. Trial-and-error computations of such a plan can be carried out step by step, as shown in Table 9.4.

TABLE 9.3 Subcontracting Costs

Quarter	Demand Forecast	Production Units	Subcontract Units	Incremental Cost at $80 per Unit ($)
1	220	130	90	7,200
2	170	130	40	3,200
3	400	130	270	21,600
4	600	130	470	37,600
5	380	130	250	20,000
6	200	130	70	5,600
7	130	130	0	0
8	300	130	170	13,600
Total				108,800

TABLE 9.4 Mixed Strategy

Quarter	Units of Demand Forecast	Regular Time Production Units	Additional Units Needed After Regular Time	Overtime Production	Additional Units Need After Regular Time + Overtime	Cost of Inventory ($)	Cost of Overtime ($)	Cost of Changing Workforce ($)	Total Cost ($)
1	220	200	20	50	−30 (−30)[c]	1500	1000	0	2,500
2	170	200	−30	—	−30[a] (−60)	3000	0	0	3,000
3	400	200	200	50	150[b] (90)	0	1000	9,000	10,000
4	600	200	400	50	350 (350)	0	1000	26,000	27,000
5	380	200	180	50	130 (130)	0	1000	33,000	34,000
6	200	200	0	—	—	0	0	19,500	19,500
7	130	200	−70	—	−70 (−70)	3500	0	0	3,500
8	300	200	100	50	50 (−20)	1000	1000	0	2,000
Total									$101,500

[a]Note that the inventory in period 2 is sixty units.

[b]If the existing inventory of sixty units is used, an increase of only ninety units is required.

[c]Negative quantities in parentheses indicate inventories, and positive quantities in parentheses denote the quantities to be produced by changing the capacity.

Our results show that the mixed strategy incurs an incremental cost of \$101,500. A smart manager would notice that subcontracting would be cheaper than hiring and lay-offs in this situation, however. The manager might also analyze the effect of constant overtime production, if permissible. These variations are left as exercises for the reader in a problem at the end of this chapter. Several combinations of pure strategies are possible. In many instances, policies are dictated by upper management. Although mixed strategies using graphical methods are not optimal, they do provide lower management with practical guidelines for day-to-day operations. In the real world, firms that have some guidelines operate much more efficiently than those that do not.

Mathematical Programming and Tabular Methods

Several versions of mathematical programming models can be formulated, depending on the complexities of the assumptions that are made. Many models have been proposed by Bowman [3], Hanssmann and Hess [10], Von Lanzenauer [28], and others for solving aggregate planning problems. A few notable examples are discussed in this section.

Transportation Model: Absorbing Demand Fluctuations with Regular Capacity and Overtime. The problem described here, originally formulated by Bowman [3], is a special case of a linear programming model. The transportation method can be used to analyze the effects of holding inventories or backordering, using overtime, and subcontracting. When more factors are introduced, such as hiring and layoffs or the cost of changing production level, the more flexible simplex method of linear programming must be used. The transportation models are relatively easy to solve with available computer routines. We also see from the structure of the model that the full power of the transportation algorithm is not needed. We elaborate on an algorithm proposed by Land [15] and on another shortcut tabular method for solving the special case of the transportation model. We find from the following example that the tabular method is a convenient worksheet for these approaches.

Example 9.2: Absorbing Demand Fluctuations
with Regular Capacity, Overtime, and Subcontracting

Warren Rogers Associates produces minicomputers that have a seasonal demand pattern. We are required to plan for the optimum production rates and inventory levels

	Supply Capacity			Demand Forecast	
Period	Regular Time	Overtime	Subcontract	Period	Units of Demand
1	700	250	500	1	500
2	800	250	500	2	800
3	900	250	500	3	1700
4	500	250	500	4	900

for the next four quarter periods. The available production capacities during regular time and overtime, as well as other cost data, are as follows:

Available initial inventory: 100 units

Desired final inventory: 150 units

Regular time cost/unit: $100/unit

Overtime cost/unit: $125/unit

Subcontract cost/unit: $150/unit

Inventory cost/unit/period: $20/unit

Table 9.5 formulates the problem in the desired tabular format. The supply consists of beginning inventory and units that can be produced using regular time, overtime, and subcontracting. Demand consists of the individual period requirements and any desired final inventory. The costs of producing and carrying inventory until a later period are entered in the small boxes inside the individual cells of the matrix. The problem can now be solved utilizing the standard transportation algorithm or a computer software package. Land [15] apparently has been working on the same problem. He suggests an alternative easier method of arriving at the solution (see Table 9.6). To facilitate easy understanding of the computations shown in Table 9.6, we define the source codes as follows: 1 = regular time production, 2 = overtime production, and 3 = subcontracted production. Columns 1 and 2 are self-explanatory; column 3 denotes the source of supply and its associated period. For example, (1, 1) means that during period 1, from source 1 (regular time production), at a cost of $100 per unit (column 4), 700 units are available (column 6). A comparison of availabilities during period 1 indicates that source 1 is the least costly, source 2 the next least costly, and so on (column 5). Naturally, if all requirements were met by the least-costly source, there would be no need to use other sources. These units can be made available for the next period if necessary, however, incurring an inventory carrying cost. The new cost is shown in column 9. Notice that the subcontractor does not charge an inventory cost. No matter how far in advance we order them, the subcontractor could deliver for the same price. During period 2, we list all available sources (1, 2, and 3) as well as the unused capacities from period 1, with the associated new cost. All sources are ranged for cost, and the iteration is repeated until demands for all period are met. The total cost of such allocation amounts to $445,750.

Tabular Method.   We suggest a much simpler approach. We deviate from the transportation procedure after the small boxes in each cell are posted with costs (see Table 9.5). We inspect each forecast period column and select the source that costs the least, the next least, and so forth. In column 1 the beginning inventory is the cheapest source; in column 2 procuring through regular time is next cheapest. Therefore, these allocations are made. Once the source is exhausted, we simply draw a dash in all cells corresponding to the source. During period 3, the allocation exhausts all current production sources. Three hundred units are obtained from the regular time production during period 1 and stored until period 3. Similarly, all other cheaper sources have been utilized, as we

TABLE 9.5 Transportation Method of Solving Aggregate Production Problem

Period	Source of Supply/Production	Period in Which Product is Forecast to Be Sold				Unused Capacity	Total Available Capacity
		1	2	3	4		
	Beginning inventory	0 (100)	20 (—)	40 (—)	60 (—)	0 (—)	100
1	Regular time	100 (400)	120	140 (300)	160 (—)	40 (—)	700
1	Overtime	125	145	165	185	0 (250)	250
1	Subcontract	150	150	150	150	0 (500)	500
2	Regular time		100 (800)	120 (—)	140 (—)	40 (—)	800
2	Overtime		125	145 (250)	165 (—)	0 (—)	250
2	Subcontract		150	150	150	0 (500)	500
3	Regular time			100 (900)	120 (—)	40 (—)	900
3	Overtime			125 (250)	145 (—)	0 (—)	250
3	Subcontract			150	150	0 (500)	500
4	Regular time				100 (500)	40 (—)	500
4	Overtime				125 (250)	0 (—)	250
4	Subcontract				150 (300)	0 (200)	500
	Demand	500	800	1,700	1,050	1,950	6,000

Note: The cost of unused capacity during regular time is $40 per unit. The solution provided inside the circles is explained by the tabular method.

TABLE 9.6 Solution to the Aggregate Production Problem Using Land's Algorithm

(1) Period	(2) Demand	(3) Supply Period	Source	(4) Cost at t	(5) Rank	(6) Available at t	(7) Used during t	(8) Available at t − 1	(9) Cost at t − 1 ($)	(10) Total Cost ($) [(4) × (7)]
1	Initial inventory 100									
	Demand 500									
	Net requirement 400	1	1	100	1	700	400	300	120	40,000
		1	2	125	2	250		250	145	
		1	3	150	3	500		500	150	
2	800	2	1	100	1	800	800	0	145	80,000
		2	2	125	3	250		250	150	
		2	3	150	5	500		500	140	
		1	1	120	2	300		300	165	
		1	2	145	4	250		250	150	
		1	3	150	5	500		500		
3	1700	3	1	100	1	900	900			90,000
		3	2	125	2	250	250			31,250
		3	3	150	5	500		500	150	
		2	2	145	4	250	250			36,250
		2	3	150	5	500		500	150	
		1	1	140	3	300	300			42,000
		1	2	165	6	250		250	185	
		1	3	150	5	500		500	150	
4	Demand 900	4	1	100	1	500	500	200	150	50,000
	Inventory 150	4	2	125	2	250	250	500	150	31,250
	Total 1050	4	3	150	3	500	300	500	150	45,000
		3	3	150	3	500		500	150	
		2	3	150	3	500		250	205	
		1	2	185	4	250		500	150	
		1	3	150	3	500				
Total cost										445,750

271

can see in Table 9.5. The cost of this program is $445,750. The total is obtained by multiplying the quantity in each cell by the corresponding cost and then adding all costs.

Modifications to the Standard Transportation Method. We assumed in Table 9.5 that products can always be stored and used during succeeding periods. A backorder situation was not considered, however. Since backordering assumes that current demand can be satisfied at a future time, appropriate cost figures can be entered in the boxes. Then the problem is solved in the usual manner. In many instances there is a cost associated with the unused regular time capacity. If such costs are known, they should be posted in the corresponding cells. In Table 9.5 notice that we have posted $40 per unit for the unused regular time production. The total cost of a program would be higher if the regular time capacity during any period were not used.

The problems illustrated so far started with initial inventory that was used during the first period. Because of uncertainties in demand, a firm may want to carry safety stocks. In such instances the safety stocks should be added to the demand, and total production requirements for the periods should be computed as follows:

$$p_t = I_t + D_t - I_{t-1}$$

Once the initial tableau is formulated with correct requirements instead of demands, the problem may be solved in the usual manner. Problems at the end of this chapter illustrate this concept. Note that the inventory requirements are implicitly included in the demand function.

The Linear Programming Method

The tabular methods assumed that production capacities can be changed within the specified upper bounds. If hiring and layoffs were employed, the model ignored the penalty costs associated with those activities. Also, desired inventories were expressed implicitly in the demand row. In a linear programming (LP) formulation, however, all these variables and their associated costs can be stated explicitly. The LP algorithm provides a solution with a mixed strategy, so that the total cost of the program is minimized.

The use of the LP model implies that a linear function adequately describes the variables for the firm studied. Hanssmann and Hess [10] developed a comprehensive LP model for production planning that was followed by many others. Excellent discussions of various LP models are found in Buffa and Miller [5], Groff and Muth [9], Lasdon and Terjung [16], Narasimhan and Gruver [20], and Shore [24]. For illustration purposes, we present here a variation of a simpler model formulated by Shore. We will start the discussion by listing many common assumptions; then we present a specific model.

Assumptions

1. Demand rate D_t is known and is assumed to be deterministic for all future time periods.
2. The costs of production during regular time are assumed to be piecewise lin-

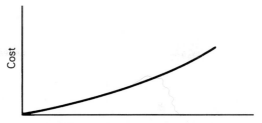

FIGURE 9.9   Regular time production rate.

ear, as exhibited in Figure 9.9. To ensure that the regular time capacities are fully utilized before using overtime, and to avoid subcontracting before using all available overtime, the following cost assumption is made:

$$c_3 > c_2 > c_1$$

where c_1, c_2, and c_3 are the production costs during regular time, overtime, and subcontracting, respectively.

3. The costs of changes in production level are approximated by a piecewise linear function, as shown in Figure 9.10.
4. Lower and upper bounds are usually specified on production quantities and on inventory levels, representing the limitations on capacities and available space, respectively.
5. A cost is always associated with inventory/backlog levels, although the unit cost could vary each period.

The Model.   The objective function and constraints of a general linear model follow. In this model, it is assumed that the manager is interested in minimizing the total costs of production, hiring, layoffs, overtime, undertime, and inventory. Minimize

$$C = r\sum_{t=1}^{k} P_t + h\sum_{t=1}^{k} A_t + f\sum_{t=1}^{k} R_t + v\sum_{t=1}^{k} O_t + c\sum_{t=1}^{k} I_t$$

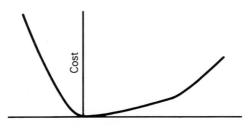

FIGURE 9.10   Changes in production level.

Subject to constraints

$$P_t \leq M_t; \qquad t = 1, 2, \ldots, k \tag{9.1}$$

$$O_t \leq Y_t; \qquad t = 1, 2, \ldots, k \tag{9.2}$$

$$I_t = I_{t-1} + P_t + O_t - D_t; \qquad t = 1, 2, \ldots, k \tag{9.3}$$

$$A_t \geq P_t - P_{t-1}; \qquad t = 1, 2, \ldots, k \tag{9.4}$$

$$R_t \geq P_{t-1} - P_t; \qquad t = 1, 2, \ldots, k \tag{9.5}$$

and all

$$A_t, R_t, I_t, P_t, O_t \geq O \tag{9.6}$$

where

r, v = cost/unit produced during regular time and overtime, respectively

P_t, O_t = units produced during regular time and overtime, respectively

h, f = hiring and layoff costs per unit, respectively

A_t, R_t = number of units increased or decreased, respectively, during consecutive periods

c = inventory costs per unit per period

D_t = sales forecast

The constraints (9.1) and (9.2) mean that the maximum production during regular time P_t and overtime O_t cannot exceed the available capacities M_t and Y_t, respectively. The third constraint expresses the inventory relationship. By defining the inventory variables as nonnegative, along with other variables in equation (9.6), we impose a no-backorder condition in the model. The constraints (9.4) and (9.5) express the hiring and layoffs when the production rate is increased or decreased during consecutive periods. Given the information required, the problem can be solved using any standard linear programming computer code.

Example 9.3

The following data were obtained from Joyce Manufacturing Company. Using the foregoing linear programming model, find the optimal production and workforce levels. Backorders are not allowed.

$$
\begin{array}{llll}
D_1 = 200 \text{ units} & M_1 = 180 & Y_1 = 30 \\
D_2 = 50 \text{ units} & M_2 = 120 & Y_2 = 20 \\
D_3 = 75 \text{ units} & M_3 = 120 & Y_3 = 20 \\
v = \$15/\text{unit} & f = \$10/\text{unit} & I_0 = 0 \\
c = \$5/\text{unit} & r = \$10/\text{unit} \\
h = \$30/\text{unit} & P_0 = 150 \\
\end{array}
$$

Solution: The problem can be formulated as follows. Minimize

$$C = 10(P_1 + P_2 + P_3) + 30(A_1 + A_2 + A_3) + 10(R_1 + R_2 + R_3)$$
$$+ 15(O_1 + O_2 + O_3) + 5(I_1 + I_2 + I_3)$$

Subject to

$$P_1 \leq 180$$
$$P_2 \leq 120$$
$$P_3 \leq 120$$
$$O_1 \leq 30$$
$$O_2 \leq 20$$
$$O_3 \leq 20$$
$$P_1 + O_1 - I_1 = 200$$
$$P_2 + O_2 + I_1 - I_2 = 50$$
$$P_3 + O_3 + I_2 - I_3 = 75$$
$$P_1 - A_1 \leq 150$$
$$-P_1 + P_2 - A_2 \leq 0$$
$$- P_2 + P_3 - A_3 \leq 0$$
$$P_1 + R_1 \geq 150$$
$$-P_1 + P_2 + R_2 \geq 0$$
$$-P_2 + P_3 + R_3 \geq 0$$

and

$$P_t, O_t, A_t, R_t, I_t \geq 0$$

The linear programming problem is also exhibited in the tableau (Figure 9.11). We can also solve the problem using any linear programming software package (we used the IBM-MPS package). The solution is as follows:

$P_1 = 170$	$P_2 = 62.5$	$P_3 = 62.5$
$O_1 = 30$	$O_2 = 0$	$O_3 = 0$
$I_1 = 30$	$I_2 = 12.5$	$I_3 = 0$
$A_1 = 20$	$A_2 = 0$	$A_3 = 0$
$R_1 = 0$	$R_2 = 107.5$	$R_3 = 0$

$$C = \$5137.50$$

In this model the inventories were assumed to be nonnegative variable. If backorder is desired, however, the variable I_t can be replaced by an expression $I_t^+ + I_t^-$ where I_t^+ and I_t^- represent inventory and backlog, respectively, during period t. The modified equation (9.3) is as follows:

$$I_t^+ - I_t^- = I_{t-1}^+ - I_{t-1}^- + P_t + O_t - D_t$$

Obj. Fcn.	10	10	10	15	15	15	30	30	30	10	10	10	5	5	5			
Deci. Var.	P₁	P₂	P₃	O₁	O₂	O₃	A₁	A₂	A₃	R₁	R₂	R₃	I₁	I₂	I₃	Sign	RHS	Remarks
	1															≤	180	
		1														≤	120	
			1													≤	120	
				1												≤	30	
					1											≤	20	
						1										≤	20	
	1			1									−1			=	200	
		1			1								1	−1		=	50	
			1			1								1	−1	=	75	
	1						−1									≤	150	
	−1	1						−1								≤	0	
		−1	1						−1							≤	0	
	1									1						≥	150	
	−1	1									1					≥	0	
		−1	1									1				≥	0	

FIGURE 9.11 Linear programming tableau for Example 9.3.

Similarly, if the desired final inventory is given, we can substitute its value in the appropriate equation. For example, suppose that the desired final inventory was $I_3 = 25$ in the foregoing example. We would simply modify the relevant equation by substituting 25 for I_3. With these modifications, we can solve such problems using an LP computer software package. Several problems involving these modifications are given at the end of this chapter.

The Linear Decision Rule

The linear decision rule was developed by Holt, Modigliani, Muth, and Simon [12] (and thus is popularly known as the HMMS rule), followed by their extensive study in the paint factory. When the costs discussed earlier can be approximated by ∪-shaped quadratic functions, the resulting decision rule formula turns out to be of simple linear form. The HMMS rule is intended to make possible routine calculations of the volume of production and size of workforce needed to be scheduled next period, be it month or week. The results are based on four cost factors: regular production costs, hiring and layoff costs, overtime and undertime costs, and inventory and shortage costs.

Regular Production Costs.   The cost of regular time production is the function

$$C_t(1) = c_1 W(t) \tag{9.7}$$

It is assumed that the cost of production is linearly related to the size of the work-force W_t in period t, as shown in Figure 9.12. Although a term involving an additional fixed cost could be added, it would not alter the optimal solution. The cost parameters here and in the following equations were determined from the company cost data.

Hiring and Layoff Costs. The cost of increasing and decreasing workforce level is assumed to be a quadratic function:

$$C_t(2) = c_2[W_t - W_{t-1}]^2 \tag{9.8}$$

where $W_t - W_{t-1}$ is the change in the workforce level from period $t - 1$ to period t. The function signifies that an increase or decrease in workforce level means the same cost for the given amount. Differences in costs can be introduced by including a constant in ex-pression (9.8). The optimal decision rule, however, will not be affected.

Overtime and Undertime Costs. For any given workforce size W_t, it is assumed that there is a desirable production rate KW_t. A lower production level means that work-ers are idle, and a higher production level signifies that overtime is required. The cost function is given by

$$C_t(3) = c_3[P_t - c_4W_t]^2 + c_5P_t - c_6W_t \tag{9.9}$$

where P_t is the production rate and W_t is the size of the workforce, as defined previously. The equation is quadratic in P_t and W_t. Note that the cost function is for a particular value of W_t and that a whole family of curves exists, one for each value of W_t.

Inventory and Shortage Costs. If the inventory level at the end of the period t is I_t, then

$$I_t = I_{t-1} + P_t - S_t \tag{9.10}$$

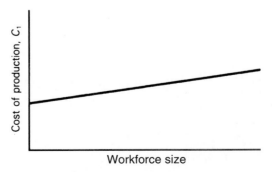

FIGURE 9.12 Regular production costs.

where I_{t-1} is the inventory at the end of period $t-1$ and S_t is the sales forecast of the actual shipments from the facility in period t. The total cost of inventory, including holding costs and shortage costs, is assumed to take the quadratic form:

$$C_t(4) = c_7[I_t - (c_8 + c_9 S_t)]^2 \tag{9.11}$$

Since the value of S_t is unknown, a whole family of curves based on each value of S_t can be drawn.

The Decision Rule. The total cost for any period t is given by

$$C_t = C_t(1) + C_t(2) + C_t(3) + C_t(4) \tag{9.12}$$

and total cost over t periods is given by the sum of individual period costs:

$$C_T = \sum_{t=1}^{T} \sum_{i=1}^{4} C_t(i)$$

It is important to note that we are attempting to minimize the expected costs. Since the exact demand is not known in advance, we deal in terms of average or expected costs. The minimization of the foregoing mathematical equation in terms of differentiation leads to the simple linear rules for W_t and P_t:

$$P_t = a_0 S_t + a_1 S_{t+1} + a_2 S_{t+2} + \ldots + g_1 W_{t-1} - h_1 I_{t-1} + c_1 \tag{9.13}$$
$$W_t = b_0 S_t + b_1 S_{t+1} + b_2 S_{t+2} + \ldots + g_2 W_{t-1} - h_2 I_{t-1} + c_2 \tag{9.14}$$

These equations are known as the production and employment linear decision rules. All lowercase coefficients are constants. Notice that each equation consists of a series of terms that include the forecasts for a given number of future period, and that each expression considers the present levels of inventory and personnel as well as the weighted forecast of sales. The optimal decision rules for the paint factory were derived by Holt, Modigliani, Muth, and Simon [12] as follows:

$$P_t = \begin{bmatrix} 0.463 S_t \\ 0.234 S_{t+1} \\ 0.111 S_{t+2} \\ 0.046 S_{t+3} \\ 0.013 S_{t+4} \\ -0.002 S_{t+5} \\ -0.008 S_{t+6} \\ -0.010 S_{t+7} \\ -0.009 S_{t+8} \\ -0.008 S_{t+9} \\ -0.007 S_{t+10} \\ -0.005 S_{t+11} \end{bmatrix} + 0.993 W_{t-1} - 0.464 I_{t-1} + 153 \tag{9.15}$$

$$W_t = 0.743W_{t-1} + 2.09 - 0.010I_{t-1} + \begin{bmatrix} 0.0101S_t \\ 0.0088S_{t+1} \\ 0.0071S_{t+2} \\ 0.0054S_{t+3} \\ 0.0042S_{t+4} \\ 0.0031S_{t+5} \\ 0.0023S_{t+6} \\ 0.0016S_{t+7} \\ 0.0012S_{t+8} \\ 0.0009S_{t+9} \\ 0.0006S_{t+10} \\ 0.0005S_{t+11} \end{bmatrix} \qquad (9.16)$$

The total cost of the program is given by

$$C_N = \sum_{t=1}^{n} \{(340W_t) + [64 \bullet 3(W_t - W_{t-1})^2] \\ + [0.20(P_t - 5.67W_t)^2 + 51.2P_t - 281W_t] \\ + [0.0825(I_t - 310)^2]\}$$

To determine the aggregate production rate, equations (9.15) and (9.16) would be used at the beginning of each month. They are easy to compute using a calculator.

Management Coefficient Model

Models discussed in previous sections attempts to capture the important trade-offs required by complex analytical structures. Difficulties arise, however, when we estimate the esoteric costs required for complex mathematical models. As an alternative, Bowman [4] developed the management coefficient model on the premise that management's past decisions can be incorporated into a system for improving present decisions. Based on his extensive consulting experience, Bowman reasoned that managers are aware of and sensitive to the variables that are important in the aggregate planning decisions. Managerial decisions might be improved by making them consistent from one time to another rather than by using optimal solution approaches, especially for problems where intangibles, such as run out cost or delay penalties, must be estimated or assumed. Using statistical regression analysis, rules can be developed that are generally of the following form:

$$P_t = a_1 S_t + a_2 W_{t-1} - a_3 I_{t-1} + a_4$$
$$W_t = b_1 S_t + b_2 W_{t-1} - b_3 I_{t-1} + a_4$$

where parameters a and b are constants derived from the regression techniques according to the past experience of the firm. These equations are based on the managerial decisions with the current workforce W_{t-1} and inventory level I_{t-1}, as well as the demand forecast S_t for the period t. Many multiple regression models may also be derived. For example, the model may account for the forecast in period $t + 1$, taking the form

$$P_t = a_1 S_t + a_2 S_{t-1} + a_3 W_{t-1} + a_4 I_{t-1} + a_5$$

Bowman [4] remarks, however, that as the number of forecast periods in the model increases, the regression gives poor results because of the high correlation between the forecast estimates. We note that as the number of forecast periods increases to eleven, Bowman's model becomes similar to the HMMS model. Using the cost equation provided by the HMMS model, however, we can predict the cost of implementing the decisions. Bowman's model also implies that the number of demand forecasts incorporated in the HMMS model is relatively unimportant. This is true in the sense that these constants are small and are uncertain. Neglecting these forecasts would not alter the solution significantly.

Direct Search Methods and Simulation

The linear decision rule assumes that a quadratic equation can be fitted to an organization's cost data. The relationships and cost functions between variables are assumed to be linear in the linear programming approach. It may be an inaccurate oversimplification, in many instances, to assume that these relationships are quadratic or linear. The linear decision rule and linear programming procedures also place restrictions on the mathematical structure necessary for the functions to be optimized, although they provide an optimum solution to the assumed model. The true cost, however, might need a higher-order equation. The mathematical complexity of the problem increases greatly when equations of higher order or step functions are present. Simulation and search methods can be used to obtain optimal solutions. Using computers, we can find many possible relationship values of the variables, select one that is acceptable, and provide a near optimal solution.

Three such models gained prominence in the 1960s. The structures of the cost equations and the methods used to obtain the optimum decision varied in these models. In 1966, Vergin [27] formulated a model that he called "scheduling by simulation." This computer simulation procedure used a search procedure to seek the minimum cost combination of values for the size of the workforce and the production rate.

In 1967, Jones [14] published a method he called "parametric production programming" (PPP). The PPP method used a computer search routine to examine various possible values for four parameters within the decision rule equations. Once the equation parameters were obtained satisfactorily, the decision rule could be applied as in the linear decision rule model. The advantage of the PPP model, however, was that it did not restrict the cost equations to any specific form, as linear decision rule or linear programming methods do.

In 1968, Taubert [26] published a procedure called the "search decision rule" (SDR) using the paint company data with the linear decision rule optimum solution as a test function. The SDR approach does not restrict the mathematical form of the cost equation and hence is superior to any linear, quadratic, or dynamic programming approach designed so far for this purpose. Taubert used a pattern search technique, which starts at a base point with a trial set of values for the workforce size (W_t) and production rate (P_t) for, say, the ten periods in the planning horizon. The model works with these twenty variables to arrive at a best combination. Small movements (variations in magnitude) are tested in a pattern around their base point, and the most promising direction is selected. A new trial point is selected by moving in the most promising direction. If this new trial point decreases the cost, it becomes the new base point, and the search continues for the new trial point. If the trial point does not improve the solution, however, the search goes back to the old base point and moves in a different direction to check whether the solution can be improved. The search terminates if no improvement can be found in any direction. The values for W_t and P_t of the current base point are recommended by the decision rule.

A flow chart of Taubert's search procedure is exhibited in Figure 9.13. Taubert points out that the pattern search method is adaptive and heuristic. For example, if the best direction to move in is the one that would be expected from the last move, the search proceeds in larger steps. If the solution does not improve, then a smaller step is taken from the base point. The adaptive search feature reduces the computer time necessary to complete the minimization process. It is important to note that the SDR approach does not guarantee optimality of solutions, although it facilitates specification of a realistic cost model. For practical purposes, however, this will be one of the satisfying solutions.

SUMMARY

The function of an intermediate-range aggregate plan is to anticipate forthcoming changes in workforce and production and to set overall goals for output rates. The overall goals may be translated into a detailed schedule, at least for the first planning period. Therefore, it is important to obtain the best-possible information for generating the plans. We have discussed many aggregate planning techniques that are suitable for either hand computation or computer-assisted methods. Table 9.7 compares the advantages and limitations of the various methods discussed in this chapter. The selection of an alternative plan is a trade-off between the desired accuracy and the cost of implementing such a program. Regardless of the method used, implementation of the plan is what matters. One problem could be obtaining necessary data. Cost data are usually obtained by using cost accounting methods, demand data from marketing and other sources, and capacity data from shop records. It is important to realize that many of the restrictions imposed on the plan are subjective. It is also essential to include as many constraints as possible when developing and evaluating alternative plans. In certain instances, very elaborate forecasting methods may be justified. For example, sophisticated methods may be used for demand forecasting, or learning curves could be used for calculating exact capacities. Although our discussion of aggregate plans centered on the manufacturing environment, plans for service systems are not much different. For some systems that supply standardized services to customers, aggregate planning may be even simpler than it is for production systems. Service industries include trucking firms, automotive services, fast-food

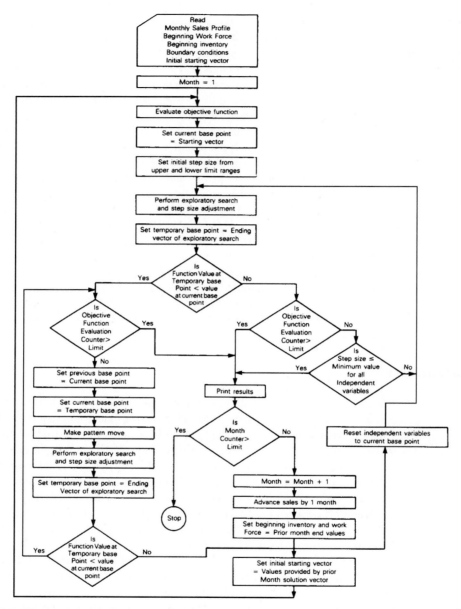

FIGURE 9.13 Flowchart of the search decision rule heuristic. From W. H. Taubert, "A Search Decision Rule for the Aggregate Scheduling Problems," *Management Science,* Vol 14, No. 6 (February 1978), pp. 343–359. Reprinted with the permission of *Management Science.*

TABLE 9.7 Comparison of Aggregate Planning Methods

Technique	Advantages	Limitations
1. Graphical methods	Simple to understand and easy to use.	Millions of solutions; chosen solution need not be optimal
2. Linear programming	Provides an optimal solution; several LP computer routines available; a popular tool in business and industry, and many managers seem to understand; sensitivity analysis easy to do; easy to add constraints on the availability of resources, warehouse space limitation and so forth; dual solutions provide shadow prices and help determine maximum allowable warehouse space, and so forth.	Mathematical functions must be linear; may not be realistic in real-world situations, although all right in most situations; ignores economic lot sizes for production runs; may plan for less than full capacity in some periods Demand is deterministic, but studies indicate it is acceptable as long as the data are updated periodically
3. Linear decision rule	Demand patterns need not be deterministic; provides an optimum solution to the given problem	Model development takes 1 to 3 months; incorporates some complex costs that are not available in a standard accounting system; ability to translate data into quadratic relationships needs very skilled personnel, and the quadratic relationship need not be true; model is insensitive to errors in cost estimates; values of variables are unconstrained; feasible solution is not guaranteed, although it is optimal
4. Management coefficients model	Attempts to duplicate managers' decision-making process; simplest, least disruptive, and easiest to implement, because it was developed from the past experience of management; when managers have limited information of forecasts, a plan based on past decision behavior is very useful for production planning	Solution need ot be optimal; assumes past decisions are good, which may not be true; takes a few personnel days to come up with a model; built on individual's behavior; any changes in personnel invalidate the model; selection of the rule is subjective
5. Search decision rules	Permits any realistic modeling by allowing all types of cost functions; alternative decisions can be tested; sensitivity analysis can be performed; modifications and variations to cost functions easily incorporated	Not absolute global optimum; takes 3 to 6 months' effort to develop a model; solution could be different depending on search routine used; cannot predict which search routine is more efficient for the given function; number of variables restricted because of response surface limitations, and cost of search cycle is expensive; requires expertise to use

stores, and banks and savings associations. Development aggregate plans for these systems poses no additional problems beyond those faced by most manufacturing systems.

PROBLEMS

1. Warren Rogers Associates has the following expected production capacity and demand:

Quarter	Capacity	Demand
1	300	300
2	400	600
3	450	300
4	550	500

The company does not accept any backorders and wishes to fulfill demand by letting inventories absorb all fluctuations. How many minicomputers must they have on hand on January 1 to meet the forecast demand throughout the year?

2. The vice-president of the Koza Company has estimated the following demand requirements for the forthcoming periods:

Period	Forecast	Period	Forecast
1	1400	5	2200
2	1600	6	2200
3	1800	7	1800
4	1800	8	1400

The operations manager is considering the following plans:

Plan 1: Maintain a stable workforce that is capable of producing 1800 units per period, and meet the demand by overtime at a premium of $50 per unit. Idle time costs are equivalent to $60 per unit. Do not build to inventory.

Plan 2: Produce at a steady rate of 1600 units per period, and accept a limited number of backorders during periods when demand exceeds 1600 units. The stockout cost of lost sales is $100 per unit. Inventory costs per period are $20 per unit.

Plan 3: Produce at a steady rate equal to minimum requirements of 1400 units, and subcontract the additional units at a $75 per unit premium.

Plan 4: Vary the workforce level, which is currently capable of producing 1600 units per period. The cost of additional workforce per 100 units is $5000, and the cost of layoffs per 100 units is $7500.

Plan 5: Vary inventory levels, but maintain a stable workforce level by maintaining a constant production rate equal to the average requirements. The company can accumulate required inventory before period 1 at no additional cost. The inventory cost per period is $20 per unit. Plot a histogram for the demand, and show the average requirements on your graph.

Discuss the merits and disadvantages of these plans. Which plan would you recommend?

3. The operations manager at Koza Company is considering the following mixed strategies. The costs given in problem 2 are also applicable to this problem.

Plan 6: Maintain the current stable workforce level, which is capable of producing 1600 units per period. Permit a maximum of 20% overtime at a premium of $50 per unit. Warehouse space contains the maximum allowable inventory to 200 units.

Plan 7:   Maintain the current workforce level, which is capable of producing 1600 units per period. Subcontract the rest of the requirements. A current inventory of 400 units is available.

Compare these strategies with the solutions obtained in problem 2, and make your recommendations.

4. Triden maintains a steady workforce of 40 persons per month. The company wants to analyze the cost of a mixed strategy employing inventories, overtime, and subcontracting as a means of absorbing demand. The cost per unit during regular time is $60, during overtime, $70; the cost of carrying is $5 per month; the cost of unused regular time capacity is $15. Find the optimum production cost for the plan, given the monthly demands and capacities shown in Table 9.8.

5. The OBS Company has a regular time production capacity of 30 units per period, and the subcontractor can supply a maximum of 50 units per period. The management forecasts the demand for the next four periods at 35, 20, 50, and 40 units. Given a production cost of $100 per unit, a subcontract cost of $130 per unit, an inventory cost of $2 per unit per period, and an unused production capacity cost of $50 per unit, provide a production plan for the company.

6. Giant Factory, Inc., found the following modified HMMS model to be very useful in aggregate planning:

$$P_t = \begin{bmatrix} 0.50S_t \\ 0.25S_{t+1} \\ 0.10S_{t+2} \\ 0.05S_{t+3} \\ 0.01S_{t+4} \end{bmatrix} + W_{t-1} + 150 - 0.50I_{t-1}$$

$$W_t = 0.75W_{t-1} + (2.0 - 0.01I_{t-1}) + \begin{bmatrix} 0.012S_t \\ 0.010S_{t+1} \\ 0.008S_{t+2} \\ 0.006S_{t+3} \\ 0.004S_{t+4} \end{bmatrix}$$

TABLE 9.8   Monthly Demands and Capacities, Problem 4

	January	*February*	*March*	*Unused*	*Capacity*
January					
Regular time					600
Overtime					300
Subcontracted					500
February					
Regular time					300
Overtime					300
Subcontracted					500
March					
Regular time					200
Overtime					300
Subcontracted					500
Demand	900	300	700		

Given a forecast demand for the next six months of 621, 415, 380, 763, 845, and 550 and an initial inventory and workforce size of 350 and 85, respectively, find the production plan for the next two months.

7. Determine the optimal production rates and workforce levels for the next four quarterly periods. The forecast demands for the next four quarters are 1600, 2100, 1800, and 1950 units (a final inventory of 300 units is desired at the end of the fourth quarter). The following information is available from the company records.

Current workforce: 600 workers

Current inventory level: 200 units

Inventory holding cost: $50/unit/quarter

Backorder cost: $100/unit/quarter

Cost of hiring: $1000/person

Cost of layoff: $1200/person

Regular time payroll cost: $5/hour

Overtime payroll cost: $7.50/hour

Each unit requires 160 person-hours to produce. Assume that each quarterly period consists of 480 regular time hours.

8. Dannon Company produces several types of slacks for retailing through department stores. The aggregate forecast (in thousands) for the next eight months is as follows:

	1	2	3	4	5	6	7	8
Forecast	100	200	200	150	200	250	100	125

Regardless of the type of material and the size, all the slacks take approximately the same amount of time to manufacture. The total cost of producing one unit during regular time and overtime is $3 and $4, respectively. With the current regular time capacity, Dannon can produce 150,000 slacks per month. Each person can produce 400 and 80 slacks per month during regular and overtime, respectively. The cost of hiring and laying off per worker amounts to $200 and $300, respectively. If the demand is not met during the month, the sales are lost. Inventory carrying cost amounts to 2% per month. The present inventory is 50,000 slacks.

a. Set up the equations for the linear programming problem.

b. Solve, using an LP code.

9. For Dannon Company, which one of the following plans would maximize their gross margin if their slacks are sold at $4.25?

Plan 1: Maintain the present workforce and use overtime for additional needs.

Plan 2: Vary the workforce to meet the needs of the month by hiring and laying off workers and work during regular time only.

Plan 3: Maintain a workforce approximately equal to the minimum requirements. Subcontract the rest at $1.10 incremental cost.

Graph the production and requirements for all the plans.

10. For the ABC Corporation example in this chapter (Example 9.1), the manager decides that he can have two hours of overtime per day. If it is not sufficient, he will add an eight-hour Saturday as overtime. Based on this information, design a strategy, assuming that it costs $50 more per unit for overtime.

11. In problem 8, if Dannon Company were able to hold back orders at 5% per month, what would be your solution?

12. In problem 4, suppose that Triden has an initial inventory of 50 units and requires 200 units of safety stock at the end of March. What is the new optimum solution?

13. Clarkson Company is a producer of microwave ovens. A schedule of their expected production capacity and demand is as follows:

Month	Capacity	Demand
1	700	650
2	700	650
3	750	850
4	780	800
5	800	750
6	800	900

The company does not accept backorders and wishes to absorb all fluctuations through their inventory. How many ovens must they have on hand on January 1 to meet the forecast demand for the first six months?

14. A producer of display panels expects the following production capacity and demand for its product:

Month	Capacity	Demand
1	400	400
2	400	450
3	450	450
4	460	500
5	480	500
6	500	500
7	550	520
8	550	530
9	550	600
10	600	650
11	550	600
12	550	600

Assuming that the company lets its inventory absorb demand fluctuations, and with a January 1 inventory on hand of 300 units, what can it expect its December 31 inventory balance to be? Assuming an inventory cost of $2 per unit per period, calculate the total and average inventory costs.

15. The operations manager at Jarrett, Inc., has received estimates for demand requirements for the next six months, as follows:

Month	Forecast
1	1000
2	1200
3	1400
4	1800
5	1800
6	1600

The following plans are being considered:

Plan 1: Maintain a stable workforce that is capable of producing 1500 units per month, and meet the demand by overtime at a premium of $50 per unit. Idle *time costs are equivalent to $60 per unit.*

Plan 2: Produce at a steady rate of 1300 units per period, and accept a limited number of backorders during periods when demand exceeds 1300 units. Stockout costs of lost sales are $100. Inventory costs per period are $25 per unit.

Plan 3: Produce at a steady rate equal to minimum requirements of 1000 units, and subcontract the additional units at a $60 per unit premium.

Plan 4: Vary the workforce level, which is at a current production level of 1300 units per period. The cost of the additional workforce per 100 units is $3000, and the cost of layoffs per 100 units is $6,000.

Plan 5: Maintain a stable workforce, and vary inventory levels by maintaining a constant production rate equal to average requirements. Inventory required before January 1 has no additional cost. The inventory cost per month after January 1 is $25 per unit. Plot a histogram for the demand, and show the average requirements on your graph.

Discuss the merits and disadvantages of these plans. Which plan would you recommend?

16. In addition, the operations manager at Jarrett, Inc., will consider the following mixed strategies. The costs in problem 15 apply to this problem also.

Plan 6: Maintain the current workforce level capable of producing 1300 units per period. Subcontract the rest of the requirements. A current inventory of 300 is available.

Plan 7: Maintain the current workforce level capable of producing 1300 units per period. Permit a maximum of 20% overtime at a premium of $40 per unit. Warehouse space constrains the maximum allowable inventory to 180 units. Subcontract the rest of the requirements at $60 per unit incremental cost.

Compare these strategies with those in problem 15, and make your recommendations.

17. Crandell Co. maintains a steady workforce of 30 persons per month. The company desires to analyze the cost of a strategy employing inventory/overtime as a means of absorbing demand. The cost per unit during regular time is $50; during overtime it is $60; the cost of carrying is $5 per month; the cost of unused regular time capacity is $10. Find the optimum production cost for the plan, given the following monthly demands and capacities: The initial inventories amount to 100 units in January; regular

time capacity is 500 units; overtime capacity is 300 units; and the demand for the next three months is 500, 700, and 900.

18. Conrad Corporation has a regular time production capacity of 50 units per month, and the subcontractor can supply a maximum of 40 units per month. Forecast demand for the next six months is 60, 40, 65, 70, 35, and 65 units. Given that production cost is $150 per unit, inventory cost is $5 per unit per month, and unused capacity cost is $75 per unit, provide a production plan for the company.

19. Determine optimal production rates and workforce levels for the next four quarters. The forecast demands are 2200, 2700, 2300, and 2500 units, and a final inventory of 500 units is desired at the end of the fourth quarter. The following information is available from the company's records.

Current inventory level: 300 units

Current workforce: 700 workers

Inventory holding cost: $75/unit/quarter

Backorder cost: $125/unit/quarter

Hiring cost: $1500/person

Layoff cost: $1600/person

Regular time payroll cost: $10/hour

Overtime payroll cost: $15/hour

Each unit requires 200 person-hours to produce. Assume that each quarterly period consists of 600 regular time hours.

20. Determine optimal production rates and workforce levels for the next six months. Forecast demands are 800, 900, 1200, 1000, 1600, and 1400 units. Final inventory desired is 250. The following information is available:

Current inventory level: 200 units

Current workforce: 300 workers

Inventory holding cost: $50/unit/month

Backorder cost: $75/unit/month

Hiring cost: $800/person

Layoff cost: $1000/person

Regular time payroll cost: $7.50/hour

Overtime payroll cost: $10/hour

Each unit requires 120 person-hours to produce. Assume that each month consists of 240 regular time hours.

21. Develop the objective function and the constraint equations for the following aggregate output problem. Assume an eight-hour, twenty-day month. Person-hour demands for the next six months are 35,000, 25,000, 40,000, 45,000, 40,000, and 30,000. The costs are as follows:

Inventory holding cost: $0.30/person-hours/month

Regular time payroll cost: $6/person-hour

Overtime payroll cost: $9/person-hour

Hiring cost: $300/worker

Layoff cost: $500/worker

The initial status consists of a current workforce of 220 workers and a current inventory of 10,000 person-hours.

22. MBI, Inc., manufactures snowmobiles. The company uses the management coefficient model for their aggregate planning;

$$P_t = 2.0S_t + 1.75W_t - 0.05I_{t-1} + 25$$
$$W_t = 0.9W_{t-1} + 0.3S_t - 0.1I_{t-1} + 10$$

At present, MBI has 150 workers and an initial inventory of 500 units. Determine the production level and workforce necessary for the next five months. The forecast demands for the next five months are 1000, 800, 1500, 1200, and 1150. Comment on the quality of the production and workforce equations based on your results. Can you come up with a better set of equations?

23. Consider the HMMS model given in this chapter. Given the following information, find the production level, workforce, and total cost for the next two periods.

Period	Forecast	Period	Forecast
1	1000	8	1300
2	1500	9	1500
3	2000	10	1600
4	1750	11	1400
5	2000	12	1300
6	1800	13	1700
7	2200	14	1600

The current workforce is 200 workers, and the initial inventory is 850 units.

24. Given the following additional data, design a mixed strategy for Example 9.1 in the text:
    a. Use a minimum overtime of 50 units per period.
    b. Increase the initial output by 10% and a constant overtime of 50 units per period.

REFERENCES AND BIBLIOGRAPHY

1. G. L. Bergstrom and B. E. Smith, "Multi Item Production Planning— An Extension of the HMMS Rules," *Management Science,* Vol. 16, No. 10 (June 1970), pp. 614–629.
2. A. B. Bishop and T. H. Rockwell, "A Dynamic Programming Computational Procedure for Optimal Loading in a Large Aircraft Company," *Operations Research,* Vol. 6, No. 6 (November-December 1958), pp. 835–848.
3. E. H. Bowman, "Production Scheduling by the Transportation Method of Linear Programming," *Operations Research,* Vol 4, No. 1 (February 1956), pp. 100–103.

4. E. H. Bowman, "Consistency and Optimality in Managerial Decision Making," *Management Science,* Vol. 9, No. 2 (January 1963), pp. 310–321.

5. E. S. Buffa and J. G. Miller, *Production-Inventory Systems: Planning and Control* (Homewood, Ill.: Richard D. Irwin, 1979).

6. R. J. Ebert, "Aggregate Planning with Learning Curve Productivity," *Management Science,* Vol 23, No. 2 (October 1976), pp. 172–182.

7. S. Eilon, "Five Approaches to Aggregate Production Planning," *AIIE Transactions,* Vol 7, No. 1 (June 1975), pp. 118–131.

8. J. H. Greene, ed., *Production and Inventory Control Handbook* (New York: McGraw-Hill, 1970).

9. G. K. Groff and J. F. Muth, *Operations Management: Analysis for Decision* (Homewood, Ill.: Richard D. Irwin, 1972).

10. F. Hanssmann and S. W. Hess, "A Linear Programming Approach to Production and Employment Scheduling," *Management Technology,* Vol 1 (January 1960), pp. 46–52.

11. A. Hax and H. Meal, "Hierarchical Integration of Production Planning and Scheduling." In M. A. Geisler, ed., *Studies in Management Sciences: Vol. 1, Logistics* (New York: North Holland, 1975).

12. C. C. Holt, F. Modigliani, J. F. Muth, and H. A. Simon, *Planning Production, Inventories and Workforce* (Englewood Cliffs, N.J.: Prentice Hall, 1960).

13. R. E. Johnson and L. B. Schwarz, "An Appraisal of the Empirical Performance of the Linear Decision Rule for Aggregate Planning," *Management Science,* Vol 24, No. 8 (April 1978), pp. 844–849.

14. C. H. Jones, "Parametric Production Planning," *Management Science,* Vol 15, No. 11 (July 1967), pp. 843–866.

15. A. H. Land, "Solution of a Purchase Storage Programme: Part II," *Operational Research Quarterly,* Vol 9, No. 3 (1958), pp. 188–197.

16. L. S. Lasdon and R. C. Terjung, "An Efficient Algorithm for Multi Item Scheduling," *Operations Research,* Vol 19, No. 4 (July-August 1971), pp. 946–965.

17. W. B. Lee and B. Khumawala, "Simulation Testing of Aggregate Production Models in an Implementation Methodology," *Management Science,* Vol 20, No. 6 (February 1974), pp. 903–911.

18. T. G. Mairs et al., "On Production Allocation and Distribution Problem," *Management Science,* Vol. 24, No. 15 (November 1978), pp. 1622–1630.

19. R. E. McGarrah, *Production and Logistics Management: Text and Cases* (New York: John Wiley & Sons, 1963).

20. S. L. Narasimhan and W. A. Gruver, "Integrated R&D Production and Inventory System," *AIIE Transactions,* Vol. 11, No. 3 (September 1979), pp. 198–205.

21. R. Peterson and E. A. Silver, *Decision Systems for Inventory Management and Production Planning* (New York: John Wiley & Sons, 1979).

22. G. W. Plossi and O. W. Wight, *Production and Inventory Control* (Englewood Cliffs, N.J.: Prentice Hall, 1967).

23. W. T. Shearon, "A Study of the Aggregate Production Planning Problem." Unpublished doctoral dissertation. Colgate Darden Graduate School of Business Administration, University of Virginia, 1974.

24. G. Shore, *Operations Management* (New York: McGraw-Hill, 1973).

25. E. A. Silver, "A Tutorial on Production Smoothing and Work Force Balancing," *Operations Research,* Vol 15, No. 6 (November-December 1967), pp. 985–1010.

26. W. H. Taubert, "A Search Decision Rule for the Aggregate Scheduling Problem," *Management Science,* Vol 14, No. 6 (February 1968), pp. 343–359.

27. R. C. Vergin, "Production Scheduling under Seasonal Demand," *Journal of Industrial Engineering,* Vol 17, No. 5 (May 1966), pp. 260–266.

28. C. H. Von Lanzenauer, "Production and Employment Scheduling in Multistage Production Systems," *Naval Research Logistic Quarterly,* Vol 17, No. 2 (July 1970), pp. 193–198.

29. U. P. Welum, "An HMMS Type Interactive Model for Aggregate Planning," *Management Science,* Vol. 24, No. 5 (January 1978), pp. 564–575.

APPENDIX 9A

Multiproduct Production Scheduling at Owens-Corning Fiberglas
Michael D. Oliff and E. Earl Burch

Owen-Corning Fiberglas (OCF), pioneer in the development of glass in fiber form and currently the world's leading manufacturer of glass fiber products, has one of its largest manufacturing facilities in Anderson, South Carolina. Anderson produces a multitude of fiberglass products that results in extremely complex production-planning and scheduling problems.

In Anderson's manufacturing process, molten fiberglass is formed, the glass is spun onto various sized spools, and this stock is used to weave fabric and to produce chopped strand mat. Approximately 20 percent of the facility's capacity is utilized to produce the mat product.

Fiberglass mat is sold in rolls in various widths and weights, treated with one of three process binders, and trimmed on one or both edges or not at all. The entire product line consists of over 200 distinct mat items. Twenty-eight of these represent over 80 percent of total annual demand and are treated as high-volume "standard" products. The remaining items are characterized as low-volume, "special order" products. Mat is primarily used in the marine industry in construction of boat hulls. It is also used as reinforcement in pipeline construction and in construction of bathroom fixtures such as bathtubs and showers.

The mat is produced on two parallel processors (Mat Lines 1 and 2). Production must be assigned and sequenced in a cost-effective manner. The processors have different capacities, line 1 having approximately three times the production capacity of line 2. Line 1 can produce mat 76 inches wide, while line 2 is limited to 60-inch material. Fiber stock is creeled in, pulled through high-speed choppers and spread over the mat chain via a forming hood. The chain carries the product past binder applicators, drying ovens, and finally, compaction rollers which add strength to the mat. The product is then trimmed and rolled into 175- to 230-pound cylinders.

The process is a high-volume one in which product demand has typically exceeded supply for approximately six months out of the year. The mat lines provide an estimated contribution to profit of over $1,000 per hour of machine time. Direct costs for machine

Reprinted from *Interfaces,* Vol 15, No. 5 (September–October 1985), pages 25–34.

downtime average $275 per hour on line 1 alone. Data suggest that maintenance costs are related to the frequency of job changeover and that losses in contribution or direct costs are incurred each time a product change is made. In addition to downtime, expensive mat waste results from each job changeover. These costs are sequence dependent and vary considerably from product to product. Such costs have ranged from $15,000 to $50,000 monthly, based on schedules that have included as many as 75 job changes and resulted in related downtimes in excess of 50 hours per month.

THE PROBLEM

The scheduling system developed for the mat line addresses the interaction between aggregate planning, lot-size determination, and ultimate job sequencing. Specifically, the system determines

1. An aggregate plan that reflects the relevant costs for workforce, overtime, and inventory;
2. Production run quantities, line assignments, and inventory levels for each standard product; and
3. Specific production sequences for standard and special order items.

For implementation, Owens-Corning Fiberglas required a flexible planning tool that could address each of the stated objectives. The approach had to be practical in terms of data maintenance; it also had to be user friendly and efficient. In addition, the company wanted an interactive model that could provide real-time response capability.

THE MODEL OVERVIEW

A discrete extension of Mellichamp's [1978] production switching heuristic (PSH) is used to smooth aggregate inventory workforce, and production levels and minimize related costs. A large-scale math program uses the derived aggregate monthly inventory levels and individual standard product demands to generate inventory levels, lot sizes, and line assignments for each of the standard items. The resulting schedule covers a three- to 12-month planning horizon. The weekly and monthly scheduling of standard and special order production is then accomplished by a third program that, given lot sizes and line assignments, minimizes the relevant setup costs. The ultimate schedule implicitly reflects the aggregate workforce costs and production costs, a property not found in standard scheduling and inventory models. The model is applicable to cases where demand is seasonal or purely random and forecasts are externally generated.

THE AGGREGATE PLAN

Aggregate production planning is the part of planning concerned with simultaneously establishing overall production, inventory, and employment levels for a given time horizon. The objective is to minimize the total of direct payroll costs, overtime costs, hiring and firing costs, and relevant inventory costs. Aggregate planning has received substantial treatment over the last three decades. Unfortunately, few of the theoretical approaches have been implemented in industry, primarily because of their conceptual or their mathematical complexity, and their unrealistic assumptions (continuous versus discrete production levels, for example).

Since heuristics are conceptually more simple than their mathematical counterparts, a heuristic based on Mellichamp's production-switching heuristic [1978] was used to determine aggregate production policy for Owens-Corning's mat lines.

The Anderson mat lines operate in any one of several shift setups. At full production, line 1 operates seven days a week, twenty-four hours a day, which requires four shifts, while line 2 runs three shifts, 24 hours a day., Monday through Friday. A shift setting will be denoted as *(i,j)*, where *i* is the number of shifts on line 1 and *j* is the number on line 2. Due to process and organizational constraints, only (4,3) (4,2) (4,1) (3,3) (3,2) (3,1) (0,0) are considered practically feasible. Each shift setting results in an expected monthly production rate given in Table 1. These rates are based on average hourly production. The total available production hours per shift, per month are multiplied by an hourly production rate (in pounds) to determine the expected monthly production level, *P(t)*.

The discrete objective function seeks to minimize the sum of direct payroll costs, overtime costs, hiring and firing costs, and relevant inventory costs over a time horizon that ranges from three to 12 months. Payroll costs are a linear function of the shift setting. Overtime costs are a linear function of the level of premium production. Hiring and firing costs are discrete multiples of changes in shift settings. Inventory costs are directly related to the difference between the average monthly inventory level and a target interval.

Owens-Corning has historically operated in a maximum of four different shift settings during any given year. Our models allow five distinct production levels $H1 > H2 > N > L1 > L2$ and two inventory controls, $A < C$.

The following rules are used:

$$P(t) = H1 \text{ if } \quad F(t) - l(t-1) + A > H1$$
$$= L2 \text{ if } \quad F(t) - l(t-1) + C < L2$$
$$= H2 \text{ if } H1 > F(t) - l(t-1) + A > H2$$
$$= L1 \text{ if } L1 < F(t) - l(t-1) + C < L1$$

$$= N \text{ otherwise,}$$

TABLE 1 Monthly shift setting and corresponding production levels.

Shift Setting	Production in Pounds per month
(4,3)	1,796,000
(4,2)	1,679,000
(4,1)	1,571,000
(3,3)	1,405,000
(3,2)	1,288,000
(3,1)	1,180,000
(0,0)	0

where *F(t), I(t),* and *P(t)* are monthly forecasts, inventory levels, and production levels (pounds), respectively.

The switching rule is interpreted as follows: Production occurs at level *H*1 if net demand for period *t,* after accounting for entering inventory *I(t−1)* and the minimum ending inventory target *A,* is greater than *H*1 (the maximum production setting). *L*2, the lowest output level, is used if net demand (accounting for entering inventory and the maximum inventory target) is less than *L*2. The remaining switches are interpreted similarly. The determination of *H*1, *H*2, *N*, *L*1, *L*2, *A*, and *C* is accomplished via an interactive simulation approach. The simulation explicitly calculates the costs for each possible rule over the specific cost region increments of 100,000 pounds (less than two days' production) are used in stepping systematically through approximately 100,000 distinct sets. The algorithm compares each possibility and chooses the rule that minimizes total costs.

The production-switching rule (PSR) is interactive in nature. OCF exercises the option to view the total set of regular and overtime production settings (in this case the seven from Table 1) and inventory target levels or any portion of the settings. Aggregate plans can be determined based upon restricted shift settings, with and without overtime, and with varying ranges for inventory targets. Starting conditions are provided, followed by the actual production plan for inventory levels, production rates, workforce settings, and overtime levels. Aggregate costs are then calculated for each planning component and explicitly given.

THE DISAGGREGATION PLAN

The disaggregate model relies on desired aggregate inventory levels *I,* as determined by the aggregate model. It also requires monthly forecasts F_{it} of individual standard product demands. Given this input, the model generates lot sizes, line assignments *X(i,r,m,p)*, and inventory levels *I(i,m)* for individual products. This information is then utilized by a scheduling heuristic to provide schedules that minimize sequence-dependent chargeovers on the parallel processors.

The formal mathematical model is composed of continuous and zero-one variables. The system minimizes total relevant changeover costs and production costs in determining the assignment and subsequent sequencing of jobs on parallel processors.

Graves [1982], Bitran and Hax [1977], and others focus on the problem with single processors or unconditional setups. Their models do not address the interaction between production costs and changeover costs. The approach of this research is similar to the hybrid method of Graves [1982] and, therefore, the hierarchical approach of Hax and Meal [1975].

Owens-Corning's specific problem entails the classification of products into two major types—standard and special order. Demand for the 170 special order products represents only 10 percent of total sales and cannot be predicted with any measure of reliability. The custom nature of these products and short customer lead times prohibit grouping. The lot-sizing decisions are then limited to the set of inventory items—the standard products. The result is imply that relevant changeover costs must be treated at a lower level of the hierarchy that includes the special orders. The nature of this treatment can be justified by the dominance of production costs over changeover costs at the lot-sizing level.

The practical disaggregation problem at OCF is address utilizing a continuous version of the mixed variable model. The decision variables are:

$X(i,1,m,p)$ = 1 if product i is produced on line 1 during month m and subperiod p.
O otherwise;
$Z(i,j,1,m,p)$ = 1 if a changeover occurs from product i to product j on line 1 during month m and subperiod p.
O otherwise;
$I(i,m)$ = the total pounds of product i in inventory at the end of month m.

These variables have a physical interpretation in the continuous realm that enables a straightforward solution and enhances computations dramatically over the discrete version. Each $X(t,1,m,p)$, whose range is $(0,1)$, represents the proportion of line 1's production during month m and subperiod p which is devoted to product i. This interpretation is used exclusively in OCF's application to supply percentages of utilized capacity and final line assignments and lot sizes that are always feasible.

The constraints (see Appendix A) are of four functional types:

1. Production balance equations facilitate demand satisfaction while forcing inventory conservation;
2. Inventory capacity equations provide the explicit satisfaction of aggregate inventory requirements as well as safety stock demands;
3. Noninterference constraints force feasible assignments of products to processors; and
4. A set of changeover equations regulates and penalizes the resulting changeovers from product to product.

Lot sizes, line assignments, and sequences are determined when the Xs (and therefore the Zs) are restricted to binary values. Given the $X(i,l,m,p)$s, lot sizes are readily available via a simple multiplication. The percentage of product i on line 1 during month m is multiplied by i's corresponding production rate per hour and then by the total hours available for the given month. Inventories are read directly as continuous variables.

Owens-Corning Fiberglas requires lot-sizing and line assignments but not sequences at this level of planning. Demand can be estimated only for the standard products, while actual sequences include many of the special orders. It makes little sense to optimally sequence again to allow for the special items. The continuous LP generates (10,000 row by 10,000 column matrix) feasible, near optimal solutions in less than 60 CPU seconds on an IBM 3081. These solutions are readily understood and implemented by management. The mixed integer version can be practically applied only in cases where all product demands are predictable on a 30- to 90-day basis and no special orders exist.

Fiberglas's use of the model is made possible by the development of a matrix generator, MAGENI, that generates the required input for any of the standard math programming packages. MAGENI, a FORTRAN program, requires input in the form of parameters that describe the desired production setting. The number of standard products, the planning horizon, and monthly estimates of individual product demands are required input. The number of subperiods and lines are set to one and two, respectively, in this application.

MPSX is used to solve the linear program, and output is read by a COBOL program, TABGEN, which generates a monthly, quarterly, or yearly master schedule. The LP, run monthly, provides OCF with specific inventory levels, lot sizes, and line assignments for the coming months. These lots are then sequenced on a weekly or biweekly basis with a host of special orders to minimize sequence dependent changeover costs.

THE SEQUENCING HEURISTIC

Given the lot sizes and line assignments for the two parallel processors determined via the linear program, MAT1 schedules these jobs and new special orders on each line to minimize sequence-dependent setups. At the disaggregate level these costs do not dominate, but as the multitude of special orders is considered, changeover costs become increasingly significant.

Fiberglas scheduled an average of 70 mat products each month during the first and second quarters of 1981. The company incurred approximately $20,000 monthly in changeover-related expenses as a result of individual costs of $300 or more per production change. These figures do not reflect increasing maintenance costs or lost efficiencies and are lower bounds on costs that may exceed $300,000 annually.

This research began with a detailed study of the production process and the costs involved in changing from one product to another. These changeover costs are highly sequence dependent but can be classified in a hierarchical fashion. Because of the process's complex nature and as a direct consequence of the data, changeovers can be accurately grouped only by family and type and not individually. The family/type segmentation of Hax and Meal [1975] is applied in this setting to transitions that share certain process-related features, rather than to products that enjoy similar demand distributions.

Changeovers are classified as fiber changes, width changes, weight changes, or as slitter changes. For each of these families there are two changeover types. Width, weight, and slitter transitions involve increases or decreases, while the two input stocks give rise to the different fiber changes.

The dominant components of changeover-related costs are direct downtime and mat waste. Fiber changes result in the most costly transitions. Considerable time and effort are expended to creel entirely new stock into the process and then establish production equilibrium. Weight changes necessitate alterations in the line speed or the amount of input stock creeled in. These transitions result in costly mat waste as the process again attempts to reach a steady state of production. Width changes result in both direct downtime and mat waste. These transitions are not as costly as the preceding ones and occur physically as the forming hood is moved to a new width. The least costly changeovers involve changes in the number of cuts in a mat panel of a given width.

As indicated, each family of changes has similar changeover costs. A distinction can be made within each family based upon the direction of the change. It is much easier to decrease weight and width than to increase them. A distinction between product changeovers is not realistic within types.

The objective function is minimize

$$Q = \sum_{ij1} C_{ij1} X_{ij1}$$

where $\quad C_{ij1}$ = cost of changing from product i to product j on line 1
and $\quad X_{ij1}$ = the number of changeovers from product i to product j on line 1.

The sequencing heuristic, MAT1, takes advantage of the above changeover classifications in determining a schedule that minimizes the average cost of sequence dependent changeovers for a seven- to 30-day time horizon. The algorithm enumerates the set of sequences that are minimum cost candidates and selects the best possible sequence in the following manner. The jobs are sorted initially by type of family. Sequencing is then done within each family to avoid fiber changes, weight increases, width increases, and slitter repositioning, that is, directional lists (minimal cost candidates) are built to prohibit the most costly possible changeovers. These lists preclude, when possible, sequences that include multiple fiber changes, weight increases and width increase. The number of minimum cost candidates depends upon the initial job setting (a new schedule may role into an existing one) and upon the composition of jobs to be sequenced. If a pool of jobs contains different fiber types, weights, widths, and cutter positions, then several candidates will be generated.

OCF inputs a batch of jobs by code, amount, and line. The model prepares a final planning document that includes sequences, complete product descriptions, expected run times, and expected changeover costs. Real-time responses to potential changes as well as to new schedule requirements are obtained daily. The FORTRAN code requires 1–3 CPU seconds of time on an IBM 3081.

SUMMARY AND RESULTS

The total system can now be reviewed (Figure 1). Given aggregate demands $F(t)$ and a basic cost structure the PSR determines monthly levels of production, inventory, overtime, and labor that minimize related costs. Aggregate inventory targets $I(t)$ are passed to the linear program that determines lot sizes and line assignments to minimize the dominant production costs of the parallel processing. These lots are then sequenced with the host of special orders via the sequencing heuristic, MAT1, minimizing changeover-related costs.

Once phase three of the model is reached, the actual demand for special orders may exceed or fall short of expected demand. In the latter case, the current standard lot sizes are increased heuristically to minimize future setups. The former situation necessitates a reduction in the given standard lots, which is again accomplished through a

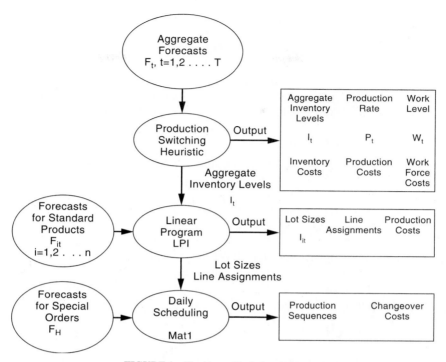

FIGURE 1 The hierarchical planning system.

straightforward heuristic. Neither of the above situations will be tolerated indefinitely. As safety stocks are depleted or capacities are exceeded, aggregate demands must be reestablished and the cycle repeated.

OCF has used the model to schedule over 20 million pounds of mat production during the past two years. The scheduling heuristic is used weekly to provide real-time cost estimates of actual and proposed schedule interruptions. The average number of monthly changeovers decreased from 70 in 1981 to less than 40 in 1982 and 1983, resulting in an estimated annual savings of $100,000 or more.

Operating efficiencies improved dramatically during this same period of time. The aggregate model has been used to predict future inventory positions as well as to plan future shift settings.

Due to the system's success, a second major production-planning model has been implemented, and a third is now being developed by Owens-Corning Fiberglas and Clemson University.

APPENDIX A: LP
PRODUCTION BALANCE

Given that $D(i,m)$ is the demand for product i during month m, let $A(i,1)$ be the production rate in pounds for product i on line 1 for one period. Then

$$I(i,\ m-1) - I(i,m)$$

$$+ \sum_1 \sum_p A(i,\ 1)X(i,1,m,p) = D(i,m) \tag{1}$$

for all i and m. These constraints force demand to be satisfied each month from sub-period production on either line or from last period's inventory.

INVENTORY CAPACITY

Given $K(m)$ is the desired total inventory level for month m determined by the aggregate model,

$$\sum_i I(i,m) = K(m) \tag{2}$$

for all m. Aggregate inventory levels are thus imposed as constraints upon the lot size and assignment decisions. Subsequent solutions will implicitly reflect the smoothing decisions made earlier at the aggregate level. Due to the stochastic nature of demand, it is required that

$$I(i,m) > M(i,m) \tag{3}$$

for all products i and months m. These constraints force accumulation of physical safety stocks $[M(i,m)]$ for each of the standard products.

NONINTERFERENCE

Only one product can occupy a processor at any given point in time. To guarantee sequencing feasibility it is required simply that

$$\sum_i X(i,1,m,p) = 1 \tag{4}$$

for all 1, m, and p.

CHANGEOVER

For each $X(i,1,\text{Im},p) = 1$, either $X(i,1m,p-1) = 1$, in which case no change occurs, or $X(i,1,m,p-1) = 0$ implying a changeover into product i takes place on line 1 at the beginning of month m and period p. If a change occurs, sequence- dependent costs are incurred and must be explicitly accounted for.

$$X(i,1,m,p-1)X(i,m,1,p)$$

$$= \sum_j Z(i,j,1,m,p) - \sum_i Z(i,j,1,m,p) \tag{5}$$

are the production changeover constraints for each $i,j,1,m,p$. It is also required that

$$\sum_i \sum_j Z(i,j,1,m,p) = 1 \tag{6}$$

for all l,m,p.

The objective function is then to minimize total costs, TC, where

$$\begin{aligned} TC = &\sum_{i,1,m,p} C1(i,1) \times X(i,1,m,p) \\ &+ \sum_{i,j,1,m,p} C2(i,j) \times Z(i,j,1,m,p), \end{aligned} \tag{7}$$

where $C1i,1)$ is the cost of one period's production of product i on line 1 and $C2(i,j)$ is the cost of changeover between product i and product j.

REFERENCES

BITRAN, G. R., and HAX, A. C. 1977, "On the design of hierarchical production planning systems," *Decision Sciences,* Vol 8, No. 1, pp. 28–55.

BITRAN, G. R.; HASS, E. A.; and HAX, A. C. 1981, "Hierarchical production planning: A single stage system," *Operations Research,* Vol. 29, No. 4 (July-August), pp. 717–743.

GRAVES, S. C. 1982, "Using Lagrangian techniques to solve hierarchical production planning problems," *Management Science,* Vol 28, No. 3 (March), pp. 260–275.

HAX, A. C. and MEAL, H. C. 1975, "Hierarchical integration of production planning and scheduling," in *Studies in Management Sciences, Rol. I, Logistics,* ed. M. A. Geisler, North Holland-American Elsevier, New York.

MELLICHAMP, J. M. and LOVE, R. M. 1978, "Production switching heuristics for the aggregate planning problem," *Management Science,* Vol 24, No. 4 (August), pp. 1242–1251.

OLIFF, M. D. 1982, "An integrated production planning model for multiproduct parallel processor environments," PhD Dissertation, Clemson University.

Chapter Ten

The
Master
Production
Schedule

INTRODUCTION

A master production schedule (MPS) represents a plan for manufacturing. When a firm uses an MRP system, the MPS provides the top-level input requirements. It develops the quantities and dates to be exploded for generating per period requirements for subassemblies, piece parts, and raw materials. The MPS is not a sales forecast, but it is a feasible manufacturing plan. It also serves as a customer order backlog system. It considers changes in capacity or loads, changes in finished goods inventory, and fluctuations in demand. A detailed MPS also determines the economics of production by grouping various demands and making lot sizes. Thus the MPS maintains the integrity of the total system backlogs, anticipated backlogs, and lower-level component requirements.

The MPS should be consistent with the aggregate production plan (APP) from which it is derived. It should consider, in detail, the unit of measure, such as pounds of steel or number of phones per period, the efficiency, and the utilization factors of the system. There are different time horizons, aggregation levels, and time buckets in MPS and APP. The relationship of the MPS to other manufacturing and control activities is shown in Figure 10.1. The output of the aggregate planning process is a set of parameters indicating aggregate inventory or backlog levels, the number of shifts to be operated, the number of employees to be hired or laid off, the anticipated amount of subcontracting, and the aggregate amount to be produced within certain time periods. The APP provides a basis for decision making regarding specific production dates, available capacity, total demand, lead time, or inventory constraints that cannot be reconciled within the company policy objectives. Although this information is necessary, it is not sufficient for the smooth functioning of a firm. What is necessary is a plan stated in terms of specific products that are to be produced in certain quantities by certain dates. The process of deriving such an MPS that is consistent with the overall APP is called disaggregation.

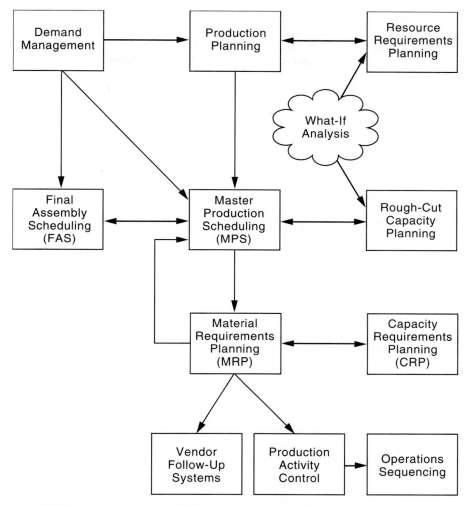

FIGURE 10.1   Relationship of MPS to other manufacturing planning and control activities.

It is important to recognize that the MPS is not a control technique or a system. Rather, it is a logical representation of information for decision making. The MPS highlights conflicts that can be solved only by people. If the MPS is done properly, the rest of the system can be harnessed to reach the desired objectives of management.

The Effect of Capacity on the MPS

The importance of an accurate and feasible schedule cannot be overemphasized. The MPS is an important input in deriving rough-cut capacity planning, as exhibited in Figure 10.1. The existing capacity and changes to it over the planning horizon become a major constraint. Considerations to capacity should include actual number of shifts scheduled, number of days scheduled in a week, overtime policy, available equipment, and workforce levels. The capacity should be expressed in terms of what is feasible, rather than in theoretical possibility estimates. In addition to scheduling hours, factors such as efficiency and utilization should be included.

When the total requirements specified by the inputs exceed available capacity, the MPS should indicate a need for corrective action. The alternative decisions may involve extending the delivery date, changing capacity, finding ways to divert parts from other activities, and so forth that may need critical inputs from management, especially when policy guidelines are inadequate.

Example 10.1

Suppose that a milling machine center has two machines. The center operates two eight-hour shifts per day, five days per week. Records show machine utilization at 95% and operator efficiency at 99%. What is the effective work center capacity per week? Is the master production schedule feasible?

Solution: Effective capacity is calculated by multiplying two shifts times two machines times eight hours per day times five days per week times 95% utilization times 99% operator efficiency. This gives approximately 150 hours of effective capacity per week.

Table 10.1 illustrates the capacity required if 200 units of product A were produced every other week and 150 units of product B were produced in the other weeks. Because the required capacities in periods 4 and 6 exceed the available capacity, the master schedule should be revised if additional capacity is not found.

Suppose that we make 100 units of product A and 75 units of product B every week instead of 200 units of A and 150 units of B in alternate weeks. The required milling machine capacities are summarized in Table 10.2. Note that the required and available capacities match. Other machine centers should similarly be checked for feasibility. The MPS may have to be revised several times before arriving at a feasible schedule.

Lead Time Constraints

The computation of lead times for MRP purposes is a complicated procedure. When end items, assemblies, subassemblies, and components are involved, the cumulative lead time, known as the *critical path lead time,* determines the earliest time that the end products could be built from the time an order is received. Orders cannot be accepted if the days remaining are less than the cumulative lead time. Tables 11.41 through 11.44 of Chapter 11 illustrate this point. One way to shorten the lead time is to carry inventory of long lead items. Unfortunately, this defeats the purpose of material requirements planning. In certain cases where a chemical process is involved, such as the distillation of beer, it is not possible to reduce the lead time.

TABLE 10.1 Capacity Requirements Plan, Milling Machine Center

| | Period | | | | | | | | | |
	1	*2*	*3*	*4*	*5*	*6*	*7*	*8*	*9*	*10*
				Lot 1, Group A 200	Lot 2, Group B 150	Lot 3, Group A 200	Lot 4, Group B 150	Lot 5, Group A 200		
Lot 1, Group A	$200 * 0.17$ $= 34.00$			67.60						
Lot 2, Group B			30.40	64.80	102.15					
Lot 3, Group B				34.00	30.40	67.60				
Lot 4, Group B						64.80	102.15			
Lot 5, Group A						34.00	30.40	67.60		
Total	34.00		30.40	166.40	132.55	166.40	132.55	67.60		

Box (between periods 7 and 8):
$$30.40 = 0.152 * 200$$
$$67.60 = 0.388 * 200$$

Gross requirements	6	7	8	9	10
Group A	200		200		200
Group B		150		150	

Gross requirement of 200 units of A in period 6 places a $0.17 * 200 = 34$ hours load on milling in period 2.

TABLE 10.2 Capacity Requirements Plan, Milling Machine Center

| | | *Period* | | | | | | | |
	1	*2*	*3*	*4*	*5*	*6*	*7*	*8*	*9*
POR									
Group A				100	100	100	100	100	100
Group B				75	75	75	75	75	
Group A, Period 4		17.0	15.2	33.8					
Group B, Period 4			32.4	51.1					
Group A, Period 5			17.0	15.2	33.8				
Group B, Period 5				32.4	51.1				
Group A, Period 6				17.0	15.2	33.8			
Group B, Period 6					32.4	51.1			
Group A, Period 7					17.0	15.2	33.8		
Group B, Period 7						32.4	51.1		
Group A, Period 8						17.0	15.2	33.8	
Group B, Period 8							32.4	51.1	
Total				149.5	149.5	149.5			

Inputs to the Master Schedule

Major inputs to the master schedule are the customer order backlog and the product sales forecast. The MPS requirements should also include (1) interplant requirements, (2) service parts requirements, and (3) distribution warehouse requirements. The backlogs are input as the firm or hard portion of the schedule, which has been committed through specific customer orders. The inputs should be very specific, such as the number of cases of beer or pounds of steel. If specified in terms of dollar value, these inputs would still need to be translated to measurable quantities or units, using conversion factors. The sales forecast provides the basis for extending the master schedule to generate the uncommitted or planned portion of inventories in anticipation of customer demand. This forecast leads directly into the order entry system. These inventories could fill seasonal peaks, promotional periods, or new-product introductions, which create capacity overloads that could affect customer service. When the orders or bookings are temporarily below production capacity, inventories help prevent underutilization and declining productivity.

Planning (Time) Periods

For production planning purposes, we want to consider intervals that are different from the forecast intervals. The objective of any system is to smooth the production process, enabling uniform production of items over the period and thus avoiding production of a month's quota during the last week of the month. The length of the planning period is a matter of convenience and compromise. Smaller periods facilitate a precise production schedule, but at the cost of extra data processing. Computing costs are more economical for longer periods, but at the loss of some precision. In auto assembly plants, for example, the orders and quantities are so large that the natural scheduling period involves a shift of production. A Boeing factory that produces 747 aircraft might find that a month is a short enough increment of time. For most manufacturing firms, a week or a fortnight

TABLE 10.3   Example of a Rolling Schedule

Initial table	0	1	2	3	4	5	6	7	8	9	10
Gross requirements	15	15	15	15	25	25	10	10	10	10	25
Table after one period	0	1	2	3	4	5	6	7	8	9	10
Gross requirements		15	15	25	25	10	10	10	10	25	30

is a satisfactory length of time. The MPS should be extended to cover at least twice the critical path lead time of the product. The total time horizon usually consists of twelve weekly periods, followed by months. As the fourth month moves into the twelfth weekly period, it must be broken down. This process is called a *rolling schedule.* An example of a rolling schedule is given in Table 10.3. The requirements for the tenth period were rolled over from the following period.

Due Date versus Need Date

The *due date* is the scheduled completion date associated with the order, whereas the *need date* is the time at which the order is actually needed. These two dates need not be the same. The need date depends on the customer requirements, whereas the due date is the result of priority planning. An MRP system has the ability to make these dates coincide at the time of order release. The system also monitors changes in the status of orders and signals the inventory planner, if necessary, to take action. Accordingly, the scheduler can expedite or deexpedite the order by scheduling it to an earlier or later date. Of course, capacity should be available to replan activities.

BILL OF MATERIAL TYPES

In production planning a bill of material (BOM) is a key input document for establishing a proper inventory control system. The BOM can best be described as a list that specifies the quantity of each item, ingredient, or material needed to assemble, mix, or produce an end product. Because inventory records (status) are an important part of MRP, an accurate BOM becomes a vital input to MRP systems. A collection of these bills—one for each assembly and one for each of its components, which are sometimes assemblies themselves—describes the flow of material through the manufacturing process in the plant. Thus a BOM also describes the relationships among parts.

Consider the Taj Mahal, with all its gorgeous marble domes and embedded precious stones, or the Leaning Tower of Pisa. Could they have been completed without a BOM? Consider, too, your grandmother's secret recipe for your favorite dish. Do you remember her ever consulting a cookbook for the dish that was gobbled up by everyone instantaneously? She probably used an informal BOM.

The BOM has several other uses. As an example, suppose that an item or part is delayed, and the manager wants to know the effects of the delay on the MPS. He or she has to know what actions are necessary to keep the shop going. A complete and accurate bill of material is needed to summarize parts requirements if an MRP system is to be used efficiently. The number and complexity of parts in most businesses today necessitates the use of computerized BOMs. It is important to note that the computations are only as accurate as the information you provide to the computer. The real value of computers in manufacturing BOMs, however, comes from three sources: (1) its ability to store massive amounts of data, (2) the speed with which information can be retrieved, and (3) the availability of software packages that organize and retrieve this information.

The relationships among parts can be represented in many ways, including the cross-classification chart, the product structure tree, and indented bills of material. In this section we illustrate these methods and show how the MPS calculations could become cumbersome even for a few products when they are extended to several planning periods.

The Cross-Classification Chart

The earliest and simplest method of representing the relationships among parts was by a matrix that is known as the cross-classification chart. It exhibits the subassemblies, parts, and raw materials that are used in each of the primary products. For example, row 1 in Table 10.4 represents product 1, which requires one unit of subassembly 4 and one unit of subassembly 6. One unit of subassembly 4, in turn, requires one unit of part 9 and two units of part 10; one unit of part 9 needs two units of raw material 12 and one unit of raw material 13. Thus the cross-classification chart shows the complete explosion of a product line.

TABLE 10.4   Cross-Classification Chart

Item	Subassembly (SA)				Part (P)				Raw Material (RM)			
	4	*5*	*6*	*7*	*8*	*9*	*10*	*11*	*12*	*13*	*14*	*15*
Finished Product												
1	1		1									
2		2										
3		1		2								
Subassembly (SA)												
4						1	2					
5					3							
6					1		1					
7						2		2				
Part (P)												
8									1			
9									2	1		
10										1	1	
11										2		3

Example 10.2

Prepare a list of the parts and raw materials required to manufacture the following:

Product	Quantity
Item 1	10
Item 2	20
Item 3	30

Solution: For clarity, we can represent the requirements for each item by means of a tree diagram. The tree diagram in Figure 10.2 shows, for example, that we need two units of raw material 12 to make one unit of part 9, that each subassembly 4 requires one unit of part 9, and finally that each item 1 requires one unit of subassembly 4, and so forth. We can compute the requirements for one branch at a time by multiplying all the quantities on that branch. The material requirements for producing ten units of item 1 are exhibited in Table 10.5. We can summarize these requirements as follows:

Raw Material	Quantity
RM12	30
RM13	40
RM14	30

Using the data in Table 10.4, we can also summarize the requirements for items 2 and 3. The overall material requirements for the three items are given in Table 10.6.

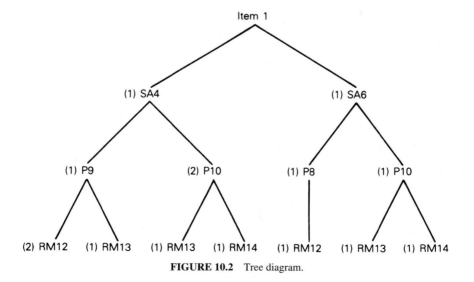

FIGURE 10.2 Tree diagram.

TABLE 10.5 Material Requirements for Ten Units of Item 1 (Example 10.2)

Part	Raw Material	Quantity	Where Used
P8	RM12	10	SA6
P9	RM12	20	SA4
P9	RM13	10	SA4
P10	RM13	20	SA4
P10	RM13	10	SA6
P10	RM14	20	SA4
P10	RM14	10	SA6

It is obvious that the calculations can become astronomical when hundreds of products (each product having many assemblies, subassemblies, and raw materials) are considered for several periods.

Product Structure Tree

The product structure tree is commonly used to display the total makeup of a particular product. An example for a bicycle is given in Figure 10.3. The highest level is zero, which represents the end item. Wheel and frame assemblies make up the next level, level 1. Similarly, tires and rim assemblies, seat, handlebars, and so forth make up level 2. The tree can be extended to the raw material level, level 4. The quantity of each subassembly or component used in producing one unit of a higher-level item is shown beside the item.

Single-Level Bill of Material

If a product was assembled from purchased parts of purchased subassemblies, then the bill of material would consist of a single level. In the bicycle example shown in Figure 10.3, if the front wheel assembly, rear wheel assembly, and frame assembly were purchased, then the assembly would be the only function. This could be represented in a single-level bill of material, as shown in Table 10.7.

This would be adequate for a single-level production process; it would not, however, accurately reflect the reality of a multilevel assembly process as seen in Figure 10.3. A multilevel assembly process would be best accomplished with the multilevel, or indented, bill of material, which is built from the single-level bills for each level.

TABLE 10.6 Overall Raw Material Requirements (Example 10.2)

Raw Material	Item 1	Item 2	Item 3	Total
RM12	30	120	330	480
RM13	40	—	360	400
RM14	30	—	—	30
RM15	—	—	360	360

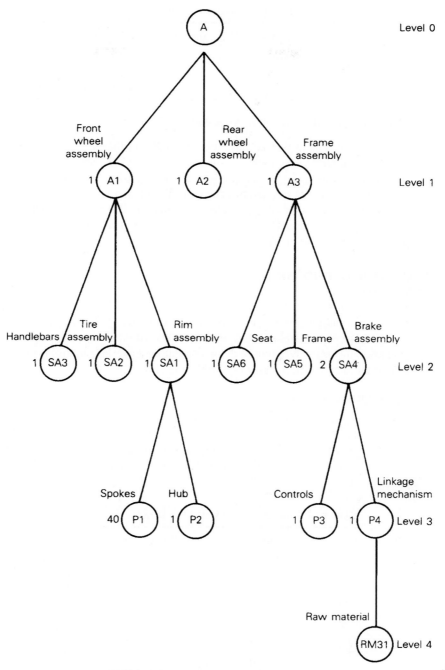

FIGURE 10.3   Product structure tree for a bicycle.

TABLE 10.7 Single-Level Bill of Material for Bicycle

Part Number	Quantity	Description
A1	1	Front wheel assembly
A2	1	Rear wheel assembly
A3	1	Frame assembly

Multilevel or Indented Bill of Material

The most convenient way to represent the BOM so that it is easy to retrieve for use in MRP and in MPS computation is the indented BOM. Table 10.8 highlights the product structure complexity of an end product. It shows concisely how many levels of components the end product has and how many different components exist on each level. It has precisely the same type of information contained in the product structure tree and in the cross-classification chart. In short, it exhibits how much of what material is required and in what order an end product is manufactured.

STRUCTURING THE BILL OF MATERIAL

Bill of material processor programs are generally used to load computer disk files when an MRP system is implemented in a firm. The BOM processor assumes that the bill is accurate and properly structured for computing material requirements. Therefore, it is important to check all features of a BOM file [14].

Desirable Features of the BOM

A bill of material should include the following features:

1. For the purpose of material requirements planning, the bill should be useful for forecasting optional product features. All items should have individual identities.

TABLE 10.8 Indented Bill of Material for Bicycle

Level	Quantity	Part Number	Description
. 1	1	A1	Front wheel assembly
. . 2	1	SA1	Rim assembly
. . . 3	40	P1	Spokes
. . . 3	1	P2	Hub
. . 2	1	SA2	Tire assembly
. . 2	1	SA3	Handlebars
. 1	1	A2	Rear wheel assembly
. 1	1	A3	Frame assembly
. . 2	1	SA4	Brake assembly
. . 2	1	SA5	Frame
. . 2	1	SA6	Seat assembly

2. The bill should facilitate statement of the master schedule planning in a small number of end items, thus reducing the total number of end item assemblies, subassemblies, and so forth.
3. The bill should be helpful in planning the release of lower- level items at the right time with valid due dates. The BOM should reflect the material flow in and out of the raw materials stock, subassemblies, and assemblies.
4. The BOM should permit easy order entry by translating customer orders into a language that the MRP system can operate efficiently, such as recognizing the model numbers or a configuration of option features.
5. The BOM should be usable for final assembly schedule purposes—for example, showing which assembly numbers and how many of the assemblies are required to build individual units of end products.
6. Finally, the BOM should provide a basis for product costing.

Definition of End Item

Given the foregoing desirable features of a BOM, one of the most difficult and confusing aspects of developing an MRP system is the definition of an end item, which varies from firm to firm.

The specific product configuration used for the MPS varies considerably among firms. For example, in the make-to-stock companies, the MPS is often stated in terms of end products. In the make-to-order companies, it is usually stated in terms of actual customer orders. If the total number of products is less than 100, then all products are included in the master schedule. When the number of products grows over 300 to 500, it becomes more difficult to deal with all products in the master schedule. The MPS may then work best with product group rather than end products. In assemble-to-order firms, an extremely large combination of products can be made from relatively few component building blocks. In such cases, the MPS is often stated in terms of options. Thus how to structure the BOM for a product becomes an important topic.

Bill of Material Structures

The three basic structures for bills of material are shown in Figure 10.4. It is important to determine the point of greatest commonality, which is represented by the narrowest part of the BOM structure. The narrowest point determines the end item for the purposes of the MPS [11].

MODULAR BILLS OF MATERIAL

Modularization consists of breaking down the bills of material of high-level items, such as products or end items, and reorganizing them into product modules. For example, consider the manufacture of bicycles. A bicycle is actually an assembly of many optional features. Because of the large number of option combinations available in each product line, a phenomenal number of possible end products exist. Table 10.9 exhibits eight different options available for a bicycle, taken from a manufacturer's catalog. With all the possible choices, it is possible to build 1024 different bicycles. Each bike represents a

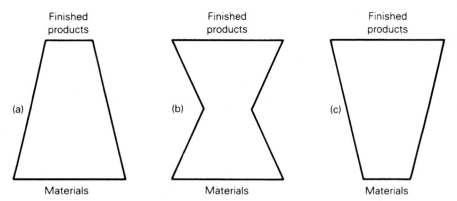

FIGURE 10.4 Three types of product structures: (a) limited number of standard items assembled from components (examples: telephones, radios); (b) many items made from common subassemblies (examples: automobiles, computers); (c) many items made from a limited number of base materials (examples: chemicals, paper).

unique combination of optional product features. Having a multitude of model designations in the sales catalog looks impressive, but the various designations merely point out differences among products belonging to the same family. Also, model identities are not fully meaningful for the purpose of forecasting and material requirements planning, because they fail to provide a precise and complete product definition. Separate BOMs for each item would be impractical and too costly to store and maintain.

To simplify master scheduling and material requirements planning, the number of models must be reduced. Instead of forecasting by finished products, we must forecast by product groups and then divide the groups into component assemblies and subassemblies. This process, known as modularization of the BOM, consists of two important steps [14]:

TABLE 10.9   Bicycle Options

Options	*Choices*
1. Wheel size	24-inch
	26-inch
2. Frame size	52 cm
	64 cm
3. Handlebars	Steel
	Alloy
4. Saddle	King: leather, vinyl
	Super: leather, vinyl
5. Brakes	Side-pull, center-pull
6. Derailleur	Twelve-speed: touring, racing
	Eighteen-speed: touring, racing
7. Chainlink finish	Silver
	Gold
8. Gear assembly	Front: two sprockets for 12-speed, three spockets for 18-speed
	Rear: six sprockets

1. Disentangling combinations of optional product features.
2. Segregating common parts from unique or peculiar parts.

Disentangling Option Combinations

Instead of maintaining bills of material for individual end products, the bills are restated in terms of the building blocks or modules from which each final product is put together. Thus disentangling makes forecasting feasible when there are numerous product variations. Many of the 1024 possible bicycle combinations, for example, may be sold only rarely. Furthermore, design changes and engineering improvements could increase bills for the file. The disentangling procedure can best be illustrated by the bicycle example.

The solution to this problem lies in forecasting each of the high- level components (i.e., major assembly options, such as wheel sizes and number of speeds) and not attempting to forecast by end products at all. Specifically, suppose that 5000 bicycles of the type in question are to be produced in a given month. If there are two choices of speeds and if the past demand averaged 65% twelve- speeds and 35% eighteen-speeds, then by applying these percentages to the bicycle speed option, we could schedule 3250 and 1750 units, respectively, as discussed in Chapter 3. Actual orders may not exactly coincide with the forecast, however, and hence safety stock would be necessary. Under this modular approach, the total number of bills of material would be as follows:

Basic bicycle	1
Wheel size	2
Frame size	2
Handlebars	2
Saddle	4
Brakes	2
Derailleurs	4
Gear assembly	2
Chainlink	2
Total	21

This total of 21 bills compares with 1024 bills if each bicycle configuration had a BOM of its own. If the manufacturer added three different colors and two types of trims, it still would give a total of 26 BOMs instead of 3 * 2 * 1024 = 6144 BOMs. Therefore, the manufacturer would forecast major assembly units, such as wheel assemblies and frames, not a specific module, such as a Royale 18-speed.

To illustrate the concept, a more simplified model will be used. First, a bill of material will be constructed for bikes with only two optional features: (1) the number of speeds and (2) styles. The customer can choose between twelve-speed and eighteen-speed and between touring and racing styles. Two options with each two choices will provide us with five BOMs (1 basic or common + 2 speeds + 2 styles). These four possible combinations are considered end items at level 0, and the options available are level

1, as shown in Figure 10.5. Descriptions of the parts are given in Table 10.10. The next step consists of restructuring these bills into modular bills.

Segregating Common Parts from Unique Parts

To restructure these bills into modules, it is necessary to segregate the level 1 components and to determine which items are common to all bike models, which are unique to a specific number of speeds, and which are unique to a specific style. Some of the items will be unassigned because they are unique to a product combination. Each item has

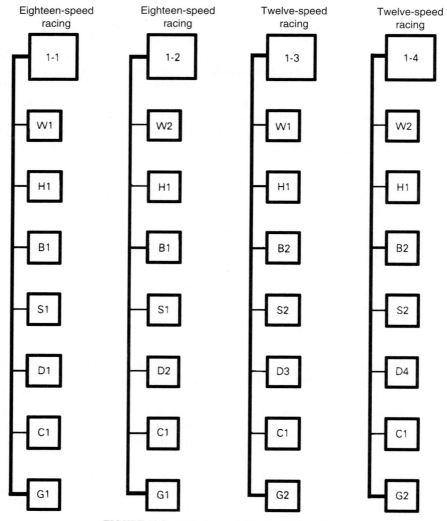

FIGURE 10.5 Bills of material for four bike models.

TABLE 10.10   Descriptions of Parts Used in Figure 10.5

W—wheel size: This option varies only when the frame size varies. For example, there is a standard wheel size for touring style W1 and a standard wheel size for racing style W2.

H—handlebars: Racing handlebars are used for all the combinations previously discussed, but there are options for other models. A suboption would be the material of which the handlebars ar made.

B—brakes: Brakes vary accoring to the number of speeds. All eighteen-speeds have side-pull brakes, B1, and all twelve-speeds have center-pull brakes, B2.

S—saddle: The saddle type also varies with the number of speeds. All eighteen-speeds have kings, S1, and all twelve-speeds have supers, S2. A suboption would be a vinyl or leather saddle.

D—derailleurs: The derailleur varies with every product option combination. Subassemblies are based on the number of component parts.

C—chainlink size: There is one standard chainlink size for all the assemblies previously discussed, although both silver and special gold finishes are available.

G—Gear assembly: The number of gear wheels waries with the number of speeds.

seven assemblies at level 1, as exhibited in Figure 10.5. For simplicity of illustration, we will deal primarily with level 1. First, we segregate all components shown in Figure 10.5 according to their use and group them into different categories, as shown in Table 10.11.

We see that items D1, D2, D3, and D4 are unique and cannot be assigned to any group option combination. For these unique components only, we carry out the same procedure at one more level (level 2) in the BOM. The breakdown of item D, the derailleurs, is shown in Figure 10.6. The derailleur assembly is composed of many common items at level 2. The level 2 items have now been elevated to level 0, as shown in Figure 10.7. These items now become a part of the manufacturing bill (M-bill) file, to be accessed whenever a particular assembly is not found in the common items of level 0. In this example a complete modularization has been achieved. The MPS will treat these bills as though we have only five (superficial) items, as described in the later discussion of S-bills. If we had been unable to decompose the four components D1, D2, D3, and D4, then we would have achieved only partial modularization. In such instances we can either forecast these items separately (resulting in more than five BOMs), or they can be assigned to more than one grouping during the modularization process. The latter approach is particularly desirable for inexpensive items, since it would prevent us from running out of stock.

Example 10.3

Pymex Clock Company manufactures several lines of clocks. Every line consists of many models and options, which can be summarized as follows:

TABLE 10.11   Components Grouped in Categories

Categories	Components
All models	H1, C1
Eighteen-speed only	B1, S1, G1
Twelve-speed only	B2, S2, G2
Racing only	W1
Touring only	W2

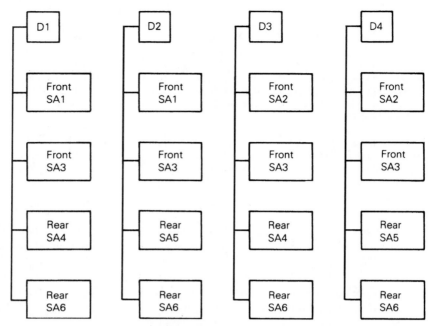

FIGURE 10.6 Subassemblies of derailleurs.

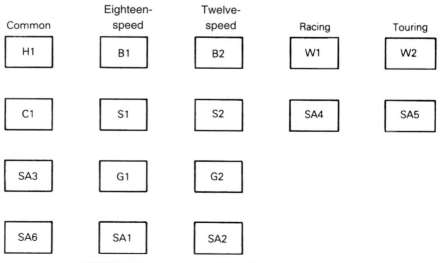

FIGURE 10.7 Complete modular (planning) bill of materials.

Components	Types
Basic clock	1
Motor assembly	2
Operating mode	2
Dial color	3
Hand color	3
Frame color	7
Reflector type	2

(a) Based on the variations available, calculate how many models Pymex can advertise in their sales catalog. (b) How many BOMs would be necessary to cover all models? (c) Under the modular approach, how many BOMs will be necessary? (d) What are the savings?

Solution: (a) The number of models that can be made is calculated by multiplying one basic clock times two motor assemblies times two operating modes times three dial colors times three hand colors times seven frame colors times two reflector types. Thus a total of 504 models can be advertised in the catalog. (b) One BOM would be necessary for every model. Therefore, 504 BOMs are needed. (c) Under the modular approach, we need only twenty BOMs. This quantity is obtained by adding the basic model to all other options available for manufacturing. (d) Total savings amounts to 504 - 20 = 484 BOMs.

Planning Bills

The modularization process elevates level 1 items and, in some instances, level 2 items to level 0. This process eliminates the former level 0 items—that is, end products. The new modular bills of material—known as planning bills—are used for forecasting and master scheduling that express the material requirements for a product without showing the final configuration of the product. Planning bills are exhibited in Figure 10.7. As explained earlier, the planning, or modular, bills area particularly useful for material requirements planning where the final configuration of the end product is extremely difficult or cumbersome to forecast.

Example 10.4

As the new production planning analyst for Pymex Clock Company, you are responsible for simplifying the system. The following table lists the four possible combinations of battery types and motor assemblies and the components necessary to make these clocks:

Battery-Operated, Regular	Battery-Operated, Heavy-duty	Electrical-Operated, Regular	Electrical-Operated, Heavy-duty
A1	A1	A1	A1
B2	C3	B2	C3
D1	D1	E2	E2
T3	T3	X4	X4
F1	G1	H1	J1

The subassembly components are as follows:

Subassembly	Component
F1	K1, M3
G1	K1, N5
H1	L7, M3
J1	L7, N5

Is it possible to achieve a complete modularization in this case? List all planning bills.

Solution: We restructure these bills into the components that belong to all models and those that are unique to specific models:

Categories	Components
All models	A1
Battery-operated	D1, T3
Electrical-operated	E2, X4
Regular	B2
Heavy-duty	C3

We find that items F1, G1, H1, and J1 are each unique to one model, but by studying their second-level components, we can list the following planning bills in each category.

Common	Battery-operated	Electrical-operated	Regular	Heavy-duty
A1	D1	E2	B2	C3
	T3	X4	M3	N5
	K1	L7		

Thus a complete modularization is achieved in this case.

M-Bills

The manufacturing bills, or M-bills, represent another technique of structuring bills of material. Note that when the planning bills were structured, we no longer identified items such as F1, G1, H1, and J1, since they were abolished during modularization. They are needed for the production control system, however, so that the sales department can place orders, the industrial engineering department can use the information for standards and product costing, and the production department can schedule manufacturing properly. Therefore, we should keep their identity for manufacturing purposes, as exhibited in Figure 10.6.

An M-bill item can be a component only of another M-bill item or of an end product. Since the components of the M-bill are included in the planning bill, it is not necessary for the computer to access M-bills for generating material requirements. These bills are coded in such a way that the MRP system bypasses them. When an order is received from a customer or warehouse, it also includes options. These options are specified in their original identities, such as F1 and G1. The order entry procedure, finding that these items are not a part of the planning bill, calls out the M-bill file and reconstructs the appropriate planning bills for manufacturing proper assemblies. Since the M-bills are not used for component requirements planning by MRP, they are segregated in a separate M-bill file. The M-bills are used for final assembly, ordering, scheduling, and costing only.

Example 10.5

In the Pymex problem, is it necessary to maintain a file of M- bills? Exhibit these M-bills.

Solution: The original items F1, G1, H1, and J1 no longer exist in the planning bills. Therefore, it is necessary to have an M-bill file, which would contain the following:

Pseudo Bills of Material (S-Bills)

Planning bills facilitate easier forecasting and requirements planning when numerous end items are involved. The modularization process also creates problems, however. When assemblies, and in some cases subassemblies, are promoted to end item level, a large number of end items remain in the MPS without a parent item. In some instances, there may be too many to work with, and the modularization must be further simplified. For example, notice in Figure 10.7 that the items are grouped in terms of options. By assigning a superficial (artificial) parent to each group—such as eighteen-speed or touring styles—we can create a pseudo bill. These pseudo bills, also known as super bills or S-bills, and the restructured bills of material are assigned a number with the suffix S, such as S-101 (see Figure 10.8). In Figure 10.8, the S-bills reduced the sixteen items of the planning bill to only five end items, thus enabling easier material requirements planning. Keep in mind that an assembly such as S-103 will never be built; the S-bill numbers are nonengineering part numbers. Using these S- bills in the BOM file, the MRP system will explode the requirements. (Note that S-*bills* and S-*numbers* are not standard terminology in the industry.) These bills are treated as end items or level zero items in the MPS.

Example 10.6

High Wheeler Bike Shop wishes to forecast the numbers and types of bikes to be sold in the upcoming summer season as input to the production process. In the past, High Wheeler sold 500 bikes, on the average, during this period. The breakdown is as follows:

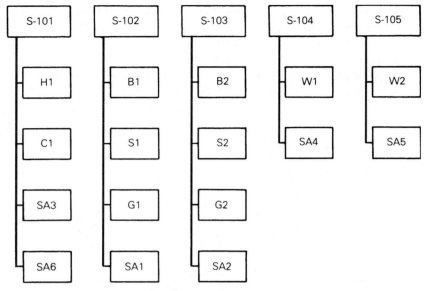

FIGURE 10.8 Super bills.

Eighteen-speed	30%
Twelve-speed	70%
Touring	60%
Racing	40%

The amount of each assembly and subassembly to be produced can be determined by referring to the S-bill in Figure 10.8. The breakdown is as follows:

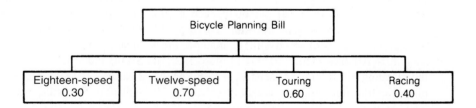

From the S-bill, we determine that 500 units of assembly S-101 (items common to all) will be produced. This will require 500 units of H1, C1, SA3, and SA6. Similarly 150 units (500 × 0.30) of S-102 (items common to eighteen-speeds) will be produced. The progression is similar for S-103, S-104, and S-105:

Common	Eighteen-speed	Twelve-speed	Touring	Racing
500	150	350	300	200

Forecasting of options within options—such as a vinyl or a leather saddle—would be done by the same method.

Kit Numbers (K-Numbers)

When there are many small, loose parts—such as fasteners, nuts, and bolts—at level 1 in the product structure, some companies use the kit number (K-number) technique. The idea behind the kit numbers is the same as that for pseudo bills. The K-numbers merely identify a group of unrelated items; as such, an engineering item does not exist. Kit numbers create a more convenient way of planning, forecasting, and master scheduling.

Example 10.7

Beanie Garage Door Opener Company manufactures several types. Although the parts could be added in the BOM, their cost does not justify the action. It is, however, important to keep track of them for requirements planning. How would you handle the situation? The three do-it-yourself models require the following loose parts:

		Quantities		
				MAXIMUM
ITEM	*REGULAR*	*DELUXE*	*HEAVY-DUTY*	*QUANTITY*
3351 Bolt	3	4	5	5
1151 Washer	3	4	5	5
4151 Nuts	3	4	5	5
101 Cotterpin	3	2	3	3
3021 Fastener	5	4	5	5
425 Clips	6	6	6	6

Solution: Find the maximum quantity of each item required, regardless of the model. For all models, include a kit containing these quantities of each item. The kits are generally identified as K-101, K-102, and so forth; the same kit number will be included in all models. The inclusion of a kit number simplifies requirements generation for forecasting. Also, the consumer is assured of receiving sufficient loose parts.

Advantages of the Modular BOM

As mentioned earlier, the modular bill accomplishes two important goals: (1) disentangling combinations of optional product features and (2) segregating common parts from unique or peculiar parts. Disentangling makes forecasting feasible when numerous product variations are encountered. Segregating facilitates the reduction of inventory investment in components that are common to optional units. The optional units must be forecast separately, thus making it necessary to carry safety stock. In all cases safety stock should be considered as a buffer against unexpected increases in demand. It must be emphasized that the bicycle example is a very simplified model, for illustrative purposes. As Orlicky [14] indicates, we do not generally modularize the BOM for bicycles. This concept can, however, be used for highly engineered items, such as trucks, automobiles, cranes, and the like.

Final Assembly Schedule versus MPS

The final assembly schedule (FAS) can serve as an MPS when the number of end products manufactured is small. These products could be simple ones such as lawn mowers, hand tools, and vacuum cleaners or complex ones such as aircraft or missiles. When the number of end items becomes very large, however, the modularization process elevates assemblies and subassemblies to higher-level items for deriving the MPS. The actual customer order entry system interacts with the MPS through the final assembly scheduling system. The FAS calls on the MRP system for components and the M-bill file, which is under its own control for options to facilitate manufacture of end items. Therefore, we can say that the MPS represents management's commitment for procurement of raw materials and production of component items in support of the FAS. The FAS, however, is constrained by the availability of components provided by the MPS via the MRP system. Thus the FAS represents management's commitment to deliver specific end products either to the customer or to the finished goods inventory. Another distinction is that the MPS could be based on anticipated customer demand, whereas the FAS generally represents actual customer demands. Therefore, the FAS needs to cover only the time span that starts when all component items are available for final assembly and ends when the items are completed.

DISAGGREGATION TECHNIQUES

The entire process of translating top-level aggregate plans to more detailed bottom-level decisions of inventory control and scheduling is referred to as *disaggregation.* The master schedule is the result of disaggregation. It specifies (1) the sizing and timing of production orders for specific items, (2) the sequencing of individual jobs, and (3) the short-term allocation of resources to individual activities and operation. Krajewski and Ritzman [9], in their survey article on disaggregation, point out the need for more concentrated research in this area, so that the various aggregate planning techniques can be implemented more easily. We can broadly categorize the existing disaggregation techniques into (1) cut and fit methods, (2) mathematical programming methods, (3) heuristic methods, and (4) others. We briefly explain the cut and fit methods, which are popular in industry. To illustrate the use of mathematical programming in disaggregation, we also present a linear programming model. In many cases, heuristic methods are tailored for special situations, as illustrated by Hax and Galovin [7], who claim that these methods are implemented in industry. Other methods, such as the "knapsack" algorithm, are limited in their applicability to real-world problems.

Cut and Fit Methods

Generally, firms try out various allocations of capacity for the products in a group until a satisfactory combination is determined. Such an approach is called a cut and fit method. For example, the master schedule shown in Table 10.12 is satisfactory in the sense that it

TABLE 10.12 Aggregate Forecast Plan

	Initial Inventory				Week					Total
		1	2	3	4	5	6	7	8	
Aggregate forecast		100	100	100	150	150	200	200	200	1200
Production		150	150	150	150	150	150	150	150	1200
Aggregate inventory	400	450	500	550	550	550	500	450	400	—
Item Forecast										
Product A		60	60	60	90	90	120	120	120	720
Product B		40	40	40	60	60	80	80	80	480
Total		100	100	100	150	150	200	200	200	1200
Master Production Schedule										
Product A		150	150	150				120	150	720
Product B					150	150	150	30		480
Total		150	150	150	150	150	150	150	150	1200
Aggregate plan capacity		150	150	150	150	150	150	150	150	1200
Deviation		0	0	0	0	0	0	0	0	0

indicates when the plant should plan to start and stop the production of individual items. The amount of capacity required to support the plan is consistent with the capacity that the aggregate planning process had indicated as appropriate. We do not know, however, whether the schedule is satisfactory in terms of the number of setups and the associated setup costs or the in-process inventory for the production line. Only the use of several alternative master schedules or the use of optimizing methods can guarantee that the MPS is feasible. Unless carefully planned, the use of aggregation may lead to infeasibilities. The product types concept is merely an abstraction that makes the aggregation process possible. Inventories and demand have physical meaning only at the end item level. When calculating product type inventories, we cannot add all items that belong to a product type; that would assume complete interchangeability of inventories among all items in a product type, which is not the case. As an example, consider a product type that consists of items A and B. Table 10.12 shows that we have adequate capacity and inventories for the given forecast and production goals of the product family. When we derive the inventories of items A and B separately, however, we find that item B is short twenty units during period 3, as exhibited in Table 10.13. The aggregate inventories are also compared with the total inventories for items A and B. The deviation is visible only when we deal with individual items, rather than with a family of products. Therefore, the MPS may have to be revised several times until a feasible schedule is obtained. This is the cut and fit method of arriving at a schedule. The same procedure can be used for checking capacity requirements.

Linear Programming Method

Krajewski and Ritzman [9] propose a linear programming version of a disaggregation model that can be used for combined aggregate-disaggregate planning in service as well as manufacturing organizations. In service organizations, we cannot store up services (inventories), and hence it becomes difficult to smooth the production rate. Disaggregation in service amounts to assigning available personnel, and possibly other resources, optimally. The general disaggregation model to be presented here minimizes the total costs of output, subcontracting, inventory, backlog, hiring, firing, overtime, and wages for T periods. Basically, this model combines aggregate planning, inventories, and scheduling decisions in one integrated problem. Minimize

$$Z = \sum_t \sum_i [C_1 X_{it} + C_2 S_{it} + C_3 I_{it} + C_4 B_{it} + C_5 H_{jt} + C_6 F_{jt} + C_7 O_{jt} + C_8 W_{jt}] \quad (10.1)$$

subject to

$$X_{it} + I_{i,t-1} - I_{it} + S_{it} + B_{it} - B_{i,t-1} = D_{it} \qquad \text{for all } i \in L \quad (10.2)$$

$$\sum_{m=1}^{li} [r_{imj} X_{i,t+m} + r'_{imj} \phi_{i,t+m} = P_{ijt} \qquad \text{for all } i \in N_j \text{ and } j \in J \quad (10.3)$$

TABLE 10.13 Inventory Status

	Initial Inventory	Week 1	2	3	4	5	6	7	8	Total
Product A										
Forecast		60	60	60	90	90	120	120	120	720
Production		150	150	150	0	0	0	120	150	720
Inventory	300	390	480	570	480	390	270	270	300	
Product B										
Forecast		40	40	40	60	60	80	80	80	480
Production					150	150	150	30		480
Inventory	100	60	20	−20	70	160	230	180	100	
Total inventory	400	450	500	570	550	550	500	450	400	
Aggregate plan inventory	400	450	500	550	550	550	500	450	400	
Deviation	0	0	0	−20	0	0	0	0	0	

$$\sum_{i \in L} P_{ijt} - W_{jt} - O_{jt} \leq 0 \qquad \text{for all } j \in J \tag{10.4}$$

$$W_{jt} - W_{j,t-1} - H_{jt} + F_{jt} = 0 \qquad \text{for all } j \in J \tag{10.5}$$

$$O_{jt} - \theta W_{jt} \leq 0 \qquad \text{for all } j \in J \tag{10.6}$$

$$\phi_{it} = \begin{cases} 1 & \text{if } X_{it} > 0 \\ 0 & \text{if } X_{it} = 0 \qquad \text{for all } i \in L \end{cases} \tag{10.7}$$

where

C_i = costs associated with variables

X_{it} = output to product (service) i in period t

I_{it} = on-hand inventory level of product i in period t (manufacturing setting only)

S_{it} = subcontracted output of product (service) in period t

D_{it} = market requirements for product (service) i in period t

l_i = production (or procurement) lead time for product (service) i (in service settings, l_i would normally equal one time period)

B_{it} = amount of product (service) put on backorder in period t

r_{imj} = number of person-hours required per unit of product (service) i at operation j in the mth period before production is finished on i (in service settings, r_{imj} is usually 0 or 1 for all i, m, j since output is normally measured in person-hours)

r'_{imj} = total setup time required by product i at operation j in the mth period before production is finished on i (in service settings, r'_{imj} is 0 for all i, m, j)

ϕ_{it} = binary variable that assigns a setup time for product i whenever $X_{it} > 0$

P_{ijt} = production output of product (service) i at operation j in period t expressed in person-hours

W_{jt} = regular person-hours assigned to operation j in period t

O_{jt} = overtime person-hours assigned to operation j in period t

H_{jt} = person-hours of labor hired for operation j at the start of period t

F_{jt} = person-hours of labor released from operation j at the start of period t

θ = proportion of the regular time workforce that can be used on overtime

L = set of all end items (services) to be controlled (in manufacturing settings)

T = length of the planning horizon

J = set of all operations where we assume there is only one type of skill at each operation

N_j = set of all products (services) that require resources at operation j

Constraint equation (10.2) represents the basic inventory balance relationship, with the added feature of the recognition of demands placed on the inventory of product i by higher-order components and subassemblies in manufacturing. In the service section, constraint equation (10.2) defines the service i backordered in period t. Constraint equation (10.3) identifies the output in terms of product or service i at operation j in period t. Constraint equation (10.4) ensures that the work planned for operation j in time period t does not exceed the personnel capacity planned for department j in period t. Hiring and releasing of personnel is represented by equation (10.5). The amount of overtime θ as a fraction of regular time person-hours is represented in equation (10.6). The term ϕ_{it} in equations (10.3) and (10.7) is a binary variable. If a setup is necessary, $\phi_{it} = 1$; otherwise, it is equal to zero.

In service sectors, variables such as X_{it}, B_{it}, and S_{it} are expressed in terms of person-hours rather than units of products. In a manufacturing setting, the X_{it} values determine the planned production of specific products from master schedule through component production, along with the personnel capacities to support these schedules. In a service setting, however, these values represent the planned output of each service. The selection of personnel capacities becomes paramount.

Example 10.8

Consider a firm that provides two different services. Given the following data, find an optimum production schedule using the linear programming technique.

	Demand Hours per Period		
	1	*2*	*3*
Service type 1	100	150	125
Service type 2	300	100	200

Available person-hours currently amount to 300, and the maximum allowable overtime is limited to 20% of the workforce. In addition, (1) setups are not required, (2) inventories or backorders are not feasible, and (3) subcontracting is not possible. We also have $C_1 = 10$, $C_5 = 20$, $C_6 = 15$, and $C_7 = 15$.

Solution: Since setups are not required,

$$\phi_{1t}, \phi_{2t} = 0 \qquad \text{for } t = 1, 2, 3$$

Since inventories and backorders are not allowed,

$$I_{1t}, I_{2t}, B_{1t}, B_{2t} = 0 \qquad \text{for } t = 1, 2, 3$$

Since subcontracting is not possible,

$$S_{1t}, S_{2t} = 0 \qquad \text{for } t = 1, 2, 3$$

The linear programming problem can now be formulated as follows: Minimize

$$C = \sum_{t=1}^{3} [10(X_{1t} + X_{2t}) + 20(H_t) + 15(F_t) + 15(O_{1t} + O_{2t})]$$

subject to

$$X_{11} + O_{11} = 100$$
$$X_{21} + O_{21} = 300$$
$$X_{12} + O_{12} = 150$$
$$X_{22} + O_{22} = 100$$
$$X_{13} + O_{13} = 125$$
$$X_{23} + O_{23} = 200$$
$$O_{11} + O_{21} - 0.2X_{11} - 0.2X_{21} \leq 0$$
$$O_{12} + O_{22} - 0.2X_{12} - 0.2X_{22} \leq 0$$
$$O_{13} + O_{23} - 0.2X_{13} - 0.2X_{23} \leq 0$$
$$X_{11} + X_{21} - H_1 + F_1 = 300$$
$$X_{12} - X_{11} + X_{22} - X_{21} - H_2 + F_2 = 0$$
$$X_{13} - X_{12} + X_{23} - X_{22} - H_3 + F_3 = 0$$

and

$$X_{it}, O_{it}, H_t, F_t \geq 0$$

The initial tableau is shown in Figure 10.9. The problem was solved using a linear programming software package. The solution is as follows:

Period (t)	X_{1t}	X_{2t}	O_{1t}	O_{2t}	H_t	F_t
1	100	100	0	0	0	100
2	229.17	125	70.83	0	154.17	0
3	150	200	0	0	0	4.17

The total cost of the program is

$$C = 10(100 + 229.17 + 150) + 10(100 + 125 + 200)$$
$$+ 15(70.83) + 20(154.17) + 15(100 + 4.17)$$
$$= \$14,750.10$$

X_{11}	X_{12}	X_{13}	X_{21}	X_{22}	X_{23}	O_{11}	O_{12}	O_{13}	O_{21}	O_{22}	O_{23}	H_1	H_2	H_3	F_1	F_2	F_3		RHS
1						1												=	100
	1						1											=	300
		1						1										=	150
			1						1									=	100
				1						1								=	125
					1						1							=	200
−0.2			−0.2			1			1									≤	0
	−0.2			−0.2			1			1								≤	0
		−0.2			−0.2			1			1							≤	0
1	1											−1			1			=	300
−1	1		−1	1									−1			1		=	0
	−1	1		−1	1									−1			1	=	0

FIGURE 10.9 Linear programming tableau for Example 10.8.

MANAGING THE MASTER SCHEDULE

In this section we discuss the basic principles and practices of designing, creating, and controlling the master schedule.

Design

Much of the discussion of selecting items to include in the master schedule has been done earlier in this chapter. In particular, the section on structuring bills of material is the primary method of determining which items will be included in the master schedule.

The types of product structure has a major influence on the master schedule. In Figure 10.4 are three different types of product structures. In Figure 10.4a, items indicated as finished goods would be included in the master schedule. In Figure 10.4b, items at the narrowest point in the structure would be in the master schedule; the final assembly schedule would still be at the finished product level. In Figure 10.4c, kits at the materials stage would be included in the master schedule, while the finished products would be at the final assembly schedule.

Other operational items, such as the planning horizon, would need to be specified at this time.

Create the Master Schedule

The master schedule is created from all sources of possible inputs, such as the forecast, customer booked orders, and inventory on hand. The previous section on disaggregation described the basic concepts of translating the aggregate plan to the master schedule.

After the master schedule is created, it should be controlled by looking at orders that have been placed by customers. The available to promise (ATP) is found to determine if future customer orders can fit within the existing master schedule. The ATP is that portion of the production that is not committed. There are at least three ways to find an ATP figure, and each provides a different result. Table 10.14 shows a forecast for Product A. The beginning inventory is just sufficient to prevent negative inventory in week 6. The line marked Orders represents firm orders from customers. Assume that this company has developed a master schedule based partly on firm orders received and partly on forecasts.

The discrete method finds the ATP by subtracting the order quantity from the production. In the first period the ATP is found by adding the beginning inventory and the production in the first period and then subtracting the orders in the first period. This would result in an ATP of 30 + 150 − 60 = 120. If the value is less than zero, then the ATP is zero. If there is no production in the second period, then the ATP is reduced by orders in the second period and any subsequent periods up to, but not including the next period in which there is production.

The ATP for periods beyond the first is determined in one of two ways. If there is no production in a period, then the ATP is zero. If there is production, then the ATP is found by taking the production in that period and subtracting orders for that period and all subsequent periods up to, but not including, the next period with production. The ATP in the second period would be 150 − 55 = 95. In the third period the ATP is zero because ATP = 150 − 50 − 75 − 70 − 50 = -95.

A second method is to find the cumulative ATP by not looking ahead (without lookahead) or with a lookahead feature. Without lookahead does not take into consideration those periods when there is no production. In Table 10.14 the first ATP (without lookahead) is found by taking the initial inventory, plus production, minus orders. In subsequent periods the ATP is found by taking the previous period ATP, adding the production, and subtracting the order in that period only. In period 2, the ATP is 120 + 150 − 55 = 215. In the third period the ATP is 215 + 150 − 50 = 315. This may be a better indicator of the actual available except in period 3, when it may be overstating the actual available. Note that there is no production in periods 4, 5, and 6. A promise of 315 in period 3 is overstated because it does not take into account the orders that will be processed in periods 4, 5, and 6.

The ATP (with lookahead) will remedy this. The ATP in period 1 is the initial inventory plus production minus orders in that period. If there is no production in period 2,

TABLE 10.14 Available to Promise

| | Initial Inventory | Week | | | | | | | |
		1	2	3	4	5	6	7	8
Product A									
Forecast		60	60	60	90	90	120	120	120
Production		150	150	150	0	0	0	120	120
Inventory	30	120	210	300	210	120	0	0	0
Orders		60	55	50	75	70	50	30	40
ATP (discrete)		120	95	0	0	0	0	90	80
ATP (without lookahead)		120	215	315	240	170	120	210	290
ATP (with lookahead)		120	215	120	120	120	120	210	290

then the ATP is further reduced by the amount of orders in all the following periods up to, but not including, the next period with production. In subsequent periods the ATP will be the previous period's ATP, plus production, minus orders for that period and all subsequent periods with no production up to, but not including the next period with production. Note that in period 3, the ATP (with lookahead) will be reduced by the orders in periods with zero production up to, but not including the next production period. The ATP in period 3 will be $215 + 150 - 50 - 75 - 70 - 50 = 120$.

In Table 10.14 note the differences in the ATP values in period 3. Without lookahead, there is an indication of 315 units available. If we now booked an order in period 3 for 315, we would consume the entire amount available and would not be able to satisfy those orders coming due in periods 3 to 6. With lookahead, the ATP is only 120, taking into account the orders to be shipped in periods 3 to 6.

Another method of controlling the master schedule is to adjust the forecast to take into consideration the orders placed by customers. The forecast is "consumed" by the actual orders that have been placed. In Table 10.15 the row Forecast (before) is the original forecast. The next row, Forecast (after), has adjusted the first row by subtracting the orders in each period. If the difference is negative, the Forecast (after) is zero.

These methods of controlling the master schedule allow the master scheduler to understand the dynamics of production and orders better. Scheduling now has better information to allow customer promises, and the scheduler can determine how the master schedule may need to be adjusted in future weeks when a new schedule is found.

MAINTENANCE OF THE MPS

The master schedule requires constant monitoring and revisions to reflect new orders, new problems, and new decisions. These functions may take one to five person-days over a period of a month [10]. The frequency of revisions depends on the type of system in operation—that is, net change versus regeneration. The details of these systems are explored in Chapter 11. The revisions are made quarterly, monthly, and weekly or daily. Quarterly revisions encompass new sales forecasts. Major revisions could also occur between the quarterly ones, but not frequently. Monthly revisions involve adding more new weeks in the future and rescheduling past due orders. Weekly or daily revisions include

TABLE 10.15 Consuming the Forecast

	Initial Inventory	Week							
		1	*2*	*3*	*4*	*5*	*6*	*7*	*8*
Product A									
Forecast (before)		60	60	60	90	90	120	120	120
Forecast (after)		0	5	10	15	20	70	90	80
Production		150	150	150	0	0	0	120	120
Inventory	30	120	210	300	210	120	0	0	0
Orders		60	55	50	75	70	50	30	40
ATP (with lookahead)		120	215	120	120	120	120	210	290

the loading of new orders and revisions on an exception basis. The key to success is to keep the master schedule up to date. The basic MPS maintenance steps [10] are as follows:

1. Load and level the capacity, using the most accurate MPS available, out through the combined lead time. Use backlogs as necessary to satisfy the forecast.
2. When new orders cannot be shipped from inventory, supply them from the first available planned lots.
3. If planned quantity is not available for a new order, schedule the order at the end of the combined lead time (freeze period) if capacity is available. If this is not feasible, extend the due date to the nearest future period that has capacity or reschedule a lower priority order in the earliest period to provide capacity.
4. If items were not met from inventory and quantities were not planned, determine whether capacity is available earlier than the combined lead time. If it is available, special handling will be required to coordinate long-lead items to meet an assembly date that is shorter than the freeze period.
5. Finally, fill up any unused capacity with standard items to obtain a full and level load for all future periods.

While keeping the MPS up to date, some basic rules must be followed for revisions.

1. *Reschedule past due orders.* Suppose that the output capacity is 500 units. If a backlog of 200 units exists, do not schedule 500 additional units without planning for additional capacity. If we lie to the MPS, it will lie back to us! There is no sense in claiming that the system is no good! The master scheduler is always blamed for the failure of the system. Unfortunately, this is the rule most frequently violated in industry.
2. *Make changes as soon as the need is recognized.* All major revisions should be done as soon as possible. This facilitates recognition of problems and issues to be dealt with—such as long lead times, capacity adjustments, and inventory adjustments—in advance.
3. *Never reschedule a component.* The component due date provided by the MPS is a firm requirement in manufacturing. If components are not available on time for some reason, such as equipment breakdown or strike, do not reschedule only the affected components. As a first priority, find other ways to get the component in time. Alternative scheduling may sometimes be feasible if safety time is built into the system or if low-priority items that use the same components can be rescheduled. If nothing is feasible, always reschedule the highest-level item, using the where-used list provided by pegging. (Details of pegging are fully explored in Chapter 11.) Otherwise, you might be left with unnecessary in-process inventory in the system.
4. *Maintain integrity.* Never lie to your MPS. Do not close your eyes and hope that problems will go away! If you overstate your MPS without additional ca-

pacity, you will end up with many past due orders. If you understate, the efficiency of the system may be impaired. A good understanding between the master scheduler and the management is very important. If you (the master scheduler) budge for everyone in the system, you might be in the wrong job.

SUMMARY

Master scheduling includes a variety of activities involved in the preparation and maintenance of the MPS. The master schedule is stated in terms of specific product configurations and usually indicates the quantities to be produced in specific time brackets (weekly) for the next twelve months or longer. The unit of production selected for the MPS varies considerably among firms. In make-to-stock companies, the MPS is stated in terms of end items. As discussed in the section on BOM structuring, it can also be stated in terms of lower-level requirements, such as major assemblies or component parts. In make-to-order firms, the actual customer orders can be used as the unit of the MPS. Assemble-to-order firms generally have a very complex problem in defining the unit. Modularization and planning bills are often a great help to the scheduler. One important criterion is selecting the units so that the total number of units in the MPS is minimized, which generally improves the forecast and reduces the administrative costs. Another important criterion of the MPS is that it represents a disaggregation of the production plan into specific items, whereas the final assembly schedule represents the production plan with specific buildable end products. The importance of keeping the MPS feasible and up to date cannot be overemphasized.

This chapter also discussed some techniques used for disaggregation. The concepts learned in this chapter will help you appreciate and understand material requirements planning better.

PROBLEMS

1. a. Using Table 10.4, compute weekly raw material requirements for the following schedule. Write a computer program if necessary.

	Period					
	5	6	7	8	9	10
Product 1	100		150		200	
Product 2		200		200		150
Product 3			300		300	

b. Assuming that the following lead times apply to these products, prepare a table to show what, when, and how much to order in a distribution requirements planning environment.

Product	Lead Time
1, 12	2 weeks
2, 13, 14	3 weeks
3, 15	2 weeks

   c. How will the answer differ if the lead times of raw materials are included for product 1? Assume that all subassemblies and fabrications require one week of lead time. Comment on your results.

2. Brown, Inc., a hoist manufacturer, offers customers a number of options, totaling 2400 configurations of hoists:

Options	Choices
Motors	10
Drums	30
Gear boxes	4
Pendents	2
Hooks	1

Smart Alex, who is being interviewed for the production control manager's job, claims that forecasting can be made much easier if the modularization process is used. To prove his point, he uses the information in Figure 10.10. In addition, he is able to describe the following items more fully:

   a. From the given data, can you prove that it is possible to obtain 2400 configurations of hoists?

   b. Illustrate the modularization process, providing examples of the planning bill, manufacturing bill, and super bill.

3. After modularization, how many end items exist in your solution of problem 2?

4. Figure 10.4 exhibits different types of product structures. What structure does the hoist example have? Why?

5. Explain how MRP, the MPS, and the FAS are related in the hoist manufacturing example.

6. Given the master schedule in problem 1, calculate the capacity requirements for the final assembly department, ignoring the lead times.

Average assembly time: 1.2 hours

Average subassembly time: 0.8 hour

Average fabrication time: 0.2 hour

The subassembly lead times are included in the assembly lead times.

7. Chethan Ballan Company currently produces several lines of sofas. Items vary, depending on the size and style. Currently there are hundreds of sofas listed in the catalog as well as in the MPS. The company is taking steps to simplify the MPS. The following is the list of options:

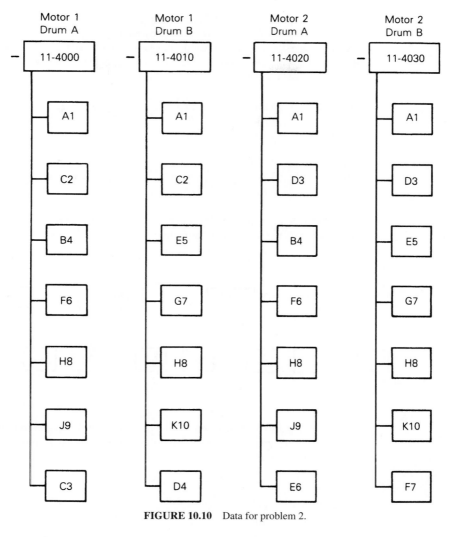

Motor 1
Drum A

Motor 1
Drum B

Motor 2
Drum A

Motor 2
Drum B

| 11-4000 | 11-4010 | 11-4020 | 11-4030 |

FIGURE 10.10   Data for problem 2.

Options	*Choices*
Widths	2
Lengths	3
Base heights	2
Arm styles	6
Rear finish	3
Trim	4
Fabrics	15

If the company has one BOM for each possible sofa, how many items will the MPS contain?

8. In the Ballan problem, if modularization is feasible, what is the least number of BOMs required? What savings would be realized by modularization?

9. Consider only the choices of width and base height (BH) combinations of sofas in Ballan Company. The level 1 items that make up the sofas are as follows. Check whether complete modularization of BOM is feasible.

Width A, BH X	Width A, BH Y	Width B, BH X	Width B, BH Y
1010	1010	1020	1020
1030	1040	1030	1040
1050	1050	1050	1050
1053	1054	1053	1054
1701	1702	1703	1704

10. If complete modularization is not feasible in the Ballan sofa problem, explore whether it is possible with the following information:

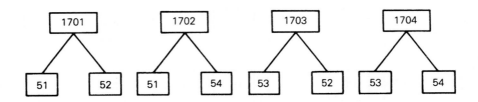

Here 51, 52, and so forth represent level 2 of items 1701, 1702, and so forth.

11. Ballan Company expects to sell 5000 sofas during the next month. The breakdown, based on the width and base heights, is as follows:

Width A:	40%
Width B:	60%
Base height X:	70%
Base height Y:	30%

Provide a forecast for sofas based on these proportions of choices.

12. Using the planning bills created for Ballan Company and the forecasts for the given options, (a) exhibit the minimum number of super bills required and (b) provide the forecasts for level 1 items, based on the super bills.

13. Lears Company markets several types of lawn mowers that require customer assembly. Depending on the type and size of the model, different sets of small, loose parts are required. To simplify the number of items on level 1, the company would like to create a kit that would satisfy the requirements of all models. Based on the following information, set up a kit number and list its contents.

	Quantities Required		
ITEMS	MODEL 1	MODEL 2	MODEL 3
Nuts 3562	3	4	2
Bolts 4531	4	4	4
Cotter pins 210	2	4	3
Washers 3251	2	2	1
Lock washers 4251	1	2	1

14. Brown Company forecasts total sales of 5000 hoists for the next month. Based on past sales information, the following breakdowns were obtained for the options:

    Motor 1: 50%

    Motor 2: 50%

    Drum A: 35%

    Drum B: 65%

    Using S-bills, indicate how many of each item would be manufactured.

15. The aggregate MPS for product group X is as follows. Check whether the MPS is feasible. Backorders are not permissible; initial inventory of the product group amounts to 200 units; and the capacity of the manufacturing facility is listed as 250 units per week.

	Weeks			
1	2	3	4	5
250	300	300	200	300

    Is the demand met?

16. Suppose that product group X in problem 15 is made up of items A and B. Given the following additional information, check whether the item MPS is feasible. Are the demands met?

	Forecasts						
						INITIAL	ALLOCATED CAPACITY
ITEM	1	2	3	4	5	INVENTORY	UNITS/WEEK
A	50	100	150	150	150	100	150
B	200	200	150	50	150	200	100

17. Using the cut and fit method, develop a master production schedule for items A and B such that the inventory constraints are met. The following data are known regarding the capacity requirements:

    Item A: 1.0 hours/unit

    Item B: 0.8 hour/unit

    Based on the capacity requirements, is your MPS feasible? If the cost of inventory amounts to $1.00 per item per month, calculate the average inventory cost. A total of 230 hours/week of capacity is available.

18. Assume that product A from problem 16 has the following firm orders from customers and feasible production schedule.

	Initial Inventory	Week				
		1	2	3	4	5
Product A						
Forecast		50	100	150	150	150
Production		200	0	200	0	200
Inventory	100					
Orders		30	70	20	50	0

Find the ATP in each of the three ways: discrete, without lookahead, and with lookahead. Determine the forecast after consumption.

19. Keysinger Corporation makes two different flash guns for cameras. The forecasts for the next three months are as follows:

	October	November	December	Initial Inventory	Final Inventory
Flash 310	1500	1500	2500	350	250
Flash 311	900	1300	1500	200	200

In addition, the following data were accumulated from various departments in the company:

	Flash 310	Flash 311
Production cost/unit	$20	$25
Inventory cost/month/unit	$0.50	$0.70
Cost of increasing capacity	$15/unit	$15/unit
Cost of decreasing capacity	$10/unit	$10/unit
Existing capacity/month	200 hours	150 hours
Assembly hours/unit	0.10	0.125

Assuming that backorders are not permitted, write the equations for the problem, using linear programming.

20. In Problem 19, suppose that the available personnel is limited to 175 and 125 hours in departments 310 and 311, respectively. Since the production involves only assembly, would this alter your equations?

21. Highwheeler Bicycle Company assembles racing and touring models. The current forecasts for the next two months are 300 touring, 500 racing; and 400 touring, 450 racing. In addition, the following information is known:

	Racing	Touring
Production cost/unit	$50	$40
Fabrication hours/unit	4	3
Assembly hours/unit	1	1
Current inventory	200	150
Inventory/cost/month/unit	$1.50	$1.25
Final inventory desired	250	200

The total labor hours last month were 2000 person-hours. The cost of increasing or decreasing the production level amounts to $5 per unit. Establish an optimum production schedule for Highwheeler. What is the total cost of your program? (*Hint:* Combine fabrication and assembly hours for solving this problem. Assume an 8-hour-per-day schedule for changes in person-hour calculations.)

REFERENCES AND BIBLIOGRAPHY

1. W. L. Berry, T. E. Vollmann, and D. C. Whybark, eds, *Master Production Scheduling: Principles and Practice* (Falls Church, Va.: American Production and Inventory Control Publications, 1979).
2. G. R. Britan and A. C. Hax, "On the Design of Hierarchical Production Planning Systems," *Decision Sciences,* Vol. 8, No. 1 (1977), pp. 28–55.
3. R. G. Brown, *Decision Rules for Inventory Management* (New York: Holt, Rinehart and Winston, 1967).
4. *Communications Oriented Production Information and Control System (COPICS),* Vols. 1–6 (White Plains, N.Y.: IBM Publications, 1972).
5. D. W. Fogarty, J. H. Blackstone, Jr., and T. R. Hoffmann, *Production and Inventory Management,* 2nd ed. (Cincinnati: South-Western, 1991).
6. Robert A. Gessner, *Master Production Schedule Planning* (New York: John Wiley & Sons, 1986).
7. A. C. Hax and J. J. Galovin, "Hierarchical Production Planning Systems." In A. C. Hax, ed., *Studies in Operations Management* (New York: North Holland–American Elsevier, 1978).
8. A. C. Hax and H. C. Meal, "Hierarchical Integration of Production Planning and Scheduling." In M. A. Geisler, ed., *Studies in Management Science: Vol. 1, Logistics* (New York: North Holland-American Elsevier, 1975).
9. L. J. Krajewski and L. P. Ritzman, "Disaggregation in Manufacturing and Service Organizations: Survey of Problems and Research," *Decision Sciences,* Vol. 8, No. 1 (1977), pp. 1–18.
10. *Master Production Schedule,* APICS Training Aid (Falls Church, Va.: American Production and Inventory Control Society, 1973).
11. *Master Production Schedule Reprints* (Falls Church, Va.: American Production and Inventory Control Society, 1977).
12. H. Mather, *Bills of Materials.* (Homewood, Ill.: Dow Jones–Irwin, 1987).
13. *MITROL: An Introduction to MRP* (Lexington, Mass.: Mitrol, Inc., 1978).
14. J. Orlicky, *Material Requirements Planning* (New York: McGraw-Hill, 1975).
15. G. W. Plossl and W. E. Welch, *The Role of Top Management in the Control of Inventory* (Reston, Va.: Reston, 1979).
16. T. E. Vollmann, W. L. Berry, and D. C. Whybark, *Manufacturing Planning and Control Systems,* 3rd ed. (Homewood, Ill.: Richard D. Irwin, 1992).
17. H. J. Zimmerman and M. G. Sovereign, *Quantitative Models for Production Management* (Englewood Cliffs, N.J.: Prentice Hall, 1974).

APPENDIX 10A

The Master Scheduler's Job Revisited
Paul N. Funk, CPIM

In my previous paper in 1987, "The Master Scheduler's Job: How To Operate While Between a Rock and a Hard Place," (see 1987 International Conference Proceedings) we explored the Master Scheduler's main responsibilities and how he/she could work his/her way out of some very difficult situations that seem to arise very often in a dynamic manufacturing environment. This paper will further explore how the job of the Master Scheduler has changed with the advent of Just-In-Time master scheduling techniques, changes in customer expectations, the huge increase in competitive pressures, and the pressure to reduce inventories and leadtimes. It will also point out what I perceive as some problems that still exist today in the master scheduling process.

THE OLD JOB

The key characteristic of the job of the Master Production Scheduler, as we discussed before, was CONFLICT. The Master Scheduler was required to be manufacturing-oriented, customer-oriented, engineering-oriented, finance-oriented, and be able to juggle the priorities as they occurred. On top of this he (or SHE—implied) was required to plan the TIMING of events which, when taken together, would achieve everyone's goals. No wonder we said that the Master Scheduler was between a rock and a hard place!

   Let's recall some of the personal characteristics of a successful Master Scheduler. He (or SHE—implied) is a thoughtful, analytic, confident individual with a great deal of product knowledge, customer knowledge, and general business sense. He is also capable of presenting himself well, preparing data quickly and verifying its accuracy, and making "ad hoc" presentations to management on short notice. This person has the ability to "sell" ideas and to negotiate compromises based on knowledge of a broad range of alternatives not known to others. He is, by and large, a proponent of formal systems and would rather plan than expedite; however, when certain situations evolve (downhill), he is not above "making deals" to get the product out the door.

CHANGES IN THE WORK ENVIRONMENT

Since that paper in 1987, the pace has definitely quickened! The competition from other American producers and from foreign producers has, in many industries, increased tenfold. This usually means that, quality and cost being equal, he that delivers in the shortest

Reprinted with the permission of APICS, Inc., "The Master Scheduler's Job Revisited," *1990 APICS Conference Proceedings* (Falls Church, VA: APICS, 1990) pp. 374–377.

lead time wins the business. This has put tremendous pressure on the Master Scheduler to turn around unusual customer requests in the shortest possible time. The luxury of several days time to analyze all the alternatives is no longer available.

Another area of tightening pressure is the new product development cycle. Master Schedulers are increasingly being called upon to contribute their expertise to the reduction of lead time required to bring new products to market. It is another "rock and hard place" situation to try to accommodate trial production runs on production equipment while still trying to achieve a monthly production/revenue target. The Master Scheduler is charged with a massive coordination effort of providing

- manpower
- materials
- manufacturing capability
- money (cash flow) and
- management of all the logistical activities

In direct conflict with the need to get the product to market is the need to continue producing enough of the current best-selling products to meet the goals of customer service. Since the popularity of your products is always a fickle thing with customers, this is a process of never-ending change. The major task of the Master Scheduler is to minimize the impact of these changes.

CHANGES IN CUSTOMER EXPECTATIONS

Because of the great proliferation in competition, and the transition from a captive U.S. market to a global market, many companies have had to change their approaches in dealing with customers. It's by and large a buyer's market. The customers are in a position to demand more and more concessions on price, delivery, special packaging, etc. Supplier companies who won't go along with these requests usually find themselves having to participate in a round of competitive bidding, where they were "locked in" as a sole source before; or they may lose in the next round of bidding if they were the majority supplier. Think about it! What would you do if you suddenly lost the business of your largest customer?

The winners in this area have enlisted the aid of their customers in helping to develop long-range plans for the benefit of both buyer and supplier. Often this has involved direct contact and the development of camaraderie between your Master Scheduler and the CUSTOMER'S Master Scheduler. Thus we see the Master Scheduler becoming a member of the SALES team for the first time in some companies.

CHANGES IN SCHEDULING TECHNOLOGY

Recall from the 1987 paper the steps in formal master scheduling in an MRP II system:

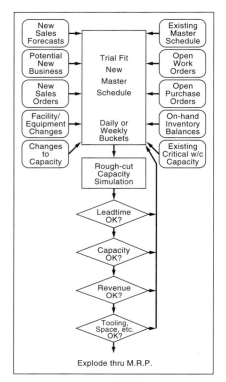

FIGURE 1

While most of the above steps are still valid and practiced today, there have been several new wrinkles added, with the advent of Just-In-Time production systems. First, contrary to the belief of many, the emphasis is on PLANNING in a J-I-T production system, not EXECUTION via Kanban cards! How do you suppose they know how many cards to put out on the floor, and where? What do they do with short runs? How do they handle unique customer requirements? Well, the good news is, the Master Scheduler's job doesn't go away when you implement J-I-T. In fact, the job becomes infinitely more important. Second, with Total Quality Control and J-I-T generating constant improvements, we can now "tighten up" the scheduling buckets from months to weeks to days to ONE HOUR! The objective is HIGH PRECISION PLANNING with NO CONTINGENCIES! That calls for some new techniques for planning, execution and feedback that our MRP II systems didn't provide. Figure 2 illustrates this.

The emphasis in precision Just-In-Time scheduling is not on the contingencies (safety stock, inventory shortages, missing tooling, missing documentation, etc.), since these are already handled by others as a part of accepting responsibility for "delighting the internal customer" in Total Quality Control. The emphasis may now be switched to what counts: improving THROUGHPUT! The best measure of throughput is Sales Dollars (shipped) Per Employee. Do you know your company's throughput yardstick? Why not find out?

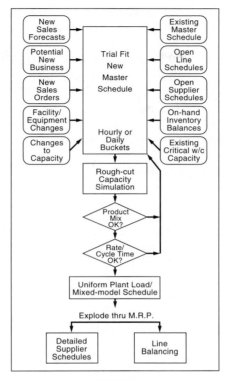

FIGURE 2

Scheduling without work orders is a difficult transition for some Master Schedulers. But once you learn the basics of line scheduling to a RATE and MIX target, the rest is easy. The key to this is the concept of Uniform Plant Loading. This can best be expressed as "make something of everything, every day." The idea is to set up a production schedule of many repeating CYCLES of the most popular products. The proportions are first determined by sales forecasts, then distilled down to weekly, daily and hourly rates. Once the required cycle time for each product has been determined, many companies often set up a schedule for 40 minutes for each hour for these most popular products, and use the technique of Mixed-Model Scheduling to create repeating cycles, like this:

- Product 'A': 10
- Product 'B': 20
- Product 'C': 5
- Product 'D': 15

- Product 'A': 10
- Product 'B': 20
- Product 'C': 5
- Product 'D': 15

The trick to making this work is total employee involvement of all departments. First, sales and marketing must agree to "time fence" guidelines for 95% of all customer orders. Constant disruption of schedules for emergency customer requirements will defeat the purpose. Engineering changes must not be implemented at random. They must be grouped and incorporated at the appropriate time (the best time is during changeovers from one product to another). Once the Final Assembly Schedule is arrived at, and agreed to by all, the objective is to achieve *linearity* to that schedule: this means the *closest possible adherence* to the Final Assembly Schedule. All of the supporting activity centers who provide material to the final assembly area must be in tune to the Final Assembly Schedule. Thus the *synchronization* of all schedules in the plant is the vehicle that brings precision to the overall process, and allows us to reduce inventory and increase throughput. Once people know they can count on a regular pattern occurring every day, the need for contingencies soon disappears. This is what we mean by schedule linearity.

Why schedule only 40 minutes of each hour this way? To allow for the processing of very small quantities for unique customer requirements. The other 20 minutes is used for this purpose. Through a concerted effort by your management and your supervision and workers, customer leadtimes can often be reduced from weeks to days or even hours!

The process of Rough-Cut Capacity Planning becomes much more meaningful when schedules are linear, because you are including a whole production line or a cellular group of machines in your calculations, not just a few "bottleneck" work centers spread throughout the factory. Visibility is provided, in a very simple form, for current work-in-process and the load against each production line, cell, or individual machine. By treating the entire factor as a cell, loads can easily be "trial fitted" into any available time slot and tested to see if the customer's needs can be accommodated within the constraints of your plant.

Note in Figure 2 how M.R.P. is still used to calculate requirements, but mostly for purchased parts and for manufactured parts which are only occasionally used. The process of *synchronization,* involving all departments who feed the final assembly area, will eventually replace the M.R.P. II Dispatch List and other priority control tools. The feeding activity centers rely on the *linearity* of the Final Assembly Schedule. They then set up their activities to produce each hour the exact number of components needed for the final assembly area. The existence of a firm, cyclical Uniform Plant Load only increases the effectiveness of M.R.P. In short, M.R.P. is used by better than 90% of J-I-T factories to generate medium-range to long-range requirements to communicate to suppliers, and to plan for plant expansion.

THE NEW JOB

Master Schedulers, this is a direct challenge! You MUST start educating yourself about J-I-T scheduling techniques. Doing things the same way you are doing them now, while it appears to be effective, will achieve the same results you are achieving now, in the future. Meanwhile, there are at least 5 to 50 competitors out there who are looking to beat your company out of its market share. How will they do that? By implementing J-I-T Continuous Improvement production. Heaven help you if you are the LAST in your industry to adopt these important manufacturing philosophies! You may not last that long.

A word about software packages for J-I-T scheduling. They are still not up to the capabilities needed for mix analyses, production rate breakdowns, rough-cut capacity planning, cycle time analysis, profitability/contribution analysis, etc. You Master Schedulers had better learn to use simple microcomputer-based spreadsheet programs for analysis. If you're not using these tools now, you are really behind the power curve. Your competitors are! And they will use the facts and plans gained by this analysis to "eat your lunch!" Also remember: people hate to look at a large spreadsheet full of numbers. Put these numbers into GRAPHS! You will be far more effective at communicating your point with pictures than with columns of numbers.

The biggest change in the Master Scheduler's job is that he (or SHE-implied) is now a more important focal point than ever before for decision-making. When you remove most of the contingencies from the production process (safety lead times, safety stock, move time allowances, queue time allowances, overtime, etc.), your scheduling must become more exact and be followed with precision. Also, the Master Scheduler is increasingly aware of the necessity for IMMEDIATE turnaround of information about alternatives. No longer is he the recipient of Sales Orders put into a shovel and thrown over the wall. He is now a working partner of Sales. Many firms have portable computers in the hands of salespeople, which they now use to dial up the factory and talk to the Master Scheduler ON-LINE! Think about that. What if you were to receive an on-line inquiry from a sales rep. How would you answer it in your current system? The Master Scheduler of today is an innovator. He has designed many "quick response" tools which he can use to get information to the field *in a hurry* to take advantage of a rapidly closing window of sales opportunity with a customer. Quick responses mean quick sales! If the Master Scheduler can provide accurate, quick turn-around of commitments to customers which can be met 99% of the time or better, he is usually worth much more than before, when the organization wasn't synchronized and achieving 75% on-time deliveries was a chore.

The Master Scheduler is increasingly seen as NOT the Materials Manager's "troop," but a working member of the staff of the Senior Management Team. There is a greater tendency to elevate him organizationally to report to a Vice-President or directly to the President. He becomes increasingly more involved in long-range planning and simulation of alternatives. He is also concerned with Performance Measurement of the entire enterprise. Master Schedulers frequently come under attack from one or more departments who perceive their needs not being met. It is up to the Master Scheduler to provide the proof that the system is working better than it did before, and defuse these arguments before they get a chance to fester and grow out of proportion to reality.

Why has the Master Scheduler's job achieved prominence? Because many bright people have taken to the position and creatively enhanced it so that their companies are dramatically more responsive to customers, to setbacks and disasters, to changes, and to market conditions. The Master Scheduler of today uses a combination of

- integrated computer systems
- microcomputer analysis tools and graphics
- MBWA (Management By Walking Around)
- superior communications skills including writing and speaking
- the judicious use of meetings and one-on-one discussions for gathering internal and external intelligence

- negotiation techniques to protect schedule integrity
- and dedication to the principle of continuous improvement

to advance his company to the position of market leader. The combination of a properly motivated, trained and dedicated Master Scheduler, working with an organization committed to linearity of schedules and constant improvement, will eventually destroy their competition.

WHERE WILL IT LEAD? THE FUTURE MASTER SCHEDULER

The Commitment to Excellence being made by companies as a competitive strategy for the 1990's will lead to a tragedy of sorts—those companies NOT making the commitment or relying on "the way we've always done it" will begin to decline, and by the end of the decade, over HALF of them will either be bankrupt or acquired by another company. You may think that this "doom and gloom" prediction means there will be less Master Scheduler jobs available. On the contrary! The surviving companies will be forced to introduce new products constantly, at an ever more accelerated pace. Their existing inefficient factories will have to be either "modernized" or closed and replaced with another factory elsewhere in the country.

Master Schedulers have an integral role to play in either of these scenarios. If an old, inefficient factor is to be "modernized," what better place to start than on Master Schedules? The Master Scheduler must introduce new techniques such as Uniform Plant Loading and Mixed-Model Scheduling. He must educate everyone on the benefits of "broadcasting" repetitive-cycle Final Assembly Schedules to internal feeding activity centers and to suppliers. He must become the *de facto* Project Manager to implement these techniques in his plant. Finally, he must pay particular attention to Performance Measurement, insuring that he reports the truth about accomplishments. If the performance is slipping back to previous levels, he must "blow the whistle" and cause the company's management to get involved to go after the next level of excellence.

If the Master Scheduler is involved in a "start-up" operation, he has the perfect opportunity to set the stage for Uniform Plant Loading, Mixed Model Scheduling, Linearity of schedules, flow of Final Assembly Schedule priority information to feeding activity centers and suppliers, and proper Performance Measurement. What he must do is to thoroughly plan all of the tools he will need to perform the three jobs of Customer Service, Change Control and Capacity Management. The emphasis must be on tools which can be used for QUICK response. Then the Master Scheduler must become the "Chief Educator." Starting with the Senior Management team, he must explain to people the value of good schedules which are not changed frequently. Once the benefits of schedule linearity are embraced by Senior Management, they must communicate to all employees that they will not tolerate major schedule disruptions unless it is for a very good reason. Armed with this commitment, the Master Scheduler can be an effective marketing weapon, manufacturing productivity enhancer, protector of the company's assets, and profit enhancer.

We have witnessed a huge explosion in the Body of Knowledge about Master Scheduling. All Master Schedulers should be learners. They need to read constantly.

They need to study other disciplines besides Materials Management. It is very important to the negotiating process that the Master Scheduler be able to talk the language of the design engineer, general accountant, cost accountant, manufacturing manager, quality manager, sales/marketing manager, customer service manager, field service manager, manufacturing engineer, materials manager, purchasing manager, President, etc. An excellent vehicle for this is APICS Certification. It should be a prerequisite for all Master Schedulers to be APICS Certified, and to have taken ALL 6 Certification Exams. The upcoming improvements in the APICS Certification program will vastly expand the Body of Knowledge as it relates to the *entire* manufacturing enterprise. Watch for the announcements on this soon. Continuous improvement is the name of the game in Master Scheduling. Make the commitment to yourself to improve 1% A DAY, EVERY DAY. That will give you about 250% improvement in your skills every year! Every Master Scheduler can do this.

America is going to get involved in a war in the 1990's. But it will NOT be fought with guns. It will be a trade war. Our objective is to **REDUCE OUR BALANCE OF PAYMENTS TO ZERO!**

This means we must build and operate our manufacturing plants at a productivity rate equal to or higher than those in the most productive countries in the world. That's a tall order. But the place to start is in Master Scheduling. The job of the Master Scheduler has never been so important.

Chapter Eleven

The Planning of Material Requirements

INTRODUCTION

This chapter examines the parts requirement problem. An end product may contain several subassemblies and parts. Each subassembly and each part takes a certain amount of production lead time. The subassemblies and parts also require raw materials that have delivery lead times. The parts requirement problem, then, is to determine the size and timing of parts and subassembly production and raw material orders.

Researchers and practitioners began working on this problem in the 1950s and 1960s. As early as 1954, in the very first issue of *Management Science,* Andrew Vaszonyi [14] described the problem and presented a matrix algebra approach to its solution. In those days the problem was also known as the explosion and netting problem in the planning of material requirements. In the late 1960s Joseph Orlicky [10] at IBM began to popularize the list processing solution known as material requirements planning (MRP). His work culminated in the outstanding book *Material Requirements Planning,* published in 1975.

It is important to distinguish between the parts requirement problem and the MRP solution. MRP has strengths and weaknesses that make it extremely attractive, but it is not the only answer. Further refinements of MRP and profitable applications require an understanding of the advantages and drawbacks of the MRP approach.

The 1970s and 1980s saw the widespread installation and acceptance of MRP in the majority of U.S. manufacturing sites. With the introduction of MRP II, manufacturing resource planning, popularized by Oliver Wight in his 1981 groundbreaking book, *MRP II: Unlocking America's Productivity Potential,* several weaknesses of MRP were addressed. Subsequent sections of this chapter address the differences in MRP and MRP II.

MRP AND MRP II

The MRP acronym is used in three ways, each of which describes a particular aspect of materials requirements. Figure 11.1 shows an overview of MRP, MRP II, and closed-loop MRP.

Materials Requirements Planning (MRP)

Materials requirements planning (MRP), sometimes referred to as MRP I, little MRP, or mrp, is the original MRP. As seen in Figure 11.1, MRP takes the output from the master schedule, combines that with information from inventory records and product structure records, and determines a schedule of timing and quantities for each item. The basic idea is to get the right materials to the right place at the right time. The major part of this chapter shows how the MRP mechanics are accomplished.

Closed-Loop MRP

As seen later in this chapter, MRP does not take into account capacity when calculating the size of production lots. The result is that the capacity may be exceeded during some periods. In addition, the production lead time will be longer than originally thought. The actual production schedule will be different from the MRP schedule, and the integrity of the MRP system will be in jeopardy. The closed-loop MRP system was developed in response to this weakness in MRP.

In closed-loop MRP, the information from the MRP and other modules is fed back to ensure that the schedules are feasible. Figure 11.1 shows that the output from MRP is fed into a capacity planning module. If the capacity plan is feasible, then the information is passed on for implementation. If the capacity plan is not feasible, then that information is fed back to the master schedule and the MRP module. Adjustments to the master schedule or MRP will result in changes to the planned order releases so that the capacity plan is feasible.

Additional loops are shown in Figure 11.1. The original MRP concept had information flowing down the system, but as can be expected, many infeasible plans were discovered only after implementation at the shop floor level. Loops were made from each module back up to feed information to the planning process. The closed-loop system was the first component of MRP II.

Manufacturing Resource Planning (MRP II)

Manufacturing resource planning is known as MRP II because the closed-loop system allows the planning of resources in the manufacturing environment. For example, as shown in Figure 11.1, the MRP module feeds information into the capacity planning module. The use of a manufacturing resource, capacity, can now be planned, and that information can be fed back through the closed loop to utilize that resource better.

It became apparent that this flow of information could be useful to other functions within a manufacturing firm. The first function to benefit was purchasing. The MRP out-

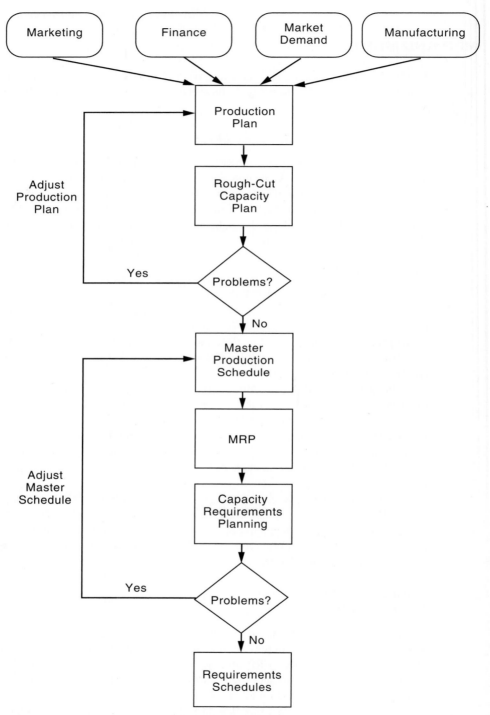

FIGURE 11.1 Overview of MRP II.

put provided the timing and quantity of materials required; thus buyers could plan their purchasing activities so that material would arrive when needed, not before nor after. This offered a tremendous improvement for buyers because they could be confident that if the MRP plan called for certain materials at a date and quantity, they could purchase that material and have it arrive at the time specified by the MRP plan. This has reduced purchasing cost due to reduced inventories and better coordination with suppliers.

Wight [18] was the first to specify the total business integration of MRP II. His intent was that all the resources of the manufacturing firm could be coordinated through MRP II. Output from the MRP plans could be used by finance, engineering, accounting, purchasing, distribution, and marketing to ensure that all functions within a firm were planning using the same numbers. For example, marketing can use the information from the MRP output to see if a new order can be worked into the existing schedule and when that order would be delivered to the customer. Marketing would no longer have to guess, but could give the customer an accurate estimate of the delivery date. This became a competitive advantage.

Class A, B, C, and D

A classification system has been devised to determine how well an MRP system has been implemented. A class D company is one in which the MRP system is only working in the computing department. Inventory records are probably poor, and few outputs from the system that are being used to manage manufacturing.

A class C company uses MRP for inventory ordering but not for scheduling. Shop scheduling is done from shortage lists. The loop has not been closed, but there is probably some inventory reduction.

A class B company may have MRP, capacity planning, and shop floor controls working. MRP is still seen as a production planning and inventory control system. They may have inventory reduction and better customer service. Some of the closed-loop system is utilized.

A class A company has MRP II in full use. The closed-loop system is working. Most important is that the firm is using the MRP II system to run the business. Finance, sales, marketing, purchasing, accounting, and engineering are receiving information from the MRP system that is useful and used. Wight [18] claims that a typical class A company should see a one-third reduction in inventory, productivity increases of 5% to 10%, and improved customer service that could result in sales increases. He cautions that every business will see different results, but a true class A company will see significant improvements.

IMPLEMENTATION OF MRP AND MRP II

The installation of an effective MRP system should not be viewed as a quick process. It may be possible to pull a computer program off the shelf and have it running the next day, but this will not result in an effective MRP system. The journey to a true class A system starts with the implementation of a basic MRP system.

Software is generally not the problem in unsuccessful installations. Software is available from many vendors and is now available for personal computers. Instead, the problems are usually associated with the use of the system and the attitudes of the people affected by MRP.

MRP requires accurate inventory records as a start. Because MRP uses inventory records to determine the number of units to make our buy, inaccuracies in such records will be translated into incorrect numbers for production scheduling. Bills of materials must also be accurate to ensure that the right parts are ordered. Procedures to ensure the accurate maintenance of records must be put in place. Engineering changes occur routinely and can have a significant affect on the bills of materials.

The most important aspect of successful MRP implementation is employee training. Everyone—from top management to shop floor workers—must be trained in how to use MRP effectively. Many stories of unsuccessful implementation are due to people not having the proper training and not utilizing MRP properly. Usually an informal production system is in place at the shop floor level: Supervisors on the shop floor may have determined, based on an informal network, which items to schedule. The problem is that the MRP system may give production schedules that the supervisors do not believe. If the supervisors do not follow the MRP schedule, then the entire sequence of schedules may be out of synch and the product will not be available when needed at a later operation.

Once the basic MRP system is in place and the informal system has been replaced with the formal MRP, the process of closing the loop can begin. As seen in Figure 11.1, the purchasing function and the capacity planning module are generally the two functions that can be pulled into the loop first. MRP software is now available that includes the loops, and as with MRP implementation, the main problem with closing the loops is getting people to use the information properly. Training is now extended to the additional functional areas that are to use the information: purchasing, engineering, and so forth.

Wight states that MRP II has three characteristics: the operating and financial system are the same, the system can do "what if" simulations, and the system is a whole company system. The closed-loop system in place is extended to include these three activities. Again, people must be able to use the system effectively and must want to use it to run the company. A system of this type is obviously one that needs top management support to be effective.

THE PARTS REQUIREMENT PROBLEM

Suppose that we have an end product A, composed of subassemblies that are made up, in turn, of parts and raw materials. We will imagine that our product A is a tricycle. The product structure, or bill of materials (BOM), for the tricycle might be pictured as a product tree. Figure 11.2 is a product tree structure showing what goes into an end item A.

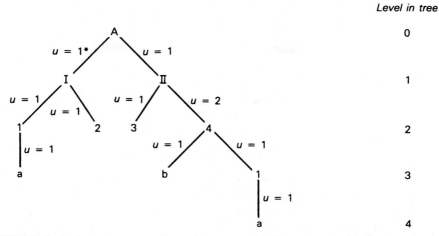

FIGURE 11.2   Product tree structure. *Note: u* is usage; *u* = 1 indicates that one unit of I, for example, goes into one unit of A.

Figure 11.3 shows the product tree for our tricycle. Such a bill of materials could also be drawn up for a bicycle, which would contain a frame, two wheel assemblies, a brake assembly, a seat, and a set of handlebars. In Chapter 10 questions of options—such as five-speed or ten-speed—will arise; for now, however, we look only at simple end products with fixed sets of subassemblies and parts.

The product structure or bill of materials can be described in several ways. The infamous Professor Zepartsdat Gozinto formulated the *Gozinto* matrix representation (see Figure 11.4). (Despite popular misconception, Z. Gozinto was not related to another famous management scientist, H. L. Gantt.) The indented bill of materials, developed by I. N. (Buckey) Dent, concisely summarizes the same information (see Figure 11.5).

Current inventory balances of parts, subassemblies, and end items are given in our inventory records, as follows:

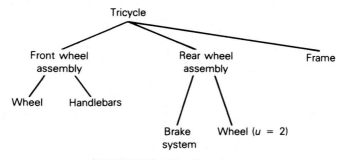

FIGURE 11.3   Tricycle product tree.

	A	I	II	1	2	3	4	a	b	into
A										
I	1									
II	1									
1		1					1			
2		1								
3			1							
4			2							
a				1						
b							1			
goes										

FIGURE 11.4 Gozinto table (rows go into columns).

	Item	On-Hand Inventory (OH)	Lead Time in Weeks (L)
End item	A	0	1
Subassembly	I	40	1
Subassembly	II	15	2
Part	1	10	3
Part	2	20	4
Part	3	15	1
Part	4	30	2
Raw material	a	10	3
Raw material	b	10	3

Figure 11.6 shows the product tree structure on a time-phased basis. The longest, or critical, path is A-II-4–1-a, with a cumulative lead time of eleven weeks.

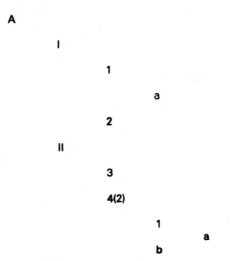

FIGURE 11.5 Indented bill of materials.

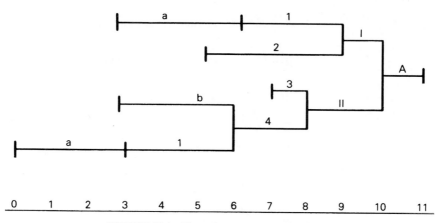

FIGURE 11.6 Time-phased product tree structure.

THE MECHANICS OF MRP

Suppose that we must produce twenty units of product A six weeks from now. Noting that only ten units of raw material a appear as an on-hand balance, how many more units of material a do we need to ensure its availability for production in end item A? As a quick reaction, we might guess that we need an extra ten units. On second thought, however, we notice that subassembly I shows a balance of forty units (already containing material a), and subassembly II shows a balance of fifteen units. Further, we already have sufficient on-hand quantities of parts 3 and 4 to produce the five additional units of subassembly II needed. Hence, fifteen units of end item A can be produced without ordering up any additional material a. The purchasing department would have cursed if we had followed our first hunch, and rightly so.

To keep track of inventory interactions from the end item to the raw material level, we can construct a set of tables based on the product tree structure. MRP computer programs, available from many vendors, perform calculations similar to those we will do with this set of tables.

Notice from Figure 11.6 that end item A requires a one-week lead time. This includes time to assemble the two subassemblies I and II into the end item A. Later we examine other elements of lead time, but it is important now to recognize that subassemblies I and II must be completed and be ready to be assembled when work begins on end item A. Suppose that the master production schedule (MPS) calls for production of twenty units of A by week 7 and 100 units of A by week 9. Because A has a one-week assembly time, this means that the order to produce twenty units of A is released at the beginning of week 6. Hence, subassemblies I and II must be ready at the beginning of week 6. With a one-week lead time, end item A should be completed by the beginning of week 7. Table 11.1 displays the master production schedule and shows the basic MRP logic.

MRP logic uses the following relationships:

TABLE 11.1 MPS for End Item A

					Week Numbers					
	0	*1*	*2*	*3*	*4*	*5*	*6*	*7*	*8*	*9*
Gross requirements (GR)								20		100
On hand at end of period (OH)	0	0	0	0	0	0	0			
Planned order release (POR)							20		100	

$L = 1.$

$$\text{Net requirements} = \text{total requirements} - \text{available inventory}$$

$$= \begin{bmatrix} \text{gross requirements} \\ + \\ \text{allocations} \end{bmatrix} - \begin{bmatrix} \text{on hand} \\ + \\ \text{scheduled receipts} \end{bmatrix}$$

Planned order release is the time that the order is released so that the units will be ready to meet the requirements.

In our example, gross requirements are for twenty units in week 7. No inventory has been allocated—that is, dedicated to an order. No previous orders have been issued for which we are expecting receipt; that is, there are no scheduled receipts.

Since on-hand inventories are zero, net requirements are simply equal to the gross requirements of twenty units in week 7. Backing off by one week for lead time, the planned order release occurs in week 6.

The master production schedule can reflect an economic order quantity decision at the end item level. Suppose that the master production schedule were based on a (Q, R) system, with $Q = 90$, $R = 30$, demand rate $d = 15$, and lead time $L = 2$ weeks. Table 11.2 demonstrates the nature of the reorder point in a (Q, R) system; the reorder point triggers an order release.

In this case the order will cover six weeks' worth of gross requirements. Gross requirements will arbitrarily be considered to occur at the middle of the week, but planned orders will be released at the beginning of the week. On-hand inventories show inventory status at the end of the week and here reflect the loss of fifteen units per week. The planned order release schedule for the end item can be considered a master production

TABLE 11.2 Time-Phased Order Point

	0	*1*	*2*	*3*	*4*	*5*	*6*	*7*	*8*	*9*	*10*	*11*	*12*
GR		15	15	15	15	15	15	15	15	15	15	15	15
OH	45	30	15	0	75	60	45	30	15	0	75	60	45
POR			90						90				

Note: Inventory projected negative in week 4. Place order to avoid negative.

$Q = 90$, $R = 30$, and $L = 2$ weeks.

Reorder point is never really used, but rather MRP logic takes over.

schedule. An order for ninety units is released at the beginning of week 2, after the ending inventory of week 1 hits the reorder point of thirty units. The order takes two weeks to process and thus arrives at the beginning of week 4, ready to be used to satisfy gross requirements in week 4. At this stage, it is useful to record our conventions for time-phasing requirements:

1. Gross requirements (GR) will occur at the middle of a period.
2. On-hand inventory (OH) will be measured at the end of a period.
3. Planned order releases (POR) will occur at the beginning of a period.

As a warm-up before tackling our MRP problem, consider a situation in which gross requirements are 70 units in week 2. On-hand inventories are 100 units, of which 50 have been allocated. A previous order has been made for 10 units, which are expected to arrive at the beginning of week 1. The normal lead time is one week. What would the MRP table look like? (AC is allocated inventory, and SR is scheduled receipt.) We would have:

	Week 0	Week 1	Week 2
GR			70
AC	50	50	
OH	100	100	
SR		10	
POR		⑩	

Net requirements = [GR + AC] − [OH + SR]

$$= [70 + 50] \quad - [100 + 10] = 10$$

By time phasing the net requirements, an order for 10 units is released at the beginning of week 1.

Now that Tables 11.1 and 11.2 are understood, we are prepared to carry our original example all the way back from end item A to raw materials. As the MPS, Table 11.1 gives the planned order order releases for 20 units of A in the sixth week and 100 units of A in the eighth week. To produce 20 units of A in the sixth week, we need 20 units of subassembly 1 and 20 units of subassembly II. These subassemblies must be ready at the beginning of week 6 so that they can be assembled into end item A during week 6. Table 11.3 displays the netting process for subassembly I. All calculations are based on the information given in Figure 11.2 and repeated in Table 11.4.

Proceeding in this fashion, we would move to subassembly II and then to parts 1, 2, 3, and 4. Part 1 causes some difficulty because it goes into subassembly I and also into part 4, as shown in Table 11.4. If we draw up a table for part 1 as it goes into subassembly I, we will then need to repeat ourselves when we prepare a table for part 4. A more convenient way to handle this is to assign lowest level codes, a code for the lowest level in the tree at which an item occurs. Part 1 occurs at level 2 and also at level 3. Hence, its lowest code will be 3. To avoid any double counting in creating tables, we need only

TABLE 11.3 MRP Table for Subassembly I

	0	1	2	3	4	5	6	7	8	9
GR							20		100	
OH	40	40	40	40	40	40	20	20		
POR								80[a]		

[a]Subassembly I has a one-week lead time

draw up the tables in order of the items' lowest level codes. Table 11.4 gives the lowest level codes for the items given in Figure 11.2. According to these lowest level codes, we can safely draw up tables in the sequence A, I, II, 2, 3, 4, 1, b, a. Table 11.5 shows the entire set of inventory interactions implied by the master schedule of 20 units of A in week 6 and 100 units of A in week 8. The following "rules of the road" should be kept in mind:

1. Planned order releases at a higher level become gross requirements at a lower level.
2. Usage rates must be noted (e.g., each unit of subassembly II requires two units of part 4).

MRP CONCEPTS AND ADVANTAGES

Several concepts can now be defined and explained against the background of the foregoing example. What are the inputs and outputs of an MRP system? Where is MRP applicable? Where does MRP fit in the overall production control environment? What are the advantages of the MRP approach?

MRP applies primarily to production systems exhibiting *parent-component* rela-

TABLE 11.4 Problem Information Summary

Indented BOM	Lowest Level Codes (LLC)		Inventory Records	
	Item	Code	OH	L
A				
I				
1	A	0	0	1
a	I	1	40	1
2	II	1	15	2
II	1	3	10	3
3	2	2	20	4
4(2)	3	2	15	1
1	4	2	30	2
a	a	4	10	3
b	b	3	10	3

TABLE 11.5 MRP Tables for the Entire Product Tree

A	0	1	2	3	4	5	6	7	8	9
POR							20		100	

LLC = 0, L = 1.
LLC = lowest level code, L = lead time.

I	0	1	2	3	4	5	6	7	8	9
GR							20		100	
OH	40	40	40	40	40	40	20	20	0	0
POR								80		

LLC = 1, L = 1.   (goes into A)

II	0	1	2	3	4	5	6	7	8	9
GR							20		100	
OH	15	15	15	15	15	15	0	0	0	0
POR					5		100			

LLC = 1, L = 2.   (goes into A)

2	0	1	2	3	4	5	6	7	8	9
GR								80		
OH	20	20	20	20	20	20	20	0	0	0
POR				60						

LLC = 2, L = 4.   (goes into I)

3	0	1	2	3	4	5	6	7	8	9
GR					5		100			
OH	15	15	15	15	10	10	0	0	0	0
POR						90				

LLC = 2, L = 1.   (goes into II)

4	0	1	2	3	4	5	6	7	8	9
GR					10		200			
OH	30	30	30	30	20	20	0	0	0	0
POR					180					

LLC = 2, L = 2.   (goes into II on a 2 for 1 basis)

1	0	1	2	3	4	5	6	7	8	9
GR					180			80		
OH	10	10	10	10	0	0	0	0	0	0
POR		170			80					

LLC = 3, L = 3.   (goes into I and 4)

(cont.)

TABLE 11.5   MRP Tables for the Entire Product Tree (*Continued*)

b	0	1	2	3	4	5	6	7	8	9
GR					180					
OH	10	10	10	10	0	0	0	0	0	0
POR		170								

LLC = 3, L = 3.   (goes into 4)

a	0	1	2	3	4	5	6	7	8	9
GR		170			80					
OH	10	0	0	0	0	0	0	0	0	0
POR		80								
SR[a]		160								

LLC = 4 and L = 3.   (goes into 1)

[a]SR = scheduled receipt; indicates the scheduled date of receipt of an order *already* released (i.e., an open order as opposed to a planned order).

tionships. The component item goes into the parent item. Manufacture of assembled products is an obvious example. If the end items can be stated in a master production schedule, an MRP system can be applied fruitfully.

The master production schedule (MPS) specifies the quantities of end items to be produced by a time period several months into the future. End items may be finished products or the highest-level assemblies from which these products can be built. The *planning horizon* for the MPS must be at least as long as the cumulative lead time for purchasing and manufacturing. In our example the critical path through the product structure was A-II-4-1-a, with a total lead time of eleven weeks. Hence, we must plan at least eleven weeks into the future to ensure that we have time to produce or order the requirements along his path.

Whereas MRP is applicable to parent-component production systems, traditional order-point techniques fail to accommodate parent-component relationships. Forecasting and order-point systems treat each component separately; but there is no need to forecast individual part requirements if we have a schedule of end item production. Sloan [13] recognized this as early as 1921, when analysts at General Motors made a "purely technical calculation" of the quantity of materials required to support the level of production necessary to yield a given number of cars. Each car has five wheels, so we need not make independent forecasts of the number of car wheels required. We may want to forecast automobile sales, however, representing *independent*, or external, demand. Once we have determined the production levels needed to support the sales forecasts, the *dependent*, or derived, requirements of wheels can be obtained by multiplication.

In forecasting, separate treatment of individual items wastes valuable information; having separate order points for individual items, however, causes excessive holding costs while increasing risks of stockout. Instead, parent-component items should be matched so that planned orders for the parent become gross requirements for the component.

TABLE 11.6   Excessive Holding in an Order-Point System

	0	1	2	3	4	5	6	7	8	9
GR		15	15	0	0	0	15	90	0	0
OH	45	30	15	15	105	105	90	0	0	0
POR			90							

$Q = 90$, $R = 30$, and $L = 2$; units held = 405.

To see the weakness of separate order points, we shall begin by studying the accusation that they cause excessive holding costs. Tables 11.6 and 11.7 show the reaction of order point and MRP to lumpy demand. The order-point system places the order much too early and hence carries inventory wastefully. Lumpy demand is not peculiar to parent-component systems, but it is certainly characteristic of them. Lot sizing at the parent level causes lumpy demand at the component level. Lumpy demand is *discontinuous* (periods of zero demand) and *irregular* (the size of the lumps varies). In an environment of parent-component relationships and associated lumpy demand, the traditional EOQ system yields excessive holding costs.

Although the order-point system causes excessive holding costs, the increased inventory levels do not contribute to increased service levels, because the inventory is held at the wrong time. The service level is the probability that inventory will be available at the time it is required. The complementary probability is the probability of stockout. Low stockout probabilities at the component level can still yield a fairly high stockout probability on all components taken together. If a parent item has, for example, ten components, and if each of these components has a probability of stockout of 0.01, then the probability of having all the items in stock ready for production of the parent item will be only $0.90 = 0.99^{10}$. Hence, parent-component relationships must be managed on a matched basis. It is not good to have the parent and component items forecast and controlled separately; even with excellent control at the individual item level, the probability of being able to produce the parent item with all components present in sufficient quantities is rather low. It is the special ability of MRP to plan for component requirements all the way from the end item level to the raw material level.

The inputs to the MRP system are (1) the master production schedule, (2) the bill of materials, and (3) the inventory status file. The output of the MRP system consists of the timing and the quantity of subassemblies, parts, and raw materials. This output can be used, first, to plan purchasing action and, second, to plan manufacturing action. A planned order release is a signal that we are going to produce a certain amount of a com-

TABLE 11.7   Reasonable Holding in an MRP System

	0	1	2	3	4	5	6	7	8	9
GR		15	15	0	0	0	15	90	0	0
OH	45	30	15	15	15	15	0	0	0	0
POR						90				

$L = 2$; units held = 135.

ponent in a specific time period. Further, the production action items also indicate that we will have needs against capacity. We see later in the text that an MRP system also enables us to plan our capacity requirements.

LOT SIZING

In our MRP discussion so far, we have used the lot-for-lot inventory lot-sizing technique: Whatever the net requirement is, we produce exactly that amount. For end item A, this meant that we planned production orders for 20 units in week 6 and 100 units in week 8. Should we combine these into one large order, thereby reducing setup costs? The answer depends heavily on the ratio of holding costs to setup costs.

The Lot-for-Lot Technique

To examine holding versus setup cost trade-offs, consider the problem presented in Table 11.8, an MRP table showing the several orders are released, each one week in advance of the net requirement. In the table, net requirements equal gross requirements minus beginning inventory as it is used up to cover requirements by period. Similarly, ending inventory from the previous week plus any order release due to be completed for the current week minus the gross requirements for the current week gives current ending inventory. The lot-for-lot (L-4-L) technique yields zero holding costs but seven separate setups, for a total cost of $1400.

The EOQ Approach

As a first at achieving better costs, we might try an economic order quantity system:

$$Q^* = \sqrt{\frac{2dS}{h}} = \sqrt{\frac{2(27)(200)}{2}} = 74$$

Table 11.9 shows the EOQ solution. The EOQ solution requires four setups, totaling $800, plus an inventory holding cost of $790, for total costs of $1590. The EOQ approach assumes a constant requirements rate. The requirements here are lumpy, however; they occur in some time periods but not in others. Moreover, not only are the requirements discontinuous, but the size of the lumps is irregular, ranging from 10 to 55 units.

TABLE 11.8 MRP Lot-Sizing Problem: Lot-for-Lot Technique

	0	1	2	3	4	5	6	7	8	9	10
Gross requirements		35	30	40	0	10	40	30	0	30	55
On hand	35	0	0	0	0	0	0	0	0	0	0
Planned order release		30	40	0	10	40	30	0	30	55	0

Holding costs = $2/unit/week, setup cost = $200, gross requirements average per week = 27, and lead time = 1 week.

TABLE 11.9   MRP Lot-Sizing Problem: EOQ Approach

	0	1	2	3	4	5	6	7	8	9	10
GR		35	30	40	0	10	40	30	0	30	55
OH	35[a]	0	44	4	4	68	28	72	72	42	61
POR		74			74		74			74	

$L = 1$; holding cost = $(44 + 4 + ... + 61) * 2$.

[a]In our holding cost calculations, we exclude the beginning inventory of 35 units because we incur the holding costs on these units regardless of any decisions we make. We include only relevant costs.

To handle the discrete nature of the problem with lumpy demand, several approaches have been advocated. At one extreme are the EOQ-based procedures, which start from a misapplication, and at the other extreme is a dynamic programming approach that yields optimal solutions.

The EOQ method was introduced in Chapter 4 as an independent inventory method. The approach was applied to MRP lot-sizing procedures even though several key assumptions of the EOQ are violated. In cases where the independent demand is the same every period, the EOQ may provide a reasonably good solution. Unfortunately, the very concept of EOQ lot sizing lumps the demand for the next level in the bill of material. Thus the lumpy demand would be a violation of the key EOQ assumption that demand is constant. In any event, the EOQ has been used in MRP systems.

Minimum Cost per Period
(or Silver-Meal) Approach

Because it is effective and can be readily understood, it is useful to start with a compromise technique: the minimum cost per period (MCP) approach, also known as the Silver-Meal (SM) method, after its developers [12].

If we combine two periods, the cost per period will be the sum of setup plus holding costs divided by the two periods. For example, combining the net requirements for periods 2 and 3 would give a cost per period of $140, calculated as $200 setup cost plus 40 units held at $2 per unit, all divided by two periods. Table 11.10 displays the MCP calculations, and Table 11.11 shows a solution by the MCP approach.

The MCP solution yields $600 in setup costs plus $310 in holding costs (ignoring required holding cost on beginning inventory), a total cost of $910. This is $490 less than the cost by the L-4-L technique.

To see the close similarity between MCP and EOQ concepts, we can develop a cost per period formulation yielding the EOQ solution. Suppose that we have $d = 27$, $h = \$2$, and $S = \$200$ (recall that d is weekly demand and h is the weekly holding cost per unit). Then the cost per period (CPP) is a function of Q:

$$\text{CPP}(Q) = \frac{Q}{2}h + \frac{S}{(Q/d)}$$

TABLE 11.10 MCP Calculations

Trial Periods Combined	Trial Lot Size (Cumulative Net Requirements)	Cumulative Cost	Cost per Period[a]
2	30	200	$200.00
2, 3	70	280	$140.00
2, 3, 4	70	280	$ 93.33
2, 3, 4, 5[b]	80[b]	340	$ 85.00
2, 3, 4, 5, 6	120	660	$132.00
(Combine periods 2, 3, 4, and 5 because cost per period is a minimum.)			
6	40	200	$200.00
6, 7	70	260	$130.00
6, 7, 8[b]	70	260	$ 86.67
6, 7, 8, 9	100	440	$110.00

[a]Cost per period = cumulative cost divided by number of periods combined.
[b]POR combines these periods and cumulative net requirements.

where Q/d specifies the number of periods covered by the lot size (e.g., $Q/d = 81/27$ covers three periods of demand). Differentiating now and setting the derivatives to zero, and solving for Q would give

$$Q^* = \sqrt{\frac{2dS}{h}}$$

Does this look familiar?

Although the MCP procedure works well, other approaches have also been developed and used, some with more success than others. Because these procedures can be found in practice and because no single procedure has been accepted as best, it is necessary to study several additional approaches.

Period Order Quantity Approach

Weakest among the other approaches is the EOQ-based period order quantity (POQ). Earlier we found the EOQ to be 74 units, with an average demand of 27 units per week. Annual requirements would be $27 * 52 = 1404$. Annual requirements divided by EOQ would be 19 orders per year ($D/Q = 19$). The time between orders would thus be 52 weeks divided by 19 orders, or 2.7 weeks between orders. Hence, POQ equals the quantity to cover P^* periods of net requirements, where $P^* = N/D/EOQ$, N is the number of

TABLE 11.11 MRP Lot Sizing Problem: MCP Approach

	0	1	2	3	4	5	6	7	8	9	10
GR		35	30	40	0	10	40	30	0	30	55
OH	35	0	50	10	10	0	30	0	0	55	0
POR		80				70			85		

TABLE 11.12 MRP Lot-Sizing Problem: POQ Approach

	0	1	2	3	4	5	6	7	8	9	10
GR		35	30	40	0	10	40	30	0	30	55
OH	35		50	10	10	0	60	30	30	0	0
POR		80				100				55	

periods in a year, and D represents annual requirements. Table 11.12 shows the POQ solution: Order three weeks' supply each order; every third time, order two weeks' supply. [In this text, we adopt the convention that POQ covers a certain number of periods of positive requirements. Facing the requirements (0, 0, 0, 20, 0, 20, 20), a three-period POQ would be 60, enough to cover seven actual periods. Hence, fewer setups will occur than EOQ would dictate if there are many periods with zero requirements.]

The total cost for the POQ solution is $980. The POQ approach performs better than the EOQ approach because it adapts to the requirements of a set of periods. In facing periods 2 through 5, the POQ approach would have one order for 80 units. The EOQ approach would have one order of 74 units, and hence a remnant of 4 units would be carried unnecessarily from period 1 through 4 until the shortfall of 6 units became evident for period 5. The advantage of POQ, then, is that it is dynamic: The order quantity changes in response to the net requirements. As we shall see later, this dynamic quality has advantages and disadvantages.

Least Unit Cost Approach

Rather than minimizing cost per period, the least unit cost (LUC) approach attempts to minimize cost per unit. Careful examination of the technique reveals its myopic nature. Table 11.13 shows the calculations, and Table 11.14 gives the planned order releases and planned inventory status. The cost of the LUC approach is $600 setup plus $390, for a total cost of $990. This total cost can also be calculated as 70 units at $4 per unit plus 80 units at $5 per unit plus 85 units at $3.65 per unit, for a total cost of $990. The myopic nature of the LUC approach stems from the range of costs from $3.65 per unit to $5 per unit. The LUC approach averages $4.21 per unit for this problem—that is, a total cost of $990 divided by net requirements of 235 units.

Least Total Cost

The least total cost (LTC) procedure is based rather fallaciously on EOQ. In the EOQ approach, the optimal solution happens to be at the point where setup and holding costs are equal. There is no reason to believe that such a result should hold in the case of lumpy demand. Indeed, our EOQ solution to the problem yielded setup costs of $800 and holding costs of $790—a reasonable balance; yet this balanced solution of $1590 is far from optimal.

The LTC approach tries to balance the holding and setup costs. A lot size is started in the first period. Starting with the next period, add the demand for that period to the lot if the cumulative carrying costs are less than or close to the setup cost. Continue adding

TABLE 11.13   LUC Calculations

Trial Periods Combined	Trial Lot Size (Cumulative Net Requirements)	Cumulative Cost	Cost per Unit
2	30	200	$ 6.67
2, 3	70	280	$ 4.00
2, 3, 4	70	No change—hence ignore	
2, 3, 4, 5	80	340	$ 4.25
(Combine periods 2 and 3 at a cost of $4 per unit)			
4, 5	10	200	$20.00
4, 5, 6	50	280	$ 5.60
4, 5, 6, 7	80	400	$ 5.00
4, 5, 6, 7, 8	80	No change—hence ignore	
4, 5, 6, 7, 8, 9	110	640	$ 5.82
(Combine periods 4, 5, 6, and 7 at $5 per unit)			
8, 9	30	200	$ 6.67
8, 9, 10	85	310	$ 3.65
(Combine periods 8, 9, and 10 at $3.65 per unit)			

(Keep raising the lot size until the cost per unit increases.)

future periods' demands until the total cumulative carrying cost exceeds the setup cost. Tables 11.15 and 11.16 indicate the solution using the data given in Table 11.8. The total cost of this solution is $600 total setup cost plus carrying cost of $310 for a total of $910.

Part Period Balancing

The part period balancing (PPB) approach is a variation of the LTC method. The PPB procedure attempts to balance setup and holding costs through the use of economic part periods. An economic part period (EPP) is the ratio of setup cost to holding cost. In our case, EPP = 200/2 = 100 units. Thus holding 100 units for one period would cost $200, the exact cost of a setup:

Cost per setup = EPP * (holding cost/unit/period)

Holding 50 units for two periods, however, would also cost $200 and could be thought of as 100 part periods. Hence, the PPB procedure simply combines requirements until the number of part periods most nearly approximates the EPP. Table 11.17 shows the calculations and Table 11.18 shows the results, with total costs of $980.

TABLE 11.14   MRP Lot-Sizing Problem: LUC Approach

	0	1	2	3	4	5	6	7	8	9	10
GR		35	30	40	0	10	40	30	0	30	55
OH	35	0	40	0	0	70	30	0		55	0
POR		70			80				85		

TABLE 11.15   LTC Calculations

Period	Demand	Periods Carried	Carrying Cost	Cumulative Carrying Cost
2	30	0	0	0
2, 3	40	1	80	80
2, 3, 4	0	2	0	80
2, 3, 4, 5	10	3	60	140
2, 3, 4, 5, 6	40	4	320	460

The addition of period 6 would make the cumulative cost exceed the setup cost of 200. Do not include period 6 demand. The lot in period 2 will be 80.

6	40	0	0	0
6, 7	30	1	60	60
6, 7, 8	0	2	0	60
6, 7, 8, 9	30	3	180	240

The addition of period 9 would make the cumulative cost exceed the setup cost of 200. Do not include period 9 demand. The lot in period 6 will be 70.

9	30	0	0	0
9, 10	55	1	110	110

End of the horizon. The lot in period 9 will be 85.

McLaren's Order Moment

The McLaren order moment (MOM) method is similar to PPB. The first step is to combine future demands into a lot and accumulate part periods until the target value is reached. A part period is one unit of inventory carried for one period. The target value is found as follows:

$$\text{OMT} = d \left(\sum_{t=1}^{T^*-1} t + (\text{TBO} - T^*)T^* \right) \qquad (11.1)$$

where

$$
\begin{aligned}
\text{OMT} &= \text{order moment target} \\
d &= \text{average requirements per period} \\
\text{TBO} &= \text{EOQ}/d = \text{time between orders} \\
T^* &= \text{largest integer less than (or equal to) the TBO}
\end{aligned}
$$

TABLE 11.16   MRP Lot-Sizing Problem: LTC Approach

	0	1	2	3	4	5	6	7	8	9	10
GR		35	30	40	0	10	40	30	0	30	55
OH	35	0	50	10	10	0	30	0	0	55	0
POR		80				70			85		

TABLE 11.17 PPB Calculations

Periods Combined	Trial Lot Size (Cumulative Net Requirements)	Part Periods
2	30	0
2, 3	70	$40 = 40 \times 1$
2, 3, 4	70	40
2, 3, 4, 5	80	$70 = 40 \times 1 + 10 \times 3$
2, 3, 4, 5, 6	120	$230 = 40 \times 1 + 10 \times 3 + 40 \times 4$
(Combine periods 2 through 5)		
6	40	0
6, 7	70	30
6, 7, 8	70	30
6, 7, 8, 9	100	$120 = 30 \times 1 + 30 \times 3$
(Combine periods 6 through 9)		
10	55	0

The first part of the MOM method is to accumulate future demands into a tentative order until the accumulated part periods equal or exceed the OMT. The EOQ was found earlier to be 74 with $d = 27$. The TBO is $= 2.74$, and $T* = 2$. The OMT target will be

$$\text{OMT} = 27 \left(\sum_{t=1}^{1} t + (2.74 - 2)2 \right) = 67$$

When the accumulated parts period equal or exceed this value, a second test is done that determines whether to include one more period in the lot:

$$h(k)D_t \leq S \tag{11.2}$$

which h is the holding cost per period, S is the setup cost, and k is the number of periods the product will be carried. If this equation holds true, then include the demand in the lot. In this example, this would be

$$2(k)D_t \leq 200$$

Tables 11.19 and 11.20 show the calculations. The total cost would be \$600 for setup and \$310 holding cost, for a total of \$910.

TABLE 11.18 MRP Lot-Sizing Problem: PPB Approach

	0	1	2	3	4	5	6	7	8	9	10
GR		35	30	40	0	10	40	30	0	30	55
OH	35	0	50	10	10	0	60	30	30	0	0
POR		80				100				55	

TABLE 11.19   MOM Calculations

Period	Requirements	Periods Carried	Part Periods	Cumulative Part Periods
2	30	0	0	0
2, 3	40	1	40	40
2, 3, 4	0	2	0	40
2, 3, 4, 5	10	3	30	70

The addition of period 5 would have the cumulative part periods exceed the OMT = 67. The second test is now done. Is $2(k) D_t \le 200$? $2(3)10 \le 200$? Yes; so include period 5 demand in the lot.

6	40	0	0	0
6, 7	30	1	30	30
6, 7, 8	0	2	0	30
6, 7, 8, 9	30	3	90	120

The addition of period 9 would have the cumulative part periods exceed the OMT = 67. The second test is now done. Is $2(k) D_t \le 200$? $2(4)30 \le 200$? No; thus do not include period 9 in the lot.

9	30	0	0	0
9, 10	55	1	55	55

End of the horizon. The lot in period 9 will be 85.

Groff's Algorithm

Groff [7] proposed a method that is similar to MOM in that it considers the addition of a future demand in a lot. A future demand is included if the following is satisfied:

$$n(n-1)D_n \le 2S/h. \tag{11.3}$$

In the example, $2S/h = 200$. Tables 11.21 and 11.22 show the calculations. The total cost would be $600 for setup and $380 carrying cost, for a total of $980.

Freeland and Colley

The Freeland and Colley (FC) method [6] also continues to add demands into a lot until $h(t)D_t > S$, where t is the number of periods that the inventory is carried. Note that this is the same as the second rule in the MOM method. Tables 11.23 and 11.24 indicate the result. The setup cost is $600 and the holding cost is $380, for a total cost of $980.

In the foregoing situations, we have seen several lot-sizing rules. The minimum cost per period rule performs well in a variety of cases and is probably the best heuristic rule to follow. It is important to recognize, however, that lot sizes for the parent item be-

TABLE 11.20   MRP Lot-Sizing Problem: MOM Approach

	0	1	2	3	4	5	6	7	8	9	10
GR		35	30	40	0	10	40	30	0	30	55
OH	35	0	50	10	10	0	30	0	0	55	0
POR		80				70			85		

TABLE 11.21 Groff Calculations

Period	Requirements	n	$n(n-1)D_n$	≤ 200?
2	30	0	0	Yes
2, 3	40	1	0	Yes
2, 3, 4	0	2	0	Yes
2, 3, 4, 5	10	3	60	Yes
2, 3, 4, 5, 6	40	4	480	No
Do not include period 6 demand in the lot.				
6	40	0	0	Yes
6, 7	30	1	0	Yes
6, 7, 8	0	2	0	Yes
6, 7, 8, 9	30	3	180	Yes
6, 7, 8, 9, 10	55	4	660	No
Do not include period 10 in the lot.				
10	55	0	0	Yes
End of the horizon. The lot in period 10 will be 555.				

come gross requirements for the component items. If the parent item were fairly far out in the schedule and the raw material item were early in the schedule, than we can visualize that lot sizes at the end product level would work themselves all the way back down to subcomponents and to raw materials. Generally, lot sizing is effective at the end product level but becomes less effective as we get down to the raw materials. At the raw material level, there is insufficient horizon, because the master production schedule may extend thirty weeks into the future, whereas the implications for raw materials may extend only ten weeks into the future. Thus we do not have a long enough planning horizon on the raw material level to do effective lot sizing.

In addition to the horizon problem, we also have the problem that the holding to setup cost trade-offs have been considered for single levels only. It would obviously be much more complicated to examine the trade-off of a lot-sizing rule at the parent level on the holding costs and setup costs all the way down through the assemblies.

Comparison of Lot-Sizing Rules

The rules described above present several different results, summarized in Table 11.25.

The order of these results are only applicable with the problem tested. A major difficulty with comparing lot-sizing rules is that each will perform differently on different problem sets.

TABLE 11.22 MRP Lot-Sizing Problem: Groff Approach

	0	1	2	3	4	5	6	7	8	9	10
GR		35	30	40	0	10	40	30	0	30	55
OH	35	0	50	10	10	0	60	30	30	0	0
POR		80				100				55	

TABLE 11.23   FC Calculations

Period	Demand	Periods Carried	Carrying Cost	> 200?
2	30	0	0	No
2, 3	40	1	80	No
2, 3, 4	0	2	0	No
2, 3, 4, 5	10	3	60	No
2, 3, 4, 5, 6	40	4	320	Yes
Do not include period 6 demand. The lot in period 2 will be 80.				
6	40	0	0	No
6, 7	30	1	60	No
6, 7, 8	0	2	0	No
6, 7, 8, 9	30	3	180	No
6, 7, 8, 9, 10	55	4	440	Yes
Do not include period 10 demand. The lot in period 6 will be 100.				
10	55	0	0	No
End of the horizon. The lot in period 10 will be 55				

Evaluation and Use of Lot-Sizing Methods

A large number of simulation studies have attempted to determine which lot-sizing method is best. A limited amount of research has focused on which methods are actually used in practice.

Simulation Study.   Nydick and Weiss [9] conducted a large number of simulation experiments on many of the lot-sizing rules that have been developed. In general, the L-4-L and EOQ rules performed very poorly. Overall Nydick and Weiss found that the PPB, GR, and MCP methods were the best, with little variation between them. When the time between orders is small, however, almost all the rules tested provided the optimal solutions.

Actual Usage.   Actual industry practice on lot sizing varies widely, as described in a survey published in 1979 in which Wemmerlov [16] interviewed thirteen MRP users in the mechanical and electronics industries. Time-variant (dynamic) lot-sizing techniques such as LTC and PPB were used by very few companies. Wemmerlov concluded that companies avoid these techniques because changes in top levels are transmitted down through lower stages, producing system nervousness, or exaggerated response at component levels to small changes at parent levels. At assembly and subassembly stages,

TABLE 11.24   MRP Lot-Sizing Problem: FC Approach

	0	1	2	3	4	5	6	7	8	9	10
GR		35	30	40	0	10	40	30	0	30	55
OH	35	0	50	10	10	0	60	30	30	0	0
POR		80				100				55	

TABLE 11.25 Comparison of Lot-Sizing Methods

Method	Total Cost
Minimum cost per period (Silver-Meal) (MCP or SM)	$ 910
Least total cost (LTC)	$ 910
McLaren order moment (MOM)	$ 910
Part period balancing (PPB)	$ 980
Groff (GR)	$ 980
Freeland and Colley (FC)	$ 980
Period order quantity (POQ)	$ 980
Least unit cost (LUC)	$ 990
Lot-for-lot (L-4-L)	$1400
Economic order quantity (EOQ)	$1590

the popular lot-for-lot technique helped maintain stability and minimized the amount of material tied up. The three most commonly used techniques were fixed period requirements, lot-for-lot, and fixed order quantity. For fixed period requirements, some companies used an ABC system, with four-week requirements for A items, eight weeks for B, and so on. Table 11.26 displays the results of the Wemmerlov survey.

In a study by Haddock and Hubicki [8] ten years after Wemmerlov, the results are almost the same. The lot-for-lot rule was the most widely used method. The next was fixed order quantity a method of merely fixing the order quantity, without regard to formal economic methods. The fixed period quantity was cited third. The user of this last method fixes the number of periods that should be included in a lot. This is different than the POQ method, which uses economic principles to determine the number of periods.

Overall, the usage of more complex methods is very limited, mainly because the more complex methods are not even included in many MRP computer software packages. Some software companies will include them as custom offerings, but there is a lack of interest in these techniques.

TABLE 11.26 Use of Lot-Sizing Rules

Technique	Number of Companies
Fixed period requirements	7
Lot-for-lot	6
Fixed order quantity	5
Economic order quantity	4
Price breaks	3
Part period balancing	2
Planner decided lot sizes	2
Least total cost	1

UNCERTAINTY AND CHANGE
IN MRP SYSTEMS

If all would go as planned, we could end this chapter right now. In the real world, however, very little goes as planned. Scheduled receipts do not come in on time, planned order sizes are changed when they are released as actual orders because of capacity constraints, changes in gross requirements dictate changes in lot sizes at subcomponent levels, and the unavailability of raw materials for one subcomponent negates the need for a fellow subcomponent because both must be ready for the parent production. Hence, we may want to change the status of an already existing open order and move its due date backward or forward in time. The uncertainty introduced in MRP systems can be neatly summarized in a chart originally drawn up by Whybark and Williams [17] and displayed here as Table 11.27. One could summarize this section by simply stating that Murphy's law applies to MRP systems; in fact, Murphy's law applies at the end item levels and its corollary applies at the component and raw material level.

The classic way to handle uncertainty is to introduce safety stock. If I need $100 to make a trip. I might add another $50 just for safety. That extra $50 is safety stock. I hope I will not need that safety stock; but if Murphy has anything to say about it, I will. In an MRP system, safety stock is normally included at the end item level; because of the parent-component matching relationships, this automatically introduces safety stock at all levels through the assembly. As an alternative to safety stock, some companies use safety lead time. The next four tables show the different effects of safety stock versus safety lead time. It is important to recognize that safety stock simply means that we do not want to see our inventory dip below a certain level. Thus, in the end item MRP rules, we place a planned order to cover a time period in which we are projecting negative inventories. With safety stock, we place the planned order to cover the time period in which we are projecting inventory below the safety stock level. Safety lead time, on the other hand, indicates that we will place the order a number of periods before it would normally have to be placed.

Tables 11.28 and 11.29 show the difference in the same time-phased order-point (TPOP) system with and without safety stock. In both cases the order quantity is ninety units and the lead time is two weeks. Without safety stock, the reorder point is thirty units; this point moves to forty units with ten units of safety stock. In the face of a fifteen-unit weekly demand rate, the safety stock system causes the planned order release to occur one week earlier.

TABLE 11.27   Categories of Uncertainty in MRP Systems

	Sources of Uncertainty	
Types	Demand	Supply
Timing	Requirements shift from one period to another	Orders not received when scheduled
Quantity	Requirements for more or less than planned	Orders received for more or less than planned

TABLE 11.28 TPOP without Safety Stock

	0	1	2	3	4[a]	5	6	7	8	9	10[a]
GR		15	15	15	15	15	15	15	15	15	15
PR					90						90
OH	45	30	15	0	75	60	45	30	15	0	75
POR			90						90		

[a]Projected negative.

PR = planned receipt.

$L = 2$, $Q = 90$, and $R = 30$.

For a time-phased order-point system, MRP logic is applied to a typical order-point situation.

Tables 11.30 and 11.31 contrast the use of safety stock and safety lead time under conditions of time-varying requirements. In Table 11.30 a safety stock of twenty units on an average weekly demand of fifteen units corresponds to roughly one-and-a-half weeks of supply. This safety stock actually causes the planned order release to occur two weeks in advance with the requirements pattern shown. A safety lead time of one week shows up in Table 11.31 as a planned order release one week early. Note that the due date on the planned order also moves ahead one week.

In his survey of thirteen MRP users, Wemmerlov [16] found much skepticism about the use of safety stock (see Table 11.32). Because safety stock causes orders to be placed earlier than absolutely necessary, practitioners felt that it distorts the true priorities. No one wants rush orders because of safety stocks!

NET CHANGE VERSUS REGENERATIVE MRP SYSTEMS

Conceptually, a *regenerative MRP system* provides the simplest approach to changes. *Regeneration* means running the entire system, exploding from the MPS through to component requirements across the board. A weekly or biweekly processing run can keep requirements updated reasonably well. The longer the time from the previous run, the more unreliable the information is, since changes invalidate some of the requirements.

Net change MRP systems avoid the gradual obsolescence of information by updat-

TABLE 11.29 TPOP with Safety Stock (SS)

	0	1	2	3[a]	4	5	6	7	8	9[a]	10
GR		15	15	15	15	15	15	15	15	15	15
PR				90						90	
OH	45	30	15	90	75	60	45	30	15	90	75
POR		90						90			

[a]Projected below 10 units.

SS = 10, $L = 2$, $Q = 90$, and $R = 30 + 10 = 40$.

TABLE 11.30   TPOP with Safety Stock and Time-Varying Requirements

	0	1	2	3	4	5	6	7	8	9[a]	10
GR		15	15	15	15	25	25	10	10	10	10
SR			90								
PR								90			
OH	45	30	105	90	75	50	25	105	95	85	75
POR						90					

[a]Projected negative without POR.

SS = 20, $L = 2$, and $Q = 90$.

TABLE 11.31   TPOP with Safety Lead Time (SLT) and Time-Varying Demand

	0	1	2	3	4	5	6	7	8	9[a]	10
GR		15	15	15	15	25	25	10	10	10	10
SR			90								
PR									90		
OH	45	30	105	90	75	50	25	15	95	85	75
POR							90				

[a]Projected negative without POR.

SLT = 1, $L = 2$, and $Q = 90$.

TABLE 11.32   Use of Safety Stock

Location of Safety Stock	Number of Companies
End item/finished good level	5
Low level items[a]	5
All levels	3
Total	13

[a]Protection against vendor supply variability.

TABLE 11.33   Standard Parent Item—Time-Varying Requirements

	0	1	2	3	4	5	6	7	8	9	10
GR		15	15	15	15	25	25	10	10	10	10
AC[a]											
SR											
PR				90						90	
OH	45	30	15	90	75	50	25	15	5	85	75
POR		90						90			

[a]AC represents an allocation—allocated inventory.

$L = 2$, $Q = 90$, and SS = 5.

ing inventory records and requirements as transactions occur. Partial explosions of items affected by changes allow the system to be continually accurate and up to date.

In the next few tables, we shall see examples of changes and updates. After getting a feel for net change systems, we can then evaluate the merits of regeneration versus net change.

In the first set of tables, we shall consider the concepts of *cashing requisitions* and *allocating inventory*. In real life, all action takes place now. In MRP terminology, now is called an *action bucket*. When a planned order release appears in period 1, it is sitting in the action bucket, and the appropriate response is to release the order. The order is released to the shop floor with a specified due date and appropriate paperwork. Part of the paperwork is an authorization (requisition) to withdraw the appropriate material for work on the item. One withdraws the material by cashing the requisition. Because the order may wait in line at the work center after being released, an uncashed requisition may exist for some time. During this uncashed time, the component's on-hand inventory is overstated, because some of the inventory is targeted for the order already released. The inventory already targeted for this purpose is called *allocated inventory*.

Tables 11.33 through 11.37 show the entire sequence of allocating inventory and cashing the requisition. Tables 11.33 and 11.34 give the standard parent-component plans with a POR in the action bucket. Table 11.35 shows the identical situation after the order has been released. Now that the order is open, a scheduled receipt becomes associated with it. Table 11.36 shows the corresponding change at the component level before the requisition is cashed. The gross requirements associated with the parent's ninety-unit POR disappear from the action bucket. Rather than deduct the ninety units from on-hand

TABLE 11.34   Standard Component Item

	0	1	2	3	4	5	6	7	8	9	10
GR		90						90			
AC											
SR											
PR											
OH	200	110	110	110	110	110	110	20	20	20	20
POR											

TABLE 11.35   Standard Parent Item after Order Release

	0	*1*	*2*	*3*	*4*	*5*	*6*	*7*	*8*	*9*	*10*
GR		15	15	15	15	25	25	10	10	10	10
AC											
SR				90							
PR										90	
OH	45	30	15	90	75	50	25	15	5	85	75
POR		0						90			

TABLE 11.36   Standard Component Item with Uncashed Requisition

	0	*1*	*2*	*3*	*4*	*5*	*6*	*7*	*8*	*9*	*10*
GR		0						90			
AC	90										
SR											
PR											
OH	200	110	110	110	110	110	110	20	20	20	20
POR											

TABLE 11.37   Standard Component Item with Cashed Requisition

	0	*1*	*2*	*3*	*4*	*5*	*6*	*7*	*8*	*9*	*10*
GR		0						90			
AC	0										
SR											
PR											
OH	110	110	110	110	110	110	110	20	20	20	20
POR											

TABLE 11.38 Parent Item POR Released with Partial Cancellation

	0	1	2	3	4	5	6	7	8	9	10
GR		15	15	15	15	25	25	10	10	10	10
AC											
SR				70							
PR								90			
OH	45	30	15	70	55	30	5	85	75	65	55
POR		0				90					

inventory at the component level, the on-hand balance remains the same and the ninety units are shown as allocated. Then, after the requisition is cashed, the on-hand balance and the allocation are both reduced by the ninety units in question.

Tables 11.38, 11.39, and 11.40 display the same allocation and cashing process, but this time with a partial order cancellation. Here the POR of ninety units has been released as seventy units, with twenty units canceled as not needed. In Table 11.39, the component's gross requirements (associated with the parent item's ninety-unit POR) disappear. The on-hand balance remains the same, but now seventy units are shown as allocated. Cashing appears simply in Table 11.40 as the balancing reduction of on-hand and allocated inventory by the seventy units of material released to the parent item.

Allocations and cashed requisitions arise in net change systems because inventory transactions are triggering the MRP program. Each inventory record must be in balance, with projected gross requirements matched by on-hand inventories, scheduled receipts on open orders, and planned receipts on properly timed PORs. If any transaction throws the record out of balance, the record must be updated immediately with revised PORs. If transaction-triggered revisions also affect a parent or component item, the principle of *interlevel equilibrium* requires that the PORs and gross requirements of the two levels be immediately realigned.

The Wemmerlov [16] survey found that five of the thirteen companies used net change while the other eight used regeneration. Of the eight using regeneration, six restrained much of the printout so that the weekly output showed only exception messages. Net change systems, as Wemmerlov points out, are genuine exception systems; hence, the regeneration users were modifying their systems to look somewhat like net change.

Regeneration users generally were disinterested in net change, because their data-processing times were acceptable and their business was not that dynamic. Regeneration

TABLE 11.39 Component Uncashed Position

	0	1	2	3	4	5	6	7	8	9	10
GR		0				90					
AC	70										
SR											
PR											
OH	200	130	130	130	130	40	40	40	40	40	40
POR											

TABLE 11.40   Component Cashed Position

	0	1	2	3	4	5	6	7	8	9	10
GR		0				90					
AC	0										
SR											
PR											
OH	130	130	130	130	130	40	40	40	40	40	40
POR											

is suitable for a stable environment in which a weekly massive batch processing run is sufficient to provide timely information. In a more volatile environment in which requirements are subject to rapid change, a net change system may be more suitable.

SYSTEM NERVOUSNESS, FIRM PLANNED ORDERS, AND TIME FENCES

Carlson, Jucker, and Kropp [2] define MRP *system nervousness* as the shifting of scheduled setups. More accurate, updated requirements can be a mixed blessing, because new, optimal schedules may disrupt previous plans. If no setup was scheduled in a particular period, expectations were raised concerning personnel scheduling and machine loading. To schedule a setup afterward is clearly undesirable.

Expanding the Carlson definition a little, we define system nervousness as an exaggerated response at component levels to small changes at parent levels. This response justifies particularly bad responses from personnel when scheduled setups must be shifted.

Forrester [5] studied the related problem of *industrial dynamics.* He found that, in a multiechelon system, small changes in demand at the retail level cause wide fluctuations in inventory levels at the factory. The more echelons there are and the longer the lead times, the wider are fluctuations.

Tables 11.41 through 11.44 show how a small change at the parent item level can cause headaches at the component level. Tables 11.41 and 11.42 display an innocent par-

TABLE 11.41   Standard Parent Item before Period 3 GR Revision

	0	1	2	3	4	5	6	7	8	9	10
GR		35	30	40	0	10	40	30	0	30	55
AC											
SR											
PR							100				55
OH	115	80	50	10	10	0	60	30	30	0	0
POR						100				55	

Parent $L = 1$; POQ (covers three periods of positive requirements).

TABLE 11.42   Standard Component before Parent Revision

	0	1	2	3	4	5	6	7	8	9	10
GR						100				55	
AC											
SR											
PR						80				55	
OH	20	20	20	20	20	0	0	0	0	0	0
POR			80				55				

Component $L = 3$; L-4-L.

TABLE 11.43   Standard Parent Item after Period 3 GR Revision

	0	1	2	3	4	5	6	7	8	9	10
GR		35	30	(40) 45	0	10	40	30	0	30	55
AC											
SR											
PR						75	(100)			85	(55)
OH	115	80	50	5	5	70 (100)	30	0	0	55 (55)	0
POR					75				85		

Parent item period 3 change from GR 40 to 45; $L = 1$.

Circled numbers show PORs and PRs without change for comparison purposes.

TABLE 11.44   Standard Component after Parent Revision

	0	1	2	3	4	5	6	7	8	9	10
GR					75	(100)			85	(55)	
AC											
SR											
PR					55	(80)			85	(55)	
OH	20	20	20 (80)	20	0	0	0 (55)	0	0	0	0
POR		55				85					

Component reaction, $L = 3$.

Circled numbers show GRs and PRs, and PORs without change.

TABLE 11.45 Standard Parent from Table 11.43 after Period 7 Revision

	0	1	2	3	4	5	6	7	8	9	10
GR		35	30	45	0	10	40	(30) 70	0	30	55
AC											
SR											
PR						115				85	
OH	115	80	50	5	5	110	70	0	0	55	0
POR					115				85		

$L = 1$; POQ (covers three periods of positive requirements).

ent-component relationship. Table 11.43 shows the same pair, except that the parent's gross requirements of forty units in period 3 have been increased to forty-five units. With that change, the POR of 100 units in period 5 is too late, because there is no longer enough projected inventory to cover period 5. The solution is to move the POR ahead to period 4. According to the lot-sizing rule, however, three weeks' supply calls for seventy-five units to cover five units of net requirements in period 5, forty in period 6, and thirty in period 7. Table 11.43 shows the parent item's projected position with the new PORs. Table 11.44 presents the troubled picture of a change in setup. This system is nervous. A five-unit change in the parent item's period 3 gross requirements leads to an immediate need for a period 1 POR at the component level. A comparison of Tables 11.42 and 11.44 reveals that the time and quantity of both PORs change at the component level.

To avoid this type of nervousness, *time fences* can be established. A fence is the shortest reasonable lead time from raw material to finished product or assembly. Within that time fence, no rescheduling is allowed except under extenuating circumstances. The master schedule must thus be fixed for the period of the time fence.

Without a time fence, we may still be able to save the day by *upward pegging,* which simply means chaining upward from the component to the parent item where it is used. *Full pegging* means chaining all the way back to the master schedule. Tables 11.45 through 11.50 show how to use upward pegging and *firm planned orders* (FPO) to solve shortage problems used by schedule changes.

Suppose that the gross requirements of thirty units in period 7 of Table 11.43 were

TABLE 11.46 Standard Component after Parent Revision

	0	1	2	3	4	5	6	7	8	9	10
GR					115				85		
AC											
SR											
PR					95				85		
OH	20	20	20	20	0	0	0	0	0	0	0
POR		95				85					

$L = 3$.

TABLE 11.47   Subcomponent after Parent Revision

	0	1	2	3	4	5	6	7	8	9	10
GR		95				85					
AC											
SR											
PR											
OH	55										
POR											

$L = 1$.

Note: The indented bill of materials shows:
   Parent
      Component
         Subcomponent

TABLE 11.48   Parent from Table 11.45 with Firm Planned Order

	0	1	2	3	4	5	6	7	8	9	10
GR		35	30	45	0	10	40	70	0	30	55
AC					75						
SR											
						(115)	(0)				
PR						75	40			85	
OH	115	80	50	5	5	70	70	0	0	55	0
					(115)	(0)					
POR					75	40			85		
						↑					

Firm planned order breaks up POQ in period 4 to relieve pressure on subcomponent.

TABLE 11.49   Standard Component Reaction to Firm Planned Order

	0	1	2	3	4	5	6	7	8	9	10
					(115)						
GR					75	40			85		
AC											
SR											
PR					55	40			85		
OH	20	20	20	20	0	0	0	0	0	0	0
POR		55	40			85					

$L = 3$.

TABLE 11.50 Subcomponent Feeling Much Better

	0	1	2	3	4	5	6	7	8	9	10
GR		55	40			85					
AC											
SR											
PR			40								
OH	55	0	0	0	0	0	0	0	0	0	0
POR		40			85						

$L = 1.$

increased to seventy units. The results of a chain reaction is shown in Tables 11.45 through 11.47.

Suppose, now, that the inventory planner suddenly detects a problem: The subcomponent item that has a procurement lead time of one week is needed right away. Facing a crisis, the planner pegs the subcomponent upward and then finds that the planned order quantity on the parent item covers periods 5, 6, and 7. The solution becomes evident. The parent item planned over can be reduced without causing a problem.

The inventory planner reduces the planned order in question by forty units (the amount of shortage in the subcomponent) and designates it as a firm planned order. The FPO generally involves a special computer command by the planner. The result is indicated in Table 11.48 by an arrow under the POR. The additional forty units are ordered in the next planning period or as a part of the next order. The resulting chain reactions are shown in Tables 11.49 and 11.50. The planned order is now called "frozen." When the MRP system replans and generates requirements, the frozen orders will not be touched.

The preceding situation illustrates a problem of coverage induced by an increase in the subcomponent requirements. The same type of problem would also arise if some items at the parent level or the component level were scrapped. Suppose that the vendor goofed or that the transportation mode was temporarily crippled. The result would create a chain reaction for which an FPO might be required. As an alternative to reducing the quantity of parent planned orders in the time buckets, it may be possible to reduce the lead time of a component or a subcomponent in some instances. For example, if we can get the component to level 1 in two weeks instead of in the regular three-week lead time, we will not encounter the problem in the subcomponent level 3. A firm planned order can resolve this problem by freezing the scheduled receipt of that particular component. Some of the problems at the end of this chapter illustrate this concept.

System nervousness goes with the MRP territory. In one way or another, it must be handled. Users who avoid net change systems or dynamic lot-sizing rules generally deal with system nervousness by trying to avoid it as much as possible. Carlson, Jucker, and Kropp [2] proposed quantifying the cost of nervousness and then trading it off against the cost of nonoptimal lot sizes. Many practitioners, however, disparage lot sizing altogether. Because setup costs and holding costs are realities, a cavalier attitude toward lot-size rules must be based on the notion that nervousness costs overwhelm lot size costs.

Because setups require both capacity and material, it is not surprising that practitioners try to avoid system nervousness. A change in setup to an earlier time slot requires that the supporting material be ready earlier and that labor and machine capacity be available.

SUMMARY

This chapter has developed the basic principles of material requirements planning and described how to use a simple master production schedule, bill of materials, and inventory status report to explode the requirements for subassemblies, parts, and raw materials.

Because setup costs and holding costs are important, the subject of lot-sizing appears in the MRP context. Such lot-sizing rules as minimum cost per period and part period balancing were developed.

Finally, this chapter explored some of the responses available in the face of change. Our survey of change-related problems was by no means exhaustive; nevertheless, it should provide a taste of the kinds of problems and solutions that appear after the MRP system begins to operate.

PROBLEMS

1. Given the following BOM, MPS, and inventory status, develop MRP tables for all items. Use the L-4-L technique. (Usage is one-for-one on all items.)

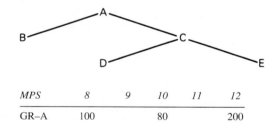

MPS	8	9	10	11	12
GR–A	100		80		200

Item	OH	LT
A	0	1
B	30	2
C	30	1
D	50	2
E	100	3

2. Given the following BOM, MPS, and inventory status, develop MRP tables for all items (ten tables in total).

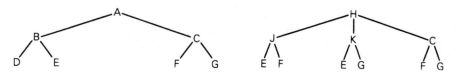

MPS	8	9	10	11	12
GR–A	100		50		150
GR–H		100		50	

Item	OH	LT
A	0	1
B	100	2
C	50	2
D	50	1
E	75	2
F	75	2
G	75	1
H	0	1
J	100	2
K	100	2

3. Set up a table using the time-phased order-point system for an independent item, using the following data:

Weekly demand: 25 units

Order quantity: 150 units

Order point: 35 units

Lead time: 1 week

Beginning inventory: 50 units

4. An item that has the following gross requirements and a beginning inventory of forty units:

	1	2	3	4	5	6	7	8	9	10	11	12
GR	30		40		30	70	20		10	80		50

Holding cost: $2.50/unit/week

Setup cost: $150

Lead time: 1 week

   a. Develop an L-4-L solution and calculate the total relevant costs.
   b. Develop an EOQ solution and calculate the total relevant costs.
   c. Develop an MCP solution and calculate the total relevant costs.
   d. Develop a POQ solution and calculate the total relevant costs.
   e. Develop an LUC solution and calculate the total relevant costs.
   f. Develop an LTC solution and calculate the total relevant costs.
   g. Develop a PPB solution and calculate the total relevant costs.
   h. Develop an MOM solution and calculate the total relevant costs.
   i. Develop a Groff Algorithm solution and calculate the total relevant costs.
   j. Develop an FC solution and calculate the total relevant costs.

5. An item has a forecasted demand of twenty-five units per week and a beginning inventory of seventy-five units. With a lead time of two weeks, an order quantity of 200 units, and a reorder point of seventy units, develop a TPOP schedule. How much safety stock is being carried? How much later would the POR occur if the safety stock were dropped?

Problems 6 through 8 are based on the following data: A parent item has projected gross requirements of twenty units per week for ten weeks and a beginning inventory of sixty units. The item has a two-week lead time and an order quantity of sixty units. The parent item has a component whose lead time is one week and whose on-hand balance is sixty units. The component is scheduled lot-for-lot.

6. Develop a set of tables to describe the parent-component relationship
   a. Before the first order is released.
   b. After the first order is released but before the requisition is cashed.
   c. After the requisition is cashed.
7. Because of capacity problems, the first order was released for thirty units, rather than for sixty units as planned. Develop a set of tables to describe the parent-component relationship.
   a. After the order release but before the requisition was cashed.
   b. After the requisition was cashed.
8. The gross requirements for period 4 are canceled; everything else remains as in the original data. Develop the appropriate parent-component tables to show the effect of this change.

Problems 9 through 12 are based on the following data. A parent item faces the following gross requirements.

	1	2	3	4	5	6	7	8	9	10
GR	0	40	30	40	10	70	40	10	30	60

The parent item has a one-week lead time, and the lot-for-lot rule is employed. Beginning inventory is twenty units. The parent item has a component whose lead time is also one week and whose starting inventory position is thirty units. At the component level, production occurs in lot sizes to cover three weeks of net requirements.

9. Develop the parent-component MRP tables to show the original planned positions.
10. At the parent level, gross requirements for period 2 are canceled. Develop the parent-component tables to show the net effect of this cancellation.
11. With the parent level gross requirements canceled for period 2, show how a firm planned order could be used to avoid a change in the timing of setups within the first four weeks.
12. At the component level, there is enough capacity to produce seventy-five units in period 1. Gross requirements at the parent level increase from forty units to fifty units in period 2. What problem arises? What solution would you recommend?
13. It is possible to show that a cost per unit formulation yields the EOQ solution when the demand rate is constant. Letting S be the setup cost, the setup cost per unit would be S/Q. To calculate the holding cost per unit, we must account for each lot covering Q/d periods of demand. Formulate the per unit cost function and solve for Q.

14. Using information system terminology, the part number master file is stored under an indexed organization using part numbers and the record key. Describe what such a system would look like.
15. Using information system terminology, the product structure master file is often implemented under a list organization. Describe what such a system would look like.
16. An item has the demand pattern shown below. Lead time is one period.

	0	1	2	3	4	5	6	7	8	9	10	11	12
GR		80	100	125	100	50	50	100	125	125	100	50	100

Holding cost: $2/period

Setup cost: $500

On hand: 0

a. Develop an L-4-L solution and calculate the total relevant costs.
b. Develop an EOQ solution and calculate the total relevant costs.
c. Develop an MCP solution and calculate the total relevant costs.
d. Develop a POQ solution and calculate the total relevant costs.
e. Develop an LUC solution and calculate the total relevant costs.
f. Develop an LTC solution and calculate the total relevant costs.
g. Develop a PPB solution and calculate the total relevant costs.
h. Develop an MOM solution and calculate the total relevant costs.
i. Develop a Groff Algorithm solution and calculate the total relevant costs.
j. Develop an FC solution and calculate the total relevant costs.
17. Consider the following problem:

	0	1	2	3	4	5	6	7	8	9	10
GR		35	30	45	0	10	40	30	0	30	55
OH	115										
POR											

With a lead time of one period and using a POQ of three periods, develop a POR schedule using an electronic spreadsheet. Set up an MRP table to experiment with changes in requirements. We give only the rudiments here and leave the rest to your own creativity:

Set the cursor at A1.
   Type in the label: Gross Req.

Set the cursor at A2.
   Type in the label: On Hand.

Set the cursor at A3.
   Type in the label: POR.

Set the cursor at A4.
   Type in the label: Sched. Rec.

Now move the cursor to B1, C1, and so forth.
   Type in the gross requirements 0, 35, 30, using column B as period 0.

Now move the cursor to B2.
   Type in the on-hand quantity: 115 units.

Now move the cursor to C2.

Type in the formula: $+ B2 + B3 - C1$.

Now replicate this formula in locations D2 through L2. Use the *relative* option.

Now enter PORs as needed in row 3.

Try again with the period 7 gross requirement changed to seventy units from thirty.

18. Use an electronic spreadsheet to replicate Tables 11.45 through 11.47 and then experiment with changes in gross requirements. We give only a general hint here: Suppose that row 5 presents gross requirements for the standard component. Then you need to move the cursor to location C5 and enter a formula such as $+C3$ where location C3 contains the POR from the standard parent. Now replicate this formula from D5 through L5 using the *relative* option. Calculate the tables with the original gross requirements and with the revised gross requirements.

REFERENCES AND BIBLIOGRAPHY

1. K. R. Bakere, "Lot-Sizing Procedures and a Standard Data Set, A Reconciliation of the Literature," *Journal of Manufacturing and Operations Management,* Vol. 2, No. 3 (Fall 1989), pp. 199–221.

2. R. C. Carlson, J. V. Jucker, and D. H. Kropp, "Less Nervous MRP Systems: A Dynamic Economic Lot-Sizing Approach," *Management Science,* Vol. 25, No. 8 (August 1979), pp. 754–761.

3. R. B. Chase and N. J. Aquilano, *Production and Operations Management, 6th Edition* (Homewood, Ill.: Richard D. Irwin, 1992).

4. D. W. Fogarty, J. H. Blackstone, Jr., and T. R. Hoffmann, *Production and Inventory Management* (Cincinnati: South-Western, 1991).

5. J. Forrester, *Industrial Dynamics* (Cambridge, Mass.: MIT Press, 1961).

6. J. R. Freeland and J. L. Colley, "A Simple Heuristic Method for Lot Sizing in a Time-Phased Reorder System," *Production and Inventory Management,* Vol. 23, No. 1 (First Quarter 1982), pp. 15–21.

7. G. K. Groff, "A Lot-Sizing Rule for Time-Phased Components Demand," *Production and Inventory Management,* Vol. 20, No. 1 (First Quarter 1979), pp. 47–53.

8. J. Haddock and D. E. Hubicki, "Which Lot-Sizing Techniques Are Used in Material Requirements Planning?" *Production and Inventory Management,* Vol. 30, No. 3 (Third Quarter 1989), pp. 53–56.

9. Robert L. Nydick, Jr., and Howard J. Weiss, "An Evaluation of Variable-Demand Lot-Sizing Techniques," *Production and Inventory Management,* Vol. 30, No. 4 (Fourth Quarter 1989), pp. 41–48.

10. J. Orlicky, *Material Requirements Planning* (New York: McGraw-Hill, 1975).

11. E. Ritchie and A. K. Tsado, "A Review of Lot-Sizing Techniques for Deterministic Time-Varying Demand," *Production and Inventory Management,* Vol. 27, No. 3 (Third Quarter 1986), pp. 65–79.

12. E. A. Silver and H. C. Meal, "A Heuristic for Selecting Lot-Size Quantities for the Case of a Deterministic Time-Varying Demand and Discrete Opportunities for Replenishment," *Production and Inventory Management,* Vol. 14, No. 2 (Second Quarter 1973), pp. 64–74.

13. A. Sloan, *My Years with General Motors* (Garden City, N.Y.: Doubleday, 1964), p. 128.

14. A. Vaszonyi, "The Use of Mathematics in Production and Inventory Control," *Management Science*, Vol. 1, No. 1 (October 1954), pp. 70–85.

15. Thomas F. Wallace, *MRP II: Making It Happen* (Essex Junction, Vt.: Oliver Wight Limited Publications, 1990).

16. U. Wemmerlov, "Design Factors in MRP Systems: A Limited Survey," *Production and Inventory Management* (Fourth Quarter 1979), pp. 15–34.

17. D. C. Whybark and J. G. Williams, "Material Requirements Planning under Uncertainty," *Decision Sciences*, Vol. 7, No. 4 (October 1976), pp. 595–600.

18. Oliver Wight, *MRP II: Unlocking America's Productivity Potential* (Essex Junction, Vt.: Oliver Wight Limited Publications, 1981).

APPENDIX 11A

A Small Manufacturer Makes a Concrete Investment in MRP
David Capel and Deborah Kakes

Managed correctly, computerization can streamline nearly any operation and open up many new possibilities. A Wisconsin maker of cement mixers found that taking advantage of technology tools to increase the availability of timely information and improve company-wide productivity was well worth its investment.

After Mixer Systems Inc. went online with an MRP II system in 1988, the company tried to sell the 23 file cabinets containing its old filing system. But there were no takers. "It was like selling buggy whips," recalls John M. Cherba, vice president of Operations at the Pewaukee, WI, cement-mixing products manufacturer. "We ended up donating them."

Like its old filing system, the company itself was falling behind its fast-moving market of the mid-1980s. Production was plodding, customers had to wait up to six weeks for replacement parts for their mixing machines, the company's sales staff could not get reliable information, and reports became long and tedious manual chores. Indeed, as other companies embraced new techniques and technologies to speed all phases of their operations, Mixer systems Inc. seemed to be wading in concrete. Top company managers decided it was time for the small company to take a bold step into the age of technology. The resulting computerization has allowed Mixer to nearly triple its business while minimizing its need for additional personnel.

But Mixer Systems' experience has amounted to more than just the hardware and software that comprised this venture into technology. Cherba's vision was to implement a solution that would help to integrate his entire organization. Just as all the components of one of Mixer's industrial mixing systems must work in harmony to be effective, Cherba wanted all the "Buzzwords" of technology—like material requirements planning (MRP) and manufacturing resource planning (MRP II)—to provide the integrated tools

Reprinted from *APICS—The Performance Advantage*, Vol. 2, No. 9 September 1992, pp. 40–41.

to assist Mixer in getting a handle on corporate information, as well as planning. Mixer has taken this philosophy to heart, taking advantage of technology tools to increase the availability of timely information and improve productivity company-wide.

Even a large company can be impacted by the enormity of investing in a comprehensive computer system and changing every facet of its operations. For a small company like Mixer Systems, which now employs 55 people and reports $12–$15 million in sales annually, so large an investment (time, more than money) can look like an overwhelming risk. But managed correctly, Cherba says, computerization can streamline an existing operation and open up many new possibilities. "A lot of small companies don't realize the amount of information and power they can have that will enable them to run their business better." In recent years, the costs of the hardware and software to run MRP and MRP II have come within the reach of small- and moderate-sized companies. At Mixer, the cost since 1986 has amounted to between $50,000 and $60,000.

BOGGED-DOWN PROCEDURES

Without the information handling provided by its computer system, Cherba says, it is unlikely that Mixer could have added nine new products while adding on only eight employees. In 1987, Mixer Systems was producing two kinds of cement mixers and a block concrete crusher, and these products had the company stretched to its limits. All the company's information files were on Kardex (a manual inventory control system), which meant that employees spent a significant amount of time creating, maintaining and searching the cards and related paperwork.

There were other problems, too. The company was always struggling to stock the right parts and get them out quickly to customers in need of service parts. Many of the parts Mixer Systems supplies are for equipment that dates back to the '70s—mixers that the company stopped making years ago. Such parts often need to be engineered-to-order or remanufactured, and the company's cumbersome Kardex procedures meant that requested parts sometimes took six weeks to get to customers.

In addition, hard, cold numbers were difficult to come by, be they for labor reporting, job estimates, or determining the actual cost of a project. "We were always behind the real world, always trying to play catch-up," Cherba recalls.

COMPUTERIZED RELIEF

After assessing the company's needs, Mixer settled on the CMI PROFIT software package, a fully integrated and interactive 4GL-based financial, distribution and manufacturing control, costing and planning system developed by M.I.S. Technology. Recently, the company upgraded the software to CMI PROFIT-IV and replaced its original hardware with IBM's RISC System/6000.

Although the original hardware and software arrived in May of 1987, it was the fall of 1988 before the system was fully operational. Cherba, having overseen transitions to computers before, attacked employee reluctance head-on. Much of the original data entry work burden was shared between employees and temporary staff. To alleviate initial fears, he set up a library system for software manuals so employees could check them out and read the materials before the implementation. This was followed up by meetings and software training sessions. When the system first arrived, a training environment was set up so employees could begin practicing using the actual programs and actual Mixer data. Within six months, an annual inventory was performed and the company's new MRP II system went 'live.'

Cherba states matter-of-factly that, before going to the computer, inventory at Mixer Systems was a guessing game. The company operated on a shortage list, in some cases; in others, employees would over-order in an effort to avoid running out. Expediting also was too common an occurrence. Now, using the information generated by the MRP II system, Mixer minimizes inventory at all levels, including raw materials, work-in-process and finished goods. "We've been very successful at minimizing inventory levels. We were struggling to track only 18,000 parts in 1987 and today we're successfully maintaining over 70,000 on the computer," says Cherba.

What makes the system so valuable to Mixer is the way it enables the company to respond rapidly to customer orders. "Service parts are a big part of our business. Our cement mixers work in hostile environments and they require constant upkeep. Since we offer service parts for the life of a mixer (which can be 30–40 years), significant tracking is required. Ninety percent of service parts orders are now filled within 10 days," Cherba explains.

Because a large portion of Mixer's business is customized, forecasting is extremely difficult. The company uses MRP to predict standard component and subassembly requirements by feeding demand through its master production schedule. When actual orders are received and booked, engineering completes the bill of material and releases the order for production. The next MRP run then catches the additional demand and schedules requirements.

Today, Cherba wonders how he was able to do his job without the indented cost reports that the system generates within a day of completing and shipping a project. This rapid feedback helps Cherba find problems that may well have gone unnoticed for weeks before computerizing. "If the gross profit does not come up to where I expect it to be, I can find out why while memories and the paper trail are still fresh. Problems can be solved once and in a timely manner. This kind of information I just didn't get before."

A TOOL FOR CHANGE

Cherba feels it's important to realize that the computer is a tool, but not a static one. Technology is always moving forward, and industry managers need to keep up with the changes with an eye on implementing continuous improvements. However, he cautions other companies to implement one step at a time and grasp each step before moving to the next. He also advises that the software must be simple to use and understand, yet provide functionality and capabilities for change and growth. Easy access to the data is essential. The hardware must provide the performance for getting the job done on time and the flexibility to grow as needs change.

"We wanted a solution to manage our business, not just a computer that needs to be managed," added Cherba.

He also knows that, for the small company, investing in a computer system can be intimidating. "It's a scarier investment for a small company. You have to look very carefully at your needs. But you'll be surprised at how many benefits result and at how confident you'll feel once the system is up and running. We aren't afraid to tackle anything, now."

PART FOUR

CONTROL ACTIVITIES

In Part III we discussed planning activities that should be undertaken before the actual implementation of production activities. In Part IV we discuss the control activities necessary to ensure that production activities go on as planned. In addition, several related topics are discussed.

Arthur Young's framework for competitive advantage is presented in Figure IV.1. This section of the text discussed the area called the production process in this model. Control activities deal with ensuring that the product is produced and shipped to the customer at the proper time.

During the development of plans, capacity is an important resource that must be recognized. In conjunction with the planning processes, capacity planning and control is undertaken to ensure that capacity can be made available if there is sufficient lead time or to ensure that the fixed available capacity is utilized at the proper time and place. This is discussed in Chapter 12.

When production is undertaken in a high-volume basis, certain techniques necessary are to ensure that the high volume is accomplished efficiently and effectively. Activities such as balancing the process and product mix scheduling are discussed in Chapter 13. The just-in-time (JIT) philosophy is a major advancement in maintaining smooth process flow with minimal inventories and waste. Arthur Young's framework indicates that JIT is a key part of the foundation of world-class manufacturing.

Job shops are used when production of items is in smaller batches. The techniques necessary in this environment are different from the high-volume process. Job shop procedures are discussed in Chapters 14 and 15. A new concept, called synchronous manufacturing, is discussed in Chapter 16. This concept provides a different view of the way to balance the product flow when bottlenecks may occur. Chapter 17 on project management concludes Part IV.

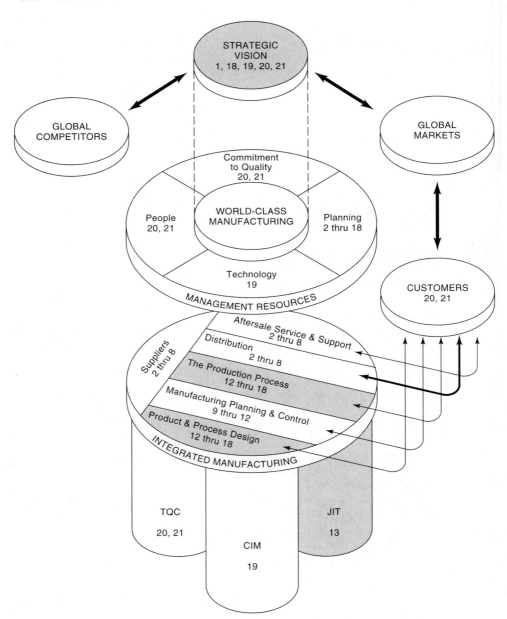

FIGURE IV.1 Arthur Young's manufacturing for competitive advantage framework. (*Source:* Copyright ®
1987 by Ernst & Young (formerly Arthur Young International. Reprinted by permission.) *Note:* numbers on this
figure indicates chapter numbers covering these subjects.

The following chapters are in Part IV:

Chapter 12, Capacity Planning and Control
Chapter 13, High-Volume Production Activity Control
Chapter 14, Job Shop Production Activity Planning
Chapter 15, Job Shop Production Activity Control
Chapter 16, Theory of Constraints and Synchronous Manufacturing
Chapter 17, Project Management Techniques

Chapter Twelve

Capacity Planning and Control

INTRODUCTION

In Chapter 9, we saw how aggregate output plans—popularly known as production plans—are formulated. Within the framework of a production plan, Chapter 10 showed how we create a master production schedule (MPS), which specifies amounts and need dates for specific items. The MPS can be exploded to determine implied loads on work centers, which generally consist of a group of machines or workers who can perform similar operations. A comparison of these implied loads and existing capacities can lead to a revision of the MPS or an increase in capacity. The capacity planning and control cycle extends from the MPS to activities at individual work centers, and the results of the capacity decision analysis are fed back to the master schedule, as illustrated in Figure 12.1. In short, capacity planning and control involves establishing, measuring, monitoring, and adjusting limits or levels of capacity to facilitate smoother execution of all manufacturing schedules, including the MPS, material requirements planning (MRP), and shop floor control (SFC).

Capacity Planning and Control Defined

Capacity planning is the process of determining the necessary people, machines, and physical resources to meet the production objectives of the firm. *Capacity* is the maximum rate at which a system can accomplish work. Consider two people washing and drying dishes after a banquet. The washer can wash 80 dishes per hour, and the dryer can dry 100 per hour. The system capacity of 80 per hour is determined by the "bottleneck,"

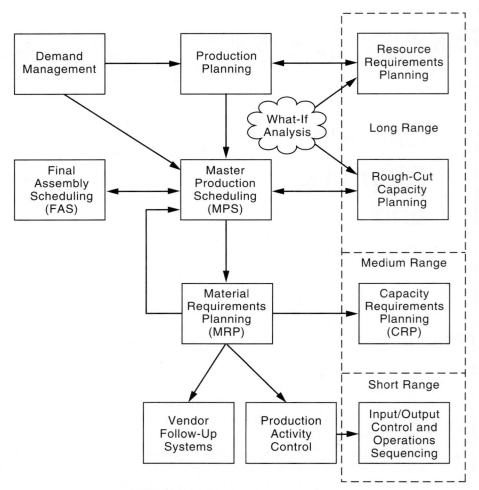

FIGURE 12.1 Capacity management techniques.

the person washing the dishes. Obviously, the person drying the dishes will be idle fairly often while waiting for the washer. But what is the load on this system? That depends on how many people attended the banquet, or how many dishes there are. At our present capacity, 400 guests would represent a five-hour load. Suppose, however, that we were renting the hall and that we had to clear out two hours after the banquet. What choices do we have? There are only two ways: reduce the load or increase the capacity. Either we allow fewer guests at the banquet or we increase the system capacity by renting a dishwashing machine or by adding extra people to the washing and drying. Thus *capacity control* can be defined as the process of monitoring output, comparing it with the capacity plan, determining if variations exceed preestablished limits, and taking corrective actions.

Machine Utilization
and Operator Efficiency

In determining effective capacity, machine utilization and operator efficiency considerations complicate the situation only slightly. Down time or repair time on equipment reduces capacity; so does labor inefficiency. Machine utilization factors indicate the percentage of scheduled time a machine actually runs. On the labor side, operation efficiency is the ratio of standard hours to actual hours ($E = S/A$). Although the load imposed on a workstation is usually calculated with standard times, workers on an incentive system often beat the standard. If it usually takes a worker only 2 minutes to perform an operation with a 2.5-minute standard, his or her efficiency rating would be 1.25. Such a worker would have an effective daily capacity of 1.25 * 8 = 10 hours. If the machine utilization factor were 90%, the effective daily capacity would be reduced to 10 * 0.9 = 9 hours.

Example 12.1

A drilling work center has three drills, three operators per shift, one shift per day, five days worked per week, and eight hours worked per shift. Records indicate that machine utilization is 95% and operator efficiency is 85%. What is the effective work center capacity per week?

Solution:

Effective capacity per week =
(3 drills)(8 hours/day)(5 days/week)(0.95 utilization)(0.85 efficiency)
= 96.9 hours/week

In a manufacturing concern, capacity planning looks ahead to predicted loads to determine whether available capacity is sufficient. Expanding the plant, purchasing equipment, and hiring personnel are options for matching long-term capacity to a long-term forecast load on the plant. Shifting the load to off-peak seasons through pricing policies would also help. In the short term, capacity can be increased through overtime and subcontracting; the forecast load can be met by making judicious delivery promises or by building up a backlog of past-due orders.

Time Horizons

In the capacity planning literature, long-, medium-, and short-term planning have meanings that are specific to materials management. True long-range planning answers the question of what the company wants to be doing five to ten years hence. Decisions about products to be offered, plants to be built, and major equipment to be purchased are the main issues. This is where manufacturing policy contributes to the company's future. Materials managers, however, are not usually involved in true long-range planning.

Hence, the term *long-range planning* in this chapter refers to one- to five-year horizon, which is long from the perspective of the materials manager.

The time horizons for capacity planning match the three major decisions to be made for materials planning. Long-range capacity planning extends a year or more into the future, where only gross estimates of capacity requirements are available. Vital long-range capacity activities include both resource requirements planning and rough-cut capacity planning.

Medium-range capacity planning generally extends as far into the future as detailed production planning data are available. The master production schedule will specify amounts and dates for all end items. When these have been chained back through subassembly and parts requirements and dates, using lead times of purchased or manufactured parts, the load on each work center can be determined. Capacity requirements planning (CRP) is the major tool used for medium-term capacity planning.

Once adjustments have been made so that we have a feasible master schedule, it is the job of the short-term capacity control system to meet schedules. Input control prevents too much work from being released to the plant, and output control ensures that work centers are producing at the expected rates. Dispatching specifies the sequence in which jobs are to be handled. The finite loading technique is used for short-term planning of actual jobs to be run in each work center, based on capacity, priority, and other relevant information about the status of the shop.

Synchronous Manufacturing

A different approach to managing manufacturing, developed by Eliyahu Goldratt and described by Umble and Srikanth [18], is known as *synchronous manufacturing*. The basic premise is that the flow of material through a manufacturing facility can be more efficient and more rapid through focusing on and managing a limited number of key constraints. In many cases the constraints are not capacity constraints but could be such things as market, material, logistical, or managerial constraints. This does not decrease the importance of a capacity constraint, known as a capacity constrained resource (CCR) in synchronous manufacturing. A CCR, if not properly managed, may cause the flow of material to deviate from the planned flow. A CCR may occur if the average load on a work center is close to or greater than the available capacity. Thus the load profiles to be developed later in this chapter have an important function in synchronous manufacturing by assisting in the location of the CCRs. Synchronous manufacturing is a way of managing capacity in the short range.

It is possible and quite likely that the load on a work center may exceed the available capacity for only several days each week. If the average load is somewhat less than the available capacity, then this work center may not be a CCR if the flow through that work center will not cause a disruption of flow through the plant. Proponents of the theory of synchronous manufacturing argue that management emphasis should be focused on true CCRs because they are the work centers that will cause disruption. The other work centers, even if they occasionally exceed available capacity, may not cause disruptions in the product flow and do not need as much attention as the CCR. Rather than develop load profiles and manage the capacity levels for all resources, as we discuss in this chapter, proponents of synchronous manufacturing suggest that only the CCR need be managed closely, and the other work centers, even if capacity is exceeded occasionally, do not need management emphasis.

In this chapter the process leading to development of load profiles is presented, while synchronous manufacturing concepts will be presented in Chapter 16.

LONG-RANGE CAPACITY PLANNING

Long-range capacity planning extends beyond the range of the master production schedule. We seek to match the long-range capacity factors (facilities, workforce, and capital equipment) and the long-range production plan. The terms *rough-cut capacity planning, resource planning, resource requirements planning,* and *long-range capacity planning* are used interchangeably for this type of approximation.

The basic idea of long-range capacity planning is quite simple. The production plan gives production rates that raise or lower inventories or backlogs. This plan is exploded through the bill of labor (bill of capacity) to give requirements on the resources. Resource requirements planning is performed on a macro level, using rough estimations of load, and a precise fit is not required. The resource requirements are then compared with resource capacities, and attempts are made to match the two. Generally, this is an iterative procedure, and these reviews lead to changes in the production plan and/or the capacities.

Long Term Resource Requirements Planning

Long-run capacity decisions are essentially sequential. Given a projection of long-term growth, what capacity should we set? Would it be better to add capacity in small increments as the growth materializes, or should we obtain economies of scale by overexpanding now to a capacity level that will be adequate for many years to come? Because of the sequential nature of the problem, dynamic programming [5] and decision tree techniques [11] have been widely advocated for capacity decision making.

Long-term capacity decisions involve facility size and location, workforce size, and capital equipment. Such "bricks and mortar" decisions tie us down for at least five to ten years into the future. Therefore, it is very important to understand the implications of long-range planning. The uncertainty inherent in our forecasts over this horizon heightens the utility of the simulation approach.

Example 12.2

Virts and Garrett [19] reported a capacity expansion simulation at Eli Lilly. Because it is simple, yet representative, we present its rudiments here. Eli Lilly had more than 1000 products, principally prescription drugs and agricultural chemicals. New-product forecasts were uncertain, especially products that had to be accepted by regula-

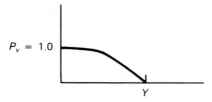

FIGURE 12.2   Yield. P_y is the probability that yield will be Y or greater.

tory agencies and customers. Not only were the product forecasts uncertain, but so were the yields from machines. Sometimes manufacturing plants were designed and even built during the development stage of new products.

To determine the number of machines needed, we could simply estimate:

$$\text{Required machines} = \frac{\text{forecasted sales}}{\text{forecasted yield/machine}}$$

From basic probability, however, we know that the expected value of the ratio of two random variables is *not* the ratio of their expected values. Therefore, simulation is in order. Basically, we use subjective probability distributions to model both sales and yields (see Figures 12.2 and 12.3). After generating sales and yields for a large sample of trials, we produce an equipment requirements curve (Figure 12.4).

Once we have the equipment requirements curve, it is relatively easy to settle on an optimal number of machines. Let the annual revenue generated by the machine when employed be R, and let the annual fixed charge for owning a maching be C. The contributions is $R - C$. Using marginal analysis, we have the payoff matrix shown in Table 12.1

Hence, we would buy, provided that we had

$$P(R - C) + (1 - P)(- C) \geqslant 0$$

or $P > C/R$.

Virts and Garrett [19] give an example of a particular product being manufactured and a process being employed for which $R = \$1,000,000$ and $C = \$571,000$. According to our analysis, we would continue purchasing machines up to the point on our equipment

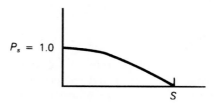

FIGURE 12.3   Sales. P_s is the probability that sales will be S or greater.

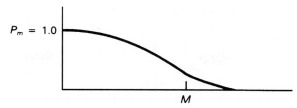

FIGURE 12.4   Machines. P_m is the probability of needing M or more machines.

requirements curve where $P = C/R = 0.571$. [Unfortunately, Virts and Garrett became confused over the concept of opportunity costs and developed the formula $P > C/(2R - C)$, which in this case would be $P = 0.4$ and an expected loss from purchasing the machine!]

TABLE 12.1   Payoff Matrix for Machine Decision

Decision	Probability P, Machine Needed	Probability 1 – P, Machine Not Needed
Buy	$R - C$	$-C$
Don't buy	0	0

Rough-Cut Capacity Planning

For long-range materials planning, we use only rough-cut procedures. After all, the long-run future is uncertain enough to make detailed planning rather pointless. Furthermore, trying out various detailed plans through simulation takes a great deal of computer storage and time. Both can be minimized by building a rough-cut model of our system.

Bill of Labor.   The model converts the five-year production plan into required capacities, which can then be compared with available capacities. What we need is a bill of labor or resources to indicate the work center person-hour requirements generated, not only by the end product but also by all its subassemblies and components. Three main simplifications allow us to estimate roughly the capacity needed for the products and thus facilitate the use of a model, termed the *bill of labor*. First, we use product groups rather than stock keeping end items as input. Just as a work center is a collection of machines and/or people with similar capabilities, so a product group is a collection of items with similar shop routings and operation times. Sometimes such a group is called a product family, but we prefer to reserve the term *family* for a collection of parts (or end products or subassemblies) that share a major setup. The second simplification concerns the use of key work centers rather than all machines. Because work centers that never cause bottlenecks are of little interest in requirements planning, we consider as key work centers those that might cause trouble. Furthermore, we want rough estimates of critical work centers, not individual machines. Finally, we choose a typical product in the product group and use its bill of materials, route sheets, and standard hours to determine capacity requirements for the planned production for the entire product group. The master production schedule associated with this group will be referred to hereafter as the *gross MPS.*

Example 12.3

Two product groups, A and B, have product trees (bills of materials) as shown here. In a company producing batteries, product group A might be watch batteries and group B might be photographic batteries.

The process sheets in Table 12.2 give the sequence of operations (route) for each item, along with setup and run time standards. In addition, our inventory records show economic order quantities for these items as A, 15; B, 10; C, 25; D, 20; and E, 30. From this information, we can calculate standard run hours in each work center:

$$\text{Standard run hours} = \frac{\text{setup}}{\text{EOQ}} + \text{run time}$$

For item C in the milling work center (operations 0010 and 0030), for example,

$$\text{Standard run hours} = \frac{0.3 + 2.7}{25} + 0.14 + 0.23 = 0.49 \text{ hours}$$

Intermediate run time tables can be created for each item by this procedure:

TABLE 12.2 Process Sheets, Example 12.3

Item	Operation Number	Work Center	Operation Description	Setup Hours	Run Hours
A	0010	1030	Assembly	0	2.00
B	0010	1030	Assembly	0	3.00
C	0010	1012	Milling	0.3	0.14
	0020	1020	Drilling	2.4	0.40
	0030	1012	Milling	2.7	0.23
	0040	1018	Grinding	1.0	0.21
D	0010	1012	Milling	0.4	0.15
	0020	1020	Drilling	2.8	0.35
	0030	1018	Grinding	2.2	0.24
E	0010	1012	Milling	0.3	0.18
	0020	1020	Drilling	2.1	0.39
	0030	1012	Milling	2.5	0.26
	0040	1020	Grinding	1.3	0.23

Item	Work Center	Work Center Standard Run Hours per Unit
C($Q^* = 25$)	Milling	0.49
	Drilling	0.50
	Grinding	0.25
D($Q^* = 20$)	Milling	0.17
	Drilling	0.49
	Grinding	0.35
E($Q^* = 30$)	Milling	0.53
	Drilling	0.46
	Grinding	0.27
A($Q^* = 15$)	Assembly	2
B($Q^* = 10$)	Assembly	3

The bill of labor, which follows readily from the intermediate run time tables, is shown in Table 12.3. Product group A requires C and D, giving total milling requirements of $0.49 + 0.17 = 0.66$.

Now, taking the production plan presented in Table 12.4, we can generate a resource requirements plan, as shown in Table 12.5, given the presently available capacity of 5000 standard run hours. We now see, for example, that producing 3000 units of A and 2000 units of B in 1995 will require $3000 * 0.66 + 2000 * 1.02$, or 4020 hours of milling time.

Table 12.5 shows that beginning in 1997, requirements exceed capacity. Our response can only be to reduce the planned production in line with the capacity constraint or to add more machines and/or people.

Resource Planning Process.   So far, we have used only gross estimates of long-run capacity requirements for making decisions regarding facilities and equipment needs. Now we turn to the more specific long-term planning process of testing the feasibility of various gross master production schedules. Although we still use work centers and product groups to gain rough approximations of load, we now want these estimates in specific time buckets for a particular master schedule.

TABLE 12.3   Bill of Labor or Resources

Work Center	Standard Run Hours per Unit	
	Product Group A	Product Group B
1012 milling	0.66	1.02
1020 drilling	0.99	0.96
1018 grinding	0.60	0.52
1030 assembly	2.00	3.00

TABLE 12.4 Production Plan for Product Groups A and B

			Units per Year		
Item	1995	1996	1997	1998	1999
A	3,000	4,000	3,000	3,000	3,000
B	2,000	2,000	3,000	3,500	4,000

The resource requirements planning concept thus facilitates balancing long-range needs and maintaining a reasonably level load on a company's resources. Resource requirements planning consists of the following steps:

1. Compute the load profile for each product group. The load profile is based on one unit of an average product.
2. Determine the total load needs on each resource for the proposed MPS. This determination is called the resource profile.
3. Simulate the effect of an alternative MPS on resource requirements, and finalize on an acceptable MPS.

Computing the Load Profile.   We can develop a product load profile by choosing a typical product in a product group. The product load profile displays the time-phased requirements of a machine load report to produce one unit of the typical product. To compute the profile, simply run one unit of the product through the MRP system, using no lot sizing and no beginning inventories for any item. The gross requirement of one unit of the typical product is exploded through all levels of the product structure in the usual fashion to generate planned order releases (PORs). The computations are made only once against each resource for each product group, and the load profile is stored in the computer for future use. Although the concept of the load profile is simple, the computations could become cumbersome for a complex product. Example 12.4 illustrates the development of a load profile for product group A (from Example 12.3) in the milling work center.

Example 12.4

Recall from Example 12.3 that product group A has a bill of materials and lead times as follows:

TABLE 12.5 Rough-Cut Requirements Plan for Milling Center (Rough-Cut Capacities)

		Standard Run Hours per Year			
	1995	1996	1997	1998	1999
Required	4,020	4,680	5,040	5,550	6,060
Available	5,000	5,000	5,000	5,000	5,000
Cumulative deficit	—	—	40	590	1,650

A $(LT = 1)$

C $(LT = 2)$ D $(LT = 3)$

To locate the gross requirement for group A, we take an arbitrary future time bucket beyond our total manufacturing lead time—for example, bucket 10. Then we would have the typical MRP time phasing shown in Table 12.6. The process file is shown in Table 12.7.

If we could schedule and load the planned order releases, we could determine where and when the per unit workload would fall, as follows:

Item A: (POR in week 9): The end item has a one-week lead time. Therefore, it requires a load of two hours during the ninth week.

Item C: (POR in week 7): Item C has a two-week lead time. We will assume that operation 0010 milling and operation 0020 drilling are performed during week 7 and that operation 0030 milling and operation 0040 grinding are performed during week 8. Standard run hours per operation for each item can be calculated in any machine center and assigned to specific time buckets, as shown in Table 12.8. For example, operation 0030 milling, which is performed during period 8 for item C, consists of a setup time of 2.7 hours, with a lot size of 25 and a run time of 0.23 hours per unit:

$$\text{Load} = \frac{2.7}{25} + 0.23 = 0.338 \text{ hours/unit}$$

Item D: Item D has three operations, with a lead time of three weeks. Assume that operation 0010 milling is performed during week 6, operation 0020 drilling during week 7, and operation 0030 grinding during week 8. Standard run hours for each machine center are assigned to each time bucket. To compute the load profile, simply add all loads in each machine center for every week, as shown in Table 12.8.

TABLE 12.6   Time-Phased Requirements (POR) Table for Product Group A

					Weeks					
	1	*2*	*3*	*4*	*5*	*6*	*7*	*8*	*9*	*10*
End Item A (lead time = one week)										
Gross requirements (GR)										1
Planned order release (POR)									1	
Item C (lead time = two weeks)										
GR									1	
POR							1			
Item D (lead time = three weeks)										
GR									1	
POR						1				

TABLE 12.7 Process Rile, Example 12.4

Item	Operation Number	Work Center	Setup Hours	Run Hours	Operation Standard Run Hours per Unit[a]
A	0010	Assembly	0	2.00	2.000
B	0010	Assembly	0	3.00	3.000
C ($Q^* = 25$)	0010	Milling	0.3	0.14	0.152
	0020	Drilling	2.4	0.40	0.496
	0030	Milling	2.7	0.23	0.338
	0040	Grinding	1.0	0.21	0.250
D ($Q^* = 20$)	0010	Milling	0.4	0.15	0.170
	0020	Drilling	2.8	0.35	0.490
	0030	Grinding	2.2	0.24	0.350

[a]This column of information is not provided by the process file but has been calculated specifically for this example. We use these standard run hours in Table 12.8: (setup hours/lot size) + run time.

Gross requirements on A in period 10
↓

TABLE 12.8 Load Profile Computations, Product Group A

					Weeks					
Machine Center	1	2	3	4	5	6	7	8	9	10
Assembly										
Load in Hours									2.000	
Milling										
Load for C							0.152	0.338		
Load for D						0.170				
Total						0.170	0.152	0.338		
Drilling										
Load for C							0.500			
Load for D							0.490			
Total							0.990			
Grinding										
Load for C								0.250		
Load for D								0.350		
Total								0.600		

Gross requirements on A in period 6
↓

	1	2	3	4	5	6
Assembly					2	
Milling Load on C			0.152	0.338		
Load on D		0.170				

Gross requirements on A in period 6 places a 0.17 hour per unit load or milling in period 2.

The Resource Profile.   What requirements are generated by a gross master production schedule? A resource profile gives a rough estimate of expected loads on key resources. To generate a resource profile, we extend the load profile for each product group in the gross master production schedule.

Example 12.5

Compute the resource profile for the milling machine center. The gross master production schedule for product groups A and B is as follows (assume 150 standard hours of existing capacity):

Gross Master Production Schedule

	6	7	8	9	10
Group A	200		200		200
Group B		150		150	

A load profile for product group B in the milling center can be calculated to match the load profile exhibited in Figure 12.5 for product group A. Multiply the load profile data by appropriate POR quantities to obtain the total resource requirements in each time bucket for every lot, as shown in Table 12.9. For example, consider lot 5 for product group A: gross MPS requirements of 200 in period 10. For the milling work center, Table 12.8 indicates that 0.152 hours per unit would be required in week 7; but 200 units of A then require 30.4 hours of milling time.

The computations for lot 5 in milling machine requirements are as follows:

Period 8: (200)(0.338) = 67.6 hours

Period 7: (200)(0.152) = 30.4 hours

Period 6: (200)(0.170) = 34.0 hours

All other requirements are similarly calculated and summarized. The totals represent the resource profile, as exhibited in Figure 12.6.

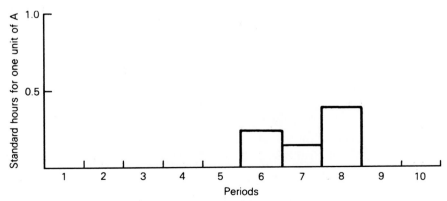

FIGURE 12.5   Product group A load profile for milling work center in real time periods.

TABLE 12.9   Resource Profile Computations, Milling Machine Center

									Period			
		1	*2*	*3*	*4*	*5*	*6*	*7*	*8*		*9*	*10*
					Lot 1, Group A 200	Lot 2, Group B 150	Lot 3, Group A 200	Lot 4, Group B 150	Lot 5, Group A 200			
			200 * 0.17									
Lot 1, Group A			= 34.00	30.40	67.60							
Lot 2, Group B					64.80	102.15						
Lot 3, Group A					34.00	30.40	67.60					
Lot 4, Group B							64.80	102.15				
Lot 5, Group A								34.00		30.40 = 0.152 * 200 67.60 = 0.338 * 200		
Total				34.00	30.40	166.40	132.55	166.40	132.55	67.60		

	6	7	8	9	10
Gross requirements					
Group A	200		200		200
Group B		150		150	

Gross Requirements of 200 units of A in period 6 places a 0.17 * 200 = 34 hour load on milling in period 2.

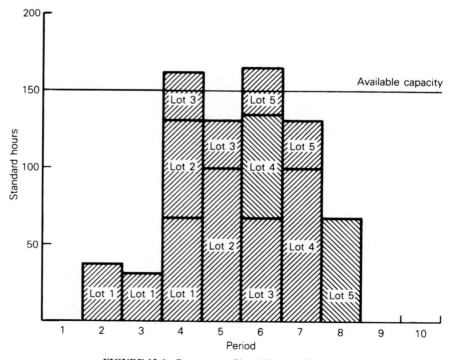

FIGURE 12.6   Resource profile, milling machine center.

Choosing an Acceptable Gross MPS.   Resource profiles are prepared especially for critical machine centers and are compared with the available capacities to gain insight into potential capacity problems. If a problem is detected, then alternative gross master schedules are used to generate new resource profiles. This information helps management decide how much additional capacity will have to be added and when.

MEDIUM-RANGE CAPACITY PLANNING AND CONTROL

In medium-range capacity planning, we generally accept the physical facilities and location as they are and add capacity by arranging alternative routings, additional tooling, subcontracting, overtime, and so on. We develop a capacity requirements plan from the master production schedule for specific end items. No longer do we use gross estimates and product groups. A detailed master production schedule can be tried out by exploding it through MRP to give planned orders. When these planned orders are added to released orders, a work center load report can be constructed for each work center in much the same way that we earlier developed a resource requirements profile (see Figure 12.7). This capacity requirements report must then be compared with available capacity at the work center.

Capacity Requirements Planning

Essentially, capacity requirements planning differs from resource requirements planning only in the level of detail considered. Resource requirements planning converts from a gross master schedule of product groups to estimates of time required on major departments, such as fabrication and assembly, or on key resources, such as a specific work center. The input to capacity requirements planning is a master schedule that shows actual model numbers, as advertised in a catalog. When exploded through the MRP system, this schedule will give capacity requirements on individual work centers.

The task of formulating a detailed master schedule consistent with the schedule for

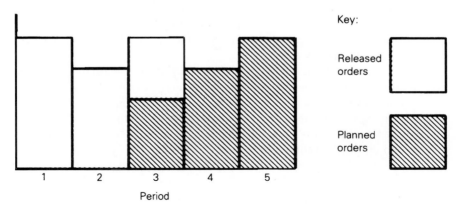

Key:

Released orders

Planned orders

Period

FIGURE 12.7   Capacity requirements report, milling work center.

product groups was studied in Chapter 10. We note that the master scheduler must disaggregate the product group plan by applying individual model popularity percentages or rule-of-thumb computations. In the case of products with major options, such as automatic versus standard transmissions, the option assemblies are often promoted to end item status in a planning bill of materials. Thus the master production schedule may include specific major assemblies rather than model numbers.

Staying with the simple case of exploding a master schedule with end item models only, we observe that the corresponding capacity requirements or load projections are calculated with no consideration of individual work center capacities.

In calculating the loads on the work centers, we have implicitly used the technique of backward scheduling, or backloading. From the due data for the end item, we back off by lead times through planned order release to estimate the timing of requirements on the work centers. Time buckets are essential. In performing this backloading, we abide by scheduling rules, such as allowing two days between operations in different departments. These rules are dealt with in detail in Chapter 15.

This process of preparing load projection without any regard to capacity has received the unfortunate name "infinite capacity loading"—unfortunate because we know that capacities are finite. Infinite capacity loading or capacity requirements planning, whatever it is called, seeks only to show the load implied by a master production schedule.

What happens if the capacity requirements plans show overloaded work centers? Our first reaction might be to shift the load around—that is, to level the load. To do this, we must take a closer look at lead times to see whether standard lead times can be compressed to allow some operations to start later and some, perhaps, earlier. The details of this process lead us to a discussion of strategies for leveling loads.

Strategies for Leveling Loads

Total manufacturing lead time is defined as the average time between the release of an order to shop floor and its delivery to stores or assembly. Usually, 10% to 20% of lead time is operation duration time—that is, setup and run time. The other 80% to 90% of lead time includes time the job is being moved, being inspected, in queue for an operation, or waiting to be moved. The estimated operation run times depend on the lot size. Interoperation time depends on the conditions in the shop, the backlog of orders on the floor, and the priority of the job.

Strategies for leveling the load depend heavily on changing lead times. Also, backward scheduling sometimes yields a start date in the past; that is, we do not have enough time to meet the due date. To change the lead time, three popular tactics are available. *Overlapping,* which reduces the lead time, entails sending pieces to the second operation before the entire lot is completed on the first operation. *Operations splitting* sends the lot to two different machines for the same operation. This involves an additional setup but results in shorter run times, since only half the lot is processed on each machine. Finally, *lot splitting* involves breaking up the order and running part of it ahead of schedule on an expedited basis. This, of course, involves extra setup but allows us to expedite a few items. We reserve lot splitting for short-term capacity planning and control, but operations splitting and overlapping properly belong to medium-range planning, where these techniques can be used to reschedule orders and level the load requirements to finite capacity.

Finite Capacity Loading for Leveling the Load at a Work Center

Finite capacity loading uses simulation procedures to schedule and load orders automatically within the given capacity constraints at each work center. Visualize a complex simulation program that reiterates between the master production schedule and capacity requirements in an attempt to piece together a jigsaw puzzle within work center constraints.

A basic feature of such a simulation program is the ability to forward schedule. Starting with the earliest possible start date for an operation, forward loading routines project the load for future periods and indicate the completion time for the entire job. For example, a planned delay in the start of an operation because of an overloaded time bucket will cause delays to all subsequent operations and hence will shift much of the load forward. Forward loading, the reverse of backloading, might then predict an order completion date beyond the due date.

In leveling the load, we might also want to perform certain operations ahead of schedule if an early time period has surplus capacity. The chief restrictions on starting operations earlier are the availability of materials and the completion of prior operations on the part or subassembly.

In the medium term, standard load times are used as the simulation attempts to level the released loads to the planned loads. Just as capacity requirements planning indicates where capacity adjustments must be made, finite loading indicates where changes should be made in the master schedule. In the next section we examine short-term scheduling, where we must control jobs at work centers and shift released loads according to priority rules. In the short term, finite loading routines help determine the timing or order releases to the machine centers.

SHORT-TERM CAPACITY PLANNING AND CONTROL

The capacity plan is finalized in the medium-range planning cycle. In the short term, most of our options have been exhausted. Our capacity adjustment alternatives are overtime, workforce reallocation, and, to some extent, alternative routings, operation splitting, and the like. The next step is to ensure that the output measured is equal to the required capacity. For this comparison, we will study input/output rate concepts at the work center level and detailed operations scheduling (short-term finite loading).

Input/Output concepts at Work Centers

Plossl and Wight [15] have popularized the bathtub model of an input/output system (see Figure 12.8). The water enters the tub at a certain rate (input), while the drain allows water out at a certain rate (output). If the input is greater than the output, the tub fills

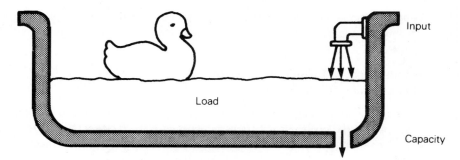

FIGURE 12.8 Bathtub model of an input/output system.

with water (load). If the output rate is greater than the input rate, the water decreases. Once a desired level of load is in the tub, the input and output can be matched to keep the load constant.

Although their model (the duck is ours) portrays the situation very well, we believe that the human body also provides an interesting example of input/output concepts. Caloric intake represents the orders released on the body. Suppose that we consume 3500 calories a day. Calories burnt in a day would be our capacity—say, 3000 calories per day. Clearly, we will build up a load—a pot belly or a backlog of fat—at a rate of 500 calories per day. Since a backlog of 3500 calories equals one pound of excess weight, we would gain one pound every seven days. Now assume that we have built up a five-pound backlog (call it our work-in-process inventory) and that we decide to go on a crash diet. By judiciously cutting out one candy bar and one piece of cake per day, we reduce our caloric intake to 3000 per day. What happens? We live with our five-pound excess. Only by reducing our intake below 3000 calories will we work off the backlog. In a manufacturing situation, the output rate must exceed the input rate to reduce the work-in-process inventory. In our biological example, a caloric intake of 2650 per day would allow us to work off the fat at the rate of 350 calories per day, so that the five pounds could be gone in ten days.

Input/output concepts help explain the dilemma of the lead time–late order cycle. A company suffering poor due date performance decides to increase lead times. If only we had two more weeks for the job, we could complete it on time! Increasing the lead time only aggravates the situation, however, because orders are released earlier, and this puts a larger load on the plant. Inevitably, queues get longer, lead times are increased further, and the revised due dates are missed. If lead times are increased, the water (input) to the tub will increase. Without a corresponding increase in output, the level of water (load) in the tub will only increase.

If increasing the lead time does not solve the problem, what will? To rephrase the question, if the water level (queue or backlog) in the bathtub is too high, how can we reduce it? We can reduce the backlog only by increasing the work center capacity or by reducing the rate of orders released to the factory. This means that we must lower our expectations of quantities to be produced, increase the work center capacity, or subcontract. In our biological example, we must jog more or eat less. Unfortunately, we cannot subcontract the jogging.

Finite Loading for Smoothing the Load at Work Centers

The input/output model focuses attention not only on the capacity or possible output rate but also on the input rate—the rate of releasing orders to the plant. Even the most finely tuned master schedule will probably produce lumpy loads at some work centers—overloads in some periods and underloads in others. Finite loading, or detailed operations scheduling, attempts to ameliorate this situation by smoothing the load at work centers.

Finite loading means simulating the factory operations to determine exactly when orders should be released, how orders will move through the plant, and what order completion dates will result. As in any simulation, we have a model of the system, measures of effectiveness, and decision variables.

In addition to such obvious factors as the number of machines, the number of shifts, and the like, the model of the plant includes priority rules that govern the flow of orders and the time each order spends in queue. Our goal is to reproduce actual job conditions in the face of the orders for the next few weeks.

Priorities determine the time an individual order will spend in queue, just as capacities determine overall queue sizes or work-in-process amounts. The order due date, the size of the order, and management's feelings about the importance of the customer all contribute to order priority. Clearly, orders with higher priority will be released to the shop first. Once released to the shop, the order must find its way out through the queues at various operations. Here, the operation priority takes over. At each work center, a priority index or a critical ratio (CR) might be calculated to decide which job should go on the machine text. Critical ratios come in various formulations, and those responsible for them always claim that theirs are the best. Most critical ratios hinge on some measure of the time remaining to due date, divided by the time or number of operations remaining.

$$CR = \frac{\text{time to due date}}{\text{required time for remaining operations}}$$

In general, critical ratios are constructed so that lower ratios represent higher or more urgent priority and ratios less than unity indicate orders that will be late. In some companies the critical ratio is further modified by a weighted measure of order priority to ensure that very important orders get high operation priorities. These aspects are fully explored in Chapter 15.

For our simulation, the measures of effectiveness are quite simple: How smooth are the input rates at the work centers, and how close are the predicted completion dates to the due dates generated by the master production schedule? To achieve good performance on these effectiveness measures, several decision alternatives are available: order release planing decisions and operations sequencing decisions.

Order Release Systems. The order release planning module of a finite loading program or system might be run every few days. Order priorities are calculated for released and planned orders. Then orders are released according to these priorities and at the latest possible start dates, as determined by the lead time offsetting routine in the capacity requirements planning module. When an order runs up against a capacity constraint at a work center, the order release planning system tries some alternatives. First, the previous and the following week are checked to see whether the operation could be handled then. Also, alternative work centers or routings might be possible. Furthermore, some orders might be released earlier to take advantage of underload situations. Early release of orders depends on material availability. As a last resort, if no alternatives can be found that allow all operations to be completed by the due date, the master production schedule will have to be changed.

Various computer software packages handle order release planning in different ways. Indeed, many companies have created their own systems. Nevertheless, the underlying procedures rest on order scheduling and operation shifting: releasing the order earlier or shifting one of its operations forward or back. The forward loading procedures mimic company operations. How do we fit the puzzle together?

Example 12.6

Jobs A and B are waiting to be released. Both will go to work center 1. Then A will continue on to work center 3 while B will go to center 4. Suppose that A has a higher priority on the first work center but that work center 3 is backlogged and center 4 is idle. Which order should be released first? Job B should be first, of course, even though it has a lower priority on the first work center.

Operation Sequencing Systems. At an even more detailed level than the order release planning system, we might run an operation sequencing system daily. The IBM Communications Oriented Production Information and Control System (COPICS) [8] uses a work sequence list rather than a standard priority list at each work center. Essentially, a work sequence list gives the work center supervisor a sequence of jobs based on coordinated priorities between work centers. A job with a low priority on the first work center might get a top priority on the second work center in the usual priority list. Yet the job will not arrive at the second work center for quite some time because of its low priority at the first center. By contrast, the work sequence list at the second work center will show jobs in an order that *combines* the usual priority at the second work center and the priority at the first work center. Hence, the list takes into account both priority and probable arrival at the second work center.

An operation sequencing system uses the work sequence concept, and such options as operations splitting, overlapping, and lot splitting are used in simulating the performance of the factory. A prime purpose of this simulation is to generate feasible, not optimal, work sequence lists for the foremen.

The complexity of finite loading does not suit everyone's taste or needs. Many companies simply do finite loading manually as problems arise, without resorting to simulation. Indeed, the simulation programs are generally designed to model the usual behavior of people who are scheduling and loading the plant.

Input/Output Control

The simplicity and usefulness of input/output control make this technique appealing. Wight [20] developed the approach of monitoring work center with input/output control reports. Released and planned orders appear as person-hour requirements in the capacity requirements plan. Since these loads may be rather lumpy, we might calculate an average rate of orders arriving at the work center and show this as our planned input at the work center, as in Table 12.10. The input in Table 12.10 comes as planned for the first three weeks. Our planned output rate, 50 hours in excess of the planned input rate, indicates an attempt to reduce the backlog to 200 units by the end of week 4. Even so, our actual output shows that we have not been entirely successful. At the end of four weeks, the released backlog has been reduced by only 170 units, to 230.

In working with input/output tables, several related tables are extremely useful. We encourage the reader to draw these up to help in creating an input/output report like Table 12.10. First, planned input versus planned output can be displayed as a planned backlog or planned queue report. Similarly, actual input versus actual output presents the most important information: what's happening to the actual queue.

Input/output control tables are useful in both production to order and production to stock. In production to stock, forecast sales become planned inputs. Planned versus actual sales, production, and inventory are then displayed in three tables, yielding a production/sales/inventory report. In production to order, the manufacturing lead time takes center stage, as *released* backlogs are closely monitored. Of course, total lead time through the plant also depends on the size of the *unreleased* backlog. The assumption must be made that total lead time can be reduced by careful management of manufacturing lead times.

Input/output tables focus attention where it should be, where we can actually make decisions. This view of the world says that backlogs and lead times can be controlled by close monitoring.

TABLE 12.10 Input/Output Report, Work Center 20

| | Standard Hours per Period | | | | | |
	1	2	3	4	5	6
Planned input	350	350	350	350	350	350
Actual input	350	350	350	350		
Cumulative deviation (actual − planned)	0	0	0	0		
Planned output	400	400	400	400	350	350
Actual output	410	395	405	360		
Cumulative deviation	+10	+ 5	+10	−30		
Backlog reduction by week						
Planned backlog	350	300	250	200	200	200
Actual backlog	340	295	240	230		

Released work backlog = 400 hours. Planned and actual backlogs show the status at the end of each period.

Example 12.7

Most of us accept the idea that some queue is necessary to act as a buffer between succeeding operations. Suppose that we have been working with an average queue size of 400 person-hours, with a standard deviation of 100 person-hours. Further, we want only a 1% chance of ever running out of work at our work center. What queue size should we maintain, assuming that the queue size is normally distributed? Using the normal tables, 99% of the curve will be to the right of the average queue size, that is, minus 2.34 standard deviations. Hence, the average is 2.34 * standard deviation = 2.34 * 100 = 234.

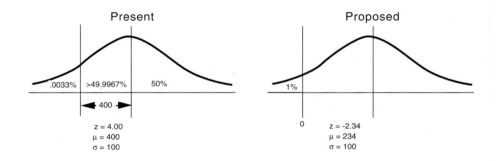

Having determined that we could be satisfied with an average queue size of 234 person-hours, compared with our current average of 400 person-hours, how do we go about reducing the queue? Simply set the planned output rate at 10 person-hours more than the planned input rate for the next sixteen weeks. This is our "diet." Of course, as the weeks go by, we may find that it takes fewer weeks than planned.

SUMMARY

This chapter has examined the capacity planning and control cycle. Production plans are checked against key resource capacities through the use of the bill of labor. Resource requirements planning and capacity requirements planning are methods for determining the capacities needed by trial master schedules. Given an infeasible schedule, adjustments to capacity or revisions in the schedule can alleviate bottlenecks.

Matching requirements to capacities becomes a juggling maneuver. Methods for compressing lead times and rescheduling orders can help meet customer needs with the capacities available. Finite capacity planning can be accomplished by planners either manually or with computer algorithsm.

Input/output control techniques help monitor the release of orders and the output rates at work centers. Reductions in backlogs occur only when output rates are greater than input rates.

PROBLEMS

1. Clark Company makes three products on three different types of equipment. The matrix of operating times and job setup times (in decimal hours), demand per month, and economical lot sizes for manufacturing are given in Table 12.11. The machine utilization factor is approximately 90%, and operator efficiency of the shop is believed to be 105%. How many of each of the machines will be needed if the plant works a forty-hour week?

2. Using the information in Table 12.12, a machine utilization factor of approximately 80%, and operator efficiency of the shop of 105%, how many of each machine will be needed if the plant works a forty-hour week?

3. A work center has the input/output table shown below. The starting backlog is 150 hours. What will be the backlog at the end of the third week?

	Standard Hours per Week				
	1	*2*	*3*	*4*	*5*
Planned input	400	400	400	400	400
Actual input	410	410	380	310	310
Planned output	420	420	420	420	420
Actual input	380	380	380	320	300

4. What would the backlog be at the end of the third week if planned output rates had been met in problem 3?

5. What will the actual backlog be at the end of week 5?

6. At a particular work center, our average queue has been 250 hours, with a standard deviation of 50 hours. Should we reduce our queue length? If so, by how much? What assumptions do we have to make?

7. At a particular work center, our average queue has been 400 hours, with a standard deviation of 75 hours. Should we reduce our queue length? If so, by how much? What assumptions do we have to make?

8. Using the following bill of materials and data, determine whether the given production plan is feasible.
   a. Bill of materials:

TABLE 12.11   Matrix for Problem 1

Equipment	*Job A*	*Job B*	*Job C*
Punch			
Setup hours	0.750		0.600
Run hours	0.040		0.060
Grind			
Setup hours		0.750	
Run hours		0.020	
Screw			
Setup hours	0.400	0.520	
Run hours	0.030	0.050	
Demand per month (units)	1500	2000	1000
Economical batch quantity (units)	300	500	250

TABLE 12.12   Matrix for Problem 2

Equipment	Job 1	Job 2
Turn		
Setup hours	0.800	
Run hours	0.070	
Press		
Setup hours		0.450
Run hours		0.020
Die		
Setup hours	0.550	0.600
Run hours	0.040	0.030
Demand per month (units)	800	1500
Economical batch quantity (units)	400	400

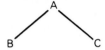

b. Route sheet:

Product or Subassembly	Work Center	Unit Hours
A	Assembly	3
B	Milling	2
	Drilling	1
C	Milling	2
	Drilling	3

c. Work center capacities:

Assembly: 250 hours/month

Milling: 150 hours/month

Drilling: 120 hours/month

d. Production plan:

	Month								LOT SIZE
PRODUCT	1	2	3	4	5	6	7	8	
A	20	25	30	35	40	45	45	45	10

If the plan is not feasible, explain the difficulty.

9. Using the following bills of materials and data, determine whether the given production plan is feasible.

a. Bills of materials:

b. Route sheet:

Product or Subassembly	Work Center	Unit Hours
A	Assembly	4
B	Assembly	4
C	Turning	1
	Tapping	2
D	Milling	4
	Tapping	4
E	Turning	3
	Tapping	4

c. Work center capacities:

Assembly: 350 hours/month

Milling: 150 hours/month

Turning: 150 hours/month

Tapping: 120 hours/month

d. Production plan:

PRODUCT				Month					LOT SIZE
	1	2	3	4	5	6	7	8	
A	30	30	30	30	30	40	40	40	15
B	15	15	20	20	25	25	25	25	15

If the plan is not feasible, explain the difficulty.

10. Using the data in problem 9, suggest a better production plan.

11. Using the following data and the technique of backward scheduling, determine whether we can meet the due dates on products Y and Z.

   a. Route sheet (times in hours):

	Product Y		Product Z	
OPERATION *NUMBER*	*WORK* *CENTER*	*TIME*	*WORK* *CENTER*	*TIME*
I	1	50	2	50
II	3	60	3	55
III	2	70	1	50

b. Work center capacities:

Work Center	Hours per Day
1	60
2	70
3	80

c. Due dates:

Job	Day
Y	3
Z	3

12. Using the following data and the technique of backward scheduling, determine whether we can meet the due dates on products A and B.
    a. route sheet (times in hours):

	Product A		Product B	
OPERATION *NUMBER*	*WORK* *CENTER*	*TIME*	*WORK* *CENTER*	*TIME*
10	1	35	2	40
20	4	35	4	50
30	3	40	1	50
40	2	50	3	55

b. Work center capacities:

Work Center	Hours per Day
1	50
2	40
3	40
4	40

c. Due dates:

Job	Day
A	3
B	3

13. Given the data in problem 12, level the load as much as possible.

14. A company makes five products on four different types of equipment. The matrix of operating times and job setup times (in decimal hours), the demand per week, and the economical lot sizes for manufacturing are given in Table 12.13. The machine utilization factor is approximately 90% and operator efficiency of the shop is believed to be 105%. How many of each of the machines will be needed if the plant works a forty-hour week?

15. Would it make more sense to change the EOQs in problem 14 rather than to purchase machines?

16. The bills of material for products A, B, and C are as follows (items 1, 2, 4, and 5 are those in problem 14); where LT = lead time:

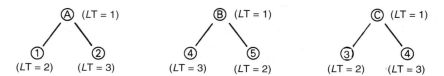

Exhibit the bill of labor for products A, B, and C.

17. Generate load profiles for product A in problem 16.

18. Given the following MPS, generate a resource requirements profile for the forge, using the data from problems 14 through 17.

| Product | 5 | *Gross Requirements per Week* | | | |
		6	7	8	9
A	1000	2000	1500	2000	
B	500	800	600	400	200
C		1100	1200	1100	1100

19. Generate load profiles for products B and C in problem 16.

20. Generate resource requirements profiles for the lathe, mill, and grinding centers, given the product information in problem 14 and the MPS in problem 18.

TABLE 12.13   Matrix for Problem 14

Equipment	Item 1	Item 2	Item 3	Item 4	Item 5
Lathe					
Setup hours	0.800			0.500	
Run hours	0.050			0.070	
Forge					
Setup hours		0.550		0.750	
Run hours		0.060		0.040	
Mill					
Setup hours	0.320	0.750			0.450
Run hours	0.010	0.020			0.015
Grind					
Setup hours			0.500		0.600
Run hours			0.050		0.060
Demand per month (units)	1000	2000	1500	500	800
Economical batch quantity (units)	250	500	300	200	400

21. One of the common uses of an input/output control report is in creating a production/sales/inventory (PSI) report. Table 12.14 shows the planned and actual production and sales data. Expand the PSI report to include planned and actual inventories.

22. At a particular work center, our average queue has been 200 hours, with a standard deviation of 30 hours. Should we reduce our queue length? If so, by approximately how much? What assumptions do we have to make?

23. At a particular work center, our average queue has been 200 hours, with a standard deviation of 100 hours. What does this indicate? What should be done?

24. Using the following bills of materials and data, determine whether the given production plan is feasible.

   a. Bills of materials:

   b. Route sheet:

Product or Subassembly	Work Center	Unit Hours
A	Assembly	4
B	Milling	3
	Drilling	2
C	Turning	1
	Drilling	4
D	Turning	3
	Drilling	3

   c. Work center capacities:

     Assembly: 300 hours/month

     Milling: 150 hours/month

     Drilling: 120 hours/month

     Turning: 120 hours/month

TABLE 12.14   PSI Report for Problem 21

			Units per Period		
	1	*2*	*3*	*4*	*5*
Production					
Planned	350	450	360	460	400
Actual	300	500	400	450	480
Sales					
Forecast	350	460	475	500	390
Actual	400	500	450	420	475
Inventory					
Starting inventory = 1500 units					

d. Production plan:

| PRODUCT | Month | | | | | | | | LOT SIZE |
	1	2	3	4	5	6	7	8	
A	20	20	25	30	40	40	40	40	10
D	15	15	15	15	15	15	15	15	15

If the plan is not feasible, explain the difficulty and suggest a better production plan.

25. Using the following data and the technique of backward scheduling, determine whether we can meet the due dates on products A, B, and C.

a. Route sheet (times in hours):

| OPERATION NUMBER | Product A | | Product B | | Product C | |
	WORK CENTER	TIME	WORK CENTER	TIME	WORK CENTER	TIME
I	2	30	1	40	1	35
II	3	30	3	30	4	20
III	4	40	2	40	2	35
IV	1	50	4	30	3	40

b. Work center capacities:

Work Center	Hours per Day
1	60
2	40
3	50
4	30

c. Due dates:

Job	Day
A	5
B	4
C	5

26. Work center 17 has the following input/output table and a starting backlog of 130 hours. What will be backlog be at the end of the third week?

| | Standard Hours per Week | | | | | |
	1	2	3	4	5	6
Planned input	300	300	300	300	300	300
Actual input	305	305	305	150	155	150
Planned output	310	310	310	300	300	300
Actual output	290	290	290	290	190	150

What would the backlog be at the end of the third week if planned output rates had been met?

27. An industrial engineer has been complaining that the workers in work center 17 (in problem 26) are lazy. As evidences, she displays data showing that they are working at half their capacity. What would be your explanation?

REFERENCES AND BIBLIOGRAPHY

1. J. H. Blackstone, Jr., *Capacity Management* (Cincinnati: South-Western, 1989).
2. R. B. Chase and N. J. Aquilano, *Production and Operations Management,* 6th ed. (Homewood, Ill: Richard D. Irwin, 1992).
3. J. T. Clark, "Capacity Management," *Twenty-Second Annual APICS Conference Proceedings,* 1979, pp. 191–194.
4. J. T. Clark, "Capacity Management—Part 2," *Twenty-Third Annual APICS Conference Proceedings,* 1980, pp. 355–341.
5. D. Erlenkotter, "Capacity Expansion with Imports and Inventories," *Management Science,* Vol. 23, No. 7 (March 1977), pp. 694–702.
6. R. Everdell, *Master Production Scheduling,* APICS Training Aid (Washington, D.C.: American Production and Inventory Control Society, 1974).
7. D. W. Fogarty, J. H. Blackstone, Jr., and T. R. Hoffmann, *Production and Inventory Management,* 2nd ed. (Cincinnati: South-Western, 1991).
8. IBM Communications Oriented Production Information and Control Systems (COPICS), Vol. 5 (1st ed.), Form No. G320–1978 (White Plains, N.Y.: International Business Machines Corporation, 1972).
9. R. L. Lankford, "Short-Term Planning of Manufacturing Capacity," *APICS Conference Proceedings,* 1978, pp. 37–68.
10. R. L. Lankford, "Input/Output Control: Making It Work," *Twenty-Third Annual APICS Conference Proceedings,* 1980, pp. 419–420.
11. J. F. Magee, "How to Use Decision Trees in Capital Investment," *Harvard Business Review,* Vol. 2, No. 5 (September–October 1964), pp. 79–96.
12. J. Orlicky, *Materials Requirements Planning* (New York: McGraw-Hill, 1975).
13. G. W. Plossl, *Manufacturing Control: The Last Frontier for Profits* (Reston, Va.: Reston, 1973).
14. G. W. Plossl and W. E. Welch, *The Role of Top Management in the Control of Inventory* (Reston, Va.: Reston, 1979).
15. G. W. Plossl and O. W. Wight, *Production and Inventory Control* (Englewood Cliffs, N.J.: Prentice Hall, 1967).
16. G. W. Plossl and O. W. Wight, "Capacity Planning and Control," *Production and Inventory Management,* Vol. 14, No. 3 (Third Quarter 1973), pp. 31–67.
17. A. Rao, *Capacity Management,* APICS Training Aid (Washington, D.C.: American Production and Inventory Control Society, 1982).
18. M. M. Umble and M. L. Srikanth, *Synchronous Manufacturing* (Cincinnati: South-Western, 1990).
19. J. R. Virts and R. W. Garrett, "Weighting Risk in Capacity Expansion," *Harvard Business Review,* Vol. 48, No. 3 (May–June 1970), 132–141.
20. O. W. Wight, "Input/Output: A Real Handle on Lead Time," *Production and Inventory Management,* Vol. 11, No. 3 (Third Quarter 1970), pp. 9–31.

APPENDIX 12A

Finite Capacity Scheduling Helps SMC Improve Delivery, Control Costs
Roberta Jung, Chris Mulherin, and Terry Riggles

SMC Pneumatics, with corporate headquarters in Japan, is the world's largest manufacturer of pneumatic components. The company's philosophy is "the closer we get to our customers, the better we can serve them." This philosophy has significant impact on the way SMC manufactures its product. Partially to be near its customers, the company has located its facilities around the globe, with manufacturing, engineering and administration facilities located on four continents and subsidiaries in all of the world's industrialized countries.

SMC structures and operates its plants so that they can provide maximum customer service and responsiveness, making use of manufacturing methods perfected over the company's 33-year history. For instance, SMC has made worldwide use of techniques popularized in Japan, such as cellular manufacturing and U-line assembly. However, SMC has not allowed tradition to prevent operations from modifying manufacturing procedures to suit local conditions. As long as overall company goals are met, local operations have the freedom to try innovative, customized solutions.

SMC's largest North American plant is in Indianapolis. Although it makes extensive use of manufacturing techniques popularized in Japan, this facility operates in an environment different from SMC's home market. To improve customer responsiveness and delivery, as well as the efficiency of manufacturing operations, the Indianapolis plant turned to a made-in-America solution, finite capacity scheduling software from Waterloo Manufacturing Software.

COMPANY BACKGROUND

In 1977, SMC established its North American headquarters in Indianapolis. Today, this facility serves as a base for manufacturing, design, sales and technical support. Business has grown so that SMC has expanded its North American operation to include manufacturing plants in Los Angeles, Toronto and Mexico City, as well as regional sales and warehouse facilities in more than a dozen other cities.

SMC Indianapolis pneumatic components are used primarily in industrial applications by customers including machine builders and providers of industrial automation equipment, as well as companies involved in the automotive, electric and electronics industries. Large customers early in the company's history included Japanese companies such as Honda, Toyota, Sony and Panasonic, which had transplanted manufacturing op-

Reprinted from APICS—The Performance Advantage Vol. 3, No 2 (February, 1993), pp. 24–27.

Cellular manufacturing and U-line assembly provide an "on-line" verification of product quality and allow accommodation of immediate changes in production demands.

erations to the United States. Today, in addition to these firms, SMC's customers include many major domestic manufacturers such as TRW, Anheuser-Busch and Eli Lilly.

A DISTRIBUTION CENTER FIRST

The Indianapolis facility started out as a distribution center. As the amount of business done by Japanese transplant firms increased throughout the 1980s, and as the company penetrated the domestic market, the volume of SMC's North American business increased dramatically. To handle this volume, and to maintain customer closeness, the Indianapolis facility evolved from executing primarily a distribution function, to doing light assembly, to performing a number of high-volume, high value-added operations that take product from raw material through to finished assemblies.

THE BUSINESS PROBLEM

Managing change is never easy, and continually increasing production requirements have presented the Indianapolis facility with many challenges. The situation is greatly compounded by SMC's basic philosophies.

Superior customer service requires that customers be able to order the type of products that best meet their needs. Superior customer service also requires that customers receive their product in a timely manner. This commitment to service results in SMC offering a large product line (more than 6,000 different pneumatic components) that includes a huge number of variations (more than 50,000). SMC's customers tend to take advantage of this diversity, so that increases in overall production volume tend to get spread out over a wide range of part numbers.

Since it would be prohibitively expensive for SMC to stock sufficient quantities of all product variants, the Indianapolis facility tends to operate in a make-to-order environment. This environment is characterized by an extremely large volume of low-quantity orders for many different part numbers of varying sizes and configurations that have differing production routings. All of these orders are characterized by very short lead times.

THE KEY TO SUCCESS

The Indianapolis plant's challenge is to maintain and continously improve customer delivery while controlling manufacturing costs. One way these goals can be reached is through good production scheduling. SMC accomplishes this task admirably in Japan. However, given the company's large share of the Japanese market, part number production volumes are much higher and customer demand is much easier to predict. Therefore, techniques used in Japan are not directly transferable to the situation in North America. The Indianapolis plant had to develop scheduling techniques that suited its own unique situation.

As production volumes grew at a rate exceeding 30 percent by 1991, scheduling for the plant was growing increasingly difficult. The problem was particularly acute in the machining area. Machining is the highest value-added process in the plant. It is also an area where change-over and setup times can consume valuable capacity.

In 1991, the production control group of the Indianapolis plant was following a scheduling procedure that made use of programs it had developed in house using the dBase data base software. The procedure required high degrees of manual data input. Given the large number of orders completed daily, data maintenance alone consumed a significant portion of the scheduling staff's time. The system also didn't explicitly consider production loads or available capacity. The best a scheduler could do with the software and data available at the time was to develop a highly arbitrary daily machine schedule. Sometimes the schedule was met, but more often it was not. Also, the procedure provided schedulers little help resolving the never-ending battle between customer service and efficiency. While SMC's philosophy mandated that customers receive product in a timely manner, wherever possible, the production department wanted to schedule similar work together to maximize machine uptime.

A SCHEDULING SOLUTION

It was clear to the Indianapolis plant staff that better scheduling was the key to improved customer delivery in an environment of high production growth where order quantities and volumes varied widely. Based on experience with their existing systems, production control staff felt that any solution had to explicitly consider available capacity and, there-

SMC continually evaluates its products for performance and durability.

fore, had to be able to schedule in a finite manner. Given the large amount of functionality desired, it was obvious that SMC would have to purchase commercially available software rather than develop a package in house. The production control group also had a preference for a PC-based solution so that staff could be as self-sufficient as possible during the implementation.

Based on input from staff in the production control group and the production department, production control began a search for commercially available software, led by the supervisor of production control. Over a two-month period, staff contacted a number of vendors and received multiple on-site demonstrations.

In September 1991, the Indianapolis facility purchased TACTIC The Scheduler's Assistant, a PC-based finite capacity scheduling software package developed, marketed and supported by Waterloo Manufacturing Software (WMS) of Twinsburg, Ohio.

The package helps to easily and effectively schedule manufacturing operations in a finite manner. The software allows "what-if" schedules to be interactively generated and allows schedulers to see the impact of potential improvements, as well as unforeseen shop floor occurrences, on customer delivery. Production control staff members were especially impressed with the software's generic data interface. They felt they could easily link the software with data in the facility's IBM mainframe as well as with its existing PC-based shop floor control system.

TRAINING AND IMPLEMENTATION

The production control group assumed responsibility for the implementation of the software and put together an implementation team. The team consisted of the production control supervisor, the production control group leader and other production control staff members as needed. Team members remained responsible for their regular duties. The team was able to quickly implement the scheduling package in the plant's machining area, the highest value-added and greatest bottleneck area of the plant, and within five weeks began receiving benefits.

The team began the implementation with one week of training and implementation assistance from WMS. This session helped the production control group understand the function and features of the software, helped identify interface issues with other systems and helped establish some of the necessary file transfer links. The team also used the training session to expose everyone even remotely involved with scheduling to the software. Team members hoped this exposure would help gain organizational acceptance for the scheduling approach.

As part of the initial week of implementation assistance, WMS helped model and enter into the software all of the production resources to be scheduled, wrote software to extract production completion data from the plant's PC-based shop floor control system and started on a program to extract routing data from a PC-based routings and standards data base. Additionally, they helped specify the necessary program to be written by the MIS department to download customer order information from the IBM mainframe.

Immediately after the training and implementation session, the team plunged into the implementation. The first step was to download the routings. Once routings were transferred to the system, staff were able to manually enter work orders and begin scheduling. A few weeks later, the MIS group completed its portion of the project and the transfer of order data was automated.

Once scheduling began, problems surfaced. These problems were not specifically software-related, but were the type that occur whenever an organization attempts to rapidly change the way it operates. The problems areas were data accuracy, training and organizational acceptance.

For the first time, the plant was actively seeking to use its routings and standards data base, and the standards information in particular was not up to the task. Problems were highlighted when the production department consistently finished work sooner or later than the software indicated. by underscoring standards problems, the software helped the plant rapidly improve its data.

Other implementation problems were training-related. When the plant actually got up and running with the software, quite a few staff members were involved one way or another with the scheduling system. While they may not have been actively scheduling, the reporting, clerical and support functions that they provided were critical to the system's overall success. The large volume of orders rapidly produced by the production department required frequent scheduling, which in turn required that the support tasks be carried out in a timely, highly accurate manner. Once the team identified training as an area of concern, it immediately documented all tasks related to the software's operation and trained staff members in these procedures, quickly and effectively solving the problem.

Only time solved the last set of problems. Even though the team had actively involved personnel from the production department in the initial software selection and training process, some staff remained skeptical that finite scheduling would be of benefit. They feared that the software would spit out schedules that would force them to set up and run work in an inefficient manner. The entire production control group alleviated these fears by continually involving others in the scheduling process and by showing them how the software could be used in a manner that incorporated their input to ensure that the best scheduling decisions were made. As better and better schedules were generated, the benefits of the software became obvious.

ONGOING USE

The Indianapolis plant has successfully been using the scheduling package in the machining area for more than a year. The production control group starts a typical day with the software by downloading orders from the mainframe. Hot orders that cannot wait until the next morning to be downloaded are entered into the system manually. Schedules are generated and dispatch lists of the day's suggested production are electronically transmitted to terminals on the shop floor. Production personnel are responsible for reporting actual work completions, twice a shift at set times, through these same terminals. This completion information, as well additional new orders, are considered in schedules generated shortly after the production reporting cut-off points.

Shop floor supervision gets a list of any orders scheduled to be finished after their due dates so that corrective action can be taken. Supervision likewise gets a list of projected machine utilization. On an as-needed basis, production supervision meets with production control to address anticipated problems highlighted by the software. These meetings make use of the software to review proposed actions such as working overtime or rerouting work from heavily loaded resources to those that are more lightly scheduled.

BENEFITS

SMC's Indianapolis facility has received numerous benefits from use of the finite capacity scheduling software package—some are easily quantifiable, others are harder to measure. One of the most obvious benefits of the software is that it has replaced an in-house scheduling system that was cumbersome and time-consuming. Shortly after implementation, the production control group was able to reduce the man-hours required to perform scheduling-related tasks by 50 percent. The software paid for itself within its first year of operation.

Operations on the shop floor have been smoothed out, also. The software has given the plant the visibility to look ahead in time and spot potential problems and make the necessary adjustments before a situation becomes critical. For instance, staff are now able to better plan overtime production and tend to get much better value for the overtime dollars spent. The production department has also been able to better organize the

"Manufacture-to-stock" and "purchase-to-stock" distribution allows quick customer service for "Just-in-Time" delivery.

shop floor and to leverage investments in capital equipment and tooling. The software has highlighted which machines on the shop floor are consistently overloaded. This information has allowed staff to better group production over machines based on part geometry, reducing setups and creating more available capacity. The software has also helped identify instances where relatively small investments in tooling can enable lightly used machine tools to be converted so that they can assume some of the load run by capacity-constrained resources.

While the plant has been able to attribute substantial cost savings to the installation of the software, the tool has also helped increase profitability. SMC's growth has been based on superior customer service and quality. However, past success is not always an indication of future performance. Rapid growth and increasing order volumes often imperil the very levels of customer service that have led to that success. The software has helped SMC avoid this trap and maintain high levels of customer service. TACTIC is currently being used by SMC's production, scheduling and customer service personnel to monitor the status of orders as they make their way through machining. When an order is in danger of finishing late, corrective action can be taken.

SMC Pneumatics has benefitted greatly from the installation of the capacity scheduling software at its Indianapolis plant. Through helping the company execute its fundamental strategy of superior customer service, the software installed has helped SMC Pneumatic maintain its enviable growth rate.

Chapter Thirteen

High-Volume Production Activity Control and Just-in-Time Systems

INTRODUCTION

Shop floor control (SFC) includes the principles, approaches, and techniques that are necessary to plan, schedule, control, and evaluate the effectiveness of production operations. Shop floor control integrates the activities of the so-called factors of production of a manufacturing facility, such as workers, machines, inventory, and material handling equipment. The SFC plan facilitates efficient execution of the master production schedule, control of processing priorities, improvement of operating efficiency through proper worker-machine scheduling, and maintenance of minimum quantities of work in process and finished goods inventories. In the final analysis, SFC should lead to improved customer service. This chapter covers the major functions of an efficient SFC and its relationships to MRP, capacity planning, and inventory status that are applicable to make-to-stock continuous-flow manufacturing for high-volume, standardized products.

High-volume production can utilize the philosophy and techniques of just-in-time (JIT). This chapter presents the basics of JIT, which has had a major impact in reducing inventories and improving quality, efficiency, and throughput time.

THE PRODUCTION ENVIRONMENT

Major differences exist in the management of production activities in make-to-order and make-to-stock firms. In make-to-order situations, due dates are important, and hence the sequencing of customer orders at various machine centers is an essential function. This involves both planning and control of activities. Make-to-stock products are generally high-volume consumer goods, such as telephones, automobiles, and wristwatches. The manufacture of standardized, high-volume items, which involves flow shops, is the subject of this chapter. Differences in the production environment could also represent technological differentiation, such as manual versus automated, chemical reactions or materials forming, and the like. The cost and quantity of production determines the degree of automation in industries. Chemicals, drugs, petroleum, and other process industries fall in the special category of high-volume production industry. Management of these industries generally requires trained technical professionals, a high degree of automation, and special materials management techniques. Production control techniques for such industries are discussed in Bensousson, Hurst, and Nasland [2]; Gruver and Narasimhan [12]; Paul [33]; and Sethi and Thompson [39]. The unique characteristics of various types of such shops are exhibited in Table 13.1.

Flow Shops

A flow shop consists of a set of facilities through which work flows in a serial fashion. The same operations are performed repeatedly in every workstation, thus requiring lower-level skilled workers. The flow shop generally represents a mass production situation, and hence the operations become very efficient. For example, an operator might be installing car doors on an automotive assembly line or assembling dials on the handset of a telephone.

In flow shops, items enter the finished goods inventory one after another, often in the same order they were input, leaving very low in-process inventories. Since the items are mostly made to stock, forecasting is a difficult job, and hence the finished goods levels carried in terms of anticipation inventories are very high. For the same reason, raw materials are carried at higher inventory levels. Machines in flow shops tend to have a special-purpose design, and hence the initial investment level is generally high for heavily automated plants.

The production control system for continuous production is called *flow control.*

TABLE 13.1   Comparison of Flow Shop and Job Shop Characteristics

Item	*Flow Shop*	*Job Shop*
Equipment	Special-purpose	General or multipurpose
Investment amount	Medium to high	Low to medium
Workers	Low-skilled	High-skilled
Product	Make-to-stock	Make-to-order
Demand volume	High	Low
Number of end items	Few	Many
Production control systems	Fixed layout and continuous flow; uses assembly line balancing	Variable flow pattern controlled by batch, order, or service needs; uses sequencing rules and simulation
Efficiency of production	Most	Least
Finished goods inventory	High	Low
In-process inventories	Low	High
Raw material inventories	High	Low
Component demand type	Dependent	Predominantly independent

Specialization, high volume, division of labor, and efficiency are built into the design of assembly lines. Hence, flow shops generally require repetitive and low skills only. The repetitive nature of the manufacturing environment also creates monotony and affects worker morale. To deal with this problem, industrial engineers and social scientists have developed job enrichment programs. Although we will not dwell on these issues in detail, they should be kept in mind when designing assembly lines.

Job Shops

Job shops follow different flow patterns through a facility in batches. The operations need not be repetitive. Job shop facilities are made up of general-purpose machines that are capable of using various tools, dies, and fixtures to perform many different jobs at the same facility. The jobs may be unique and hence may never be repeated. Variations in sequencing of jobs on different machines, processing time requirements, priorities, and number of operations make management of job shops very challenging. These factors also necessitate highly skilled workers. Unfortunately, this leads to a production system that is not very efficient. Job shops accumulate large amounts of in-process inventories. Many end items are made to customer specifications, requiring low amounts of raw materials and low end item inventories. The production control system for a job shop is called *order control*. Hospitals, restaurants, and service stations are classic examples of job shops in the service sector. Management of job shops will be addressed in later chapters.

Intermittent Flow Shops

An intermittent flow production shop is useful when high production volumes are needed on a periodic basis. Examples include the production of refrigerators, air conditioners, and heat pumps. Such systems are known as *batch processing*. This chapter also discusses several techniques for managing such systems. When batches of several items are to be manufactured in a job shop environment, the process is dominated by serial production. The volume may not justify special-purpose machines because of recurrent use. For production and inventory system purposes, such shops can be treated as flow shops. These systems are treated in Chapter 14, which covers job shop activity planning.

CONTROLLING CONTINUOUS PRODUCTION

The major flow shop problem is to attain the desired production rate, such as 60 cars per hour or 600 telephones per day, with maximum possible efficiency. The total work content of the job is divided into elementary operations, and these operations are grouped together at workstations. The job moves successively, and in many instances continuously, from one station to another. All workstations are occupied with jobs that are at different stages of completion. The speed of the assembly line is controlled by the required output rate, the space between stations, and the time requirements for each workstation. By controlling the speed of the conveyor or the cycle time, we can essentially control the output rate on the production line.

Table 13.2 exhibits all operations involved in the assembly of a calculator, which requires 120 seconds of total work. Suppose that we allocate operations 1 through 7 to station A; operations 8 through 14 to station B; and, finally, operations 15 through 20 to station C. The workload at stations A, B, and C would be 43 seconds, 43 seconds, and 34 seconds, respectively. The maximum output of such a system could only be 83.7 units per hour, even though individual output rates of 83.7, 83.7, and 105.8 units per hour at stations A, B, and C, respectively, are possible. This can be explained by the following. Stations A and B each have a total of 43 seconds of work, and hence a job arrives at station C every 43 seconds. Station C has only 34 seconds of work, and hence the station will be idle for $43 - 34 = 9$ seconds during every cycle.

For an efficient layout and production operation, therefore, all stations should be loaded with equal amounts of work. Several aspects of manufacturing make it difficult to assign equal work assignments to all stations. Generally, most operations must be done in a particular sequence, which is specified by a precedence diagram. Also, the capacity of equipment and the efficiency of people on the assembly line differ. The interaction of these aspects complicates the balancing of workload among stations. Generally we must ask the following questions: How do we determine the sequencing of operations? How do we determine the ordering of stations? How many stations are necessary? How well balanced will the flow shop be? The answers to the first two questions are given by the

TABLE 13.2   Calculator Assembly

Station	Operation	Duration t_i (seconds)	Σt_i for Station	Maximum Output Rate per Hour
A	1	5		
	2	14		
	3	6		$\dfrac{60 \times 60}{43}$
	4	3		
	5	8		
	6	5		$= 83.7$
	7	2	43	units
B	8	5		
	9	7		
	10	12		$\dfrac{60 \times 60}{43}$
	11	2		
	12	7		
	13	4		$= 83.7$
	14	6	43	units
C	15	3		
	16	4		$\dfrac{60 \times 60}{34}$
	17	9		
	18	10		
	19	3		$= 105.8$
	20	5	34	units
Total		120	120	

process sheet released by the manufacturing process engineer. Very few choices may exist to alter the operation times or assembly sequence. The answer to the last question, and other questions, depends on the required production rate, feasible cycle times, and available facilities. A brief discussion of these issues follows.

Precedence Diagram

A precedence diagram specifies the order or sequence in which the activities must be performed. As an example, Figure 13.1 shows the precedence relationship for the calculator assembly. Each circle is a node, and the numbers inside the circles identify particular operations. The number outside the circle represents the duration of the operation. We have specified them in seconds; it is also popular to specify them in hundredths of minutes. Arrows indicate which operations should be completed first. For example, operation 14 cannot be started until operations 3 and 7 are completed.

Cycle Time

Cycle time (c) is directly related to the production rate of the assembly line:

$$c = \frac{\text{productive time}}{\text{demand}}$$

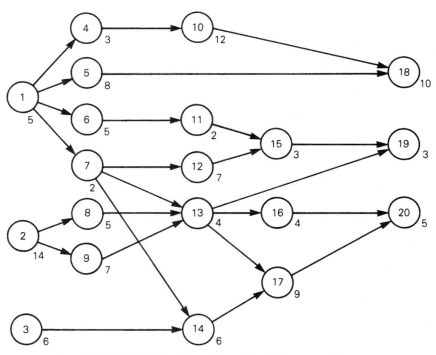

FIGURE 13.1   Precedence diagram for calculator assembly.

For example, if 2400 calculators are required per day, then the cycle time, based on an eight-hour day, would be $(8 \times 60 \times 60)/2400 = 12$ seconds. This cycle time tells us that the items must be spaced at least twelve seconds apart on the assembly line and that each station has that amount of time. Referring to Table 13.2, we notice that operation 2 exceeds twelve seconds. Thus, a cycle time of twelve seconds is not feasible in this case. Therefore, we can state that the cycle time should be larger than the longest duration, t_{max}. It is also evident that c need not be greater than the total time Σt_i to complete all operations. Therefore,

$$t_{max} \leq c \leq \sum_i t_i$$

where t_i represents the duration for operation i.

Number of Work Stations

The total number of workstations (n) must always be an integer. The value of n is dependent on c, and vice versa. That is,

$$n = \frac{\sum t_i}{c} \quad \text{or} \quad c = \frac{\sum t_i}{n}$$

Table 13.3 exhibits several alternative cycle times and production rates for many possible numbers of stations. The results are based on perfect balance, which may not be feasible in all cases. A layout is said to have a perfect balance when the total work content in all stations are equal, and only a limited number of combinations satisfy the precedence constraints. For example, it is not possible to obtain exactly 17.1 seconds per station for a seven-station layout, since elementary operations are not divisible.

TABLE 13.3   Relationship between Cycle Time and Production Rate

Number of Workstations (n)	Cycle Time (c = $\Sigma t_i/n$)	Daily Production	Remarks
1	120	240	
2	60	480	
3	40	720	
4	30	960	
5	24	1200	
6	20	1440	
7	17.1	1685	Not feasible
8	15	1920	
9	13.3	2165	Not feasible
10	12	2400	

Balance Delay

The quality of the solution is measured by the balance delay (d) equation:

$$d = \frac{\left(nc - \sum t_i\right)100}{nc}$$

The ratio indicates the inefficiency of the system, which is induced by the idle time $nc - \sum t_i$. In our calculator example, the ratio would be

$$d = \frac{[(3)(43) - 120]100}{(3)(43)}$$
$$= 6.98\%$$

The proposed solution does not meet the required production of 720 units per day. Management may explore the use of overtime or the addition of another station or may find other ways of increasing the productivity of assembly operations. Although this layout resulted in an imbalanced solution, we see later that it is possible to achieve a perfect balance for this example. In a perfectly balanced layout situation, $nc = \sum t_i$, and hence the balance delay will be equal to zero.

We see that cycle time (c) is a function of the required output rate and the total number of workstations (n). We also know that the required basic output rate depends on demand, available inventory, and capacity of the facility. Therefore, we see that many aspects of production are interrelated and, hence, that the design of an assembly line is a complex problem. Assuming that a suitable production schedule is stated, a feasible cycle time and an acceptable number of workstations can be determined using one of the several balancing methodologies discussed in this chapter.

SEQUENCING AND LINE BALANCING METHODOLOGIES

Line balancing concepts have been implemented in a number of industries, such as automobile, telephone, and consumer electronic goods manufacturing. These are generally large-scale problems involving perhaps up to 100 tasks or more, with ten to fifteen or more stations. Line balancing methods and techniques include (1) linear programming, (2) dynamic programming, (3) heuristic methods, and (4) computer-based sampling techniques. The size of the problems (number of operations) that can be successfully and economically tackled by optimizing methods such as linear programming and dynamic programming is limited. Further, complexities inherent in manufacturing, as described later, make heuristics and other computer-based techniques very attractive for solving large-scale line balancing problems. In this section, we explain the basic concepts of a heuristic method that was developed by Kilbridge and Wester [24].

Heuristic Techniques in Line Balancing

A heuristic, or rule of thumb, is a shortcut solution procedure that searches for a satisfactory, rather than optimal, solution. Analogous to the human trial-and-error process, heuristics reach acceptable solutions to problems for which optimizing solutions are not feasible or are too costly to employ. There is no way to guarantee that an optimal solution will be found with the heuristic techniques; rather, the best of the arrangements considered has been found.

The Trial-and-Error Technique.   The trial-and-error technique merely groups work elements so as not to violate the cycle time, starting with precedence diagram and activity times. For example, suppose that 1400 units of the product shown in Table 13.2 are required per day. This translates to approximately twenty seconds of cycle time with six workstations, as shown in Table 13.3. The ease of rearrangement of existing facilities, economics, and quality reasons lead the management to consider two assembly lines, each producing 720 units. This layout will have a cycle time of forty seconds, with three workstations in each assembly line.

Assuming that activities may be combined within a given zone without violating the precedence relationship, we can designate work zones/stations on the precedence diagram. When we obtain the theoretical minimum number of workstations, the optimum assignment is made. The solution need not be unique. One such solution is exhibited in Figure 13.2. In this case a perfect balance was achieved, since all work zones have exactly the same workload of forty seconds. Hence, the balance delay is 0%. This can also be shown by use of the balance delay formula discussed earlier:

$$d = \frac{(nc - \sum t_i)100}{nc}$$

$$= \frac{[(3)(40) - (120)]100}{(3)(40)} = 0$$

Comparison of Balancing Techniques

The previous discussion has been limited to the basic concept of line balancing. Numerous heuristic procedures are available to find good solutions to the problem of line balancing. Linear programming, dynamic programming, and other mathematical models can be used to develop solutions.

Although we have not dealt with any analytical techniques, we would like to remark on a notable algorithm by Held, Karp, and Sharesian [17]. Their algorithm combines the best of heuristic methods and dynamic programming. We refer to this algorithm as the *combined heuristic.* Mastor [30] compared the effectiveness of these techniques on the basis of the computational time required to provide a feasible solution and the resulting balance delay for the solutions. Held et al. [17] report that the combined heuristic procedure yielded the least balance delay. Since the combined heuristic procedure used dynamic programming, the size of the problem that could be economically

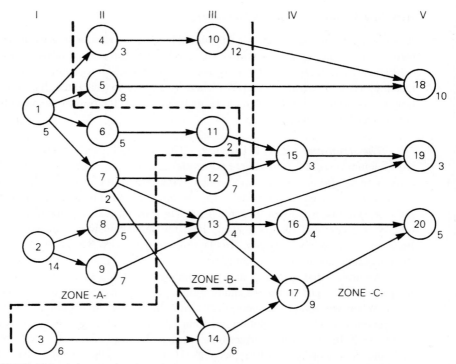

FIGURE 13.2 A three-workstation solution by the trial-and-error procedure for the problem shown in Table 13.2.

solved was a limitation. Mastor concluded, however, that the magnitude of computational time was still moderate.

Simpler heuristic techniques are also available, including such methods as the "longest-task-time" heuristic and the "most-followers" heuristic. Readers are referred to Gaither [10], for example, for more information.

RELATIONSHIP TO AGGREGATE PLANNING

It is important to realize that assembly line balancing is a component of overall scheduling. It was indicated earlier that cycle time is a function of many variables, such as the output rate, the speed of conveyors, the distance between stations, and, of course, the number of stations. We also know that the required output rate is designed by aggregate planning. The aggregate plan, in turn, depends on demand, inventory, and the capacity of the available facility. Thus we see that many of these factors are interrelated and that the design of an assembly line is a complex problem, as illustrated by the following example.

Example 13.1

The demand for calculators is seasonal. During the school opening and Christmas seasons, the demand is heavy. The Nikity Company needs a plan to produce calculators according to the following schedule:

Month	Production Rate
August	2400
September	1920
October	1200
November	1920
December	2400
January	960

Solution: To accommodate this production schedule, we would need two parallel lines, each with five stations, for August; two parallel lines, each with four stations, or a single line with eight stations, for September; and so forth, as exhibited by Table 13.3. It is important to note that frequent changes in the number of workstations lead to a loss of productivity, since these changes would lead to reassignment of activities, which generally requires a learning period. Rather than changing the production rate drastically every month, the firm might want to smooth the schedule or explore other strategies, as explained in Chapter 9.

BATCH PROCESSING TECHNIQUES

In assembly line situations, we have assumed that the same item is manufactured continually. In many instances, however, several items belonging to the same product family are produced in the same line. A setup may or may not be required between batches of production. A problem involving a setup is called an economic lot size problem (ELSP) in the literature [8, 27]. Hax and Meal [16] developed a method involving a major setup for the family and a minor setup for individual items in the family. These methodologies find the economic batch quantities by analyzing the trade-off between setup and inventory costs. In addition to the economic batch sizes, if we want to determine their best sequence along with capacity considerations, mathematical programming techniques can be used to solve these problems. If we include the uncertainty in demand, however, we would need a dynamic method of sequencing and scheduling items. In this section we illustrate some simple but effective batch processing techniques.

The Runout Time (ROT) Method
with Lot Size Considerations

The runout time (ROT) method is based on the depletion times for different items in the family. For example, a family may contain linens, pillow cases, and fitted sheets with the same design. The ROT of an item is expressed as the ratio of the current inventory to the demand forecast for the period [34]:

$$\text{ROT}_i = \frac{\text{current inventory of item } i}{\text{demand per period for item } i}$$

With the ROT method, the ROT for each item in the family is computed, and the items are ranked in the ascending order of their ROTs. The item with the lowest ROT is scheduled for production first, followed by the item with the next lowest ROT. This process is repeated for the remaining items. As discussed in Chapter 12, problems involving a setup generally specify the lot size, setup time, and run time. Based on the given data, we can calculate the required standard hours per unit for every item in the family. The following example illustrates the ROT method.

Example 13.2

Table 13.4 gives the data on current inventory, production lot sizes, standard hours per unit, and the forecast of demand for all items in a product family. Using the ROT method, determine the sequence of production. The available production capacity is stated as eighty hours. Analyze the effect of capacity on your schedule.

Solution: As a first step, we calculate the runout time for each item by dividing the current inventory by the forecast demand per week (see Table 13.5). This process provides the sequence D, B, A, C. Given the available capacity of eighty hours, the effect of capacity on the sequence is analyzed in Table 13.6. We thus find that the capacity is inadequate to produce the desired units of all items in the family.

The Aggregate Runout Time Method

In many cases we may be able to remove the lot-size restrictions. In some other cases, however, we may be forced to delineate from the lot-size considerations. For instance, in Example 13.2, although we found ourselves short of the required capacity, we might be

TABLE 13.4   Data for Example 13.2

Item	Standard Hours per Unit	Lot Size	Forecast per Period	Current Inventory	Machine Hours per Order
A	0.10	100	35	100	10
B	0.20	150	50	120	30
C	0.15	100	40	130	15
D	0.20	200	60	80	40
Total					95

TABLE 13.5 ROT Determination, Example 13.2

Item	Current Inventory	Demand per Week	ROT (weeks)	Sequence
A	100	35	2.86	3
B	120	50	2.40	2
C	130	40	3.25	4
D	80	60	1.33	1

forced to produce all items in a family so that a shortage of any single item does not occur. Such problems can be solved by the aggregate runout time (ART) method [5]:

$$\text{ART} = \frac{\begin{array}{c}\text{Machine hours inventory} \\ \text{for all items in the family}\end{array} + \begin{array}{c}\text{total available} \\ \text{machine hours}\end{array}}{\begin{array}{c}\text{machine hour requirements forecasted} \\ \text{for all items in the family}\end{array}}$$

We extend Example 13.2 to illustrate this method.

Example 13.3

Suppose that the firm wants to prevent any anticipated stockouts caused by the lack of capacity. How would we devise a new schedule for Example 13.2?

Solution: As a first step, we compute the available and required capacities and determine the value of ART (see Table 13.7). We obtain

$$\text{ART} = \frac{69.5 + 80}{31.5} = 4.746$$

As a second step, we calculate the schedule, assuming the same ART for all items in the family (see Table 13.8). The gross and net requirements are calculated using the following formulas:

TABLE 13.6 Capacity analysis, Example 13.2

Scheduled Sequence	ROT	Lot Size	Machine Hours Required	Remaining Capacity
D	1.33	200	40	40
B	2.40	150	30	10
A	2.86	100	10	0
C	3.25	100	15	−15

TABLE 13.7 ART Determination, Example 13.3

Item	Standard Hours per Unit	Forecast per Period	Machine Hours for the Forecast	Current Inventory	Machine Hours Inventory
A	0.10	35	3.5	100	10.0
B	0.20	50	10.0	120	24.0
C	0.15	40	6.0	130	19.5
D	0.20	60	12.0	80	16.0
Total			31.5		69.5

$$\text{Gross requirements} = \text{forecast per period} \times \text{ART}$$

$$\text{Net requirements} = \text{gross requirements} - \text{current inventory}$$

Using the net requirements, and with the given available capacity of eighty hours, we can check the capacity requirements, as exhibited in Table 13.9. The use of the ART formula facilitated the production of all items in the family, thus avoiding any shortages of any individual items in the family.

PROCESS INDUSTRY SCHEDULING

Process, or continuous-flow, industries have a common characteristic of numerous outputs from few inputs. For example, a refinery takes crude oil as the input and produces kerosene, jet fuel, and various grades of gasoline as outputs. Many process industries have an inverted bill of materials, as was shown in Figure 10.4c. Due to the cost and time involved, particularly the capital intensity, changeovers occur very infrequently. Table 13.10 indicates the characteristics of process industries and the scheduling implications.

One possible scheduling application is to determine the lot sizes when changing production. Consider a small refinery that produces three products: jet fuel (JF), unleaded regular gasoline (UR), and unleaded high-octane gasoline (UH). Common inputs are used to produce any one of the three products. The scheduling problem is to determining the lot size of each product to produce before changing over production to the next product. Unless some simplifying conditions are considered, this type of problem can be very complex. One simplification is to produce each product in sequence, each

TABLE 13.8 Schedule Requirements, Example 13.3

Item	Standard Hours per Unit	Forecast per Period	ART	Gross Requirements	Current Inventory	Net Requirements
A	0.10	35	4.746	166	100	66
B	0.20	50	4.746	237	120	117
C	0.15	40	4.746	190	130	60
D	0.20	60	4.746	285	80	205

TABLE 13.9   Capacity Requirements, Example 13.3

Item	Standard Hours per Unit	Net Requirements	Machine Hours Required	Remaining Capacity
A	0.10	66	6.6	73.4
B	0.20	117	23.4	50.0
C	0.15	60	9.0	41.0
D	0.20	205	41.0	0

the same fraction of the annual demand. The question then is to determine the number of cycles each year, resulting in a lot size. Table 13.11 indicates the parameters of this sample problem.

Assuming 250 days per year, the daily gallons are found from the annual gallons. If there were one cycle per year, then the process would first be set up to produce 10,000 gallons of jet fuel, then 20,000 gallons of unleaded regular, and then 5000 gallons of unleaded high octane.

Krajewski and Ritzman [28] develop a process for determining the best plan. Total production days per year will be

$$T_i = N_i t_{si} + D_i t_{pi}$$

for product i, where

N_i = number of cycles per year

t_{si} = setup time

D_i = annual demand for product i

t_{pi} = time to produce one unit of i

The total days for the example from Table 13.11 is 219. The annual cost of any plan would be the sum of the setup cost and holding cost:

$$C_i = \frac{D_i S_i}{Q_i} + \left(\frac{Q_i}{2}\right)\left(1 - \frac{r_i}{p_i}\right)H_i$$

TABLE 13.10   Process Industry Characteristics and Scheduling Implications

Characteristics	Scheduling Implications
Standard design of product and process.	No need for routing.
Produce to inventory.	Can use economic order quantities in scheduling lot sizes.
Process steps are tightly coupled.	Scheduling is accomplished through gating decisions at the raw material input stage.

TABLE 13.11   Refinery Scheduling

Output	Annual Gallons D_i	Daily Gallons r_i	Production in Daily Gallons p_i	Process Time/Gallons (days) t_{pi}	Setup Time (days) t_{si}	Holding H_i	Setup S_i
JF	10,000	40	150	0.0067	1.0	1.00	200
UR	20,000	80	200	0.005	0.5	0.50	100
UH	5,000	20	100	0.01	0.5	0.75	100

For a cycle of one per year, the total cost of our example would be $8567. As the number of cycles increases, the setup costs will decrease but the holding costs will increase. Table 13.12 shows the results for cycles of four to six. The lowest cost solution would be the $N = 5$, which means that there would five cycles per year.

JUST-IN-TIME (JIT)

Just-in-Time (JIT) is often thought to be a technique for reducing inventories. That is only partly correct. JIT can be considered in two ways: first, as a philosophy of waste reduction, and second, as a set of techniques for the reduction of inventory and waste. JIT is just one of the many improvement philosophies that are now in vogue; such as total quality management (see Chapter 20), world-class manufacturing, and zero inventories.

TABLE 13.12   Cost at Various Cycles per Year

	N = 4		
Product	Q = D/N	T	Total Cost
JF	2500	70.7	$1717
UR	5000	102	$1150
UH	1250	52	$ 775
Total		224.7	$3641

	N = 5		
Product	Q = D/N	T	Total Cost
JF	2000	71.7	$1733
UR	4000	102.5	$1100
UH	1000	52.5	$ 800
Total		226.7	$3633

	N = 6		
Product	Q = D/N	T	Total Cost
JF	1667	72.7	$1811
UR	3333	103	$1100
UH	833	53	$ 850
Total		228.7	$3761

JIT Definitions

As a philosophy, JIT's primary goal is the elimination of waste in the production system. Anything that does not add value to the product in the system is waste. Rework and scrap are obvious wastes and should be eliminated. Less obvious as a source of waste is inventory. Consider inventory between work centers: There is no value added by allowing this inventory to sit there, and therefore it is considered waste. The name just-in-time epitomizes the concept of reduced inventory: Get the material to the next work center or customer just in time for the next production step. If this is done, then inventory between production stages is reduced.

JIT has, as its basics, the concept that the right part should be at the right place at the right time. The objectives of JIT are to eliminate waste, to improve quality, to minimize lead times, to reduce costs, and to improve productivity.

The key elements to successful JIT are the following:

- Housekeeping is the organizing of the workplace for higher productivity.
- Quality improvement through process improvement is necessary so that there are no interruptions in the flow due to defective material.
- Reduced setup times allow smaller lots.
- Preventive maintenance is practiced to avoid unexpected interruptions.
- Incremental inventory is reduced to force problems into the open.
- Workers are cross-trained to allow higher efficiency of the workforce.
- A level schedule is maintained so that flow is easier to balance throughout the process.
- Operations are balanced to allow even flow and to prevent inventory between work centers.

Process Flow and Layout

The design of the production process and the plant layout are critical components for ease of flow in a JIT environment. Several common techniques, such as group technology and dedicated lines, will allow uninterrupted product flow. Rearranging layout to allow for a minimum amount of flow time between workstations will also speed the flow by allowing a transfer batch separate from the process batch to be moved to the next process operation. If the work centers are immediately adjacent, the transfer batch could be as little as one.

In operations with workers adjacent on assembly benches, the transfer could easily be one. In an example from Hewlett Packard, a section of the bench between each worker was marked with masking tape in a rectangle equal to the size of the unit being assembled. This area is known as the kanban area. When the first worker was done, the unit was placed in the kanban area between that worker and the next. The second worker could then take the unit from the kanban area, do the assigned operations, and place it in the next kanban area adjacent to the next assembly worker. If a unit occupied a kanban space between a worker and the subsequent worker a worker could not start on a unit until that area was clear. This prevented inventory buildup and forced workers to ensure that they completed the assigned assembly operations correctly because the next worker would find any incorrect operations.

Setup Time Reduction

Setup time reduction is an important aspect for successful implementation of JIT. Consider the classic economic order quantity (EOQ) inventory formula introduced in Chapter 4. If the setup cost is considered a surrogate for the setup time, then a reduced setup cost will result in smaller EOQ value. Assume that the setup cost (S) is initially $500 per setup, the annual demand (D) is 10,000 units, and the holding cost (H) is $4 per year per unit. The EOQ value would be

$$Q = \sqrt{\frac{2DS}{H}} = \sqrt{\frac{2 \times 10,000 \times 500}{4}} = 1581$$

and total cost, TC would be

$$\text{TC} = \frac{SD}{Q} + \frac{HQ}{2} = \$6325$$

This would be about forty days' worth of inventory.

If this were one machine in the process, then the lot would need to be 1581 to be economical. If the setup cost were to be cut in half to $250, the new Q would be

$$Q = \sqrt{\frac{2DS}{H}} = \sqrt{\frac{2 \times 10,000 \times 250}{4}} = 1118$$

and total cost would be

$$\text{TC} = SD/Q + HQ/2 = \$4472$$

or about twenty-eight days' worth of inventory. A significant reduction in the setup cost is needed to reach a smaller lot size. A most demanding standard for setup time is known as single minute exchange of dies (SMED), meaning that the setup should be accomplished in less than ten minutes.

In our example above, if the setup cost were reduced to $50, then Q would be

$$Q = \sqrt{\frac{2DS}{H}} = \sqrt{\frac{2 \times 10,000 \times 50}{4}} = 500$$

and total cost would be

$$\text{TC} = SD/Q + HQ/2 = 2000$$

or about twelve and a half days' worth of inventory.

At a lot size of 500, there would be twenty setups per year, still a large number. Say that we want to have a lot that equals one day's worth of demand, given 250 working days per year. We need to find the setup cost that would accomplish this. Rearranging the EOQ formula,

$$S = \frac{HQ^2}{2D}$$

Since Q is to equal one day's demand, then $Q = 10,000/250 = 40$. In the formula above, then

$$S = \frac{4 \times 40 \times 40}{2 \times 10,000} = \$0.32$$

The total cost would be about \$160. This would require an extremely fast setup time! But consider the payoff: Total annual cost is reduced to \$160 from \$6325. We would be willing to pay a large amount to reduce this setup. Smaller lots would reduce the inventory significantly because only one day's worth of inventory is produced at any one time. Any product produced on this machine could be produced in small lots for customer demand. Several researchers have studied the cost-benefit trade-off of including the setup reduction cost; see Billington [3] and Porteus [35] for more advanced studies.

Uniform (Level) Flow

Uniform, level flow is a key to successful JIT. The master scheduler must take the demands for products and start to batch them into daily quantities so that the final assembly schedule cycles through the products. Consider three products A, B, and C that have production requirements per day as shown in Table 13.13.

A uniform flow will allow for uniform work center flows and loads. In Table 13.13 the requirements per day could be produced in batches equal to the daily requirements. This would result in imbalance at the work centers if the subassemblies needed for product A are different from B and C. To balance the load, the production can be done in smaller lots per cycle. We can find the number per cycle by dividing the daily require-

TABLE 13.13   Production Cycles

Product	Requirements per Day	Number Required per Cycle
A	200	2
B	300	3
C	400	4
Total	900	9

ments by the largest common denominator of 100. A cycle would then consist of nine units, with two A's, three B's and four C's. One possible example is

A   B   C   B   C   B   C   A   C

which would result in relatively even load to the supplying work centers.

Pull Production and Kanban

In the previous situation, note that the subassemblies and parts required for the final assembly schedule will need to be pulled in small batches from the supplying work centers whenever they are required; hence, "pull" is associated with JIT systems. MRP systems are considered to be "push" systems because a batch will be pushed to the next work center when it is finished even if it is not needed there yet. A fully implemented JIT system has so little inventory that the production process forces a pulling of material through the system in response to the final assembly schedule.

One of the most publicized methods is the kanban system, which means "card" or "visible record" in Japanese. Toyota was one of the first and most important developers of the kanban system. In this method, production is only undertaken when a kanban is available. A kanban becomes available when the subsequent process uses a container of parts and places the kanban associated with that empty container into a receiving post at the supplying process. The important determinant in this system is the number of containers (and hence kanbans) that should be between two processing steps.

Fast-food restaurants such as McDonald's are essentially run by kanban systems. In this type of kanban, some signal is sent to the supplying process. Consider the holding area for hamburgers and cheeseburgers that usually consists of a bin for each type. As the inventory is depleted in a bin, some form of signal is used to tell the production area to produce more of that type. A manager may also signal a production start based on his or her observation of the people entering or about to enter the line (i.e., a bus rolls into the parking lot) in relation to the inventory in the bins.

Single-Card Kanban.   A single-card kanban system uses a withdrawal or "move kanban" to trigger production. In Figure 13.3, work center A produces a part that is moved into the storage area in containers that hold a certain number of parts. Work center B pulls a container from storage when it needs to do work and may return an empty container to the storage area. When B pulls the container, the move kanban is removed from the container and placed in the card rack at A. The move kanban in A indicates an authorization for A to produce another container of parts. A may only produce a container when it has a move kanban.

It is possible to have more than one move kanban in this system. Each kanban represents a container of parts; the more kanbans, the more inventory in the storage area. The number of kanbans can be found from

$$n = \frac{DL(1+\alpha)}{a}$$

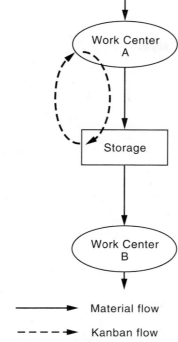

Material flow

Kanban flow

FIGURE 13.3   Single-Card Kanban System.

where

       D = demand per unit of time
       L = lead time
       a = container capacity
       α = safety stock factor

For example, if the lead time for a container is one hour, a container holds ten parts, and the demand per hour is twenty, then there should be $n = 20(1)(1 + 0)/10 = 2$ kanban. Any positive α will force a third kanban to hold safety stock. If no safety stock is needed, then in this example the two containers will be filled just in time for the next processing center to use them. Krajewski and Ritzman (28, p 601) indicate that Toyota's typical container holds no more than 10% of a day's requirements.

Two-Card Kanban.   A two-card or dual-card system has a move (withdrawal) kanban and a production kanban. Figure 13.4 shows the flow of material and kanbans. Say that Work center B is using a container of parts that are next to B. A worker pulls the move kanban from the container and goes to the storage area. A full container is found, the production kanban is removed, and the move kanban is placed in the container. This authorizes the worker to move the container from storage to B. The production kanban is

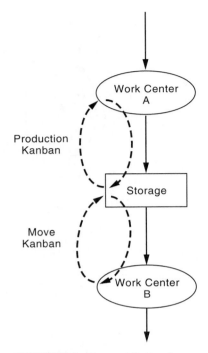

FIGURE 13.4 Two-card Kanban System.

then placed in a rack next to work center A as an authorization to A to produce another container of parts. This production kanban is placed in a full container of parts and moved to storage.

We can see that the kanban system allows a pull of material through the work centers without centralized coordination of work center scheduling.

JIT Interdependencies

Implementation of JIT is commonly accomplished by reducing batch sizes so that problems are forced into the open. Quality problems are often the first to surface. With large lots, a few (or even many) defective units cause few problems because they may be placed aside and the next unit used. When the lot size is small, however, a defective unit will cause disruption in the production process. This discovery is actually advantageous and allows problems to be solved. This is a basic concept of the implementation process. To ensure that production is not disrupted, quality must be extremely high.

Some JIT advocates have argued that a system should be either JIT or MRP, not both. Karmarkar [23], however, argues that JIT and MRP both serve useful purposes in the appropriate settings. In continuous-flow operations, a JIT system alone may work effectively. In a repetitive, batch type of system, then MRP and JIT can work together. MRP can be used for materials coordination, materials planning, and purchasing; JIT is used to control the flow on the shop floor.

Implementation Issues

As with any new system, implementation must be done purposefully so that success is assured. We cannot just announce that JIT will be implemented, today, and hope that the workforce, supervisors, and managers will embrace it. Implementation of JIT is accomplished through a continuous improvement process. Lot sizes are reduced, problems must be located, quality must be improved, and setup times and costs must be reduced to make the smaller lots economically feasible.

Workforce management is critical to the success of JIT. Education and training are required to ensure that all understand the objectives of the JIT system. Improved skills will allow workers to improve the process in ways that are not readily noticed by supervisors and managers. An important aspect is to allow workers to bring forth observations and improvements in the manufacturing process.

JIT Purchasing

A popular strategy for implementing JIT is to start the process within a manufacturing facility and reduce inventory and speed the flow through the plant. At some point it becomes obvious that those who supply materials to the plant must participate in the process to gain the full advantage. If large batches arrive at the dock only to be moved one by one through the plant, total inventory may not decrease much; the raw material and purchased parts inventory is still large. The quality imperative may not be forced upon suppliers because there still may be defective units in the large batches. Defective units that come out of the stockroom are then disruptive to the production process.

A now-common approach is to work back to the suppliers to develop JIT purchasing. Suppliers are instructed to deliver small batches more frequently than before. If the supplier is local, daily batches may be appropriate. To have a supplier desire to deliver in a JIT manner, a firm should have a requirements schedule that is relatively certain for some period into the future. Suppliers can then schedule their production to match the requirements of the buyer.

Some firms have implemented programs known as "dock-to-stock" or "dock-to-line" plans to bypass the usual stockroom-inspection process to speed the delivery of products to the shop floor. The supplier delivers a shipment that is immediately delivered to the shop floor in appropriate containers. The stockroom only verifies the units delivered; it does not inspect them.

Suppliers need to meet certain certification requirements before they can deliver in this format. A key component is the ability to have close to perfect quality. A monitoring program in conjunction with a supplier will ensure that the supplier is shipping only good products. Obviously, if there is no inspection in the stockroom, then the quality must be high when delivered directly to the shop floor. Participation in this type of program comes only after a supplier is certified to be able to do.

The benefits of JIT are significant to both the supplier and the buyer. The buyer gets to move the product directly onto the shop floor for immediate use, reducing the amount of stockroom space needed. Quality must be high, and the supplier is considered a partner with the firm and may be first to be considered for supplying other parts and future contracts.

When a supplier takes on the JIT philosophy, a JIT program will exist in that plant. When JIT was first introduced in the United States and firms were asking suppliers to deliver on a daily basis, some suppliers did not really undertake JIT. Instead, they merely stockpiled inventory in a convenient location (some were known as JIT warehouses) so that delivery could be made faster and in small lots. This, of course, is not really a JIT system. A supplier should undertake JIT throughout an operation so that the benefits can be realized there also. Who pays the supplier's costs? The buyer pays, and if the supplier can reduce waste and cost, then the reduced costs can be passed to the buyer through reduced prices.

Expected Outcomes

Successful JIT implementation should result in significant benefits to a business. When first considered, the primary benefit was reduced inventory levels in work-in-progress inventory between work centers. This reduced inventory should improve quality in production processes and materials and cut down on line stoppages. Reduced inventory and improved quality will result in reduced cycle time. Less space will be needed on the shop floor for inventory, so new layouts that cut down on the total amount of floor space can be designed. If JIT is pushed into the supplier stream, there will be reduced inventory in the stockrooms.

All these actions result in reduced waste and inventory holding costs, which may make a firm more competitive. Smaller lot sizes will allow more responsiveness to customer needs; special orders can be accommodated without significant setup cost. With smaller setup times, machines can be switched to produce another product with little disruption to the shop floor. The reduced cycle times should allow firms to produce custom orders faster and should give a firm a competitive advantage.

Consider the following highway analogy when thinking about JIT. Say that a line of cars is traveling at 65 mph but are spaced at six and a half car lengths. Consider the space between the cars to be inventory. If the first car in line slows down a little to look at the scenery, the driver of the second car may see this slowdown and put on the brakes. The space between the first two cars (inventory) may be consumed, but a chain reaction occurs back through the entire chain of cars when a red brake light flashes on. Some cars will slow down or speed up in wild reaction, but the drivers do not need to be constantly vigilant because the next car is a safe distance away. This illustrates lots of inventory, but also potential swings in speed. More importantly, however, a driver may not frequently check the condition of the car's brakes and taillights because there is a wide safety margin in this type of driving.

During rush hour in a city, however, cars will travel at 65 mph but will be within two feet of each other! This forces drivers to be alert at all times. Any noticeable decline in braking power will force drivers to go immediately into their closest brake shop for repair or else court disaster on the highway. The driver at the head of the line quickly learns (through flashing headlines or beeping horns) that slowing down to gaze at the scenery is not appreciated during rush hour. The result is a constant speed (level production load) for the first car (finished goods), but the cars will be able to drive down the highway safely as long as all cars are maintained the drivers are alert. The distance be-

TABLE 13.14   Classification of SFC Techniques

	Make-to-Stock	Make-to-Order
High volume	Assembly line balancing Process planning techniques Just-in-time	Intermittent flow shops scheduling techniques
Low volume	ROT, ART Batch/intermittent processing techniques	Job shops planning and control techniques

tween the cars (inventory) is reduced significantly so that drivers will get home sooner than if all drivers maintained six and a half car lengths.

SUMMARY

In this chapter we discussed the functions of shop floor control in a make-to-stock manufacturing environment. Table 13.14 shows the useful techniques according to manufacturing environment, which can be broadly classified as continuous flow shops, intermittent flow shops, and job shops. Controlling a flow shop involves successfully manipulating cycle time, number of stations, and idle time without violating precedence relationships.

Just-in-time (JIT) systems have been introduced in this chapter as another means to control the shop floor. Implementation of JIT requires significant time and effort to ensure success, but many benefits have been reported. Inventory can be reduced, quality can be improved, and waste can be reduced through JIT.

We also dealt with the ROT and ART methods for scheduling and sequencing intermittent or batch processing environments. As these plans are being implemented, there may be cases where wide fluctuations in demand occur for certain items, requiring adjustments to the run quantities. It is also important to realize that planning and control of flow shop activities are integral parts of aggregate planning and master scheduling. The management of make-to-order systems, known as job shops, is discussed in later chapters.

PROBLEMS

1. Rago, Inc., manufactures radios. The company wishes to make approximately 290 radios per day. The following table lists all basic tasks performed along the assembly line:

Task	Operation Time (min.)	Immediately Preceding Tasks
A	0.20	—
B	0.30	A
C	0.40	A
D	0.60	B, C
E	0.80	—
F	0.80	E
G	0.60	D, E

H	1.20	F, G
I	1.20	H
J	1.00	I
K	1.90	J

The shop operates five days per week and two shifts per day. The company provides two coffee breaks of ten minutes each during an eight-hour day.

a. Group the activities into the most efficient arrangement.

b. What is the cycle time of your arrangement?

c. What is the balance delay percentage?

2. The Easy Corporation manufactures light fixtures. Subassemblies are produced and packed in cartons. The durations of operations and their sequence are given in the following table:

Task	Operation Time (min.)	Immediately Preceding Tasks
A	1.1	—
B	0.4	A
C	0.5	A
D	1.2	A
E	1.1	A, D
F	0.3	B, C
G	0.4	D
H	1.1	F
I	0.8	E, G
J	0.7	H, I
K	0.3	J
L	1.2	K

Draw the precedence diagram for the assembly. What is the total time of manufacture for this light fixture?

3. The facility given in problem 2 is operating on an eight-hour, two-shift, five-day schedule. Each shift consists of two ten-minute coffee breaks. The assembly schedule calls for approximately 700 units per day. If the shop operates at 95% efficiency, how many workstations should we need to obtain the desired production rate? Group the activities into appropriate workstations using the trial-and-error technique. Calculate the balance delay.

4. The following table exhibits the assembly time assigned to the respective workstations. The efficiency percentages of workers presently allocated to the corresponding workstations are also exhibited.

Station	Production Time (sec.)	Efficiency Percentage	Operator
1	25	95	A
2	30	100	B
3	20	90	C
4	35	110	D
5	25	90	E

a. What is the cycle time of this layout?

b. If the shop operates an eight-hour shift, five-day work schedule, what is the average

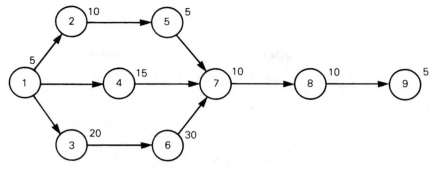

FIGURE 13.5   Precedence diagram for problem 5.

hourly production rate? Assume that the company provides two fifteen-minute rest periods every day.

   c. What is the balance delay percentage?
   d. If these workers can be assigned to any station without any loss of efficiency, can you increase the production rate by switching them among stations?
   e. What is your new hourly production rate, and what percentage productivity increase was accomplished?

5. The necessary activities and their durations in seconds required to assemble the Krown Blunder are shown in Figure 13.5. Given a cycle time of thirty seconds, balance the assembly line stations using the trial-and-error technique. How many stations would you have in your layout? Assuming that the company provides two fifteen-minute breaks during an eight-hour shift, what production rate would you expect from the assembly line? Would your answer vary if the efficiency of the workers is stated as 95%?

6. Find a five-station balance for the assembly line shown in Figure 13.6, using the trial-and-error technique. (The performance times for the tasks are given above the nodes.) What is the balance delay percentage in your layout?

7. Find a five-station balance for the assembly line shown in Figure 13.7, using the trial-

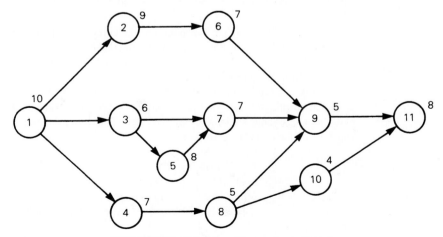

FIGURE 13.6   Precedence diagram for problem 6.

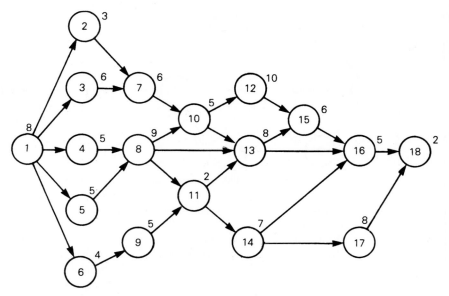

FIGURE 13.7 Precedence diagram for problem 7.

and-error technique. (The performance of corresponding tasks is also shown in the precedence diagram.)

8. Taylor Company needs to produce items A and B. The production times, lot sizes, current inventories, and weekly forecasts are as follows:

Item	Production Time (hrs.)	Current Inventory	Weekly Forecast	Lot Size
A	0.25	100	60	200
B	1.00	50	40	100

Assuming that 120 hours are available for scheduling these items, find the production quantities for items A and B using the runout time method.

9. Ignoring the lot sizes and using the aggregate runout time method, find the production quantities for items in problem 8.

10. Barry Mojena Corporation bottles three different types of wines. Since retailers order them periodically, they are made to stock for satisfying the demands. The company operates two eight-hour shifts, five days per week. The filling times, current inventories, and demand forecasts are as follows:

Type, Size	Filling Time per 100 Bottles (min.)	On-hand Inventory (in thousands)	Lot Sizes (in thousands)	Demand Forecast per Week (in thousands)
Rosé, 16 oz.	8	70	40	20
Rosé, 24 oz.	10	60	40	30

Chablis, 16 oz.	8	50	40	35
Chablis, 24 oz.	10	40	40	25
Chablis, 1 qt.	12	30	40	20

Using the runout time method, schedule the winery for bottling these varieties.

11. If the Barry Mojena winery has only 330 hours of filling time left next week, determine the sequence for bottling these wines, using the aggregate runout time method.

12. White Radar Company produces components for electronic equipment. These components are made to stock to meet anticipated customer demands. Currently, sixty-six hours of machine capacity are available for the next week. Using the aggregate runout time method, establish the sequence for the given data:

Items	Machine Hours per Unit	Lot Size	On-hand Inventory	Demand Forecast
W	0.07	300	240	200
X	0.10	200	225	150
Y	0.05	500	400	250

Calculate the projected inventory at the end of two weeks.

13. Suppose that White Radar Company has only forty hours of machine time available in the planning period and that lot sizes can be modified without any sacrifice in capacity. Assuming that the data given in problem 12 are still valid, establish the sequence of items to be produced, using the aggregate runout time method. Project the anticipated inventories at the end of two weeks.

14. The following formula establishes the economical production run quantities when a family of items is to be produced:

$$EOQ = \sqrt{\frac{2 \Sigma A \Sigma S}{(1 - \Sigma d / \Sigma p)}}$$

where ΣA = sum of yearly dollar volumes of all items in the family
$\quad \Sigma S$ = sum of all setup times in dollars
$\quad \Sigma d$ = sum of daily usage in dollars for all items
$\quad \Sigma p$ = sum of daily production rates in dollars for all items

Given the following data, find the dollar values of items to be produced:

Items	A (in dollars)	S (in dollars)	d (in dollars)	p (in dollars)
1	3,000	3	12	90
2	6,000	5	20	120
3	1,200	5	6	110
4	15,000	3	60	100
5	1,500	3	5	95

Assume that the cost of a major setup is $100.

15. Assume that demand for a certain product is 3600 per year, holding cost is $20 per unit, setup cost is $1000, and there are 250 days worked per year.

    a. Use the EOQ approach to determine the order quantity and total cost.

    b. The objective is to reduce the setup cost to a point that results in one week's worth of demand in each lot size ordered. What setup cost is required, and what is the total annual cost?

    c. What would you be willing to spend each year to reduce the setup from the original to that found in part b?

16. Refer to the problem illustrated in Table 13.11. We will cut the setup cost to one-fourth of the original setup cost for each product. Determine the number of cycles per year that would result in the lowest cost.

17. Three products X, Y, and Z have daily demands of 120, 200, and 360. Develop a cycle of production.

REFERENCES AND BIBLIOGRAPHY

1. A. L. Arcus, "COMSOAL: A Computer Method of Sequencing Operations for Assembly Lines." In Elwood S. Buffa, ed., *Readings in Production and Operations Management* (New York: John Wiley & Sons, 1966).

2. A. Bensousson, E. Hurst, and B. Nasland, *Management Applications of Modern Control Theory* (New York: North Holland-American Elsevier, 1974).

3. P. J. Billington, "The Classic Economic Production Quantity Model with Setup Cost as a Function of Capital Expenditure," *Decision Sciences,* Vol. 18, No. 1 (1987), pp. 25–42.

4. E. S. Buffa and J. G. Miller, *Production-Inventory Systems: Planning and Control,* 3rd ed. (Homewood, Ill.: Richard D. Irwin, 1979).

5. R. B. Chase and N. J. Aquilano, *Production and Operations Management,* 6th ed. (Homewood, Ill.: Richard D. Irwin, 1992).

6. E. M. Dar-El (Mansoor), "MALB—A Heuristic Technique for Balancing Large-Scale Single Model Assembly Lines," *AIIE Transactions,* Vol. 5, No. 4 (December 1973), pp. 343–356.

7. E. M. Dar-El, "Mixed Model Assembly Line Sequencing Problems," *OMEGA,* Vol. 6, No. 4 (1978), pp. 313–323.

8. S. Elmaghraby, "The Economic Lot Scheduling Problems (ELSP): Review and Extensions," *Management Science,* Vol. 24, No. 6 (February 1978), pp. 587–598.

9. D. W. Fogarty, J. H. Blackstone, Jr., and T. R. Hoffmann, *Production and Inventory Management,* 2nd ed. (Cincinnati: South-Western, 1991).

10. N. Gaither, *Production and Operations Management,* 5th ed. (Fort Worth, Tex.): Dryden Press, 1992).

11. Walter E. Goddard, *Just-in-Time: Surviving by Breaking Tradition* (Essex Junction, Vt.: Oliver Wight Limited Publications, 1986).

12. W. A. Gruver and S. L. Narasimhan, "Optimal Scheduling of Multistage Continuous Flowshops," *INFOR,* Vol. 19, No. 4 (November 1981), pp. 319–330.

13. L. Gutjahr and G. L. Nemhauser, "An Algorithm for the Line Balancing Problem," *Management Science,* Vol. 11, No. 2 (November 1964), pp. 308–315.

14. P. G. Gyllenhammer, "How Volvo Adapts Work to People," *Harvard Business Review,* Vol. 55, No. 4 (July-August 1977), pp. 102–113.

15. R. W. Hall, *The Implementation of Zero Inventory/Just-in-Time* (Falls Church, Va.: American Production and Inventory Control Society, 1988).

16. A. Hax and H. Meal, "Hierarchical Integration of Production Planning and Scheduling." In

M. Geisler, ed. *TIMS Studies in the Management Sciences, Logistics* (Amsterdam: North Holland, 1975).

17. M. Held, R. M. Karp, and R. Sharesian, "Assembly Line Balancing—Dynamic Programming with Precedence Constraints," *Operations Research,* Vol. 11, No. 3 (May–June 1963), pp. 442–459.

18. W. B. Helgeson and D. D. Birnie, "Assembly Line Balancing Using the Ranked Positional Weight Technique," *Journal of Industrial Engineering,* Vol. 12, No. 6 (November–December 1961), pp.394–398.

19. T. R. Hoffmann, "Assembly Line Balancing with a Precedence Matrix." *Management Science,* Vol. 9, No. 4 (July 1963), pp. 551–562.

20. E. J. Ignall "A Review of Assembly Line Balancing," *Journal of Industrial Engineering,* Vol. 16, No. 4 (July–August 1965), pp. 244–254.

21. J. R. Jackson, "A Computing Procedure for a Line Balancing Problem," *Management Science,* Vol. 2, No. 3 (April 1956), pp. 261–271.

22. L. A. Johnson and D. C. Montgomery, *Operations Research in Production Planning, Scheduling and Inventory Control* (New York: John Wiley & Sons, 1974).

23. U. Karmarkar, "Getting Control of Just-in-Time," *Harvard Business Review* (September–October, 1989), pp. 122–131.

24. M. D. Kilbridge and L. Wester, " A Heuristic Method of Assembly Line Balancing," *Journal of Industrial Engineering,* Vol. 12, No. 4 (July–August 1961), pp. 292–298.

25. M. D. Kilbridge and L. Wester, "The Balance Delay Problem," *Management Science,* Vol. 8, No. 1 (October 1961), pp. 69–84.

26. M. D. Kilbridge and L. Wester, "A Review of Analytical Systems of Line Balancing," *Operations Research,* Vol. 10, No. 5 (September–October 1962), pp. 626–638.

27. M. Klein, "On Assembly Line Balancing," *Operations Research,* Vol. 11, No. 2 (March–April 1963), pp. 274–281.

28. L. J. Krajewsky, and L. P. Ritzman, *Operations Management,* 2nd ed. (Reading, Mass.: Addison-Wesley, 1990).

29. E. M. Mansoor (Dar-El), "Assembly Line Balancing—An Improvement on the Ranked Positional Weight Technique," *Journal of Industrial Engineering,* Vol. 15, No. 2 (March–April 1964), pp. 73–77.

30. A. A. Mastor, "An Experimental Investigation and Comparative Evaluation of Production Line Balancing Techniques," *Management Science,* Vol. 16, No. 11 (July 1970), pp. 728–746.

31. W. L. Maxwell, "The Scheduling of Economic Lot Sizes," *Naval Research Logistics Quarterly,* Vol. 11 (1964), pp. 89–124.

32. J. O. McClain and L. J. Thomas, *Operations Management* (Englewood Cliffs, N.J.: Prentice Hall, 1980).

33. R. J. Paul, "A Production Scheduling Program in the Glass Container Industry," *Operations Research,* Vol. 27, No. 2 (March–April 1979), pp. 290–302.

34. G. Plossl and O. Wight, *Production and Inventory Control* (Englewood Cliffs, N.J.: Prentice Hall, 1967).

35. E. L. Porteus, "Investing in Reduced Setups in the EOQ Model," *Management Science,* Vol. 31, No. 8 (1985) pp. 998–1010.

36. W. A. Sandras, Jr., *Just-in-Time: Making It Happen* (Essex Junction, Vt.: Oliver Wight Limited Publications, 1989).

37. N. A. Schofield, "Assembly Line Balancing and the Application of Computer Techniques," *Computers and Industrial Engineering,* Vol. 3 (1979), pp. 53–69.

38. Richard J. Schonberger, "Some Observations on the Advantages and Implementation Issues of

Just-in-Time Production Systems," *Journal of Operations Management,* Vol. 3, No. 1 (November 1982), pp. 1–11.

39. S. P. Sethi and G. L. Thompson, *Optimal Control Theory: Application to Management Science* (Boston: Martinus-Nijhoff, 1981).

40. H. A. Simon and A. Newell, "Heuristic Problem Solving: The Next Advance in Operations Research," *Operations Research,* Vol. 6, No. 1 (January–February 1958), pp. 1–10.

41. M. K. Starr, *Operations Management* (Englewood Cliffs, N. J.: Prentice Hall, 1978).

42. N. T. Thomopoulos, "Some Analytical Approaches to Assembly Line Balancing Problems," *Production Engineer* (July 1968), pp. 345–351.

43. N. T. Thomopoulos, "Mixed Model Line Balancing with Smoothed Station Assignments," *Management Science,* Vol. 16, No. 9 (May 1970), pp. 573–603.

44. F. M. Tonge, "Summary of a Heuristic Line Balancing Procedure," *Management Science,* Vol. 7, No. 1 (October 1960), pp. 21–39.

45. F. M. Tonge, "Assembly Line Balancing Using Probabilistic Combinations of Heuristics," *Management Science,* Vol. 11, No. 7 (May 1965), pp. 727–735.

46. T. E. Vollman, W. L. Berry, and D. C. Whybark, *Manufacturing Planning and Control Systems,* 3rd ed. (Homewood, Ill.: Richard D. Irwin, 1992).

47. L. Wester and M. Kilbridge, "Heuristic Line Balancing: A Case," *Journal of Industrial Engineering,* Vol. 13, No. 3 (May–June 1962), pp.139–149.

APPENDIX 13A

Case in Point: Buick City Genuine JIT Delivery
Mehran Sepehri

Buick City marks the most extensive application of Just-In-Time (JIT) parts delivery in the U.S. to date. Buick City is General Motors' transformation of its Flint, Michigan production facility into a state-of-the-art automobile assembly complex. Buick City's goal is to become a competitive assembler of world-class products that will provide complete customer satisfaction. A radically-different management approach is utilized to manufacture cars with minimum inventory and manufacturing waste. The highlights of this approach include:

- A Just-In-Time program for parts deliveries that results in in-plant inventory levels ranging from less than an hour to a maximum of 16 hours' worth of parts;
- Separate delivery schedules for more than 4,000 parts from a total of more than 600 suppliers;
- The use of 85 point-of-use receiving dock doors, resulting in dock-to-line delivery distances of no greater than 300 feet;
- The use of returnable containers for all parts used in the plant;
- Auto assembly through "synchronized production," which minimizes inventory levels and makes maximum use of floor space;

Reprinted from P&IM Review with APICS News Vol. 8, No. 3 (March 1988), pp. 34–36.

- The use of a "flow-through" parts terminal which will sort incoming parts before moving them to the main plant;
- State-of-the-art materials handling equipment, including an assembly AGVS and a total of 222 handling and manufacturing robots; and
- Relationships with suppliers and union members that are a drastic change from historical relationships in the U.S. auto industry.

The JIT program entails establishing separate delivery schedules for more than 4,000 individual parts, considering their pick-up and transit times. There is a 20-minute window during which a particular part must be delivered, otherwise the line may be shut down.

Buick City contains about 1.80 million square feet of floor space, which is about half the size of other comparable GM plants. Approximately 70 percent less floor space in Buick City is dedicated to materials storage than the typical auto plant. Major reductions are accomplished in labor costs of materials, damage to parts, and repair operations. The most important benefit has been improvement in quality. The operators are responsible for doing their jobs correctly while the work is in their particular work station. There are no repair people at the end of the line.

Buick City contains 85 dock doors spread around the perimeter of the complex. The approach is to install docks as close as possible to the point where the parts will be used in the manufacturing process. Using synchronized production at Buick City, many parts, including metal stampings, are delivered to the line in the order that they are added to the car. Since suppliers are electronically linked to the scheduling computer, each part supplier knows the build sequence for the next 10 days (the first five days of which are frozen). The facility uses rail as well as truck deliveries. Robotic unloading is used in specific areas.

Robots are used to load and unload trucks at docks for seats, transmissions, and engines. At one dock that robotic loading and unloading is used, a robot interfaces with an overhead conveyor after unloading seats to their point of use on the assembly line. A second robot is used to transfer empty seat containers from the conveyor to a truck after the empties are cycled back to the dock.

Most containers are bar coded to decrease the opportunity for error in data entry.

Bar coding is a basic premise that much of JIT is built on. It was actually decided as a milestone to implement JIT and automation. All containers and most of the components are bar coded for automatic receiving and dispatch. Bar coding is coordinated with the suppliers so it is readable by both the factory and its suppliers. Most of the automation equipment was bought from a major U.S. manufacturer, but were programmed and integrated in-house.

Buick City is using seven different standard containers, and is headed toward 100 percent utilization of returnable containers. Due to the deregulation of trucking and railroads in 1980, the parts are delivered more efficiently and inexpensively than before.

The heart of automation and Just-In-Time in Buick City is a sophisticated computerized control system. The system receives real-time information about assembly equipment, including the robots, parts movement, and operational performance. The system is also connected to Material Requirements Planning (MRP) and inventory control systems at the factory. The system recalculates and submits orders to automated equipment continuously. A centralized control room is manned to oversee the performance of the factory. The control room can override the computerized control system and dispatch repair or assistance crews to operations.

General Motors undertook a major training program in an attempt to educate Buick City's suppliers and workforce. In a three-phase supplier program, the philosophy and objectives of the program were discussed; then, the suppliers prepared plans to accomplish those objectives; finally, GM representatives traveled to the suppliers' factories and aided in the implementation of improvement plans. The majority of the suppliers were supportive of the project, due to their interest in establishing and maintaining a long-term relationship with GM and its JIT plan. Each employee also received 120 hours of training. Through training and communication, all employees were made to feel a part of the team in the "team concept" of auto manufacturing. Managers were also trained since the JIT approach was completely new to them, too.

There's a sense of mission and firm determination to succeed. Thus, no more "business as usual." The Buick City concept utilizes state-of-the-art technology, banks heavily upon creative and innovative methods, and fosters a high level of employee involvement in decision-making.

The success of the Buick City concept depends heavily upon the maintenance of very high quality standards and the elimination of waste. Buick City management has identified 11 areas which it regards as vital to Buick City's success.

The first, **Employee Involvement,** is the underlying catalyst for both the long-and short-term success of Buick City's assembly approach and divisional viability. Key elements of employee involvement include joint union-management planning, early training programs for start-up, ongoing training within Buick City, and improved communication.

The second area, **Material Management,** is predicated on Just-In-Time management systems. The absence of significant inventory levels requires high-quality performance as well as reduction of the need for non-value-added effects and floor space. The ability to implement a JIT program is made possible by Buick City's location relative to its suppliers.

The Buick City concept will allow a consolidation of organizations within the plant, all under one plant manager. This consolidation will increase the **Span of Control**, the third area vital to Buick City's success. An improvement in communication is expected with the realization of this type of organization.

Synchronized Production, the fourth area, means having each operation build or fabricate at the same rate and in the same sequence each day, thus reducing inventory.

Plus or minus Zero Performance to Schedule (the fifth area) means producing on each shift, on each operation, the exact quantity that has been scheduled. For example, if the plant build is 600 jobs per shift, then each operation from blank to finish should build 600 units.

The sixth area, **Efficient Facility Utilization,** is the ability to use every square foot of space to its maximum efficiency.

Planned Maintenance, the seventh area, not only takes advantage of scheduled maintenance during down periods, but also includes an attitude that doesn't tolerate breakdowns.

The eighth area, **Problem Visibility Resolution,** is more readily accomplished when inventories are maintained at low levels. The ability to cover up problems is virtually eliminated.

Small Specialized Work Units (area nine) provide an identity with the specific system or sub-assembly, providing ownership of performance and helping foster a team spirit.

Minimum Set-Up Requirements (the tenth area) allows not only the ability to easily accommodate product model mix, but also provide reduced run size because the tooling change is easily accomplished.

The eleventh area, **Process Improvements,** will be continuous. The first day of production was the first day to further eliminate waste. Far more than a manufacturing and assembly operation, Buick City is a concept of total involvement of production processes, the materials utilized, and the people who work there.

Chapter Fourteen

Job Shop Production Activity Planning

INTRODUCTION

The job shop is the most widely used production organization. It represents a variety of industries that produce goods as well as service industries such as restaurants, repair shops, hospitals, and colleges. In high-volume production systems, idle facility time and excessive in-process inventories can be controlled by the proper use of assembly line balancing techniques. In a job shop, however, every order is different, and each job could be unique. Workstations are grouped together according to their functions. The job shop production control system should schedule the incoming orders in a way that does not violate the capacity constraints of individual workstations or processes. The system should check the availability of materials and tools before releasing an order to departments. It should establish milestones or due dates to measure progress against need dates and lead times for each job. It should check work in progress as the jobs progress through the shop and should provide feedback on plant and production activities. The system should also provide work efficiency statistics and should collect operator times, thus satisfying payroll and labor distribution objectives.

In general, an effective production activity control system should cover many activities, such as planning, executing, monitoring, and control, as shown in Figure 14.1. In doing so, the system interfaces with the overall objectives of the production plan, the inventory systems, and other activities in manufacturing.

In this chapter we see how various functions of the production activity system are executed as the job shop manager attempts to run the shop in a balanced and efficient manner, like assembly line production. If the system is unable to execute the overall objectives, what choices do we have? What are the possible corrective actions we can take to assure that the job is completed on time? To answer these and other questions, we organize this subject into three stages: (1) planning, (2) execution, and (3) monitoring and

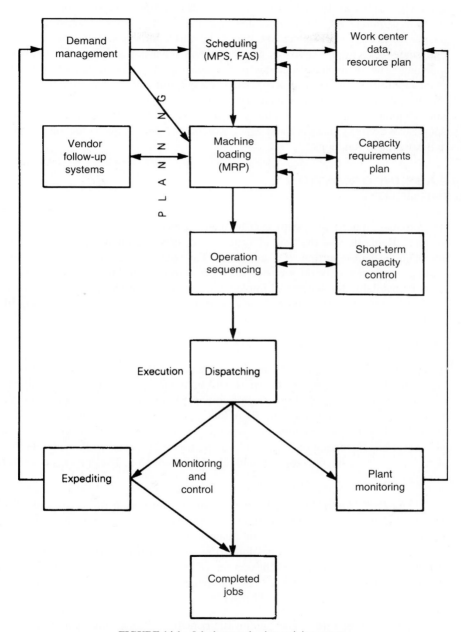

FIGURE 14.1 Job shop production activity system.

control. The planning stage consists of scheduling, machine loading, and operation sequencing. (In Figure 14.1) comparable activities in a high-volume production system are given in parentheses.) This chapter discusses the planning stage of the system, which interfaces with demand management, material control, and capacity management. The second stage—execution—involves the production releasing activity, also known as dispatching. The final phase includes production reporting and status control as well as plant monitoring and control activities. The execution and control phases of the system are the subject of the next chapter. To fully understand how these functions are carried out in a job shop, it is important for us to know what information is available from which sources. We briefly discuss these aspects before we delve into the details of production activity system functions.

DATABASE INFORMATION REQUIREMENTS

The production activity system may be manual, mechanized, or a combination of these systems. Regardless of the type of system used, it is important to set up the framework for planning and reporting. We are all familiar with the expression GIGO (garbage in, garbage out). Not only do we need *accurate* information, but *all* relevant information must be available for successful implementation of any production activity control system. The production system database can be broadly categorized into planning files and control files [1].

The Planning Files

The planning files consist of (1) the part master file, (2) the routing file, and (3) the work center master file. Each work center should be uniquely identified (e.g., lathe #15), and its capacity should be given in terms of units per hour or per day, its utilization factor in terms of periods it is active, and its output efficiency in comparison to specified time standards. Routing information indicates the plan of an order's flow through the shop. It also describes the operation that should be performed on each work center and the standard for how long the operation should take for completion. The contents of each file are as follows.

The Part Master File.   The purpose of the part master file is to store all relevant manufacturing and inventory data related to a part or item in a single location. It contains the part number and description; the manufacturing lead time; and on-hand, allocated, and on-order quantities. It also has lot-sizing requirements. As explained in Chapter 11, the allocated quantities refer to the amount of material assigned to a specific future production order but not yet cashed. The lot-sizing requirements are determined by one of the lot-sizing techniques described earlier.

The Routing File.   The routing file contains all important operations required to manufacture an item. Each record consists of a specific operation number, operation description, setup hours, run hours, and codes to identify the type of operation. A separate record is kept for each operation, and all are arranged in the exact sequence in which the item is produced. Special instructions, such as alternative operations and subcontracting, may also be included in the data file. In determining how elaborately defined a file should be, keep in mind that a well-defined manufacturing process is a potential control application, even if only in the form of a simple routing.

The Work Center Master File.   All data relevant to a work center, such as capacity, efficiency, utilization and so forth, are contained in the work center master file. Every work center is uniquely identified. The capacity data include the number of shifts worked per week, the number of machine hours utilized per shift, and the number of labor hours used in the work center. The master file also contains the average queue time of the work center and an identification of alternative work centers if the specific work center becomes overloaded.

The Control Files

The control files perform both monitoring and control functions once the job is released for production. They measure the actual progress made against the plan. The control file consists of all information pertaining to a particular order: (1) the shop order master file and (2) the shop order detail file. The contents of these files are as follows.

The Shop Order Master File.   All data related to a shop order, such as order quantity, reference data, priority, status information, and cost data, are stored in the shop order master file. Each shop order is uniquely identified in the file. Included in the file are total number of units completed, quantities scrapped, and actual quantity disbursed for any given order quantity specific to the order. The shop order master file also contains the due date on which the order is scheduled for completion. Of course, MRP systems may revise these data files during replanning. Priority may be specified in terms of a critical ratio. In case of shortages of raw material or interruption of operation, the balance due will also be recorded. The balance due quantity obviously affects the quantity disbursed, unless actions such as alterative routing or subcontracting are taken.

The Shop Order Detail File.   All information that pertains to the planning, scheduling, actual progress, and priority related to an operation of a shop order is contained in the shop order detail file. This file is similar to the master file except that the master file data refer to the entire order, whereas the detail file consists of all information necessary for each operation. Each operation is uniquely identified by a number. The progress data include reported actual setup hours, run hours and quantity disbursed as each operation is recorded. The detail file also includes the operation due date or the lead time remaining prior to completion of the operation, which permits dynamic calculation of the critical ratio.

The Shop Planning Calendar

When the production control system is computerized, use of the regular Gregorian calendar creates scheduling difficulties, since months have varying numbers of days and the pattern of holidays is irregular. Therefore, it becomes necessary to develop a shop planning calendar. This calendar is also referred to as the *shop calendar* or the *planning cal-*

endar. It consists of two-digit week designations (00–99) and three-digit day designations (000–999), resulting in a 100-week or 1000-day scheduling horizon, as exhibited in Figure 14.2.

SCHEDULING

The purpose of scheduling is to optimize the use of resources so that the overall production objectives are met. In general, scheduling involves the assignment of dates to specific jobs or operation steps. As mentioned earlier, many jobs on the shop floor compete simultaneously for common resources. Machine breakdowns, absenteeism, quality problems, shortages, and other uncontrollable factors further complicate the manufacturing environment. Hence, the assignment of a date does not ensure that the work will be performed according to the schedule. Developing reliable schedules for completion of jobs on time requires a method or discipline to determine the sequence in which scheduled work will be performed.

A good scheduling approach should be simple, unambiguous, easily understood, and executable by the management and by those who must use it. The rules should set tough but realistic goals that are flexible enough to resolve unexpected floor conflicts and allow replanning, as priorities may continually change. When people trust and use these rules, scheduling becomes a reliable and formal means of communication.

Many scheduling techniques can be employed to schedule a job shop. The type of technique used depends on the volume of orders, the nature of operations, and the overall job complexity. The selection of the technique also depends on the extent of control required over the job while it is being processed. For example, we would try to minimize or eliminate idle time in costly machine operations, and we might want to minimize the cost of in-process inventories at the same time. Scheduling techniques can generally be categorized as (1) forward scheduling and (2) backward scheduling. In practice, a combination of forward and backward scheduling is often used.

Forward Scheduling

Forward scheduling, which is also known as set forward scheduling, assumes that procurement of material and operations start as soon as the requirements are known. The events or operations are scheduled from this requirements point of view. Forward scheduling is used in many companies, such as steel mills and machine tools manufacturers, where jobs are manufactured to customer order and delivery is requested as soon as possible. Forward scheduling is well suited where the supplier is usually behind in meeting schedules. The set forward logic generally causes a buildup of in-process inventory, which cost money. Figure 14.3 provides an example of forward scheduling.

Backward Scheduling

In backward scheduling, which is also known as set backward scheduling, the last operation on the routing is scheduled first. Then the rest of the operations are offset one at a time, in reverse order, as they become necessary. Finally, by offsetting the procurement time, the start date is obtained.

Sunday	Monday	Tuesday	Wednesday	Thursday	Friday	Saturday
					1 New Year's Day	2
3	4 001	5 002	6 003	7 004	8 005	9
10	11 006	12 007	13 008	14 009	15 010	16
17	18 011	19 012	20 013	21 014	22 015	23
24 31	25 016	26 017	27 018	28 019	29 020	30

January

Sunday	Monday	Tuesday	Wednesday	Thursday	Friday	Saturday
	1 021	2 022	3 023	4 024	5 025	6
7	8 026	9 027	10 028	11 029	12 030	13
14 Valentine's Day	15 Presidents' Day	16 031	17 032	18 033	19 034	20
21	22 035	23 036	24 037 Ash Wednesday	25 038	26 039	27
28						

February

FIGURE 14.2   Shop planning calendar.

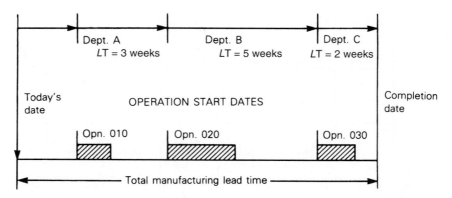

FIGURE 14.3 Forward scheduling. *Note:* The leadtimes are measured forward from today's date as pointed by the arrows.

Set backward scheduling is used primarily in assembly-type industries. After determining the required schedule dates for major subassemblies, the schedule uses these required dates for each component and works backward to determine the proper release date for each component manufacturing order. Backward scheduling minimizes the in-process inventory. It works well in the MRP environment and is used for establishing shop order start and due dates using the lead time offset. Figure 14.4 provides an example of backward scheduling.

Simple and Block Scheduling Rules

Simple scheduling rules roughly estimate the total operation times *(O)*, move times *(M)*, wait times *(W)*, queue times *(Q)*, and transit times *(T)* for each job in number of days and schedule them approximately on a calendar for computing the completion times. Exhibited in Table 14.1, these rules generally are not detailed enough to be implemented. Block scheduling rules are even worse. They estimate operation times in weeks for every department and express the scheduled completion dates in week numbers. Table 14.2

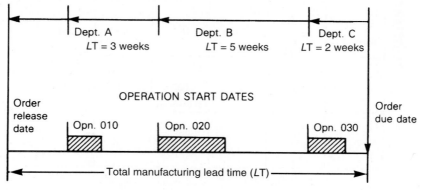

FIGURE 14.4 Backward scheduling. *Note:* The lead times are offset from the due date as pointed by the arrows.

TABLE 14.1   Simple Scheduling Rules

1. Compute operation time to the nearest day.
2. Allow one day for successive operations in the same department, two days if they are in different departments.
3. Allow two days for inspection.
4. Allow two days to withdraw material from stockroom and one day to deliver completed material to stockroom.
5. Allow five extra days for certain specified bottleneck operations.

portrays the block scheduling rules. Example 14.1 illustrates simple and block scheduling rules [13].

Example 14.1

XYZ Company is required to quote a completion time for a possible job to a preferred customer. The job has four operations, with estimated durations of operations as follows:

Operation	*Description*	*Duration (days)*
01	Make patterns	5
02	Pour casting (Bottleneck)	2
03	Grinding	15
04	Milling and assembly	10

Table 14.3 illustrates how completion times are derived. The codes indicate the type of activity. The simple scheduling rules allot a certain number of days for each activity in the department, whereas the block scheduling rules round the days into the approximate number of weeks, thus providing a much worse schedule. (Compare fifty-six days for simple scheduling versus seventeen weeks, or eighty-five days, for block scheduling rules.) Note that these rules provide examples for the forward scheduling technique. Many problems may arise while using forward or backward scheduling. For example, in forward scheduling, the completion date may fall beyond the due date; in backward scheduling, the start date may turn out to be a past date. In such cases, the schedule becomes infeasible. Also, we did not pay any attention in our scheduling to the available capacities of facilities. Such problems may be resolved by the judicious use of capacity control techniques.

TABLE 14.2   Block Scheduling Rules

1. Express operation times in weeks; combine short operations.
2. Allow one week of inspection time for piece parts, two weeks for equipment.
3. Allow one week for releasing order, withdrawing material from stockroom, and delivering completed products into stockroom.
4. Allow one extra week for certain specified bottleneck operations.

TABLE 14.3 Simple and Block Scheduling, Example 14.1

Operation	Code	Simple Scheduling		Block Scheduling	
		Days Allowed	Day Number	Weeks Allowed	Week Number
Release date			394		80
Withdraw material		2	396	1	81
Opn. 01	O	5	401	1	82
	MWQ	2	403	1	83
	I	2	405	1	84
Opn. 02	O	2	407	1	85
	MWQ	2	409	1	86
	I	2	411	1	87
Bottleneck		5	416	1	88
Opn. 03	O	15	431	3	91
	MWQ	2	433	1	92
	I	2	435	1	93
Opn. 04	O	10	445	2	95
	MWQ	2	447	1	96
Inspection and delivery	I, D	3	450	1	97
Time to complete			56 days		17 weeks

Note: O refers to the operation time, I is the inspection time, and MWQ refers to the move, wait, and queue times, as described in Chapter 15. The sequence of activities within the operation is not important because we are in the planning range.

SHOP LOADING

As orders are released to the shop according to a schedule, individual jobs are assigned to work centers. The process of determining which work center receives which jobs is known as *loading*. Loading procedures are categorized as infinite or finite. In finite loading procedures, the jobs are assigned by comparing the required hours for each operation with the available hours in each work center for the period specified by the schedule. In infinite loading, jobs are assigned to work centers without regard to capacity. The procedures for generating load reports were discussed in Chapter 12. In the job shop environment, capacity requirements and availability are rarely equal. When the number of jobs is limited, the jobs can be allocated to machines where the operations can be done more efficiently, and hence the loading becomes manageable. When the same job can be performed by many work centers, however, with varying time periods (for example, job A can be done at work center 1 in ten hours, whereas it needs only eight hours in work center 2), the choice of work centers becomes very complex. When we consider the costs of operations and available capacities, it becomes even more complex. In addition to the trial-and-error technique described in Chapter 13, graphical and analytical techniques such as charts, the index method, the assignment method, and the transportation method of linear programming can be applied. Each of these techniques is suited to particular situations.

The Gantt Load Chart

Several charts, graphs, tables, and boards are available for projecting machine center loads in a department. They can be either manual or computerized. The Gantt chart is a visual aid that is commonly used in job shops. It is also used in maintenance and service industries. Figure 14.5 shows a Gantt load chart. In the figure, the department has four machine centers, and each machine center may have more than one machine. The cumulative hours assigned to each machine center are plotted on the chart, which then displays the relative workloads in the system. When a center is overloaded, it is easy to identify the problem areas and to develop corrective action by reassigning workloads to alternative machines.

Gantt charts are simple to devise and easy to understand. Many commercial schedule boards are available, using magnetic strips, pegs, plastic inserts, and other devices. There are several limitations to the Gantt chart, however. For example, the sequence of operations is not considered in detail. Also, since the machines are grouped together, the wait times of individual jobs and the idle times of machines are not apparent. In addition, the charts do not reflect maintenance and breakdown times of machines, operator performance variables, and other details.

The Assignment Method
of Linear Programming

Consider the problem of assigning three jobs to three machines. Jobs A, B, and C, for example can be done on machines I, II, and III in 3! = 6 different ways, as exhibited in Table 14.4. Similarly, problems involving ten jobs may be done in 10! ways. These are typical combinatorial problems, and all possible solutions may need to be enumerated to find an optimal solution. Thus enumeration methods are impractical for large problems. The assignment method, which is a special case of linear programming, is better suited to such problems. The assignment method allocates jobs to work centers, workers to jobs, salespeople to territories, contracts to bidders, and the like. The procedure can minimize

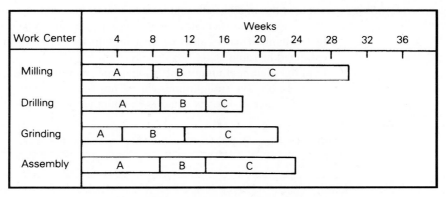

FIGURE 14.5   Gantt load chart for one department.

TABLE 14.4 Possible Ways of Assigning Three Jobs to Three Machines

Machines			Assignments			
I	A	A	B	B	C	C
II	B	C	C	A	A	B
III	C	B	A	C	B	A

such objectives as cost, time, and efficiency or can maximize profit or some other single-criterion objective functions. The assignment procedure known as the *Hungarian method* is presented here.

This solution procedure involves forming a square matrix and systematically developing relative costs of assigning projects to work centers, which is called *matrix reduction.* An optimal solution is obtained when at least one zero appears in each row and each column.

As an example, consider Andy's Auto Shop, where the manager needs to assign three different jobs to Tom, Dick, and Harry. Based on the workers' past experience, the required duration for completing a job varies among them. The manager would like to minimize the total hours spent by all three employees, since their hourly wages are the same. The effectiveness of assigning workers to various jobs can be presented in a tabular form, known as the *effectiveness matrix,* which is presented in Table 14.5. The Hungarian method transforms the effectiveness matrix into the desired matrix, with one zero in each column and each row, by adding or subtracting constants to columns or rows without changing the set of optimal assignments.

This method requires that the number of jobs equals the number of work centers. Otherwise, a dummy row or a column would be added to the matrix to make it a square matrix. A zero cost coefficient is assigned to the dummy to identify which job would be eliminated. If a particular assignment for a worker is not feasible, then either the corresponding cell would be blocked out or a high cost would be assigned. The solution procedure consists of the following three steps:

Step 1: Subtract the smallest element in each row from every element in that row. Then subtract the smallest element in each column from every element in that column. This is called the matrix reduction. The resulting matrix is exhibited in Table 14.6.

Step 2: Find the minimum number of straight lines required to cover all zeros in Table 14.6b. The straight lines can be horizontal or vertical. If the maximum number of lines is equal to the number of columns or rows in the

TABLE 14.5 Hours Required by Three Employees on Three Different Jobs

Employee	X	Y	Z
Tom	11	12	17
Dick	7	11	21
Harry	5	8	15

TABLE 14.6   Matrix Reduction, Step 2

(a) Row Reduction				(b) Column Reduction			
	X	Y	Z		X	Y	Z
Tom	0	1	6	Tom	0	0	0
Dick	0	4	14	Dick	0	3	8
Harry	0	3	10	Harry	0	2	4

given problem, then an optimal solutions has been obtained. Otherwise, go to Step 3. We need only two lines to cover all zeros in Table 14.6b. We have a situation in which we can assign any job to Tom and in which job X can be done by anyone. Suppose that we assign job X to Dick and job Y to Tom; then Harry will not have an optimum assignment. Regardless of how we reassign these jobs, the situation remains the same. We want to move any one of the zeros to another position on the matrix in order to remove this deadlock situation. Therefore, we go to Step 3.

Step 3:   Find the smallest uncovered number in the new matrix exhibited in Table 14.6b. In our example it is 2. Subtract this number from every row that is not covered with a horizontal line (see Table 14.7a). Since this gives us some negative elements in the first column, add the number to the first column. The resulting matrix is shown in Table 14.7b.

Because we need three lines to cover all zeros in Table 14.7b, the problem is solved. It must be possible to assign a zero for each row-column combination, as shown by squares in Table 14.7b. The solution to the assignment can be found by referring to the original problem and adding the corresponding times associated with the optimal solution. A total of thirty-two hours are necessary to complete all jobs. Suppose that we still needed fewer than three lines to cover all zeros in Table 14.7b, then we would repeat Steps 2 and 3 as often as necessary until an optimum solution was found.

What if you cheat in Step 2 by, for example, just drawing three horizontal lines and looking for an optimal assignment? Unfortunately, you will not find an optimal solution, which just shows that cheating does not pay! You must do it again, correctly.

There are limitations to this procedure. First, it uses a static approach. When jobs arrive continually, however, loading decisions may have to be checked continually. Also, there is no provision for checking capacity in this method; hence, the most efficient machine may be overloaded while other machines are lightly loaded. We might need a

TABLE 14.7   Matrix Reduction, Step 3

(a) Removing Deadlock			(b) Final Solution Matrix		
0	0	0	2	0	[0]
−2	1	6	[0]	1	6
−2	0	2	0	[0]	2

Gantt chart or some other mechanism to balance the load among machines. In addition, an important assumption in this method is that several choices (machines, workers, or processes) exist for each job. This may not be true in many cases, and this procedure is not very valuable when choices do not exist. Although the transportation method of linear programming can overcome these difficulties, it has very restrictive applications.

The Index Method

The index method overcomes many of the problems inherent in the assignment method. The index method is better suited, in fact, for shop loading purposes. This heuristic procedure uses opportunity cost, time, or some other single criterion as a variable. The index number represents the ratio of opportunity cost or time to the value of the best assignment.

As an example, consider a situation in which seven jobs, A to G, need to be completed. The capacity, in hours, needed for each job at every interchangeable machine center is as follows:

	Alternative Work Centers		
JOB	*1*	*2*	*3*
A	25	35	70
B	50	60	40
C	300	200	450
D	180	160	120
E	60	90	150
F	90	45	60
G	75	220	250
Available capacity	140	235	250

It is desirable to assign a job to the work center that takes the shortest amount of time or has the lowest cost. It may not always be possible to assign jobs to the most efficient work center because of capacity limitations. If we assign job A to work center 1, job B to work center 3, and so on, we would have a shortage in work centers 1 and 2 and excess capacity at work center 3, as shown in Table 14.8. The hours assigned to each machine are shown in parentheses, and the total assigned capacities are compared with the available work center capacities. Since all jobs cannot be done by the most efficient work center, some jobs need to be shifted to the next most efficient center. The steps involved in the index method are as follows:

Step 1:   Find the lowest process time for each job among the alternative work centers. Divide the lowest process time of each job into the other alternative process times and obtain index numbers. Show the index numbers next to the process times, as in Table 14.8.

Step 2:   Allocate the jobs with the lowest index numbers to the corresponding

TABLE 14.8 Calculation of Index

| | Alternative Work Centers | | | | | |
| | 1 | | 2 | | 3 | |
Job	Hours	Index	Hours	Index	Hours	Index
A	(25)	1.00	35	1.40	70	2.80
B	50	1.25	60	1.50	(40)	1.00
C	300	1.50	(200)	1.00	450	2.25
D	180	1.50	160	1.33	(120)	1.00
E	(60)	1.00	90	1.50	150	2.50
F	90	2.00	(45)	1.00	60	1.33
G	(75)	1.00	220	2.93	250	3.33
Assigned	160		245		160	
Available	140		235		250	
Excess	−20		−10		90	

work centers. If sufficient capacity exists, the problem is solved. Otherwise, go to Step 3. We already know that the available hours and assigned hours do not match in our example; therefore, we proceed in Step 3.

Step 3:   Shift some jobs to the next most efficient center—that is, to the center with the next lowest index number. After moving Job A to Center 2, we find that center 1 has adequate capacity for completing Jobs E and G, as shown in Table 14.9. Center 2 is in a worse situation, however, needing thirty-five more hours. Therefore, we repeat Step 3 to shift some jobs from Center 2 to Center 3.

We notice that Job F can be moved to Center 3, which has the next lowest index number. The solution found in Table 14.10 is feasible, and so we need not continue the

TABLE 14.9 Index Method, Step 3: Moving Job A to Center 2

| | Alternative Work Center | | | | | |
| | 1 | | 2 | | 3 | |
Job	Hours	Index	Hours	Index	Hours	Index
A	25	1.00	(35)	1.40	70	2.80
B	50	1.25	60	1.50	(40)	1.00
C	300	1.50	(200)	1.00	450	2.25
D	180	1.50	160	1.33	(120)	1.00
E	(60)	1.00	90	1.50	150	2.50
F	90	2.00	(45)	1.00	60	1.33
G	(75)	1.00	220	2.93	250	3.33
Assigned	135		280		160	
Available	140		235		250	
Excess	5		−45		90	

TABLE 14.10 Index Method, Repeat Step 3: Moving Job F to Center 3

	Alternative Work Center					
	1		2		3	
Job	Hours	Index	Hours	Index	Hours	Index
A	25	1.00	(35)	1.40	70	2.80
B	50	1.25	60	1.50	(40)	1.00
C	300	1.50	(200)	1.00	450	2.25
D	180	1.50	160	1.33	(120)	1.00
E	(60)	1.00	90	1.50	150	2.50
F	90	2.00	45	1.00	(60)	1.33
G	(75)	1.00	220	2.93	250	3.33
Assigned	135		235		220	
Available	140		235		250	
Excess	5		0		30	

iterative procedure any longer. In some instances we may be forced to move a partial load to another center. For example, suppose that we need to shift only twenty hours of job A from center 1 to center 2. The additional hours needed at center 2 can be computed as (20) (index) = (20)(1.4) = 28 hours. If a setup time is included, however, these hours would vary.

There are limitations to the index method. In complex situations, especially when several job splits are included, several iterations would be needed to complete the loading. Although this heuristic solution procedure does not always provide an optimal solution, it generally gives reasonably good loads.

SEQUENCING

We know that scheduling provides a basis for following jobs as they progress through succeeding manufacturing operations. Machine loading is a detailed capacity control technique that highlights daily or weekly overloads and underloads. We still need to determine the priorities of operations at each machine to meet schedule dates of individual jobs, however. In many factories, expediters, known as stock chasers, also follow jobs through the shop, attaching red tags to indicate urgent jobs. Expediting is the real production control system in these factories. The expediter works with the shop floor supervisor to complete jobs that are due. In due course the expediter discovers shortages and rushes important jobs through production by establishing priorities. Alas, very soon almost all jobs have red tags, and then the expediter is told what the real job priorities are! Fortunately, in most cases such situations can be prevented by using a reliable MRP system in conjunction with priority rules for sequencing jobs on the floor.

Sequencing specifies the order in which jobs should be done at each center. For example, suppose that ten patients are assigned to a medical clinic for treatment. In what order should they be treated? Should the first patient to be served be the one who arrived first or the one who needs emergency treatment? Sequencing methods provide such detailed information. These methods are referred to as priority rules for dispatching jobs to work centers.

PRIORITY RULES
FOR DISPATCHING JOBS

Priority rules are used for preparing dispatch lists of jobs or lots in job shops. They provide simplified guidelines for the sequence in which the jobs should be worked when the machine center or facility becomes available. Numerous rules have been developed; some are static and others are dynamic. The rules are especially applicable for intermittent and batch processes with independent demands. The priority rules attempt to minimize mean flow time, mean completion time, and mean waiting time and to maximize throughput and so forth. Several simulation experiments have been conducted to compare the performance of priority rules [11]. In this section we discuss some well-known rules and their effectiveness in sequencing.

Priority System

Many properties are desirable in a good priority system. The system should be relative, and it should specify the order in which the jobs should be processed—first, second, third, and so on. A priority system should be dynamic, so that the priority rule permits regular updating of priorities if necessary. This is particularly true for items with long lead times whose demand is uncertain. Finally, a priority system should truly reflect the due dates. Properly maintained MRP systems can provide priority planning information that meets these specifications and can supply valid input to a priority control system. A discussion of how an MRP system can be used in job shop situations is explored by Teplitz [30].

Priority Rules

The following priority decision rules are commonly used:

FCFS (first come, first served): By this rule, jobs are scheduled for work in the same sequence as they arrive at the facility. This is used particularly in service firms, such as automotive repair shops, barber shops, and restaurants.

EDD (earliest due date): This rule sequences the jobs waiting at the facility according to their due date, and they are processed in that order. This does not guarantee that all jobs will be completed on time, as illustrated by Example 14.2.

SPT (shortest processing time): This rule, which is also known as the shortest operation time rule, selects first the job with the shortest operation time on the machine. Many simulation experiments have demonstrated that this rule minimizes in-process inventories. This is accomplished, however, at the expense of keeping the bigger jobs longer.

LPT (longest processing time): This rule selects first the job with the longest operation time on the machine.

TSPT (truncated shortest processing time): This rule sequences jobs according to the SPT rule, except for jobs that have been waiting longer than a specified truncation time. Those jobs go to the front of the waiting line in some specified order (using, for example, the FCFS rule).

LS (least slack): This rule selects first the jobs with the smallest slack. Slack is defined as the number of days remaining before the due date minus the duration of the job.

COVERT (cost over time): This rule computes the ratio of expected delay cost *(C)* to processing time *(T)*. The job with the largest ratio is selected first.

Example 14.2

AMX Company has received the following jobs and wishes to use priority decision rules for sequencing. All dates have been translated according to shop calendar days; assume that today is day 120.

Job Number	Production Days Required	Date Order Received	Date Order Due
117	15	115	200
118	10	120	210
119	25	121	185
120	30	125	230
121	17	125	150
122	20	126	220

Assuming a five-day work week, determine the sequence in which the jobs should be performed according to each of the seven priority rules. For the TSPT rule, assume that jobs cannot be delayed more than sixty-five days.

Solution: The sequences according to the priority rules are as follows:

Priority Rule	Sequence
FCFS	117, 118, 119, 120, 121, 122
EDD	121, 119, 117, 118, 122, 120
SPT	118, 117, 121, 122, 119, 120
LPT	120, 119, 122, 121, 117, 118
TSPT	118, 117, 121, 119, 120, 122
LS	121, 119, 117, 122, 120, 118
COVERT	121, 122, 117, 118, 119, 120

The first four sequences, as given by the FCFS, EDD, SPT, and LPT rules, are obvious. For the TSPT rule, we specified that jobs cannot be delayed more than sixty-five days. If none of the jobs in the SPT sequence violates the constraint, the sequence provided by rules SPT and TSPT will be identical. Let us now check the wait time of the jobs according to the SPT sequence:

Job Sequence	Duration (days)	Date Order Received	Start Date	Wait Time (days)
118	10	120	120	—
117	15	115	130	15
121	17	125	145	20
122	20	126	162	36
119	25	121	182	61
120	30	125	207	82

The SPT sequence results in a delay of eight-two days for job 120, which is more than the sixty-five days specified. Therefore, the TSPT rule would schedule job 122 after job 120. The slack for each job is computed as follows:

Job Number	Duration (days)	Days Remaining	Slack (days)	Sequence
117	15	85	70	3
118	10	90	80	6
119	25	64	49	2
120	30	105	75	5
121	17	25	8	1
122	20	94	74	4

If, by chance, one of these jobs were already being processed, the rest of the jobs would be arranged using the least slack rule. The COVERT rule computes possible delays for individual jobs, using rules such as FCFS. If one or more jobs are delayed, the ratio is computed, as shown in Table 14.11. Then we can sequence the jobs such that the largest C/T is completed first. We have explained the concept of the COVERT rule in its simplest form. In its original version, C/T ranges from zero to one. The priority rules are fully described by Carroll [9] and Buffa and Miller [5].

TABLE 14.11   COVERT Priority Rule Ratio Calculation, Example 14.2

Job Number	Duration (T)	Completion Date	Due Date	Delay	Cost (C)	C/T
117	15	135	200	No	—	0
118	10	145	210	No	—	0
119	25	170	185	No	—	0
120	30	200	230	No	—	0
121	17	217	150	Yes	680	40
122	20	237	220	Yes	500	25

Dynamic Sequencing Rules

The following rules are often used for sequencing:

DS (dynamic slack): When the LS rule is used repeatedly at each machine center for sequencing jobs, it is known as the dynamic slack rule. This rule, however, does not consider the duration of the job.

DS/RO (dynamic slack per remaining operation): This rule computes a ratio for each job waiting. The ratio is obtained by dividing the total slack time available for the job by the number of operations remaining (including the current machine). Obviously, the job with the smallest ratio is scheduled first. The DS/RO rule suffers from the same shortcoming as the DS rule.

CR (critical ratio): This dynamic priority rule constantly updates priorities according to most recent conditions. It has been found effective in MRP for review and revision of the existing schedule. The rule develops a comparative index of any job in relation to others at the same facility. The CR is designed to give priority to jobs that have the most urgently needed work to meet the shipping schedule. The CR is the ratio of the time period left prior to the shipping date to the time period needed to complete the job:

$$CR = \frac{\text{need date} - \text{today's date}}{\text{days required to complete the job}}$$

$$= \frac{\text{days remaining}}{\text{days required}}$$

As a job gets further behind schedule, its CR becomes smaller, and jobs with low CR take precedence over others.

Example 14.3

AMX Company has a list of jobs to be started on day 358. The jobs' durations and due dates are as follows:

Job Number	Duration (days)	Due Date
150	25	360
151	17	372
152	35	367
153	19	377
154	29	370
155	10	390

Using the critical ratio technique, find the sequence of jobs to be done.

TABLE 14.12   Critical Ratio Calculation, Example 14.3

Job Number	Duration (days)	Due Date	Days Remaining	Critical Ratio	Sequence
150	25	360	2	2/25 = 0.080	1
151	17	372	14	14/17 = 0.824	4
152	35	367	9	9/35 = 0.257	2
153	19	377	19	19/19 = 1.00	5
154	29	370	12	12/29 = 0.41	3
155	10	390	32	32/10 = 3.2	6

Solution: We can calculate the days remaining by subtracting today's date from the due date. Assuming that no job has been started, we can say that the time remaining for each job is equal to its duration. If a certain portion of the work was completed on a job, we would merely subtract that amount from the duration of that job. The critical ratio is found by dividing the days remaining by the days required (see Table 14.12). Ratios range from zero to one for jobs that are behind or on schedule. Smaller ratios signify more urgent jobs. Ratios that are greater than one represent noncritical jobs. By the CR rule, the sequence of jobs is 150, 152, 154, 151, 153, and 155.

We have described many rules for dispatching jobs. These rules can be useful for generating the start dates of the jobs or operations on a particular machine as the job progresses. Obviously, dynamic rules are superior to static rules. The effectiveness of these rules in the job shop environment was studied by Conway, Maxwell, and Miller [11], Nanot [21], and others. The critical ratio technique has been the most highly acclaimed by industry practitioners, including Plossl and Wight [25], Orlicky [24], and others.

MATHEMATICAL PROGRAMMING, HEURISTICS, AND SIMULATION

Several techniques have been presented for scheduling, loading, and sequencing in job shops. We have pointed out that many jobs in job shops are unique and that they move through work centers as specified by the route sheets. The sequences of operations often are different. In many instances, however, the sequences of operations are the same, but the orders may require varying loads in different work centers. Note that the job shop is different from an assembly line environment, in which most operations, their durations, and the sequence are essentially the same. Many mathematical programming and heuristic techniques for job shop production control can be found in the literature [15], and several efficient heuristic techniques have been developed to solve special cases encountered in job shop situations. These solution techniques can be classified as (1) series machines, (2) parallel machines, and (3) series-parallel machines. They are referred to as "*N* job, *M* machine problems" or simple *N/M* problems in the scheduling and sequencing literature. We now discuss the solution procedures for some configurations that are popularly known as flow shops.

N Jobs, One Machine

This configuration consists of several jobs waiting to be processed by a single facility:

Jobs   Machine

Many of the priority rules discussed in this chapter can be used to solve problems in this category. They provide the sequence in which the jobs are to be processed by a single facility.

N Jobs, Two Machines in Series (No Passing)

In this case, each job has to go through two facilities in the same order. "No passing" means that no job is allowed to pass any other job while the first job is waiting between facilities and that no job can be started before the previous job is completed.

Jobs   Machines

Johnson's rule [16] can be used to obtain an optimal solution to this problem. The rule minimizes the total make span. Given a set of jobs and their associated operation durations in corresponding machines, Johnson's rule consists of the following steps:

Step 1: Select the shortest operation duration.

Step 2: If the shortest duration requires the first machine, schedule the job in the first available position in the sequence. If the shortest duration is on the second machine, schedule the job in the last available position in the sequence.

Step 3: Remove the assigned job from further consideration and return to Step 1 for the next job.

Example 14.4

Bracken Company tests laboratory specimens. The testing department must perform two consecutive operations for each job. The following table lists jobs and their corresponding durations of operations:

| JOB | Duration (hours) | |
	MACHINE 1	*MACHINE 2*
A	3	6
B	5	2
C	1	2
D	7	5

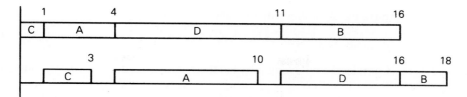

FIGURE 14.6 Make span for the optimum sequence, Example 14.4.

1. Use Johnson's rule to set the sequence of processing the shipments.
2. Use a Gantt scheduling chart to determine how much time is required to complete all jobs listed.

Solution: The shortest duration is one hour on machine 1. Schedule job C first in the sequence and remove it from further consideration. Now the shortest duration is two hours on machine 2. Schedule job B last in the sequence and remove it from further scheduling. Since the next smallest duration of three hours is on machine 1, schedule job A as the second job in the sequence. Finally, job D is scheduled in third place. We now have the optimal sequence: C, A, D. B.

The Gantt chart for the sequence is shown in Figure 14.6. The chart is constructed by scheduling job C first on machine 1. Job C on machine 2 can be started as soon as it is completed on machine 1. The succeeding job, job A, can be started on machine 1 as soon as job C is completed there. Similarly, upon job A's completion on machine 1, it can be started on machine 2, provided that the previous job has been completed. Otherwise, it will wait in the in-process inventory until the previous job is finished.

The make span, or throughput time, for these jobs is given by the schedule chart as eighteen hours. It is not hard to explain how this algorithm works. The smaller jobs are scheduled on machine 1 first, so that machine 2 is not kept waiting too long. Once the machines are busy, the situation reverses at the end of the sequence. Machine 2 has all small jobs left, which again helps complete all jobs quickly.

N Jobs, Three Machines in Series (No Passing)

The optimal solution for a general case is quite complicated. If either or both of the following conditions are met, however, the solution is given by the *N*/3 Johnson's rule [16]:

1. The smallest duration on machine 1 is at least as great as the largest duration on machine 2.
2. The smallest duration on machine 3 is at least as great as the largest duration on machine 2.

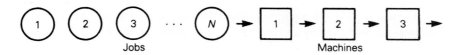

We explain the algorithm with the following example.

Example 14.5

Consider the following jobs and their processing times at corresponding machines:

| | *Duration (hours)* | | |
	MACH 1	MACH 2	MACH3
JOB	t_{i1}	t_{i2}	t_{i3}
A	13	5	9
B	5	3	7
C	6	4	5
D	7	2	6

Using Johnson's rule, find the optimal sequence.

Solution: Since both conditions of Johnson's rule are met, we can apply the algorithm. First, form a new matrix, as follows:

Job	$t_{i1} + t_{i2}$	$t_{i2} + t_{i3}$
A	18	14
B	8	10
C	10	9
D	9	8

Now, using Johnson's rule for the $N/2$ problem, we get the optimal sequence: B, A, C, D.

Essentially, Johnson's rule converts an $N/3$ problem into an $N/2$ problem, provided that certain conditions are met. Even if these conditions are not met, the rule still provides a near optimal solution, as we soon see.

N Jobs, M Machines in Series (No Passing)

When several jobs have to be processed through many facilities, finding an optimal sequence requires a combinatorial search procedure.

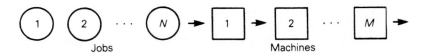

An efficient heuristic procedure, suggested by Campbell, Dudek, and Smith [8], is known as the CDS algorithm. The CDS algorithm extends the $N/3$ Johnson's rule to a general N/M problem and provides a near optimal solution.

Example 14.6

The following table provides jobs and their durations on respective machines:

| | | Duration (Hours) | | |
JOB	MACH 1	MACH 2	MACH 3	MACH 4
A	3	1	11	13
B	3	10	13	1
C	11	8	15	2
D	5	7	7	9
E	7	3	21	4

The algorithm generates $m - 1 = 4 - 1 = 3$ two-machine (M1, M2) solutions. The sequence that yields the minimum make span is chosen.

Solution 1: From the foregoing table, consider only the durations on the first and last machines:

Job	M1	M2
A	3	13
B	3	1
C	11	2
D	5	9
E	7	4

Applying Johnson's rule, we obtain the sequences A, D, E, C, B.

Solution 2: From the original data, add the durations of the first two machines for M1 and add the durations of the last two machines for M2 for every job:

Job	M1	M2
A	4	24
B	13	14
C	19	17
D	12	16
E	10	25

Again applying Johnson's rule, we obtain the sequence A, E, D, B, C.

Solution 3: In this solution, the process times for M1 and M2 are obtained from the original data by adding the first three and the last three durations, respectively, for each job. This concept can be extended to any number of machines in a series.

Job	M1	M2
A	15	25
B	26	24
C	34	25
D	19	23
E	31	28

Using Johnson's algorithm, we get the sequence of A, D, E, C, B.

We can now find the make span for the solutions generated, as shown in Figure 14.7. Since the sequences obtained in solutions 1 and 3 have the smallest make span—seventy-two hours—we would probably choose A, D, E, C, B.

N Jobs, *M* Machines in Parallel

In this configuration, the N available jobs may be processed by any one of the M machines available. A simple yet effective heuristic solution is given by the LPT rule [2].

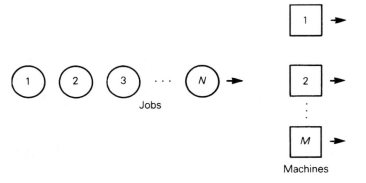

Example 14.7

The following table provides the data for an $N=8/M=3$ problem. Assign the jobs using the LPT rule.

Job	Duration
A	11
B	18
C	4
D	13
E	2
F	7
G	5
H	3

Solution: As a first step, using the LPT rule, we obtain the sequence B, D, A, F, G, C, H, E. The second step consists of assigning jobs to each machine in an order such that the

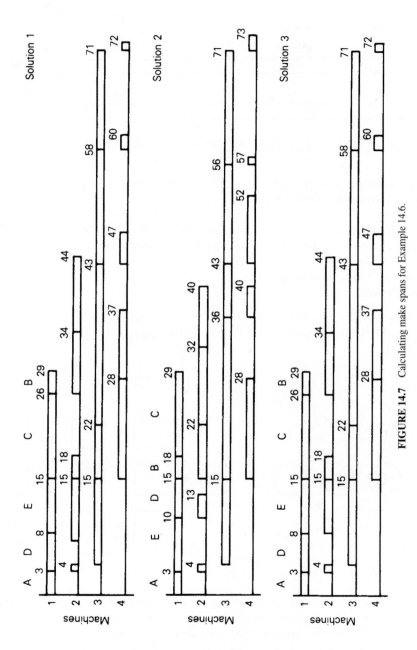

FIGURE 14.7 Calculating make spans for Example 14.6.

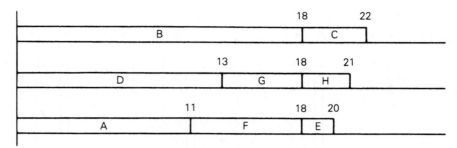

FIGURE 14.8   LPT sequence for Example 14.7.

least amount of total processing is already assigned. Ties are arbitrarily resolved. The solution is shown in Figure 14.8

    This procedure assumes that all parallel machines have identical capacities. Several elaborate algorithms for minimizing make spans are discussed in Baker [2].

N Jobs, *M* Machines in Series and Parallel

In this case, the jobs may be processed by any one of the machines at stage I. Once the operation at stage I is completed, the job is then served by any one of the machines at stage II.

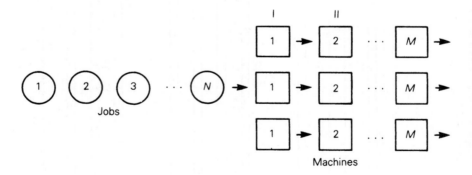

This is probably the most difficult type of scheduling problem to solve. Zangwill [33] presented a mathematical programming model to solve multiproduct, multifacility production and inventory problems. As indicated earlier, however, mathematical programming techniques for solving such problems are not yet practical. So far, computer simulation techniques have been found to be the most effective tools.

WORKER SCHEDULING
IN SERVICE OPERATIONS

Scheduling service operations poses a challenge because many of these operations serve customers six or seven days a week. The daily demand, as well as the demand during the day, could vary; this is not generally the case in manufacturing operations where the output can generally be uniform. Typically, the demands in service operations are met with full-time, part-time, and temporary workers. Most of the available service scheduling algorithms can be classified into (1) days off, (2) shift, and (3) tour [32]. Days-off algorithms specify what days specific employees work and what days they are off during a given period. Shift scheduling algorithms determine a set of employee work schedules across a daily planning horizon. Tour scheduling algorithms deal with the combined problem of days off and shift scheduling over a planning period. Algorithms for solving these problems range from some simple ones that can be done on the back of an envelope to others where a sophisticated computer programs are employed for generating detailed schedules.

Tibrewala, Philippe, and Browne [31] developed an algorithm where each employee gets two days off every week. It is apparent that everyone cannot get all weekends off, although all employees would prefer that. Burns and Carter [6] provide an exact lower bound on the number of workers required to meet the daily demand that ensures that each employee receives at least A out of B weekends off. During the remaining (B − A) weeks, they receive two days off on weekdays, although not necessarily two consecutive days. Burns and Koop [7] propose a modular approach for scheduling multi-shift tour scheduling problems. These algorithms assume that a fixed demand is encountered week after week. We now present the Burns and Carter Algorithm.

Burns and Carter Algorithm

The Burns and Carter algorithm consists of four steps. The first step determines the maximum numbers of workers required to meet the given weekly demand. The second step assigns weekends off for those workers who are not scheduled to work during the weekend. The third step identifies pairs of days on which workers can be given days off. The fourth and final step assures that every worker gets two days off during any given weekly period.

> *Step 1:* *Determine the workforce size.* The algorithm determines the minimum number of workers necessary to meet given weekly demand. The larger of the following three lower bounds determines the workforce size.

> L1: weekend demand constraint:

$$\frac{(B-A)}{B} * W \geq n \quad \text{or} \quad W \geq \frac{Bn}{(B-A)}$$

where $n = \max(n_1, n_7)$ and W is an integer rounded up.

L2:   total demand constraint:

$$5W \geq \sum_{j=1}^{7} n_j \quad \text{or} \quad W \geq \frac{1}{5} \sum_{j=1}^{7} n_j$$

L3:   maximum daily demand constraint:

$$W \geq \max_j (n_j) \quad j = 1, 2, \ldots, 7$$

Size of workforce $W = \max[L1, L2, L3]$

Step 2:   *Assign weekends off to* $W - n$ *workers.* As the weekend requirements n_1 and n_7 are met, the remaining $(W - n)$ employees get their weekend off. The algorithm simply gives the weekend off to the first $(W - n)$ workers during the first weekend, the next $(W - n)$ workers during the next weekend, and so forth in a round-robin fashion. Therefore, it is necessary to find n pairs of days to give off for those workers who were working on either one or both of the two consecutive weekends. Those n pairs are identified in step 3.

Step 3:   *Identify n pairs of days off.* This step assures that every employee gets exactly two days off every week. First, the daily weekday slack $s_j = (W - n_j)$ for each $j = 2, 3, \ldots, 6$ and the weekend slack $s_j = W - n_j - (W - n)$ for $j = 1$ and 7 are calculated. The following procedure chooses n day-off pairs iteratively:

(a) Find day k where $s_k = max_j(s_j)$
(b) Find any $s_j > 0$ where $j = k$. If none found, then $j = k$.
(c) Days (k, j) forms an off-day pair while days (k, k) forms a nondistinct off-day pair.

Step 4:   *Assign days off for n workers.* The final step assigns days off so that every employee gets two days off every week. Each employee can be classified into one of the following four types for determining how many days and when they should be given off.

Types	Weekend 1	Week 1	Weekend 2
T1	Off	No days off needed	Off
T2	Off	One day off needed	On
T3	On	One day off needed	Off
T4	On	Two days off needed	On

A pair of days for each T4 type is assigned first because those employees are scheduled to work on both weekends. Then pairs are split between T2 and T3 types such that T3 gets the earlier day and T2 gets the later day of the week. Note that type T1 does not need any days off because those employees already have two consecutive weekends off. The algorithm ensures that every worker gets two days off during every one-week period in addition to the constraint that they be given A weekends off during a period of B weekends.

Example 14.8

Given $A = 3$, $B = 5$, and daily requirements, find the minimum number of workers required. Provide a schedule for the week for each worker.

Day (j)	Su	M	T	W	Th	F	Sa
Requirements	3	4	5	6	7	7	3

Solution:

n = max (3, 3) = 3

L1 = [5 * 3/2] = 8

L2 = 35/5 = 7

L3 = 7

Therefore, W = 8.

Weekend slack $s_j = W - n_j - (W - n) = 8 - 3 - (8 - 3) = 0$

Weekday slack $s_j = W - n_j$ is tabulated below.

Day (j)	Su	M	T	W	Th	F	Sa	
Requirements	3	4	5	6	7	7	3	
Slack	0	4	3	2	1	1	0	Select M/T pair
	0	3	2	2	1	1	0	Select M/T pair
	0	2	1	2	1	1	0	Select T/W pair
	0	2	0	1	1	1	0	

We need $n = 3$ pairs, and we were able to obtain three distinctive pairs. Table 14.13 presents the working days (•) and days-off (X) for each worker. Workers 1 and 2 belong to type T1 and hence they do not need any more weekdays off. Workers 3, 4, and 5 belong to type T2, whereas workers 6, 7, and 8 belong to type T3. Therefore, workers 3, 4, and 5 get the later day of the pair (that is, T, T, and W, respectively), while workers 6, 7, and 8 are assigned the earlier day of the pair (that is, M, M, and T, respectively), as shown in Table 14.13. The same procedure is repeated for the remaining weeks.

In addition to the required days off, special provisions such as the right mix of

TABLE 14.13   Solution to Example 14.8

Worker	Type	Sa	Su	M	T	W	Th	F	Sa	Su	M	T	W	Th	F	Sa	Su
1	T1	X	X	•	•	•	•	•	X	X	•	X	•	•	•	•	•
2	T1	X	X	•	•	•	•	•	X	X	•	X	•	•	•	•	•
3	T2	X	X	•	X	•	•	•	•	•	X	•	•	•	•	X	X
4	T2	X	X	•	X	•	•	•	•	•	X	•	•	•	•	X	X
5	T2	X	X	•	•	X	•	•	•	•	•	X	•	•	•	X	X
6	T3	•	•	X	•	•	•	•	X	X	•	•	•	•	•	X	X
7	T3	•	•	X	•	•	•	•	X	X	•	•	•	•	•	X	X
8	T3	•	•		X	•	•	•	X	X	•	•	X	•	•	•	•

Note: (X) days off and (•) working days

skills, seniority, personal preferences, vacations, and sudden changes in requirements can also be accommodated by many commercially available software packages on personal computers. Both *Schedule Master* by Schedule Master Corporation, and *Who Works When* by Newport Systems are suitable for small- and medium-sized businesses [28]. In many instances, a manual check can further improve the schedules generated by the computer.

SUMMARY

This chapter has focused on job shop activity planning. We have discussed many aspects of scheduling, loading, and sequencing. In addition, we have pointed out the importance of scheduling in service operations. Service operations are inherently more difficult to plan because the forecasted demand could easily vary due to weather, competition, or other uncontrollable factors. We cannot, however, ignore the importance of proper planning. Although we have dealt with only one algorithm in this chapter, several excellent references are listed at the end of this chapter. As service operations become more and more important to our economy, further research will be forthcoming in this area. The details and the extent of planning depend on the situation, but planning alone does not get the job done. Executing the job involves issuing actual shop orders, following through on the job progress, and delivering goods to the customer on time. These factors are fully explored in the next chapter.

PROBLEMS

1. Job A has the following four operations:

Operation	Description	Duration (hours)
010	Casting	15
020	Milling	28
030	Drilling	18
040	Finishing	8

The due date on the planning calendar is day 235.

Using the simple scheduling rules in Table 14.1, find the following:

a. The completion date, using forward scheduling; today is day 198 on the planning calendar. Assume a five-day work week.

b. The starting date, using backward scheduling.

2. For problem 1, apply the block scheduling rules specified in Table 14.2 and find the following:

a. The completion date (week number), using forward scheduling.

b. The starting date, using backward scheduling; the due date is week 47. Note that 235 days is 47 weeks.

3. Operations required for completing jobs A, B, and C are given by the route sheet information in Table 14.14. Draw a Gantt load chart for each machine center.

4. A scheduler in Mary Mont Hospital needs to assign four technicians to four different jobs. Estimates of the time to complete every job were provided by each technician and are summarized as follows:

	Hours to Complete Job			
Technician	1	2	3	4
A	20	36	31	17
B	24	34	45	12
C	22	45	38	18
D	37	40	35	18

a. Using the assignment method of linear programming, how would you assign jobs to technicians to minimize the total work time?

b. Assuming that the estimates are fairly accurate, can these jobs be completed in three eight-hour days? If not, give a schedule for completion of all jobs.

5. Suppose that, in problem 4, it is not desirable to assign technician A to job 1 and technician D to job 3. How will this affect your results?

6. A scheduler has five jobs that can be done on any of four machines, with respective times (in hours) as shown here. Available capacities in each machine center are also given.

	Machines Job			
	1	2	3	4
A	50	60	80	70
B	100	120	110	70
C	100	80	130	60
D	80	70	40	30

TABLE 14.14   Route Sheet for Problem 3

	Job A		Job B		Job C	
Operation Number	Work Center	Time (hours)	Work Center	Time (hours)	Work Center	Time (hours)
I	2	30	1	40	1	35
II	3	30	3	30	4	20
III	4	40	2	40	2	35
IV	1	50	4	30	3	40

E	75	100	70	120
Available capacity	70	120	70	120

Determine the allocation of jobs to machines that will result in minimum hours.

7. Prepare a Gantt chart schedule based on the criterion of the shortest total processing time per order, using problem 3 data.

8. Using the techniques learned in this chapter, solve problem 9 in Chapter 12.

9. The following jobs are waiting to be processed at the same machine center. Jobs are logged as they arrive.

Job	Due Date	Duration (days)	Cost
A	313	8	100
B	312	16	100
C	325	40	200
D	314	5	50
E	314	3	75

In what sequence would the jobs be ranked according to the following decision rules: (a) FCFS, (b) EDD, (c) SPT, (d) LPT, (e) TSPT, (f) LS, and (g) COVERT? All dates are specified as manufacturing planning calendar days. Assume that all jobs arrive on day 275. No job is allowed to wait more than fifty days.

10. Suppose that today is day 300 on the planning calendar and that we have not started any of the jobs given in problem 9. Using the critical ratio technique, in what sequence would you schedule these jobs?

11. Job B has the following four operations:

Operation	Description	Duration (hours)
01	Casting	10
02	Reaming	21
03	Milling	14
04	Finishing	6

Apply the simple scheduling rules specified in Table 14.1 and find the following:
a. The completion date, using forward scheduling.
b. The starting date, using backward scheduling; the due date is 310.
Assume a total procurement lead time of ten days. Today is day 257.

12. For problem 11, apply block scheduling rules specified in Table 14.2 and find the following:
a. The completion date, using forward scheduling.
b. The starting date, using backward scheduling; the due date is 320. Today is day 257. Assume a five-day work week.

13. The operations required for completing the jobs A and B are given by the route sheet information in Table 14.15.
a. Draw a Gantt schedule chart for each machine center.
b. Calculate the percentage of the time that the machines are idle prior to the completion of operations in individual work centers.

14. The operations required to complete jobs X, Y, and Z are given by the route sheet infor-

TABLE 14.15 Route Sheet for Problem 13

	Job A		Job B	
Operation Number	Work Center	Time (hours)	Work Center	Time (hours)
010	2	30	1	40
020	3	30	3	30
030	1	40	2	50

mation in Table 14.16. Draw a Gantt load chart for each machine center. Assume two machines at center 2.

15. A manager at Paul's Laboratories must assign three engineers to three different duties. Estimates of the time to complete each duty for every engineer have been determined from historical data and are summarized as follows:

	Hours to Complete Job		
Engineer	1	2	3
A	10	12	16
B	12	24	20
C	14	20	24

a. Using the assignment method of linear programming, how would you assign jobs to the engineers to minimize the total work time?
b. Assuming that the estimates are fairly accurate, can all three duties be performed completely in two eight-hour days? If not, give a feasible schedule for completion of all jobs.

16. The scheduler at CPA Company needs to assign four accountants to four different projects. Estimates of the time to complete every job were determined by the scheduler and are summarized as follows:

	Hours to Complete Job			
Accountant	1	2	3	4
A	16	29	34	19
B	11	43	32	23
C	17	36	43	21
D	17	33	40	36

TABLE 14.16 Route Sheet for Problem 14

	Job X		Job Y		Job Z	
Operation Number	Work Center	Time (hours)	Work Center	Time (hours)	Work Center	Time (hours)
Release order		40		40		40
010	2	80	1	60	2	120
020	1	40	2	100	1	160

a. Using the assignment method of linear programming, how would you assign projects to minimize the total work time?

b. Assuming that the estimates are accurate, can these jobs be completed in three eight hour days? If not, give a feasible schedule for completion of all jobs.

17. For problem 16, suppose that it is not desirable to assign accountant A to job 4 and accountant D to job 2. Will this affect your results; If so, how? (*Hint:* Block off unwanted squares or assign very high costs to those squares and recompute the problem.)

18. A scheduler has four jobs that can be done on any of five machines, with respective times (in hours) as shown here. Determine the allocation of jobs to machines that will result in minimum hours.

		Jobs		
Machines	1	2	3	4
A	60	110	70	50
B	70	40	100	80
C	70	80	100	120
D	100	70	80	110
E	30	120	60	70

(*Hint:* Add a dummy column with zero values for job 5 and proceed as usual. One of the machines will be allocated to a dummy job in the final solution; that is, the machine is idled.)

19. Using the index method, find the allocation of the following six jobs to four machines that will result in minimum hours.

		Machines		
Job	I	II	III	IV
1	40	50	60	70
2	100	40	80	30
3	50	60	20	100
4	110	80	100	120
5	120	60	70	80
6	60	30	80	50
Available capacity	100	80	120	40

20. The following jobs are waiting to be processed at the same machine center:

Job	Due Date	Duration (days)	Cost per Day
010	260	30	100
020	258	16	100
030	260	8	200
040	270	20	75
050	275	10	100

In what sequence would the jobs be ranked according to the following decision rules: (a) FCFS, (b) EDD, (c) SPT, (d) LPT, (e) TSPT, (f) LS, and (g) COVERT? All dates are specified as manufacturing planning calendar days. Assume that all jobs arrive on day 210. No job is allowed to wait more than fifty days.

21. The following jobs are waiting to be processed at the same machine center.

Job	Date Order Received	Production Days Needed	Date Order Due	Cost of Delay ($)
A	110	20	180	500
B	120	30	200	1000
C	122	10	175	300
D	125	16	230	500
E	130	18	210	800

In what sequence would the jobs be ranked according to the following rules: (a) FCFS, (b) EDD, (c) SPT, (d) LPT, (e) TSPT, (f) LS, and (g) COVERT? All dates are according to shop calendar days. No job is allowed to wait more than seventy days. Today on the planning calendar is day 130.

22. Suppose that today is day 150 on the planning calendar and that we have not yet started any of the jobs in problem 21. Using the critical ratio technique, in what sequence would you schedule these jobs?

23. Dotmat Data Processing Company estimates the data entry and verifying times for four jobs as follows:

Job	Data Entry (hours)	Verify (hours)
A	2.5	1.7
B	3.8	2.6
C	1.9	1.0
D	1.8	3.0

In what order should the jobs be done if the company has one operator for each job? Using a Gantt chart, show how long it will take to complete all four jobs.

24. Six jobs are to be processed through a two-step operation. The first operation involves preparation and the second involves painting. Processing times are as follows:

Job	Operation 1 (hours)	Operation 2 (hours)
A	10	5
B	7	4
C	5	7
D	3	8
E	2	6
F	4	3

Determine a sequence that will minimize the total completion time for these jobs. Using a Gantt chart, find the make span time.

25. Consider the following jobs and their processing times at the three machines. No passing of jobs are allowed.

Job	Machine 1 (hours)	Machine 2 (hours)	Machine 3 (hours)
A	6	4	7
B	5	2	4
C	9	3	10
D	7	4	5
E	11	5	2

Using Johnson's rule, find the sequence in which the jobs are to be processed.

26. Given the processing times for three consecutive operations: (a) find the optimal sequence for machines 1 and 2; (b) find the optimal sequence for machines 2 and 3; and (c) find the optimal sequence for the $N=5/M = 3$ problem. If all sequences are the same, we have an optimal sequence for the $N/3$ problem. Verify the solution.

Job	Machine 1	Machine 2	Machine 3
A	3	6	7
B	5	3	2
C	1	2	4
D	6	8	9
E	7	5	4

27. Suppose that, in problem 25, a fourth operation was just added, with processing time on machine 4 given as 5, 7, 15, 3, and 4 hours, respectively, for jobs A through E. Using the CDS algorithm, find a near optimal solution. What is the make span time?

28. A list of jobs and their respective job durations is given here: The jobs can be processed by any one of three identical machines. Using the LPT rule, sequence these jobs.

Job	Duration
A	17
B	25
C	50
D	10
E	15
F	9
G	35
H	28
I	5
J	13

29. Given $A = 3$, $B = 5$, and the daily demand from Sunday through Saturday as 3, 5, 5, 5, 6, 7, 3, determine the total number of full-time workers necessary to meet the given demand. Using the Burns and Carter algorithm, provide a schedule indicating what days each worker gets off and what days each works.

REFERENCES AND BIBLIOGRAPHY

1. *APICS Training Aid—Shop Floor Control* (Falls Church, Va.: American Production and Inventory Control Society, 1979).

2. K. R. Baker, *Introduction to Sequencing and Scheduling* (New York: John Wiley & Sons, 1974).

3. K. R. Baker, R. N. Burns, and M. W. Carter, "Staff Scheduling with Days-Off and Workstretch Constraints," *AIIE Transactions,* Vol. 11, No. 4 (December 1979), pp. 286–292.

4. S. E. Bechtold, M. J. Brusco, and M. J. Showalter, A Comparative Evaluation of Tour Scheduling Models," *Decision Sciences,* Vol. 22, No. 4 (September–October 1991), pp. 683–699.

5. E. S. Buffa and J. G. Miller, *Production-Inventory Sustems: Planning and Control,* 3rd ed. (Homewood, Ill.: Richard D. Irwin, 1979).

6. R. N. Burns and M. W. Carter, "Workforce Size and Single Shift Schedules with Variable Demands," *Management Science,* Vol. 31, No. 5 (May 1985), pp. 599–607.

7. R. N. Burns and G. J. Koop, "A Modular Approach to Optimal Multiple-Shift Manpower Scheduling," *Operations Research,* Vol. 35, No. 1 (January–February 1987), pp. 100–110.

8. H. G. Cambell, R. A. Dudek, and M. L. Smith, "A Heuristic Algorithm for the *n* Job, *(m)* Machine Sequencing Problem," *Management Science,* Vol. 16, No. 10 (June 1970), pp. 630–637.

9. D. C. Carroll, "Heuristic Sequencing of Single and Multiple Component Jobs," Unpublished Ph.D. dissertation, Sloan School of Management, MIT, 1965.

10. E. G. Coffman, ed., *Computer and Job Shop Scheduling Theory* (New York: John Wiley & Sons, 1976).

11. R. W. Conway, W. L. Maxwell, and L. W. Miller, *Theory of Scheduling* (Reading, Mass.: Addison-Wesley, 1976).

12. S. E. Elmaghraby, "The Machine Sequencing Problem-Review and Extensions," *Naval Research Logistics Quarterly,* Vol. 15, No. 2, pp. 205–232.

13. J. H. Greene, ed., *Production and Inventory Control Handbook* (New York: McGraw-Hill, 1970).

14. W. C. Healy, "Shift Scheduling Made Easy," *Factory,* Vol. 127, No. 10 (October 1969), pp. 87–91.

15. L. A. Johnson and D. C. Montgomery, *Operations Research in Production Planning, Scheduling, and Inventory Control* (New York: John Wiley & Sons, 1974).

16. S. M. Johnson, "Optimal Two- and Three-Stage Production Schedule with Setup Times Included," *Naval Research Logistics Quarterly,* Vol. 1, No. 1 (March 1954), pp. 61–68.

17. V. A. Mabert and A. R. Raedels, "The Detail Scheduling of Part-Time Workforce: A Case Study of Teller Staffing," *Decision Sciences,* Vol. 8, No. 1 (January 1977), pp. 109–120.

18. P. M. Mangiameli, S. L. Narasimhan, and D. West, "A Heuristic Algorithm for Scheduling Service Operations in a Dynamic Demand Environment." Proceedings of the National DSI Conference held in San Diego, November 17–21, 1992.

19. R. McNaughton, "Scheduling with Deadlines and Loss Functions," *Management Science,* Vol. 6, No. 1 (October 1959), pp. 1–12.

20. L. G. Mitten, "Sequencing *n* Jobs on Two Machines with Arbitrary Time Logs," *Management Science,* Vol. 5, No. 3 (April 1959), pp. 293–298.

21. Y. R. Nanot, "An Experimental Investigation and Comparative Evaluation of Priority Disciplines in Job Shop-Like Queuing Networks." Unpublished Ph.D. dissertation, UCLA, 1963.

22. S. L. Narasimhan, "A Survey of Workforce Scheduling in Service Operations." Paper presented at the National CORS/TIMS/ORSA Conference, Vancouver, Canada, May 8–10, 1989.

23. S. L. Narasimhan and S. S. Panwalker, "A Heuristic Solution Procedure to Process Industry," Paper presented at the ORSA/TIME Annual Meeting, Detroit, April 1982.

24. J. Orlicky, *Material Requirements Planning* (New York: McGraw-Hill, 1975).

25. G. W. Plossl and O. W. Wight, *Production and Inventory Control* (Englewood Cliffs, N.J.: Prentice Hall, 1967).

26. A. A. B. Pritsker, L. W. Miller, and R. J. Zinkl, "Sequencing *n* Products Involving *m* Independent Jobs on *m* Machines," *AIIE Transactions,* Vol. 3, No. 1 (1971), pp. 49–60.

27. A. O. Putnam et al., "Updating Critical Ratio and Slack Time Priority Scheduling Rules," *Production and Inventory Management,* Vol. 12, No. 4 (1971), pp. 51–73.

28. W. Rash, Jr., "Who, What, When, and Why Not?" *Byte,* June 1990, Vol. 15 Iss. 6. pp. 85–87.

29. M. K. Starr, *Operations Management* (Englewood Cliffs, N.J.: Prentice Hall, 1978).

30. C. J. Teplitz, "MRP Can Work in Your Job Shop," *Production and Inventory Management,* Vol. 19, No. 4 (Fourth quarter 1978), pp. 21–26.

31. R. Tibrewala, D. Philippe, and J. Browne, "Optimal Scheduling of Two Consecutive Idle Periods," *Management Science,* Vol. 19, No. 1 (September 1972), pp. 71–75.

32. J. M. Tien and A. Kamiyama, "On Manpower Scheduling Algorithms," *SIAM Review,* Vol. 24, No. 3 (July 1982), pp. 275–287.

33. W. L. Zangwill, "Deterministic Multiproduct, Multifacility Production and Inventory Model," *Operations Research,* Vol. 14, No. 3 (1966), pp. 486–507.

CHAPTER 14A

Computer Simulation Drives Innovation in Scheduling
Michael B. Thompson

Shop floor scheduling of a manufacturing facility is a complex task that is done today largely by manual means. This complexity arises from several factors. First, there are a large number of possible combinations to consider. Second, manufacturing facilities are very dynamic, and unforeseen events that occur invalidate previously developed schedules. Third, the scheduling process is time consuming and there is normally not time to develop and evaluate more than one version to see if a better schedule could result in improved shop performance. Coordinated rule-based planning scheduling, in combination with computer simulation, can help overcome the aforementioned problems.

BACKGROUND

In most cases, scheduling does not become a concern until some fundamental *planning* issues are resolved. For example, the planning decisions that must precede any scheduling decision include: the selection of the product or products to be manufactured; the scale of production volume; the manufacturing processes that will be used; the required quantities of equipment, tooling and personnel; and the facility layout. Computer simula-

Reprinted from APICS—The Performance Advantage February 1993, Vol. 3, No. 2, pp. 31–34.

tion has been traditionally used to assist in providing the answers to these planning decisions.

Scheduling, the time-sequenced allocation of resources to manufacture products, begins once these planning decisions have been made and the set of available resources has been determined. Finite capacity scheduling is the process of assigning manufacturing tasks to limited capacity resources while not violating any of the resource constraints.

The single most difficult component of any scheduling problem is the number of possible combinations of schedules that exist for a given number of orders and available resources. For example, "the sequencing of 12 orders through six operations has $12!^6$ or more than 10^{52} possible schedules in a simple job shop without even considering any alternate machines."[1]

Generally, manufacturing scheduling problems, in terms of scheduling theory, are known as sequencing problems. A processing sequence is imply the order in which jobs are processed through the machines. The time duration for each task or operation is normally predetermined by using time standards. For developing a schedule, the time duration for each task is normally not a variable, and therefore the sequence is the essence of the schedule. Schedulers have used dispatching as a technique to rank the order of jobs waiting for a machine to be processed. Dispatching rules have evolved to provide the criteria by which jobs are prioritized as they enter machine queues. At the lowest level of detail, you can think of scheduling in terms of two critical decisions that must be made:

1. When do I launch an order? (induct a new order into the system)
2. What does each resource work on next? (maybe nothing)

SIMULATION AND SCHEDULING

While computer simulation is rapidly increasing in manufacturing applications, it has been used in the design and analysis of systems for more than 20 years. Simulation modeling involves the creation of a mathematical model of a manufacturing system or facility so that the model can be experimented with, on a computer, to gain information and understanding of the system. Experimenting with a model, as opposed to the real system, has several distinct benefits:

- With a model, the element of time can be accelerated so that long-term effects on the system can be understood.
- The effects of capital acquisitions, process changes and scheduling rule changes can be evaluated prior to their commitment.
- Experimentation with the actual system is not often possible, and even when it is possible, it isn't cost effective and most likely will be disruptive.

An overwhelming majority of the simulation models that have been created to date could be classified as "throw away" models. This means that the model was developed, validated, and experimented with and after the information it provided was understood, the model was put on the shelf. These throw away models are really analytical models. This means they are used for the design and analysis of the facility but not for the operation and control of the facility.

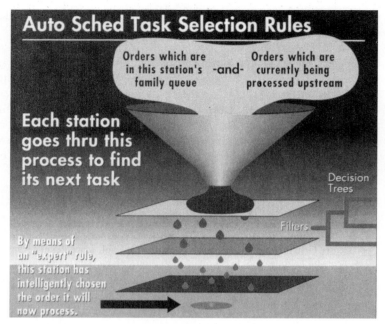

FIGURE 1   Task selection utilizing user-defined rules for each resource.

Simulation-based scheduling takes the analytical model and expands its usefulness, accuracy and, in some cases, its degree of detail. Models that are used for scheduling and control are called operational models. The accuracy of the operational model is verified by comparing the model's predictions of factory performance parameters such as throughput, cycle time, equipment utilization and bottlenecks, with the actual measurements of the factory operation. The model is calibrated by carefully evaluating and adjusting parameters such as time standards for setup and processing, batching rules and task selection rules so that a high degree of confidence can be placed in the model.

SIMULATION-BASED SCHEDULING

Simulating a factory is the process of proceeding through time (simulated time) while imitating the moment-by-moment operation. "Proceeding through time" means synchronizing the simulated clock and making all the appropriate changes to the factory model parameters for each increment. Such changes must be consistent with real changes that would occur in the factory in real time. Each change in the factory model corresponds to an event. Many types of events may occur, such as order releases, operator shift changes and material pickup or delivery. Perhaps one of the most significant kinds of events, in terms of the impact on the schedule, is an *operation completion event*. This occurs when a machine or an operator completes the processing of a material load for one of the operation steps in the load's process plan. At this point, the machine or operator selects the next operation, considering all the possible operations and constraints. This selection process is a large component of the total scheduling problem.

Each time a machine, workstation or operator completes an operation, comes on shift, or is idle and a new load enters its work queue, it must decide if any work should be done. Each resource must select its next task. The selection process is based on a task selection rule (also referred to as a dispatching or scheduling rule). The literature is full of studies aimed at formulating and measuring the effectiveness of dispatching rules for job shop scheduling.

Conway's[2] classic study of 39 different dispatching rules established Shortest Processing Time as the favored rule and it has become the de facto standard against which proposed rules are measured.

The method of task selection is a critical and important difference between commercially available simulation and scheduling software. Most systems provide a collection of standard dispatching rules which the user may specify for each machine. Examples of dispatching rules are: shortest processing time, critical ratio, earliest due date, etc. While this is certainly a user-friendly approach, the author has learned firsthand that this method is not flexible enough to handle real world shop floor scheduling problems. The traditional dispatching rules are one dimensional—only a single criterion is used to determine which load should be selected next. In addition, traditional approaches only provide a view of the loads that are currently waiting in queue and do not provide visibility of upstream loads or of downstream congestion.

Now available is a far more powerful approach, which provides the capability for plant scheduling personnel to specify powerful, user-defined rules for each resource that are not limited to a single criterion or a single view of a load. Auto-Simulations Inc. has implemented this concept in its AutoMod II with AutoSched. AutoSched is designed to solve the primary problems associated with scheduling and floor management, and assists in controlling manufacturing facilities. Reduced inventory and lead times result in lower operating costs and higher profits. The AutoSched user can create a customized schedule that fully considers the user's real world constraints.

These rules can be thought of as a series of filters into which the potential tasks are fed. The filters screen out more and more loads from the set of possible choices until either one load or none remains to be selected. Each filter is a test that the loads must pass to be considered for final selection or further filtering. The rules provide decision tree, sorting and filtering capability in a simple to apply manner. There are no limits to the number of filters that a rule may comprise.

DEVELOPING RULES

Before we discuss the process of developing scheduling rules, the management of the factory must answer the question, "What is a good schedule?"

While this seems like it should be an easy question to answer, it is often far more difficult to quantify than one might think. Over the years I have been told the attributes of a good schedule are:

- Minimum lead time
- On-time completion of all orders

- Minimum work-in-process and finished goods inventory
- Maximum resource utilization
- Minimum or no overtime
- Minimum cost in terms of the routing through alternate machines

While these are all worthy goals, the fact of the matter is that some of these are inherently in conflict. For example, let's say that we just purchased a new flexible machining center that cost the company $250,000. The plant manager mandates that this machine will be utilized in excess of 90 percent of the available time so that the investment is maximized. To utilize equipment that much, it is safe to say that there must be a queue of work in front of that machine constantly so that when the machine completes an order there is a new one immediately waiting to be worked on. To be safe, a fairly large number of orders will be kept in queue just in case unforeseen situations arise. The utilization goal directly conflicts with lead time, inventory, and possibly on-time completion goals, because loads requiring services by the new machining center must wait in the large queue.

REACHING COMPROMISE

A factory is much like a balloon. These goals must be considered together so that the best overall compromise can be achieved. By purely maximizing a single goal, you might be accomplishing it at the expense of other goals. Therefore, "a good schedule" can be accurately defined as one that conforms to management's goals and objectives. These goals and objectives must be developed understanding the inherent conflicts and the dynamics of their relationships. These goals and objectives can be stated in terms of:

- Measure due date performance (the measure of early or lateness)
- Measure of throughput (the number of products completed per unit time)
- Measure of lead time (the time orders stay in the system)
- Measure of inventory levels (WIP and finished goods)
- Measure of resource utilization (the productive time a resource spends)

Computer simulation is an excellent tool to use to establish aggressive schedule performance measures.

Once the goals and objectives have been established to quantify schedule performance, we can begin to think about how to develop scheduling rules. Rules should be developed off-line with a representative set of data that is indicative of the real scheduling problem. There is a high level strategy, which I have titled coordinated rule-based scheduling. The concept is to make a shop flow even if it is a job shop. This means it is necessary for all the machines, workstations and operators (resources) to operate in concert. It is seldom effective to have each resource attempt to optimize its own utilization. The general strategy is this:

- First, identify the critical machines, stations, or operators in the factory. A critical resource is one that will increase the overall throughput of the facility once

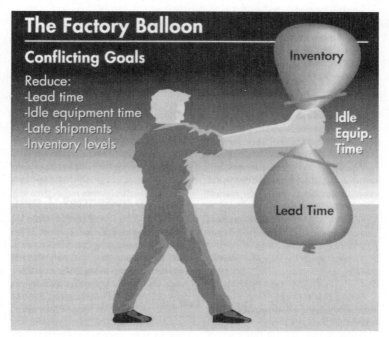

FIGURE 2   Goals must be considered together to reach the best compromise.

its constraint is relaxed. Analyze the production requirements (loading) to determine those with the greatest load. When using a simulation model to identify the critical resources, they are the resources that have the highest utilization in combination with the largest measures of queue contents, and time spent waiting in queue.

- Critical resources must be kept busy processing by keeping manageable levels of work in their queue and choosing tasks smartly to minimize setups and to avoid unnecessary setups. If necessary, critical resources should look upstream in an attempt to wait a reasonable amount of time for a better choice rather than just taking what is available in the current queue. They should also look downstream in an attempt to keep things flowing through the shop by choosing jobs that will not just get bogged down in a large downstream queue.
- Identify server stations—those with low to moderate load that can play the role of servers to the critical stations to ensure the critical stations have work and the work they have minimizes setup time. They should always look downstream to feed work to the critical resources smartly.
- Run the model and analyze the resulting schedule performance measures. Pay particular attention to critical resources and look at why they were not fully utilized. In addition, determine if new critical resources have surfaced. If a trend or cause is identified that a rule change could help, enhance the rule and run the model again.

- Iterate with the previous step until you are satisfied with the schedule performance.
- Last, employ the rule in an on-line mode.

Using this strategy, Norman[3] improved schedule performance by 25 percent over Conway's shortest processing time rule. The author has experienced schedule performance improvements of up to 60 percent better than previous methods.

PRE-PLANNED SCHEDULING

There are two modes in which coordinated rule-based scheduling can be applied to the manufacturing scheduling problem. The first and most commonly used is the periodic scheduling mode. Periodic scheduling implies developing a finite capacity schedule from an accurate status of the shop floor. The pre-planned schedule is then provided to the shop floor for implementation. If an unforeseen event occurs in the real world that wasn't considered in the original schedule, a new one is prepared with the appropriate constraints adjusted and provided to the shop floor. This concept, coupled with today's high performance scheduling software, allows planners to create "work around" schedules with ease and accuracy not previously available.

One fundamental problem with pre-planned schedules arises if the rate of unforeseen events is excessive. Periodic scheduling under such conditions is not as feasible an alternative due to the number of iterations that would be required during the scheduling period. A powerful extension to simulation-based scheduling is real-time dispatching.

REAL-TIME DISPATCHING

Rather than create a new schedule each time an unplanned event occurs, the software can run in a mode where the simulation runs in parallel with real time. This method requires the model to receive detailed messages from the shop floor control system as the appropriate events occur on the shop floor. These are actual events rather than simulated events. The model updates the status on the internal data structures as the event messages are received. When an event message is received that necessitates task selection, as described earlier, the model performs the task selection and sends a message back to the shop floor system specifying the next job to be run. Machine failure events are received in the same manner as regular operation start and completion events. The model will react accordingly and schedule the factory efficiently, even in a dynamic environment. Even with the capacity of real-time dispatching, there is still the need to do pre-planning schedules and traditional capacity analysis. Therefore, one method does not preclude the other.

CYCLE-TIME REDUCTION

Simulation-based scheduling with coordinated rules is being used today. A major international semiconductor manufacturer recently reported that, as a result of this technology, average cycle time was reduced from 25 days to five days and average work-in-process inventory was cut to one fifth the level of 18 months before. "It was not just the software that allowed us to make these improvements, it was the confidence that the software provided management that allowed them to cut inventories to previously unheard of levels."

Today we are all looking for ways to be ever more responsive to customer needs, to improve the time to market and to utilize the resources that are currently available to the maximum. This technology provides a means to increase the productivity of factories without adding personnel or equipment.

The key requirements are: An on-line shop floor control system that tracks all of the orders in process, accurate time standards, some special tailoring to gain access to existing data bases, and time allocated to developing and analyzing rules so that the schedules will reflect the way your company wants to do business.

REFERENCES

1. Rickel, J., "Issues in the Design of Scheduling Systems," *Expert Systems and Intelligent Manufacturing,* Michael D. Oliff Editor, Elsevier Science Publishing Co. Inc., New York, New York, 1988.
2. Conway, R. W., Maxwell, W. L., and Miller, L. W., *Theory of Scheduling.* Addison Wesley, Reading, Mass., 1967.
3. Norman, T. A., "Tailoring Dispatching Rules for Job Shop Scheduling," A White Paper Developed in the Department of Computer Science at Brigham Young University, Provo, Utah, 1989.

Chapter Fifteen

Job Shop Production Activity Control

INTRODUCTION

Chapter 14 dealt with many important aspects of planning, such as scheduling, loading, and sequencing. The rules learned in that chapter can be used to calculate how long it will take to complete any job and to specify start and want dates. Remember, however, that rules and computer systems cannot cut chips and will not get jobs through the shop. A basic function of any planning system is to identify the necessary materials, tooling, and labor skill levels. Once this is done, the control functions must coordinate the production effort and communicate the results effectively to all concerned. This results in good customer relations and savings for the firm.

DISPATCHING

Dispatching is a production control function that is performed by a dispatcher—a production control person who coordinates with the manufacturing department. The dispatcher maintains a file of all open orders related to his or her department, whether or not they are released. This file is known as the *dead load file*. Once a job is released to the dispatcher's department, that order is kept in the *live load file*. Between the time the order reaches the department and the time the material arrives, the dispatcher issues a dispatch list, which authorizes the manufacturing department to produce the item. Subsequently, a shop packet is issued, consisting of detailed drawing specifications, a bill of materials, a route sheet, a shop order, tickets for materials and tools, and any other information deemed necessary. The dispatcher is also responsible for maintaining an accurate and up-to-date shop order file, using forms or computer terminals.

Dept.:	310					Date:	10/1/94	
Work Center:						Capacity:	40 hrs./wk.	

Part No	Start Order No	Opn No	Description	Qty	Hours	Operation Start	Dates Due	Order Due Date
A112	110	0100	Drilling	100	10.0	10/1/94	10/4/94	10/9/94
C315	98	0040	Milling	100	10.0	10/5/94	10/6/94	10/13/94
B512	117	0020	Reaming	400	20.0	10/5/94	10/10/94	10/15/94
							Priority	
		Total Standard Hours			40.0			

FIGURE 15.1   Sample dispatch list.

The dispatch list, which is the key document for priority control, is generally prepared by computer. This list consists of all jobs that are available to run. They are ranked by priority, using rules such as the critical ratio technique. The dispatch list is the vehicle used for formal communication; a sample is exhibited in Figure 15.1. Many commercial devices are available to assist the dispatcher in preparing the dispatch list. The simplest device, of course, is a Gantt load chart. In addition, modern electronic communication equipment, such as barcode devices, modems and data collection terminals are also available.

CORRECTIONS TO SHORT-TERM CAPACITY

The dispatch list is generally prepared for a short period of time. We should never schedule (input) more than what we can complete (output) within the time frame specified by the dispatch list. If a capacity shortage is detected, corrective actions must be taken. The most common corrective actions [2] are (1) scheduling overtime, (2) selecting alternative routings, (3) reallocating workforce, (4) operation overlapping, (5) lot splitting, (6) order splitting, and (7) subcontracting. Some of these tactics are exhibited in Figure 15.2, and we discuss them briefly here.

Alternative Routings

Generally, we schedule jobs through the most efficient machines and methods so that the due dates are met. This approach might lead to overload situations at some critical work centers, however. By routing some jobs to alternative machines, we might be able to gain additional capacity, even though it may not be the most efficient procedure. The trade-offs involved in alternative routing should be fully explored. The route sheet may already indicate alternative routings for accommodating emergency situations.

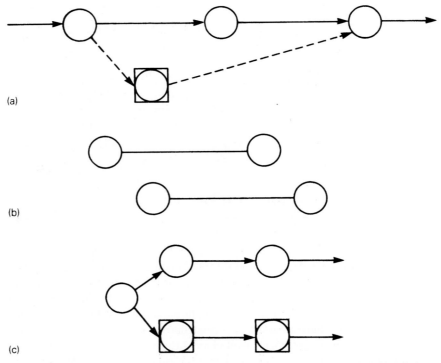

FIGURE 15.2 Corrective actions: (1) alternative routings, (b) overlapping, and (c) lot splitting.

Overlapping

Overlapping involves sending part of the first lot to the next machine center immediately, so that the second machine center can get a head start. Then the lot-sizing rules may be followed as usual. This approach reduces the total lead time of the job. Overlapping may be designated as standard for certain operations, especially when additional setups can be avoided.

Lot Splitting

Jobs are completed in lots, and they generally are not moved to the next workstation until the entire lot is completed. In many instances, it is possible to send a partial lot ahead of time, thus preventing idle time in the succeeding machine centers. Decreasing idle time essentially increases capacity. For example, suppose that ten additional pieces are scrapped during an operation but that an order quantity of fifty is being produced in the previous operation, which cannot be expedited. As an alternative, the order quantity of fifty can be split into ten and forty, and ten pieces can be expedited ahead of time. This reduces the delay for the order that had ten excess scraps and effectively prevents idle time of machines. Note that overlapping generally occurs at the beginning of a job, whereas lot splitting can occur at any time prior to completion of the job.

Example 15.1

Consider the following jobs on a dispatch list.

Job	Opn 10, Milling (hours)	Opn 20, Drilling (hours)	Due Date
A	40	24	140
B	80	16	141
C	24	40	146

Assuming a forty-hour work week, determine whether it is possible to meet the due dates. If not, what alternative actions can the manager take to complete the jobs on time? Assume today is day 129.

Solution: Assuming that the machines are available right away, the schedule chart would be as exhibited in Figure 15.3.

The manager finds from the process sheet that the first operation of job A can be done by an alternative machine, but it will take fifty-six hours. Based on this additional information, she develops the schedule chart shown in Figure 15.4.

She finds that jobs A and B meet the due dates but that job C would be a day behind. Realizing that the process sheet is an important source of information, she finds out that part of job C can be sent ahead to a drilling machine. The lot splitting and overlapping not only eliminate idle time on the drilling machine but also result in completing all jobs on or before the due dates. Note that lot splitting and overlapping generally contribute to additional materials handling costs.

Order Splitting

Large orders can be split and run on several machines simultaneously, thus reducing the processing lead time. Such decisions are generally made during activity planning stages, and each split order is rescheduled as a separate order.

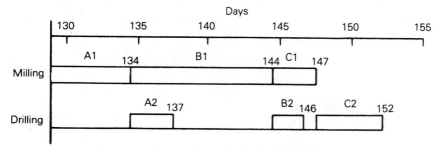

FIGURE 15.3   Schedule chart for Example 15.1

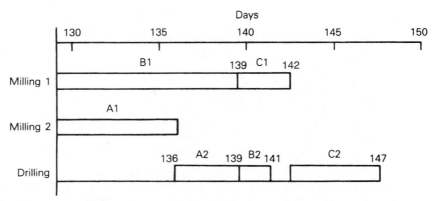

FIGURE 15.4 Revised schedule chart for Example 15.1.

PRODUCTION REPORTING AND STATUS CONTROL

Manufacturing feedback to the production schedule is done by production reporting. Production reporting helps manufacturing take corrective actions on maintenance of valid priorities of on-hand and on-order jobs. The reporting covers delays in scheduled performance, the efficiency of the shop, the productivity of the workers, and the utilization of capacity. The information is derived from schedules, job authorizations, and job documents, such as the move ticket and the labor ticket, the scrap and rework reports, inventory receipts, and material usage reports, as shown in Table 15.1. The reporting also includes operation completions, order closings, and necessary information to assist the payroll department.

FACTORS AFFECTING THE COMPLETION TIME OF JOBS

Even though the dispatch list is issued and the operational sequence is defined, the completion of jobs on time is not guaranteed. Many unplanned and unforeseen circumstances arise in the real world. Machines can break down. A key operator can be absent. A rush order from a preferred customer may force us to reschedule, and when we try to reschedule, we might find that we do not have enough capacity or that our lead time is too high

TABLE 15.1 Status Reports and Information Sources

Information	Sources
Delays	Schedules
Capacity utilization	Job authorizations
Schedule performance	Job documents
Machine efficiency	
Operator productivity	

to satisfy so valuable a customer. Reporting the delay does not solve the problem. Therefore, it is important for us to understand how to reduce the lead time of jobs and to learn how to accommodate many emergency situations. A detailed discussion of manufacturing lead time and in-process inventories (queues) will help us in mapping a strategy for coping with such situations [7].

Example 15.2

Friendly Company uses a block scheduling rule for calculating estimated completion dates. Usually, the durations of operations are rounded up to the nearest week. One week is allowed for releasing the order and withdrawing material from the stock room. Two weeks are permitted between operations and one week for final inspection and delivery. Calculate the feasible promise dates for jobs A and B, for which the route sheet is as follows:

	Job A		Job B	
		Time		*Time*
Operation	*Work Center*	*(hours)*	*Work Center*	*(hours)*
I	3	160	3	40
II	4	120	2	200
III	1	200	4	80

This is week 28 on the planning calendar. Given the due dates for jobs A and B as week 44 and week 38, respectively, can we complete them on time?

Solution: Assuming a five-day, forty-hour work week, the number of weeks required for every operation in all departments can be calculated. Using the block scheduling rule, we estimate the completion dates for jobs A and B as week 46 and week 42, respectively. Thus according to the schedule (Table 15.2), jobs A and B cannot be completed by the due date.

TABLE 15.2   Block Schedule for Example 15.2

	Job A		Job B	
Operation	*Weeks Allowed*	*Week Number*	*Weeks Allowed*	*Week Number*
This week		28		28
Release date	1	29	1	29
I	4	33	1	30
Interoperation	2	35	2	32
II	3	38	5	37
Interoperation	2	40	2	39
III	5	45	2	41
Inspect and deliver	1	46	1	42

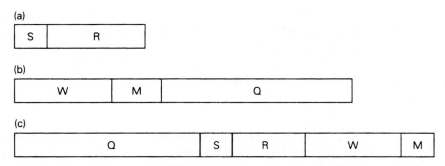

FIGURE 15.5 Elements of manufacturing lead time: (a) operation duration, (b) interoperation time, and (c) lead time.

Manufacturing Lead Time

The manufacturing lead time of a job is the interval between its release to the shop floor and its delivery to stores or higher-level operations. Although lead times are made up of several time elements, they fall into two major categories: (1) operation duration and (2) interoperation time. Operation duration is the time the job actually spends on the machine. It consists of setup time and run time, which is also referred to as production time (see Figure 15.5):

$$\text{Operation duration} = \text{setup time} + \text{run time}$$
$$= S + R$$

Interoperation time refers to the interval between the completion of one operation and the start of the subsequent operation. It is defined as

$$\text{Interoperation time} = \text{wait time} + \text{move time} + \text{queue time}$$
$$= W + M + Q$$

Thus total lead time can be defined as

$$\text{Total lead time} = \text{operation duration} + \text{interoperation time}$$
$$= S + R + W + M + Q$$

Operation Duration. Operation duration is the actual production time of the job. It depends on the lot size. The necessary data for calculating operation times are given by the manufacturing route sheet, which is also known as the process sheet. The calculation of operation duration was dealt with in detail in Chapter 12.

Interoperation Time. Interoperation time is the time the job waits in the work center queue after the completion of any operation and before the start of the next operation on the same job or order. In high-volume assembly lines, work flows from station to station continuously, and hence the accumulation of in-process inventory between work centers is minimal or negligible. The interoperation times in assembly lines usually are very short. In the job shop environment, however, the work centers are farther apart. Depending on the work content of jobs, the loads among machine centers vary considerably. Overloads at work centers increase interoperation time. The interoperation time is the primary determinant of the lead time. In fact, up to 85% of manufacturing lead time is interoperation time. Therefore, the total lead time can be reduced by cutting the excess from the interoperation time [15].

Queue Time.   Queue time is the time that a job spends at a machine waiting to be worked on because there are other jobs ahead of it. Recall that the priority of each job is specified by a sequencing rule. Regardless of the shuffling of the sequences of jobs at work centers, however, the total work load at each center remains the same. The queue time of an individual job is affected by the priority assigned to it.

Move Time.   Move time, also referred to as transport time, is the actual time a job spends in transit between operations. Move time depends on the location of the two work centers involved. Time values for it can be provided in a matrix format.

Wait Time.   The term *wait time* is usually applied to the time a job spends waiting before being transported to the next operation and after completion of the prior operation. Note that all elements of lead time except move time are work center-dependent. It is not necessary to specify all time elements for every work center. It is, however, desirable to specify all elements as accurately as possible for critical work centers where close control is needed. In other instances, all elements can be lumped as a single value, such as one day or one week. Once a time estimate has been specified, it should be monitored for accuracy periodically.

Example 15.3

Even though the block schedule indicates that the jobs can not be completed on time (see Example 15.2), Friendly Company wants to review the situation in detail. Since the shop is not usually busy at this time of the year, the manager decides that he can get by with one week between operations. Draw a finite forward schedule chart based on this additional information, given the due dates as weeks 44 and 38 for jobs A and B, respectively, and this is week 28 on the planning calendar.

Solution: Table 15.3 summarizes all available data pertaining to the route sheet and the machine center capacities. Next, we develop a Gantt schedule chart. Since jobs A and B need machine 3 simultaneously, we calculate the critical ratios (CR) for these jobs:

$$CR(A) = \frac{44 - 28}{\dfrac{160}{40} + \dfrac{120}{40} + \dfrac{200}{40}} = \frac{16}{12} = 1.33$$

$$CR(B) = \frac{38 - 28}{\dfrac{40}{40} + \dfrac{200}{40} + \dfrac{80}{40}} = \frac{10}{8} = 1.25$$

TABLE 15.3   Load and Capacity Data, Example 15.3

Operation	Job A			Job B		
	Work Center	Load (hours)	Capacity (hours/week)	Work Center	Load (hours)	Capacity (hours/week)
I	3	160	40	3	40	40
II	4	120	40	2	200	40
III	1	200	40	4	80	40

Since job B has a smaller critical ratio, we schedule it on machine 3 first. It is followed by job A, as exhibited in Figure 15.6. In the figure, BI, for example, represents the first operation of job B. As they are completed on machine center 3, job B is moved to machine center 2, whereas job A is moved to machine center 4. Note that we permit one week between operations. Then job A goes to machine center 1, whereas job B goes to machine center 4 for the final operations. We see that jobs A and B are completed by weeks 45 and 41, respectively. Unfortunately, Friendly Company is still unable to complete these jobs before their respective due dates.

Queue Length (In-Process Inventory)

Many managers ask what the optimum length of a queue is. They all know that a queue serves as insurance against idle time. Queues absorb changes in efficiency, randomness in job length, scrap, product mix, and changes in job arrival time. Queues will fluctuate as a result of (1) over- and understatements of the master production schedule, (2) excess or inadequate capacity, and (3) changes in product mix. Therefore, in a job shop situation it is very difficult to specify or attain an optimum queue length. In practice, however, it is possible to graph the distribution of queue length as the number of work hours waiting at intervals. Some typical distributions are shown in Figure 15.7. Diagram (a) portrays a situation with a desirable queue. The distribution of queue length is approximately normally distributed. It indicates that various factors influencing the queue size occur randomly. Note that the queue size is rarely zero, amounting to a very small idle time. Diagram (b) indicates the situation in which a persistent backlog exists. The queue size is always above ninety hours. If this center produces thirty hours a work per day, then the

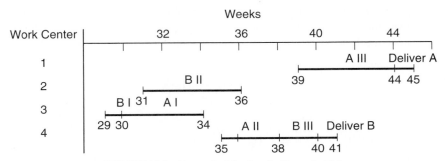

FIGURE 15.6   Gantt schedule chart for Example 15.3.

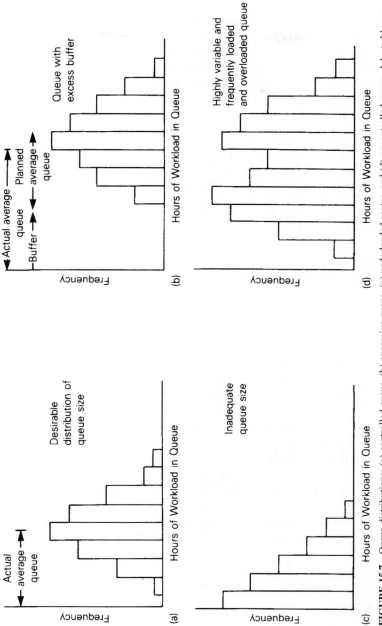

FIGURE 15.7 Queue distributions: (a) controlled queue, (b) excessive queue, (c) underloaded center, and (d) uncontrolled queue. Adapted by permission from *COPICS*, Volume 5, First edition, p. 16. © 1972 by International Business Machines Corporation.

manufacturing lead time can be reduced by three days (90/30) by reducing the average queue size by ninety hours, still resulting in a negligible amount of idle time. Reducing the unnecessary buffer would reduce the lead time of all future orders. The queue represented by diagram (c) is short. It could contribute to a persistent underload situation. Finally, diagram (d) portrays a work center with a higher variable queue length. It indicates frequent overload situations [7]. Using such diagrams, it is possible to calculate the approximate amount of desirable queues for particular machine centers.

Example 15.4

Suppose that a work center has an average queue size of 600 hours, with a standard deviation (σ) of 80 hours. The manager is a very conservative individual, who does not want to run out of work more than one in 200 chances. The capacity of the work center is 160 hours per week. Assuming that the queue size is normally distributed, what queue size should we recommend? By how much can we safely afford to reduce the queue size? What alternatives do we have to attain the desired queue level?

Solution: The queue out risk is specified as one in 200, or $\alpha = 0.005$ probability. Using normal probability tables, we find the value of $z = -2.57$. The desired average queue length (μ) can then be calculated using the following formula:

$$z = \frac{0 - \mu}{\sigma}$$

That is,

$$2.57 = \frac{-\mu}{80}$$

or

$$\mu = (80)(2.57) = 206 \text{ hours}$$

The excess queue size = 600 − 206 = 394 hours (see Figure 15.8). The buffer is approximately equal to two weeks of workload. The excess queue size can be reduced by increasing the capacity by using overtime or by reducing the amount of work input to the work center. Alternative routing and subcontracting are other options.

TECHNIQUES FOR ALIGNING COMPLETION TIMES AND DUE DATES

The importance of delivering goods on time cannot be overemphasized. When some jobs are ahead of due dates and others are behind, corrective actions are taken with the help of dispatch lists. Jobs that are ahead of schedule can easily be delayed until a future date,

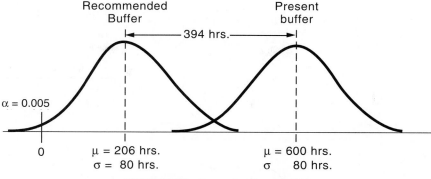

FIGURE 15.8   Queue size, Example 15.4.

but jobs that are behind schedule need attention. Essentially, we need to complete these jobs in less-than-normal lead times. Some techniques for reducing the lead times are discussed in this section.

Lead Time Control

Lead time can be reduced by decreasing the interoperation time. Although the interoperation time consists of wait time, move time, and queue time, it is no secret that queue time is the most difficult one to control. The queue length or the queue time can be reduced by increasing the capacity of critical work centers or reducing the workload to specific work centers. In the short term, capacity can be increased by using overtime or additional shifts or by subcontracting operations. Some techniques discussed for short-term capacity corrections, such as depicted in Figure 15.9, as overlapping and lot splitting, are also applicable here. Operation splitting is another way to reduce the lead time. For example, if the operation is labor-oriented, several workers can be assigned to the job. Or, if the operations are machine-controlled, they may be performed on several machines. It is important that the scheduling department be informed of the new routing and its impact on lead times and loads at affected machine centers. Additional setups may be incurred when several machines are employed instead of one. Unfortunately, that is the price we pay to reduce the lead time and keep our customers happy.

Example 15.5

Suppose that the supervisor in Example 15.3 manages to obtain additional forty-hour capacities at machine centers 1 and 2. Can we complete the jobs on time?

Solution: Based on the new information, we first tabulate the machine center loads and capacities, as shown in Table 15.4.

Recall that the due dates for jobs A and B are weeks 44 and 38, respectively, and that this is week 28 on the planning calendar. We know that jobs A and B compete for machine center 3 simultaneously, and hence we calculate their critical ratios:

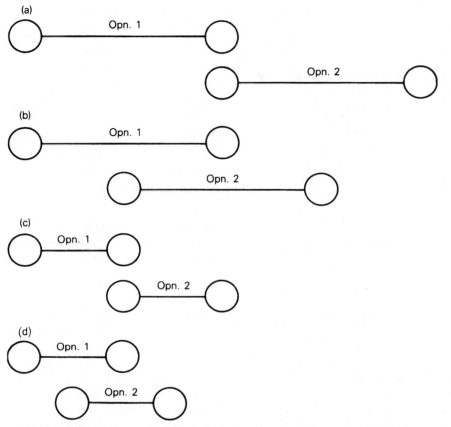

FIGURE 15.9   Techniques for reducing lead time: (a) standard production routing, (b) overlapping, (c) doubling the number of shifts, and (d) shift doubling and overlapping.

$$CR(A) = \frac{44 - 28}{\dfrac{160}{40} + \dfrac{120}{40} + \dfrac{200}{40}} = \frac{16}{9.5} = 1.68$$

$$CR(B) = \frac{38 - 28}{\dfrac{40}{40} + \dfrac{200}{80} + \dfrac{80}{40}} = \frac{10}{5.5} = 1.82$$

Job A has a smaller critical ratio, and hence we schedule job A at machine center 3, followed by job B. Upon their completion, they are moved to machine centers 4 and 2, respectively, as shown in Figure 15.10.

We see that jobs A and B are completed by the middle of week 42; that is, job A can be done before its due date, but job B is still late by approximately four weeks.

TABLE 15.4   Load and Capacity Data, Example 15.5

Operation	Job A			Job B		
	Work Center	Load (hours)	Capacity (hours/week)	Work Center	Load (hours)	Capacity (hours/week)
I	3	160	40	3	40	40
II	4	120	40	2	200	80
III	1	200	80	4	80	40

Example 15.6

The manager at Friendly Company finds out that lot splitting and overlapping can be done on job A. Based on this information, determine whether it is possible to meet the due dates in Example 15.5. If not, can you suggest an alternative method to solve this problem?

Solution: We notice in Example 15.5 that splitting job A does not hasten the completion of job B in any work center. Instead of using the critical ratio technique, we try the SPT rule at machine center 3. The detailed schedule chart is exhibited in Figure 15.11. We see that Jobs A and B can be completed by weeks 44 and 38, respectively. This example shows the complexity of solving job shop problems. Hence, many commercial packages resort to simulation for obtaining a satisfactory schedule.

Queue Control

When queue sizes are too large, the associated cost of in-process inventories is high. Large inventories not only increase shop floor congestion, they also increase materials handling costs. Since queues control the lead time, the reduction of queue size provides better customer service through decreased lead time. Overall, this increases productivity. Queue control consists of (1) measuring the present queue size, (2) establishing the optimum queue size, and (3) adjusting the capacity of machines or controlling the input to the work center [15].

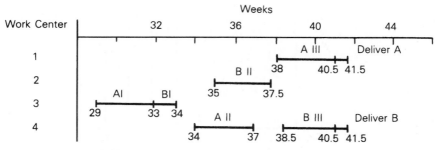

FIGURE 15.10   Gantt schedule chart for Example 15.5.

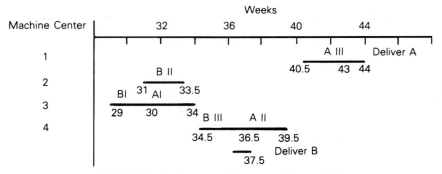

FIGURE 15.11 Gantt schedule chart for Example 15.6.

Measuring Present Queue Size. Queue size fluctuates for various reasons, as discussed earlier. By measuring queues regularly, they can be plotted, as was shown in Figure 15.7. For example, the following are the buffer inventories in a work center for the past thirty weeks:

450, 500, 700, 540, 320, 750, 650, 870, 930, 400,

730, 550, 610, 850, 890, 930, 650, 710, 550, 850,

660, 750, 690, 350, 450, 630, 620, 640, 650, 580,

We can group these inventories into different classes and draw a frequency distribution diagram, as exhibited in Figure 15.12. The calculation of mean and standard deviation, as found in most elementary statistics books, can be calculated as follows:

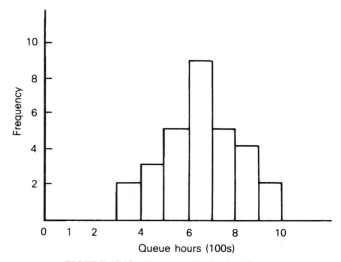

FIGURE 15.12 Frequency distribution diagram.

Range of Queues (hours)	Queue Size (x_i)	Frequency (f_i)
300–399	350	2
400–499	450	3
500–599	550	5
600–699	650	9
700–799	750	5
800–899	850	4
900–1000	950	2

$$\mu = \frac{\Sigma f_i x_i}{\Sigma f_i} = \frac{19,700}{30} = 656.67 \text{ hours}$$

$$\sigma = \sqrt{\frac{\Sigma x_i^2 f_i}{N} - \frac{(\Sigma x_i f_i)^2}{N^2}}; \text{ where } N = \Sigma f_i$$

$$= \sqrt{455,833.33 - 431,211.11} = 156.91 \text{ hours}$$

Establishing the Optimum Queue Size.   Suppose that the manager does not want to exceed the queue out risk more than 1% of the time. We can calculate the maximum buffer necessary to meet the goals:

Required Buffer $= z\sigma$

$$= (2.33)(156.91)$$

$$= 365.6 \text{ hours}$$

Therefore, the excess queue, or the unnecessary buffer, is equal to $656.67 - 365.6 = 291$ hours.

Controlling Queue Size.   The final phase consists of reducing the buffer in a planned manner. The input/output control technique described earlier (Chapter 12) is a useful tool for controlling queues. Suppose that the planned input/output table for the work center is as shown in Table 15.5. We can anticipate a buffer size of 600 hours at the end of nine weeks. Even though the buffer size is below average, it is above the required queue size of 365 hours. Suppose that we decided to reduce the queue to the optimum value in approximately ten weeks. We should examine the planned jobs (inputs) to see whether it is possible to divert some of them to alternative machine centers. Essentially, we will be reducing the inputs to the overloaded work center. The resulting planned input/output table might be as shown in Table 15.6.

If we are unable to divert some of the loads to other machine centers, we might plan to increase the output by scheduling overtime for the next ten weeks. Essentially,

TABLE 15.5   Planned Input/Output Table

	Backlog	Week								
		1	*2*	*3*	*4*	*5*	*6*	*7*	*8*	*9*
Planned input	—	75	100	80	90	85	70	90	70	80
Planned output	—	80	80	80	80	80	80	80	80	80
Buffer	580	575	595	595	605	610	600	610	600	600

we are increasing the capacity of the work center. The revised input/output table might be as shown in Table 15.7.

Thus the strategy is either to decrease the input or to increase the output at the particular work center to correct the overload situation. Once the optimum buffer size is achieved, we should monitor the work centers periodically to make sure it stays that way always.

Short Interval Scheduling

Short interval scheduling (SIS) is a scheduling and control technique that has met with some success in a variety of manufacturing and service industries, such as fabrication, repetitive manufacturing, warehouses, banks, and mail-order houses. The SIS technique is especially suited to machine-controlled operations. It assists supervisors in periodically auditing worker productivity at short intervals to examine whether the workers are meeting the set output objectives [18]. The first step in SIS is to determine the short interval of time to be used, which varies according to the application. It could be an hour in keyboarding or an entire day in the maintenance of large machines. The next step is to schedule a reasonable amount of work that should be completed in the short interval. This step obviously requires established work standards for defining the unit of work, such as the number of pieces per hour or the number of hours and minutes to complete a certain amount of work. Then forms are developed for the supervisor to use in monitoring and taking corrective actions, if needed.

A sample form is exhibited in Figure 15.13. The SIS reporting system highlights problem areas and focuses necessary attention on the bottleneck operations, which invariably results in improved productivity.

TABLE 15.6   Planned Input/Output Table After Reducing Queue Size

	Backlog	Week								
		1	*2*	*3*	*4*	*5*	*6*	*7*	*8*	*9*
Planned input	—	75	50	75	60	50	50	50	40	60
Planned output	—	80	80	80	80	80	80	80	80	80
Buffer	580	575	545	540	520	490	460	430	390	370

TABLE 15.7   Revised Input/Output Table

					Week					
	Backlog	1	2	3	4	5	6	7	8	9
Planned input	—	75	100	80	95	85	70	90	75	70
Planned output	—	105	105	105	105	105	105	105	105	105
Buffer	580	550	545	520	510	490	455	440	410	375

SCHEDULING IN FLEXIBLE MANUFACTURING SYSTEM

A flexible manufacturing system (FMS) consists of a set of computerized machine tools connected by an automated material handling system as described in detail in Chapter 19. The FMS can manufacture a variety of customized products with varying demand rate while increasing the utilization of the manufacturing facility. The key elements include (1) automatically programmable machines, (2) automated tool delivery and changes, (3) automated material handling including the loading and unloading of parts to and from the machine, and (4) a centrally coordinated control mechanism. An FMS represents a significant investment in hardware, software, and personnel training. Askin and Standridge [3, p. 126] succinctly describe an ideal FMS: "The part types assigned to the FMS should have sufficient production volumes to make automation attractive but insufficient to justify dedicated production lines." Such a system poses a formidable challenge to everyone involved in the design, manufacture, and delivery of items. Production planning in an FMS environment becomes a very complex task. Approaches to conceptual representation of FMS and analysis of production planning include computer simulation, queuing, mathematical programming, heuristic algorithms, and hierarchical control methodologies. Jaikumar and Van Wassenhove [13] propose a three-level hierarchical model for decisions relating to the manufacturing processes occurring at several stages.

Work Center: 310 Description: Milling		Period: From 4/1 to 4/5											
		Monday 4/1				Tuesday 4/2				Wednesday 4/3			
Machine Number/ Operator		8t 10	10t 12	1t 3	3t 5	8t 10	10t 12	1t 3	3t 5	8t 10	10t 12	1t 3	3t 5
Machine 1 Cap. 100/hr Kodale, R.	Scheduled Actual Variance	100 100 0	100 100 0	100 90 −10	100 100 0								
Machine 2 Cap. 80/hr Kazmere, T.	Scheduled Actual Variance	80 80 0	80 90 10	80 90 10	80 70 −10								
Machine 3 Cap. 110/hr John, P.	Scheduled Actual Variance	110 105 −5	100 105 5	110 105 −5	100 105 5								

FIGURE 15.13   Short interval scheduling: sample performance work sheet.

Recognizing the computational burden associated with their model Sodhi, Askin, and Sen [23] propose a four-level hierarchical model for production planning in the flexible manufacturing (machine) system environment. We next describe the four levels of their model as illustrated by Figure 15.14.

Level 1: Part Selection

The selection of items to be produced on the FMS over the aggregate-term planning period is made in the first level. The demand for the item is converted into individual components or parts, and their corresponding production volumes are calculated for each

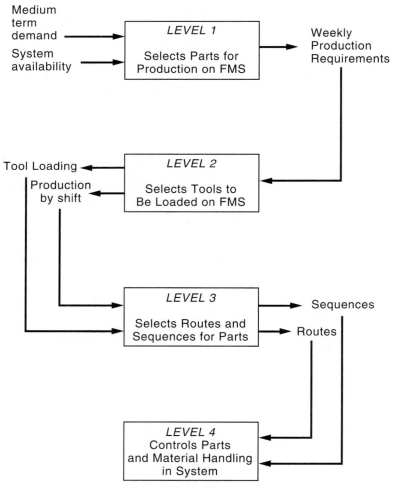

FIGURE 15.14 A hierarchical model for FMS. From M. S. Sodhi, R. G. Askin, and S. Sen, "A Hierarchical Model for Control of Flexible Manufacturing Systems," p. 2 Report No. 51, Department of Industrial and Manufacturing Systems, University of Rhode Island, Kingston, R.I. (October 1991).

item. The decision is based on costs relating to the savings in producing a part on FMS as opposed to other sources and available capacities. The FMS is assumed to be a part of a larger manufacturing facility so that off-loading of some demand to other conventional machines become feasible.

Level 2: Loading

At level 2, determination of the shift-wise production level and tooling of FMS necessary for manufacturing each part is made for the period defined by Level 1. By changing tools and fixtures required by various machines, the FMS can frequently be reconfigured to meet the production requirements over the planning period. A mathematical model has been proposed by Sodhi, Askin, and Sen [23] to address this problem. The model includes details such as limited availability of tools needed at more than one location, machine times, and other related variables.

Level 3: Route Assignment

Level 3 makes route selection decisions while reducing the excessive material handling, which is accomplished by the selection of a combination of alternative routings for each part that minimizes the transportation time or distance. A provision for additional tools and fixtures at key machine centers can often increase routing flexibility and reduce excessive material handling costs.

Level 4: Dispatching

The selection of proper dispatching rules can influence the productivity of the manufacturing facility by reducing idle times on machines as well as by reducing work-in-process inventory. In Level 4, dispatching rules are incorporated for entry of parts (order release) into the system and for detailed sequencing of part movement (operational control) requests. Parts are released in proportion to their planned production levels at fixed intervals so that a fixed amount of in-process inventory is maintained in the system. In general, material handling equipment can be assigned to these parts of a first come, first served basis.

SOLUTION METHODOLOGY

Utilizing these details a four level hierarchical model has been proposed by Sodhi, Askin, and Sen [23]. The input for each part includes its demand, operational sequences, durations at each alternate machine, and material handling equipment necessary to support the operations.

At Level 1 of the hierarchy, the objective function consists of operating costs, subcontracting costs, and inventory holding and shortage costs. The linear model ensures that adequate machine capacity is available on each machine type in each period for the given schedule while material handling equipment for the part mix does not exceed the available resources. Historical estimates of transport times for part movement can be utilized in the model. Level 2 of the hierarchy utilizes the flexibility of the system to process different part mixes at different times to obtain shift-wise tooling and production requirements from the part mix provided by Level 1. Setup costs are a significant factor in manufacturing operations. Based on the system capabilities, operating policies, and associated costs, tools are assigned to individual machines. The objective of this level is to minimize the total cost of deviation from the period 1 production plan suggested in Level 1. Assuming that tools and fixtures are available as required in each period, the total costs caused by the shortages and inventories are minimized in the FMS. Level 3 minimizes the level of transportation time or distance traveled. This is achieved by increasing the routing flexibility of the system by possibly providing additional tools at these machines. The results of Level 3 are passed to the shop floor, and dispatching rules are used for detailed scheduling at Level 4. Further details on the implementation of the procedure and an example are found in Sodhi, Askin, and Sen [23].

SUMMARY

In this chapter we described many aspects of dispatching and monitoring the progress of jobs through the shop floor. The relationships among major planning, execution, and control activities are exhibited in Table 15.8. In the short term, resolving overload situations is important. When the available capacity is not adequate, the master schedule should be revised. Low-priority jobs should be rescheduled for a later date. Overload situations can also be overcome by scheduling anticipated jobs in the immediately following periods. As illustrated earlier, the scheduling of job shops is a complex and challenging task. Many commercial software packages are available for use. General Electric's

TABLE 15.8   Relationship among Major Planning and Control Activities

Function	Purpose	Type of Planning/ Control Activity	Capacity Control Techniques
Scheduling	Overall output plan	Rough-cut plan	Add land and/or facilities, capital equipment, workforce
Machine loading	Calculate order priority; adjust release dates; assign specific jobs to work centers	Capacity requirements plan	Make versus buy; alternative routing; subcontract; reallocate workforce; additional tooling
Operation sequencing	Calculate operation priorities; sequence jobs for dispatching; identify start/finish dates for each job on each machine	Short-term capacity plan	Determine overtime needs; alternative routing; change workforce; operation overlapping; lot splitting; subcontract operations
Dispatching	Issue shop order, move ticket; labor, material and tool orders	Start the ball rolling	Input/output control
Status reporting and control	Report delays; salvage rework and production counts; labor and material utilization	Expediting and lead time control	Reduce interoperation times by overtime, operation splitting, operation overlapping
Plant monitoring	Monitor facilities; report machine utilization; generate productivity reports	Feed back information to database	Replan priorities and capacities; update data files

GJSCH$* can be used on a time-share basis [8] *GJSCH$ can handle up to seventy-five work centers and any number of jobs. Sandman Associates have developed a computer simulation approach called Q-control for scheduling job shops [19]. These programs provide several feasible schedules. The manager can choose the one that best satisfies the company's needs. Several references have been provided at the end of this chapter for more details on FMS.

PROBLEMS

1. The dispatch list contains the following jobs, due dates, and capacities of machines. Allow two days between operations; today is day 130 on the planning calendar. Assume that consecutive jobs can be started on a machine without any delay.

Job	Opn. 10, Reaming (40 hours/week)	Opn. 20, Finishing (40 hours/week)	Due Date
A	40	16	140
B	80	20	145
C	64	24	151

   a. Ignoring the due dates, find an optimal schedule and the completion date. (*Hint:* Use Johnson's rule.)

   b. Using the EDD rule, can we complete these jobs on time?

2. A supervisor finds that a previously scheduled job can be deexpedited and hence that she can get an additional forty hours of reaming during the second week. Based on this information, can the due dates in Problem 1 be met? Use EDD rule.

3. With the increase of forty hours in reaming capacity during the second week, the supervisor finds that she still cannot meet all due dates. The penalty for not meeting the due dates amounts to $200 per job. She finds that Job C can be split into lots. However, an additional materials handling cost of $50 per split is incurred. Is it economically feasible to split the lot? What is the completion date for your solution?

4. In Problem 1, suppose that an alternative machine is available for reaming but that it will take forty-eight hours for Job A. Can we meet all due dates? Use the EDD rule.

5. With an alternative machine center for Job A, the supervisor finds that she still cannot meet all due dates. Delays translate to $50 per job per day, but they can be prevented by lot splitting Job C, at an additional cost of $50 for materials handling per split. What is the optimum number of splits for this problem? What is the completion date for your solution?

6. If we apply the critical ratio technique in Problem 1, how will it be different from the EDD schedule? Which is the more reasonable solution for this problem? Why?

7. The supervisor finds that the shop does not have much in-process inventory to finish in the department. If she allowed only one day between operations, could all jobs in problem 1 be completed on time? If not, which job creates the problems?

8. The buyer has rescheduled Job C for day 152. With one-day interoperation time, can we meet all due dates in Problem 1?

9. Work center 364 has an average queue size of 400 hours, with a standard deviation of 60 hours. The manager is willing to take a chance of being out of work 1% of the time.

Assuming that the queue size is normally distributed, what queue size would you recommend? By how much should the manager reduce the queue size?

10. The capacity of the work center in Problem 9 is eighty hours per week. If management
permits a maximum of 25% overtime, how long will it take to reach the required queue
length? Show the existing and the desired queue sizes in a normal probability distribution diagram.

11. The following are the buffer inventories (hours) in a work center for the past fifteen
weeks:

450, 500, 700, 540, 320, 750, 650, 870,

930, 400, 730, 550, 610, 850, 890.

Draw a frequency distribution diagram. Calculate the mean and standard deviation of
the queue size. If the manager is willing to take a 0.5% chance of running out of work,
what queue size would you recommend? (*Hint:* Assume that the queues are normally
distributed. Since the sample size is less than thirty, use the student's t distribution tables).

12. In Problem 11, how long will it take to reach the desired queue size if the capacity of
the work center is 120 hours per week?

13. Several priority rules are being considered for dispatching jobs in a job shop. The following table lists the status of all jobs to be processed:

Job	Days before Due Date	Remaining Work (days)	Remaining Number of Operations	Order of Arrival
A	30	25	2	2
B	29	22	3	5
C	25	18	1	1
D	20	15	2	3
E	28	21	1	4

Using the following priority rules, determine the sequence of processing jobs:
a. FCFS
b. LPT
c. SPT
d. Earliest due date
e. Minimum slack per operation
f. Critical ratio technique

14. Which one of the situations illustrated in Figure 15.7 do the data in Problem 11 describe? Explain.

15. The queue sizes for the past fifty weeks in the milling department of Extron Company
are as follows:

100, 50, 100, 150, 200, 200, 150, 0, 0, 180, 170, 250, 250,

300, 500, 100, 200, 0, 110, 350, 310, 210, 400, 300, 100, 150,

400, 0, 0, 500, 200, 300, 400, 200, 250, 350, 450, 210, 400,

500, 200, 300, 500, 200, 0, 0, 100, 200, 100, 100

Draw a frequency histogram using the data for (a) first twenty weeks, (b) first thirty weeks, (c) first forty weeks, (d) last forty weeks, (e) all fifty weeks, (f) last thirty weeks, and (g) last twenty weeks. Which one of these diagrams in Figure 15.7 does each of these described situations portray?

Use the following data for problems 16 through 21. The route sheets and due dates for the following jobs are given in Table 15.9. Today is day 141 in the planning calendar. Possible modifications to operations are shown by X marks in the table.

The machine center capacities and current backlogs are as follows:

Work Center	Capacity (hours/week)	Current Backlog
1	80	20
2	40	40
3	40	40
4	80	20

Assume that the capacities of alternative machine centers are not included above.

16. Develop a dispatch list for each machine center, using the critical ratio technique. Assume one week for transport between operations and two days for final inspection and transport to the delivery department. Since the use of alternative machine centers, overlap of operations, and job splits is costlier, can we meet the schedule without utilizing them? Assume a five-day work week.

17. For some operations, alternative machine capacity is available up to forty hours per week, as shown in the table. These machines take 50% more time, however. Does this change your results in problem 16? The interoperation time is five days.

18. As you know, overlapping can sometimes do the trick. The succeeding operation can be started after 25% of the work is done on the machine center. What effect does overlap of operations have on the results of problem 16? The interoperation time is five days.

19. In problem 17, we gave the additional available capacities. In some instances we can

TABLE 15.9   Data for Problems 16 through 21

Job (Due Date)	Operation No.	Work Center	Duration (hours)	Alternative Machine Center	Overlap	Job Splits
A (190)	0010	1	80	X		
	0020	2	60	X		
	0030	3	24			
	0040	4	40	X		
B (200)	0010	3	80			
	0020	2	40	X		
	0030	1	120		X	X
	0040	4	160	X	X	
C (210)	0010	1	120	X		X
	0020	3	80			
	0030	4	80			
	0040	2	120		X	X

split portions of the operation and allocate them to an alternative machine, but it takes 50% more time. Can this help us meet the due dates?

20. If none of the alternatives given in Problems 17, 18, and 19 are helpful, can a combination of these alternatives help resolve the situation? State any additional assumptions necessary to solve the problem.

21. Repeat Problem 16 with two days for interoperation time instead of one week (five days).

REFERENCES AND BIBLIOGRAPHY

1. E. E. Adam and R. J. Ebert, *Production and Operations Management,* 2nd ed. (Englewood Cliffs, N.J.: Prentice Hall, 1982).
2. *APICS Training Aid—Shop Floor Control* (Falls Church, Va.: American Production and Inventory Control Society, 1979).
3. R. G. Askin and C. R. Standridge, *Modeling and Analysis of Manufacturing Systems* (New York: John Wiley & Sons., 1993).
4. J. F. Bard, "A Heuristic for Minimizing the Number of Tool Switches on a Flexible Manufacturing Machine," *IIE Transactions,* Vol. 20, No. 4 (1988), pp. 382–391.
5. M. H. Bulkin, J. L. Colley, and M. W. Steinhoff, "Load Forecasting, Priority Sequencing, and Simulation in a Job Shop Control System," *Management Science,* Vol. 18, No. 2 (October 1966), pp. 29–51.
6. R. B. Chase and N. J. Aguilano, *Production and Operations Management,* 6th ed. (Homewood, Ill.: Richard D. Irwin, 1992).
7. *Communications Oriented Production Information and Control Systems (COPICS),* Vols. 1–6 (White Plains, N.Y.: IBM Publications, 1972).
8. *General Electric Company Users Guide to General Shop Schedules (GJSCH$)* (General Electric Company, 1970).
9. V. Godin and C. H. Jones, "The Interactive Shop Supervisor," *Industrial Engineering,* Vol. 20, No. 11 (November 1969), pp. 16–22.
10. J. H. Greene, *Production and Inventory Control* (Homewood, Ill.: Richard D. Irwin, 1974).
11. F. S. Gue, *Increased Profits through Better Control of Work in Process* (Reston, Va.: Reston, 1980).
12. A. C. Hax and H. Meal, "Hierarchical Integration of Production Planning and Scheduling." In M. Geisler, ed., *TIMS Studies in Management Science, Logistics* (Amsterdam: North Holland, 1975).
13. R. Jaikumar and L. N. Van Wassenhove, "A Production Planning Framework for Flexible Manufacturing Systems," *Journal of Manufacturing and Operations Management,* Vol. 2 (1989), pp. 52–78.
14. A Kusiak, ed., *Modeling and Design of Flexible Manufacturing Systems* (Amsterdam: Elsevier, 1986).
15. N. P. May, "Queue Control: Utopia or Pie in the Sky," *APICS Conference Proceedings,* 1980, pp. 358–361.
16. G. W. Plossl, *Manufacturing Control* (Reston, Va.: Reston, 1973).
17. G. W. Plossl and O. W. Wight, *Production and Inventory Control* (Englewood Cliffs, N.J.: Prentice Hall, 1967).
18. W. J. Richardson, *Cost Improvements, Work Sampling and Short Interval Scheduling* (Reston, Va.: Reston, 1976).

19. W. E. Sandman and J. P. Hayes, *How to Win Productivity in Manufacturing* (Dresher, Pa.: Yellow Book of Pennsylvania, 1980).
20. S. C. Sarin and C. S. Chen, "The Machine Loading and Tool Allocation Problem," *International Journal of Production Research,* Vol. 25, No. 7 (1987), pp. 1081–1094.
21. M. R. Smith, *Manufacturing Controls* (New York: Van Nostrand Reinhold, 1981).
22. M. S. Sodhi, A. Agnetis, and R. G. Askin, "Tool Addition Strategies for Flexible Manufacturing Systems," *International Journal of Flexible Manufacturing Systems,* Vol. 6, No. 4, (1994), pp. 287–310.
23. M. S. Sodhi, R. G. Askin, and S. Sen, "A Hierarchical Model for Control of Flexible Manufacturing Systems," Report No. 51, Department of Industrial and Manufacturing Systems, University of Rhode Island, Kingston, R.I., October 1991.
24. K. E. Stecke and I. Kim, "A Study of FMS Part Type Selection Approaches for Short-Term Production Planning," *International Journal of Flexible Manufacturing Systems,* Vol. 1, No. 1 (1988), pp. 7–29.
25. R. Suri and C. K. Whitney, "Decision Support Requirements in Flexible Manufacturing," *Journal of Manufacturing Systems,* Vol. 3, No. 1 (1984), pp. 61–69.

APPENDIX 15A

The Flexible Factory Revisited
Robert U. Ayres and Duane C. Butcher

Anyone who shops discount electronics stores knows that consumer products have recently undergone a kind of combinatorial explosion. In the old days a camera was a box with a button. These days a store displays 40 or 50 camera models, each offering a different combination of features and modes: auto flash, pre-flash, fill-in flash, fill-in preflash, backlight control, self-timer, double self-timer, remote control, interval shooting, double exposure, continual shooting, bulb mode, zoom mode, portrait mode, landscape mode, "intelligent" shooting with fuzzy logic.

One of the most interesting aspects of the trend toward increased complexity and diversity in manufactured objects probably occurs to few discount-store browsers. How are all these objects made? The techniques of mass production achieve their celebrated economies only if they are used to make many copies of a few designs. Is there an economical way to make a few copies of many designs? If so, what is it?

A few years ago many manufacturers thought they knew the answer to this question. They would have said that the solution lies in applying to the problems of manufacturing the computer's power and flexibility, and they would have proceeded to rattle off a list of computer-based innovations in manufacturing technology, including computer-aided design, robots, computer-controlled numerical machines, automated guided vehicles, automated storage-and-retrieval systems, and flexible manufacturing cells. The unexpressed assumption was that all of these innovations would somehow add up to an

Reprinted from *American Scientist,* Vol. 81 No. 5 (September–October 1993), pp. 448–459.

integrated manufacturing system, which would be just as good at producing many product variants as the old assembly line was at producing one product.

This assumption turned out to be optimistic. In 1986 the Nobel laureate Robert Solow of the Massachusetts Institute of Technology remarked that "we see computers everywhere except in the productivity statistics." By then computers had been applied to a host of applications, with consistent gains in speed and accuracy, but they had somehow failed to produce measurable increases in overall productivity. Solow's observation suggests not that computers are inefficient in manufacturing applications, but that work is changing in ways that are poorly understood and probably not measurable by the old yardsticks. Productivity itself may now be a less important statistic than flexibility.

The disappointingly slow progress of computer-integrated manufacturing, or CIM, has the same paraodoxical quality. American industrialists are belatedly recognizing that CIM is a completely new approach to manufacturing, not just an automated version of the old approach. It requires the wholesale reorganization of work, something to which there are major technological, educational and cultural barriers. To complicate matters, CIM's benefits are not always the same as those offered by older production technologies. Indeed, some of the most important advantages would not even appear on books kept by traditional accounting methods.

Although CIM has still made only minor inroads in U.S. manufacturing, it is already clear that wholehearted adoption of this technology can dramatically improve a firm's competitive position. What profits a firm, however, may not profit the nation. Mass production set up a feedback loop whereby improvements in productivity tended to ratchet up demand, making possible further gains in productivity. It is not clear whether the new manufacturing technology includes a comparable engine for economic growth.

This article discusses these and other issues, relying largely on the results of a four-year international study of CIM completed in 1990 under the auspices of the International Institute for Applied Systems Analysis (IIASA) in Laxenburg, Austria. The IIASA study was prompted by the conviction that the world of discrete manufacturing—the production of individual objects such as toasters or cars—is experiencing an industrial revolution. IIASA attempted to define the existing situation with regard to CIM technologies, and to find out why computerized automation has spread more slowly in discrete manufacturing than in continuous processing, the production of goods, such as petrochemicals, that flow continuously through the production process.

The study concluded that CIM offers a solution to the current manufacturing dilemma, but it warns that its adoption will be complicated by cultural and economic problems that are rarely discussed in engineering textbooks.

CHANGES IN THE MARKETPLACE

Today's manufacturers sell their products in markets that have changed dramatically since World War II. Immediately after the war, U.S. manufacturers were catering to a relatively homogeneous domestic market with huge unspent demand. The integration of the world's economy and the emergence of Japan and other "Asian Tigers" as manufacturing powerhouses dramatically increased competition in the domestic and foreign markets. At the same time, the domestic market for many consumer goods approached satu-

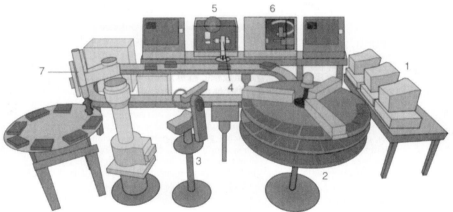

FIGURE 1. Flexible manufacturing systems are representative of the level of integration in advanced manufacturing facilities. The adoption of such technologies has been complicated by the radical changes they imply—not only a wholesale reorganization of work, but new ways of assessing the economic benefits of production technology. An FMS is a group of machines linked by a computer network. Unlike a traditional assembly line, it can automatically machine and assemble a family of product variants, allowing design changes to be implemented easily. The instructional system shown here was developed at the University of Wisconsin–Madison under the supervision of Jerry Sanders of the Department of Industrial Engineering. It is capable of going "from art to part," or machining parts to fit geometries defined by a computer-aided design system. Orders for parts or assemblies are entered in the master cell-control computer (*1*). Raw materials and partially finished parts are stored on an automatic storage-and-retrieval system, or ASRS (*2*), which is also computer-controlled. As manufacture begins, the ASRS rotates into position. A robot (*3*) selects and picks up the desired workpiece from a template and deposits it on a pallet on the conveyor belt. The pallet stops in front of a second robot (*4*), which picks up the workpiece and places it in a lathe (*5*) or mill (*6*) for machining. The door of the machine tool automatically opens as the cutting path for the part is downloaded to the tool. After the cutting operation is completed, the workpiece is moved to another machine tool or returned to the ASRS for stoage. Several workpieces may be in transit or undergoing machining at one time. In their turn, finished parts are called off the ASRS and travel on the conveyor to the rotating table at the base of the third robot (*7*), which assembles and inspects with the aid of its vision systems and a force-sensing "wrist."

543

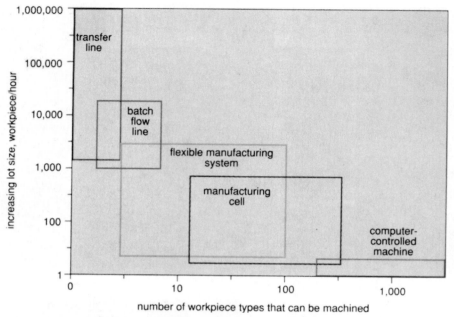

FIGURE 2. Modern manufacturing systems use five basic production technologies, which make different trade-offs between productivity (expressed here in terms of hourly output) and flexibility, or the variety of products that can be made on demand. The transfer line, or assembly line, consists of custom-made automation machines and a continuous transportation system. It achieves a higher yearly production rate than the other technologies, but at the cost of flexibility. At the other extreme is the numerically controlled machine, which can be programmed to produce almost any desired surface contour. It is very flexible, but productivity is much lower, largely because of the setup time needed to switch from one product to another. The manufacturing cell consists of several computer-controlled machines that perform various machining operations on the same workpiece, which is transferred from one to another by a special setup device or robot. Flexible manufacturing is similar but adds a transportation system. Finally, a batch flow line is a flexible system that makes parts that have been ganged to minimize setup time. As this illustration suggests, each technology has its place: The best technology for the job depends on the number of parts to be made. (Adapted from Rembold, Nnaji and Storr 1993.)

ration. By the end of the 1970s, one out of two people in the U.S. had an automobile, and 99 percent of American households had television sets, refrigerators, radios and electric irons.

To hold their own, manufacturers must now cater to narrow demographic segments in scattered regional markets. They do so by offering many customized products instead of a few standarized ones. Seiko sells 3,000 watch models, and IBM produces more than 55,000 models of the Selectric typewriter. Philips Electronics, which produced about 100 different color-television models in 1972, put 800 on the market in 1988.

Because product features have become the key to market differentiation, each new model tends to be more complex than the previous one. An admittedly extreme example of this trend is Xerox's top-of-the-line photocopier. In addition to the mechanical components needed to make an image of a page, it includes 30 microprocessors linked by local-

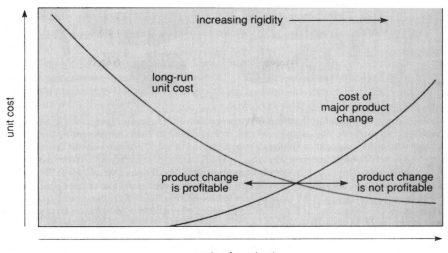

FIGURE 3. Today's market conditions have exposed a weakness of mass-production technology, which is sometimes called the productivity dilemma. The more parts a mass-production system makes, the lower the cost of producing each part. But to realize these economies of scale, manufacturers must make inflexible capital investments, which then become impediments to changing the product design. Mass-production technology thus stifles both innovation and customization of products and features. In the volatile and segmented markets of recent years, the ability to offer a variety of short-lived products often confers a significant competitive advantage.

area networks that monitor operations and make adjustments for wear and tear on the mechanical parts.

At the same time that products have become more various and more complex, design and production cycles have become shorter and shorter. Automobile models with major design changes used to be introduced only once every 15 or 20 years, but Toyota and Honda can now bring new models to market in two or three years. Motorola reportedly has cut the time it takes to build and ship an electronic pager from three weeks to two hours.

Because of these market trends, a manufacturing company can succeed only if it is able to produce a wide variety of complex products in a short time with little waste. These demands, in turn, have placed tremendous pressure on manufacturing engineering, a discipline that has declined steadily in prestige in this country since World War II. One result: Only one American-owned firm produces television sets domestically and none produces videocassette recorders, even though both the TV and the VCR were invented here.

THE DECLINE OF MASS PRODUCTION

It should surprise no one that a production technique that dates back to the 1920s is inappropriate to the conditions of the 1990s. Yet the obsolescence of mass production has sunk in only slowly, as its liabilities have surfaced in what are known within industry as the "productivity paradox" and the "quality problem."

The productivity paradox, also sometimes called the Abernathy dilemma after the late William J. Abernathy, a Harvard Business School professor, is that the need to cut costs conflicts directly with the need to introduce new products. In a mass-production system, production costs are minimized if standardization is driven as far as possible. Extreme standardization allows manufacturers to extract the maximum benefit from hard automation that is dedicated to particular tasks and from a specialized labor force. Large investments in hard automation and workforce specialization are barriers to change, however, because the capital invested in them cannot easily be converted for use in manufacturing new products.

Computer-integrated manufacturing offers a way around the productivity paradox, in that it relies on computers and microprocessor-controlled devices that can be programmed to produce a mix of low-volume products. CIM systems aim to produce lots of one at the same unit cost as lots of 10,000. The capital invested in them is flexible rather than fixed and easily applied to the manufacture of new or altered products.

The second drawback of mass production—the quality problem—has been chewing at the edges of industrial productivity for years. As manufacturing companies are uncomfortably aware, in recent years the cost of quality control have accounted for an increasing fraction of production costs. In fact, poor quality is symptomatic of many workplace problems, such as adversarial relations with parts suppliers and an alienated workforce. For this reason, many would argue that the quality problem is "prior to" the automation problem. Unless a company has addressed quality and brought it under control, the introduction of a CIM system is bound to fail. Computerizing a system that produces rejects will yield only automated confusion.

On the other hand, under the right conditions, computerized assembly systems can reach levels of perfection unattainable by manned assembly systems. Computers are notoriously better than people at complex, high-speed or repetitive tasks. Proponents of CIM often speak of the "lights-out factory," the "paperless process" or a product "untouched by human hands." The point of these idealizations is not simply that CIM reduces labor costs but that it reduces human error. Whatever its subtleties, the quality problem has been a major factor driving the adoption of CIM.

A company ignores these fundamental shifts in the economics of production at its peril. A cautionary example is provided by the U.S. auto industry. The industry's strategy of gradual "managed" innovation and "planned obsolescence," which once helped it to extract the maximum possible return from its capital investments, have now become a recipe for failure. Standardized products differentiated primarily by style—the strategy introduced by General Motors under Alfred Sloan—are increasingly difficult to sell, even with the advanced marketing techniques developed in the U.S.

A surprising counter-example is provided by Beretta, the Italian manufacturer of small arms, which has been in business since 1800. According to a fascinating case study by Ramchandran Jaikumar of Harvard Business School, productivity at Beretta has increased almost 500 times since 1800. The early gains in productivity clearly owed a great deal to design standardization. More recently product diversity has been increasing, and yet productivity is still improving. This time the gains are owed to a dramatic improvement in quality. Whereas in 1800, 80 percent of the small arms Beretta produced had to be reworked, by 1990 only 0.5 percent were scrapped or reworked. The improvement in quality, in turn, is attributable largely to Beretta's advanced position on the road to full computer integration.

ISLANDS OF AUTOMATION

The astute reader may be wondering why we have not yet defined CIM. Simply put: CIM is not a machine that arrives conveniently in a shipping crate, and no two CIM systems will be exactly alike, even if designed to make the same product. Moreover, no firm has yet achieved a degree of integration that could fully justify the term "computer-integrated manufacturing."

IIASA described CIM as "the application of computers and micro-electronics to supplement human decision-making in manufacturing." To many practitioners, however, CIM is the use of the computer to bring about a sort of planned synergy among many discrete technological modules.

Thus, although microprocessors originally were applied piecemeal to many iso-

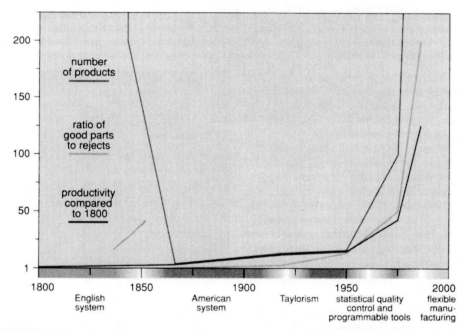

FIGURE 4. Tremendous gains in productivity have come to companies such as Beretta, the Italian small-arms manufacturer, that have changed their process-control strategies as manufacturing technology has advanced. Between 1800 and 1867 the "American system," which emphasized product standardization and the interchangeability of parts, supplanted the "English system," which used new power-driven machines but relied on old methods of production; the number of products offered by Beretta plummeted. Since then the number of products has steadily increased; recent productivity gains, then, are owed to dramatic quality improvements, reflected in the ratio of good parts to rejects. Two additional trends are noteworthy: The number of workers per machine dropped from 13.3 to 1 between 1800 and 1985, and the proportion of off-line workers in the workforce grew from 13 percent to 67 percent. Overall direct employment, meanwhile, dropped from 300 to 30. In 1987 the company introduced flexible manufacturing systems, and shortly thereafter computer-aided design was integrated with computer-controlled machines.

FIGURE 5 A flexible manufacturing cell at Allen-Bradley, a unit of Rockwell International, assembles printed circuit boards for the company's industrial-control products and systems at a plant at Twinsburg, Ohio. The industrial market demands small lots of many different circuit boards, making this an ideal application for flexible manufacturing. As many as 10 different panel designs may flow through the center simultaneously, each using a different combination of assembly operations. Also, the fact that most components are surface-mounted—rather than inserted through pre-drilled holes, as in traditional "through-hole" circuit-board assembly—requires highly precise positioning but allows the manufacturer to reap advantages from flexible assembly. A few steps are shown here. After a solder paste has been screen-printed onto the panel through a stencil, high-speed machines place components on it, aligning their terminations with the solder traces.

lated production problems, CIM is not limited to the machining and assembly processes within the firm. CIM systems designers have moved steadily toward the ultimate goal of bringing every segment of a company's operations within the system's embrace, and telescoping those operations so that tasks that were once done sequentially can now be done simultaneously. For their part, manufacturing companies are coming to realize that the organizational structures and management practices typical of postwar American industry are incompatible with the radical reintegration of work that CIM seeks to achieve. Many companies are struggling to change the corporate culture at the same time that they are wrestling with the new production technology.

Examples of transitional technologies that have enabled CIM include programmable machine tools, industrial robots, automated guided vehicles and automated storage-and-retrieval systems. On the factory floor these technologies can sharply reduce the handling of workpieces, producing startling reductions in error and defect rates and greatly increasing productivity. The technologies central to CIM, however, integrate larger segments of the production process.

Two key technologies are CAD/CAM, computer-aided design and manufacturing, and FMS, or flexible manufacturing systems. CAD/CAM strives to link design workstations directly to manufacturing equipment. Digital representation of designs can be electronically converted to coded instructions to drive, for instance, metal-cutting machines on the factory floor. Flexible manufacturing systems are used for batch production of families of relatively similar parts. A typical system consists of sequentially arranged, programmable manufacturing cells linked by a common control-and-transport system.

No doubt CAD/CAM and FMS will one day be seen as transitional technologies as well, as CIM designers expand systems to include purchasing, marketing, sales and administration under a single closed-loop control system. All functions are interdependent, and can be connected in a dynamic and continuous interaction with modern communication and database management.

THE TELESCOPING OF WORK

This abbreviated history conceals the degree of reorganization that adoption of a CIM technology often requires. The history of CAD/CAM and of FMS provides examples of both the difficulties reorganization presents and the opportunities it creates.

In the heyday of mass production, industry solved the design problem much as it solved the production problem—by dividing design into many specialized tasks, which were then done in sequence. When the designers completed their work, the blueprints were "tossed over the wall" to the manufacturing department, which was required to find the cheapest way to produce the design it had been given.

CAD systems and CAM systems initially developed quite independently, observing the traditional separation of engineering and manufacturing functions in industry. Newer systems have tried to integrate these functions through the use of concepts called design for manufacturing and design for assembly, which incorporate a variety of manufacturing and assembly constraints directly into the design process. More holistic approaches, variously called simultaneous engineering, concurrent engineering or life-cycle design, bring together teams of engineers and experts from all phases of the product's life cycle.

Simultaneous engineering has two benefits: reduced production costs and shortened design and production cycles. Many studies have shown that over 70 percent of the cost of manufacturing an item is fixed when its design if frozen. Under the old system, manufacturing engineers had little leverage over the production costs of a new product. Under the new system, production costs are taken into account during design, and much greater economies are possible. Simultaneous engineering also substantially reduces the lead time between conceptual design and commercial production. In today's fast-paced markets, shorter lead times can be an even greater competitive advantage than lower unit costs.

The first flexible manufacturing systems (FMS) were designed and built in the late 1960s and early 1970s. When the systems were installed, costs were often greater and benefits less than had been anticipated. The new systems were more complex and had a greater variety of pathological behaviors than the machines they replaced. Workers familiar with manually controlled machines or stand-alone computer-controlled numerical machines were not able to operate or troubleshoot FMS systems without special training. Yet many firms, particularly in the U.S., failed to provide training, apparently because managers underestimated the difficulty of the transition.

The installation of an FMS system is ideally accompanied by the reorganization of the supply chain as well as by the retraining of the workforce. An FMS system makes the

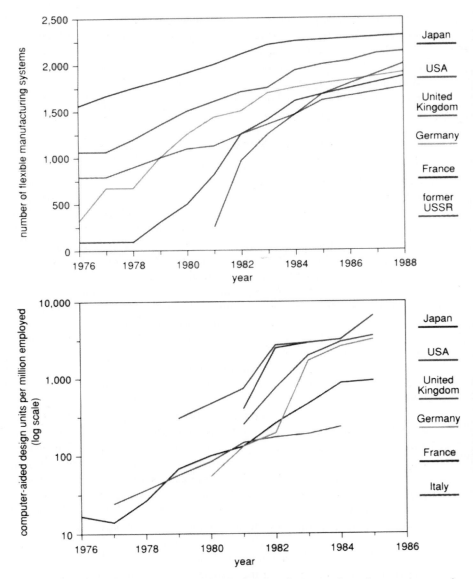

FIGURE 6. Although only 1 or 2 percent of U.S. manufacturing companies make extensive use of computer-integrated manufacturing (CIM) today, surveys indicate that adoption of key CIM technologies is accelerating. Shown here are results from surveys by the International Institute for Applied Systems Analysis, indicating the diffusion of flexible-manufacturing and computer-aided-design (CAD) systems in major manufacturing countries. CAD is one of the few CIM technologies in which the U.S. maintains a considerable lead over other countries, a state of affairs consistent with the traditional U.S. emphasis on engineering innovation over production technology. (Adapted from Iouri Tchijov, "The Diffusion of FMS," a chapter in Ayres and Haywood 1991.)

manufacturing process much more vulnerable to interruption. Because at least 70 percent of the materials used in a modern factory are brought in from outside, flexible manufacturing will be productive only if it is paired with a subcontracting network capable of delivering high-quality parts and components exactly on schedule. In Japanese-style production, the final assembler works intimately with suppliers and may even share ownership. At the very least the parties eschew the adversarial relations typical of suppliers and manufacturers in the U.S. and much of Europe. A Massachusetts Institute of Technology study of the automobile industry estimated that in the mid-1980s cooperative supply networks gave Japanese automotive manufacturers an advantage of about $2,000 per car over U.S. manufacturers.

TAYLORISM AND LEAN PRODUCTION

The technocentric picture we have presented so far omits a key question: How should the manufacturing enterprise be managed? The era of mass production was also the era of "scientific management," a method developed and promoted by Frederick W. Taylor and adopted most enthusiastically by Henry Ford. The burden of recent experience is that scientific management is as obsolete as the assembly line. Indeed the major impediment to the revamping of American industry may be the scleroses of this management philosophy, rather than any technological barrier.

As Harvey Brooks of Harvard University and Michael Maccoby have pointed out, Taylorism is based on familiar but questionable assumptions about the nature of work. The assumptions are: that the most complex manufacturing enterprise can be organized into a hierarchy of independent functions, tasks and subtasks; that boundaries between management levels and functions should be well-defined and tasks should be formally codified and reduced to "work rules"; that the most efficient way to manage work is to train each worker to do one task or subtask; that there is a single best way to accomplish each task, which can be discovered by time-and-motion studies; and that the organization should be directed from the top, by managers who have absolute control of every aspect of the business.

The first critiques of Taylorism came from organized labor, which attacked task-level optimization as exploitative and inhumane. The labor movement has long since adapted Taylorism to its own purposes, however, and now uses detailed job descriptions and work rules as a means of job protection. Yet the original criticism was not without merit. Taylorism fails to extract the best from workers in part because it is based on restrictive notions of their abilities.

Two other criticisms of Taylorism have also been made. Economists and sociologists have pointed out that hierarchical organizations tend to obstruct the flow of information needed to coordinate their activities. Middle-level managers have more incentives for impeding the flow of information than for facilitating it. Taylorism also assumes that information need only flow from the top of the organization downwards. Yet significant elements of the organization's expertise reside at all levels of the hierarchy, and some of the most critical expertise resides at the lowest level.

The third criticism strikes at the very heart of Taylorism. In the 1940s and 1950s mathematicians working with linear programming models noted that the optimization of

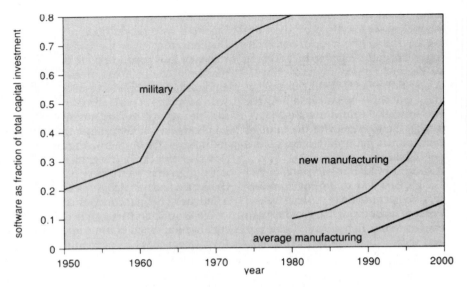

FIGURE 7. Flexibility of a modern manufacturing system derives from the incorporation of key system elements in software rather than in hardware. Investment in software by U.S. manufacturing industries is growing as a fraction of capital investment—a trend that follows the pattern that was seen as the military invested in sophisticated products and systems in the 1960s and 1970s—and much new manufacturing is software-intensive. The economic implications of the increasing capital investment in specialized software are poorly understood, but this graph probably understates the trend, since in-house software development is often not counted as a capital investment. Software requirements are expected to grow disproportionately with the growth of computer control in manufacturing.

individual steps in a process does not necessarily result in the optimization of the process as a whole. Industrial managers have been very slow to recognize the implications of this finding. Nevertheless, the painstaking attempt to reduce work to its elements and to optimize at the task level is fundamentally flawed.

Taylorism also incorporates several hidden assumptions that have been less criticized because they have been less clearly understood. One of these assumptions is that the manufacturing problem is static. The hierarchical organization of work makes sense only if product designs are stable and production technologies are enduring. Product designs have become ephemeral and production solutions fleeting, but organizational structures have remained the same. In today's fluid business environment, the resistance of the familiar organizational pyramid to change is a major liability.

A second hidden assumption is that the single best way to improve productivity is to maximize the product output per unit of labor input. Large American manufacturing firms have been obsessed with reducing direct labor costs and have neglected opportunities to reduce overhead and other indirect costs, such as defect rates. The prevailing accounting methodology, which was developed in the 1920s during the ascent of Taylorism, fails to give adequate weight to quality improvement and to reducing costs that add no value—such as inventory, inspection, rework and warranty servicing. Only recently has this deficiency been addressed in depth by the accounting profession.

An alternative to Taylorism, called Japanese-style production, or lean production,

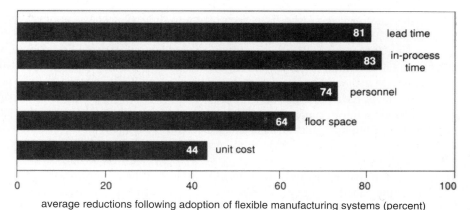

average reductions following adoption of flexible manufacturing systems (percent)

FIGURE 8.   Benefits of flexible manufacturing for a manufacturing company can include substantial reductions in turnaround time, labor costs, floor space and unit costs. IIASA surveyed three applications of FMS systems—metal cutting, metal forming and welding assembly—and reported the average reductions in allocations of space, personnel and time required for manufacturing that resulted from the adoption of FMS in each sector; these are collapsed here into an overall average. The payback time for investments in FMS equipment was fastest—3.1 years—in metal forming; the average payback time for the three applications was 3.8 years.

emerged in the 1970s. The elements of lean production are familiar to readers of the business press. They include: stable, cooperative relationships with suppliers of parts and components; flexible rather than hard automation; an emphasis on quality control; the drastic reduction of work-in-progress and stored inventory; the integration of product design and production planning; a motivated and adaptable workforce; and a flattened organizational structure.

We would contend that these elements have been more important to the success of Japanese manufacturing than trade protectionism or government investment. Lean production may in fact be more important to the future of American business than CIM. At any rate, the evidence suggests that unless a company first adopts the basic tenets of lean production, conversion to CIM is likely to fail; cultural change must precede rather than follow—or, at least, it must accompany—technological change.

Although the practices of lean production do not depend in any fundamental way on the use of computers, they are not inconsistent with CIM technologies and can even be facilitated by them. For example, Japanese products are typically designed by teams that include production engineers and representatives of part suppliers. CAD/CAM systems could facilitate teamwork by allowing team members to exchange information electronically. Production expertise might even be incorporated in software advisors residing on CAD workstations, so that it would be available on demand.

HUMAN-INTEGRATED MANUFACTURING

The truth is that converting to lean production may be harder for American firms than installing CIM systems. The difficulty of cultural change is nowhere more apparent than in the case of the workforce. Japanese and German manufacturers are said to consider the worker their most important fixed asset, but this is a sentiment not traditionally shared by

American managers, who have been taught to think of labor as an unstable and costly input to the production process.

CIM brings the American attitude toward labor to a crisis. For many firms, one attraction of CIM is that it increases productivity and reduces errors by eliminating labor. But CIM also requires workers to be much more multi-skilled than in the past. Workers who might once have managed a high-speed lathe must now be able to manage or even to program three or more computer-controlled numerical machines, and they may also be authorized to make maintenance and procurement decisions. Under these circumstances, it can be counterproductive to think of workers as variable inputs rather than as valuable assets.

It is interesting that European and Japanese attitudes toward CIM are very different from the American attitude. A representative of Toyota has said that Toyota doesn't even use the term CIM, because "the computer is only a tool that we use in our process, not really the heart of our system of manufacturing." Indeed, European and Japanese firms fear that computerized systems may displace skilled labor prematurely, with the result that interpretive and cognitive skills will be irretrievably lost. They worry that these losses might outweigh any gains a CIM system might offer.

In short, Japanese and European firms recognize that although the computer is more flexible than the lathe or the punch press, people are more flexible than the computer. Many overseas firms are moving not toward CIM but toward what some call human-integrated manufacturing. Human-integrated manufacturing recognizes that computers are useful tools, but emphasizes that computer integration begins and ends with people. The purpose of the computers is ultimately to allow the people to work together more effectively.

Why have attempts by U.S. companies to create a motivated workforce met with indifferent success? Perhaps only partly because of the managerial culture of Taylorism. Organized labor has also been obstructive. In the end, however, the difficulty may lie even deeper. Mass production led to the steady de-skilling of an increasingly itinerant workforce, a process that has proved difficult to reverse. We may simply be more deeply committed to the culture of mass production than other countries.

THE DIFFUSION OF CIM

How widespread are CIM technologies, and how fast are they spreading? Somewhat surprisingly, we don't really know the answers to these questions.

Because it can be very difficult to quantify the degree of integration a CIM system achieves, observers typically measure the diffusion of CIM elements rather than of CIM systems. Most surveys focus on metal cutting and do not ask about the automation of other aspects of the business. Most surveys also ask whether a firm or plant owns a particular CIM technology, not how much of the firm's manufacturing capacity has been converted to the technology.

Nonetheless, it is clear that CIM is still not in widespread use. It is clear that at most, only 1 or 2 percent of U.S. manufacturing companies have approached full-scale use of FMS and CAD/CAM, let alone CIM systems. The IIASA study identified 107 plants in the U.S. (mostly large) that employ all or nearly all of the elements of CIM, and

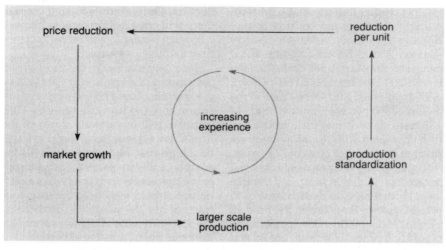

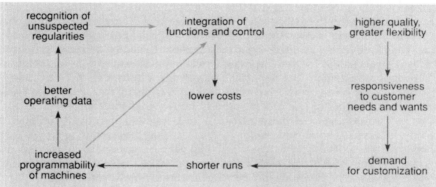

FIGURE 9. Mass production has served as an engine of economic growth driven by economies of scale (*top*). It remains to be seen whether the "economies of scope" provided by flexible manufacturing can drive economic growth in the same way. In a mass-production system, the unit cost of production is driven downward as production volume goes up. Manufacturers can lower retail costs, permitting more people to buy goods, and they can employ more workers, providing expanding payrolls that can be spent on more products. Increased sales fund investment in even higher-volume production lines, which reduce costs further, and so on. A different spiral (*bottom*) can be envisioned for the new production technologies. Computerized manufacturing systems might allow a manufacturer to produce more product variants for a given capital investment, allowing manufacturers to offer a greater variety of products, and stimulating the expansion of markets responsive to quality and diversity rather than to cost. Increased sales would then fund the adoption of more sophisticated automation, and further diversification, and so on. The sharing of capital investment among product variants might replace lower unit costs as the force driving economic growth.

where integration is quite advanced. On the other hand, a survey of manufacturing companies conducted by the U.S. Census in 1988 showed that of the 100,000 manufacturing "units" in the U.S., more than 40 percent are using one or more elements of CIM technology.

Integration at the lowest level lags behind that at higher levels. Machine groups on the shop floor are less likely to be able to exchange information than are the organizational units of multiplant, multi-product corporations. In particular, it is still uncommon for CAD systems to directly drive FMS, computer-aided process planning, or other production systems.

Is diffusion likely to continue at this slow pace? The late Yuri Tchijov of IIASA used the institute's data on FMS and a "diffusion model" to predict the growth of this technology. He estimated that the number of installed FMS systems will increase by 11 percent per year in the 1990s, and that growth will then slow to 3.6 percent per year. This would mean that there would be 5,000 FMS systems worldwide by the year 2010. Of these 1,060 would be in the U.S. and 1,200 would be in Japan.

Tchijov's predictions are based on an S-curve diffusion model, the traditional way of predicting the adoption of a new technology, but we doubt whether it has much relevance to CIM. For one thing, the model assumes that the technology is a stable entity. In reality, CIM systems grow and change from the moment they are installed. The protean nature of the technology may render the S-curve meaningless.

There are many reasons for thinking CIM will be adopted more quickly than Tchijov predicts. CIM systems, which are reaping the benefits of exponential growth in computing power and communication bandwidths, are improving rapidly. Many major firms, including Nissan Steel, Digital Equipment Corporation, IBM, NEC and Hewlett-Packard, are now competing to sell "complete" CIM packages. For these and other reasons we would be very surprised if at least 25 percent of U.S. manufacturing firms were not claiming to utlize CIM by the turn of the century.

CIM AND THE COMPANY

How does CIM affect the profitability of a company? Because so few manufacturing companies are making extensive use of CIM technology, data on its economic impact are still sparse and difficult to interpret. Nonetheless, some general observations can be made.

Most studies indicate that CIM technologies produce astounding reductions in unit labor costs. One study showed that in European firms, CAD increased the output per draftsman between 200 percent and 6,000 percent, with an average gain somewhere in the region of 500 percent. Another study of 137 cases showed that the introduction of FMS systems led to an average reduction in unit labor costs of 77 percent.

But CIM is by no means only about labor saving. Studies in Japan and the U.S. have shown repeatedly that desires to achieve greater flexibility in the product mix, to reduce product development and production cycles, to improve quality and to reduce materials costs are more important motivators in the move toward CIM than is the desire to reduce labor costs.

CIM also clearly increases the productivity of the capital invested in automation. When an FMS system replaces a conventional assembly line, for example, one of the greatest benefits is capital savings. Conventional manufacturing systems use machine tools rather inefficiently. In a typical small-scale U.S. job shop, machine tools actually engage in productive work a scant 6 percent of the time. Large-scale, high-volume producers keep the tools busy only 22 percent of the time. The utilization rates for the numerical machines in an FMS system, on the other hand, already approach 40 percent and may be as high as 60 percent by the end of the decade.

These results are startling, but the shifting demands of the market have made both labor productivity and capital productivity uncertain measures of a firm's competitiveness. Other benefits of CIM, such as shorter design and production cycles, less set-up time, smaller inventories and lower reject rates, may actually be more important. A company that is fast out of the blocks, for example, has a major advantage in a market characterized by the rapid evolution of product designs.

The adoption of CIM will also affect the business environment in subtler ways that are harder to quantify but that could eventually be equally important to the bottom line. America's traditional leadership in product innovation may become less and less relevant. As product life cycles contract, the competitive advantage will shift from innovative firms to responsive firms. Short turnaround times will make imitation progressively easier and the advantage conferred by innovation ever more fleeting. Only manufacturing competence will offer real protection, because it is much harder to imitate competence than to imitate a product.

For many managers, the most unsettling aspect of migration to CIM will be the key role played by software and decisions about software. Because the CIM software will incorporate much of the firm's "knowledge base," writing it will be a high-risk, high-stakes task. Large, complex software projects have a frightening propensity to go awry. Nor will it be possible to avoid the risk by contracting out the work; the purpose of the software, after all, is to capture the in-house manufacturing expertise.

Software is also something of an accounting problem. An increasing percentage of industry's fixed capital consists of specialized software. But many firms are forced by accounting rules to treat in-house software development as an expense rather than as an investment. As a result, economists don't really know what fraction of total capital costs is invested in software, although it is clear that the fraction is growing. In general the economic significance of the migration of capital from hardware to software is poorly understood.

CIM AND THE U.S.

If enthusiastic adoption of CIM makes individual companies more competitive, will wholesale adoption of CIM make the U.S. more prosperous? For most of this century an important engine of growth was the economies of scale achieved by mass production. Reduced prices and larger payrolls led to increased demand, which paid for further capital investment, which further reduced unit costs, and prices, in an indefinitely rising spiral.

CIM is unlikely to work in this way for several reasons. CIM increases labor productivity dramatically, but at the same time labor productivity has become a dubious measure of economic growth. Direct labor costs now account for only 10 to 20 percent of total factory costs, on average, and many observers predict that by the second decade of the next century, direct labor will account for as little as 3 percent in many manufacturing sectors. There is little room for further savings here.

Capital productivity has not experienced a similar increase, but it is difficult to measure, and its influence on the economy is unclear. The optimistic view is that increases in capital productivity will become the new engine for economic growth. The notion is that increasingly more flexible automation will progressively decrease the cost difference between customized and standardized products.

The difficulty is that it is unclear whether mass manufacture of customized goods will permit a relative reduction in prices. If relative prices of manufactured items do not continually spiral downward, will the economic growth engine eventually stall? Or could a changing panoply of goods act much as diminishing prices did, continually stimulating the demand that keeps the economic engine running?

A further difficulty is that it is still too early to predict CIM's effect on employment levels. CIM will certainly decrease the number of unskilled jobs available in the manufacturing sector, but its effect on aggregate employment is less clear. Improved capital productivity will be meaningless unless it is accompanied by the high levels of income—hence employment—needed to sustain demand.

One minor but interesting side effect of CIM may be to alter the familiar shape of the business cycle. Traditionally, periodic efforts to build or reduce inventory have provided much of the impetus for the business cycle. CIM systems should allow companies to fill customers' orders without costly inventories, and therefore allow the firms to match production to fluctuating demand without hiring or firing workers. The result may be flatter employment statistics; companies will probably lay off fewer workers during an economic downturn, but they will probably also hire fewer new workers during an upturn.

CIM AND THE WORLD ECONOMY

The impact of computer-integrated manufacturing on the world economy is even less certain than its impact on the U.S. economy. This can be appreciated by considering just one of the many questions that could be asked about CIM's impact on the world economy: How will CIM affect the competitive position of less-developed countries?

If CIM remains the preserve of developed countries, the prospects for the less-developed countries are grim. Cheap labor is the only advantage many of these countries offer to manufacturers, but labor costs are increasingly unimportant. Once CIM is widespread, manufacturers will have little incentive to produce in low-wage areas for export to their home countries. Instead they will have strong incentives to invest in productive capacity in the market to which they expect to sell. This trend is already visible in the automobile and consumer-electronics sector, where there has been a substantial increase in investment in the U.S. by European and Japanese producers. Under these conditions, it is unlikely that less developed countries will be able to emulate the "Asian tigers," for whom lowcost labor provided a springboard to export-led growth.

At least one scenario offers a glimmer of hope, however. Countries such as India and China, which have large internal markets, may be able to attract foreign investors who would bring CIM technology with them. If the technology can be successfully transplanted (and the struggles of U.S. companies suggest that this might be difficult), these countries might be able to fuel economic gorwth by replacing expensive imports with cheaper, locally produced goods. Perhaps smaller nations could exploit the same strategy by joining together to form regional common markets.

CONCLUSION

In this article we have deliberately tried to move beyond a strictly technical description of CIM and to look instead at its implications for corporate culture and for the economy. The American tendency is to assume that technology is a panacea whose impact is always positive. This assumption has never been true, but at this late date there is perhaps less excuse for technological boosterism.

The advantages of CIM are not in doubt, nor is the eventual migration of industry from mass production to flexible manufacturing. Without CIM, manufacturing concerns will be unable to meet the market demand for an everchanging array of complex, well-made goods. But CIM is a constantly evolving matrix of solutions to an exceptionally complex set of manufacturing problems. There is no single best way of solving manufacturing questions—and in the case of CIM, precious few precedents. The more clearly these problems are understood, the more likely the economic dislocations and human suffering that accompanied the transition to mass production can be avoided during the transition to CIM.

THE STUDY

The IIASA study resulted in four volumes: *Computer Integrated Manufacturing: Revolution in Progress; Computer Integrated Manufacturing: The Past, the Present & the Future; Technology & Productivity: The Challenge for Economic Policy; and Computer Integrated Manufacturing: Economic & Social Impacts.* All four were published by Chapman & Hall in 1991.

BIBLIOGRAPHY

ABERNATHY, WILLIAM J. 1978. *The Productivity Dilemma.* Baltimore: The Johns Hopkins University Press.

ABERNATHY, WILLIAM J., KIM B. CLARK and ALAN M. KANTROW. 1983. *Industrial Renaissance: Producing a Competitive Future for America.* New York: Basic Books.

AYRES, ROBERT U. 1984. *The Next Industrial Revolution.* Cambridge, MA: Ballinger.

AYRES, ROBERT U., and WILLIAM HAYWOOD. 1991. *The Diffusion of Computer Integrated Manufacturing Technologies: Models, Case Studies & Forecasts.* New York: Chapman & Hall.

BURT, D. N. 1989. Managing suppliers up to speed. *Harvard Business Review* 7/8:127–135.

CLARK, K. B., and T. FUJIMOTO. 1989. Overlapping problem solving in product development. In *Managing International Manufacturing,* ed. K. Ferdows. Amsterdam: Elsevier.

CLARK, K. B. 1989. Project scope and project performance; The effect of parts strategy and supplier involvement on product development. *Management Science,* 35(10):1247–12623.

COHEN, STEVEN, and JOHN ZYSMAN. 1986. *Manufacturing Matters.* New York: Basic Books.

COLE, W., A. BOS and R. D. SANDERS. 1991. Management Systems as Technology: United States, Japanese and National Firms in Brazil. Paper presented at the Annual Meeting of the Southwestern Economics Association.

EBEL, K. H., and E. ULRICH. 1987. Some workplace effects of CAD and CAM. *International Labour Review* 126(3):351–370.

EBEL, KARL H. 1990. *Computer-Integrted Manufacturing: The Social Dimension.* Geneva: International Labor Office.

ETTLIE, J. 1988. *Taking Care of Manufacturing.* San Francisco: Jossey-Bass.

Management methods keep Japan ahead in CIM. 1988. *FMS Magazine,* July, p. 149.

GUNN, T. G. 1987. *Manufacturing for Competitive Advantage.* Cambridge, MA: Ballinger.

HICKS, DONALD A., ed. 1988. *Is New Technology Enough? Making and Remaking U.S. Basic Industries.* Washington, DC: American Enterprise Institute for Public Policy Research.

KUMPE, T., and J. BALDWIN. 1988. Manufacturing: The new case for vertical integration. *Harvard Business Review* 3/4:75–81.

MCALINDEN, S. 1989. *Programmable Automation, Labor Productivity and the Competitiveness of Midwestern Manufacturing.* Ann Arbor, MI: Industrial Technology Institute.

Manufacturing Technology. 1988. *Current Industrial Reports: 1989.* Washington, D.C.: U.S. Dept. of Commerce, Bureau of the Census.

National Research Council. 1986. *Toward a New Era in U.S. Manufacturing: The Need for a National Vision.* Washington, DC: National Academy Press.

PIORE, M., and L. SABEL. 1984. *The Second Industrial Divide.* New York: Basic Books.

REMBOLD, U., B. O. NNAJI and A. STORR. 1993. *Computer Integrated Manufacturing and Engineering.* Reading, MA: Addison-Wesley.

TCHIJOV, I. 1989. *FMS in Use: An International Comparative Study.* WP-89-45. Laxenburg, Austria: IIASA.

WOMACK, JAMES P., DANIEL T. JONES and DANIEL ROOS. *The Machine That Changed the World.* 1990. New York: Macmillan.

Chapter Sixteen

Theory of Constraints and Synchronous Manufacturing

INTRODUCTION

A Brief History

In the late 1970s Eliyahu Goldratt, an Israeli physicist, began to present his ideas on production scheduling. He developed a proprietary computer black-box software program known as optimized production technique (OPT). This software was sold to companies with no information about the theory or the methodology of OPT. The promise was that the schedules developed would take capacity into account and would make more efficient use of capacity-constrained resources to maximize the throughput. Purchasers of the OPT software reported some successes and some failures, but the overall perception was that the secret algorithms prevented more widespread acceptance.

In 1984 Goldratt and Cox published a novel, *The Goal* [6], which presented some of the concepts underlying OPT. This was followed in 1986 by *The Race* [7], which further explained the concepts. OPT, the software, was abandoned by Goldratt about that time, but he continued the education and marketing of his ideas.

Umble and Srikanth [11] presented a detailed look at these concepts, then known as synchronous manufacturing, in 1990, and claim that the term was coined in 1984 at General Motors. With more widespread understanding, synchronous manufacturing concepts have been adopted by many more companies.

In the late 1980s Goldratt refined his ideas into what is now known as the theory of constraints, an expansion of his original OPT concepts. He has since made the concepts more widely known through seminars and the publication of *What Is This Thing Called Theory of Constraints* [9], which includes a management philosophy on improvement based on identifying the constraints to increasing profits.

Basic Concepts

The basic concept of synchronous manufacturing is simple: The flow of material through a system, not the capacity of the system, should be balanced. This results in materials moving smoothly and continuously from one operation to the next; and thus lead times and inventory waiting in queues should be reduced. Improved use of equipment and reduced inventories can reduce total cost and can speed customer delivery, allowing a company to compete more effectively. Shorter lead times improve customer service and give a company a competitive edge.

In synchronous manufacturing, the bottlenecks are identified and used to determine the rate of flow. To maximize flow through the system, bottlenecks must be managed effectively. Called capacity constrained resources, these bottlenecks led Goldratt to expand the idea of managing constraints. The theory of constraints expands the concept to include market, material, capacity, logistical, managerial, and behavioral constraints.

THEORY OF CONSTRAINTS

The basis of the theory of constraints is that every organization has constraints that prevents it from achieving a higher level of performance. These constraints should be identified and managed to improve performance. Usually only a limited number of constraints exist, and they are not necessarily capacity constraints. When a constraint has been broken, identify the next constraint and improve that, thus continuing the process of improvement.

Five Steps

At any time, usually very few constraints prevent the improvement of performance. A five-step process works at one constraint at a time.

1. Identify the system's constraints.
2. Describe how to exploit the system's constraints.
3. Subordinate everything else to the above decision.
4. Elevate the system's constraints.
5. If in the previous steps a constraint has been broken, go back to step 1.

Step 1 is to identify the system's constraints and to prioritize them according to their impact on the goal. Although there may be a lot of constraints at any time, usually only a few really constrain the system at that moment.

Step 2 is to determine how to exploit those constraints to improve performance. Once we believe that there are only a few constraints that are limiting performance, then all other resources are not constraints. Hence Step 3 is included to ensure that the other resources are subordinated to the constraints. There is no reason to spend extra time managing resources that are not constraints to improved performance.

Step 4 states that the constraints must be elevated so that action is taken to reduce their impact and improve performance. When this is accomplished, we cannot stop, because there is a natural tendency to slip back to the old ways. Thus Step 5 is included to make sure that we move forward and find the next constraint.

There is a strong parallel here with the continuous improvement processes that are espoused in total quality management (see Chapter 20). Both processes suggest that the way to proceed with improvement is not through major changes but through small, incremental changes to a limited set of problems. Once a problem is solved, another will be found that needs to be solved. This continuous process of improvement should not stop. The cumulative effect of small improvements is major improvement, which will result in increased customer satisfaction, reduced costs, and improved profits.

THE GOAL
AND PERFORMANCE MEASURES

The Goal

The first sections of this chapter mention the goal, which according to Goldratt, is to make a profit. The financial performance measures of importance are net profit, return on investment (ROI), and cash flow. Operations measurements are throughput (T), inventory (I), and operating expense (OE). Throughput is defined as the rate at which the system generates money through sales (not production). Inventory is all the money the system invests in purchasing things that the system intends to sell. This does not include labor and overhead. Operating Expense is all the money the system spends in turning inventory into throughput. This includes all the other expenses, including direct labor. Net profit is, then, throughput minus operating expense ($T - $ OE). Return on investment is throughout minus operating expense, divided by inventory:

$$\text{ROI} = \frac{T - \text{OE}}{I}$$

These measures are defined this way to focus decision making on activities that will improve the goal of making a profit. Because it has the greatest impact on net profit and ROI, throughput is elevated to the most important measure. Once we have focused on throughput, we can determine the constraints that are preventing throughput from increasing and thus increasing profit.

Goldratt's Rules

Goldratt's motto is "The sum of the local optimums is not equal to the global optimum." He suggests nine global rules that differ significantly from conventional wisdom.

1. Do not balance capacity, balance the flow.
2. The level of utilization of a nonbottleneck is determined by some other constraint.
3. Utilization and activation of a resource are not the same.
4. An hour lost at a bottleneck is an hour lost for the entire system.

5. An hour saved at a nonbottleneck is a mirage.
6. Bottlenecks govern both throughput and inventory.
7. The transfer batch does not have to equal the process batch.
8. A process batch should be variable.
9. Priorities can be set only by examining the system's constraints.

These nine rules form the foundation for synchronous manufacturing. Rule 6 states that the overall throughput of a system is determined by the bottleneck. Note that the other resources, which are not bottlenecks, have no effect on the overall throughput, unless they are poorly scheduled and cause the bottleneck to be underutilized. If the bottleneck has no work available, then that production time is lost and cannot be made up; thus the bottleneck determines the throughput and must be the best managed (Rule 4).

Conventional wisdom suggests that machine centers should be fully utilized to keep labor working. If they are nonbottlenecks, however, the production on those nonbottlenecks will only end up in inventory and will not increase the throughput of the system (Rule 5). The idea, then, is to synchronize the flow of material through the system.

CAPACITY

Bottlenecks and Nonbottleneck Resources

A bottleneck resource is one whose capacity is equal to or less than the demand placed on it. A nonbottleneck is one whose capacity is greater than the demand placed on it. Goldratt designates bottleneck resources with X and nonbottleneck resources with Y.

These two building blocks, X and Y, result in five basic interactions. Assume that both X and Y have forty hours available per week and that a product that flows through these two resources uses ten minutes per unit on X and six minutes per unit on Y. It is easy to see that the throughput on X is 240 ($40 \times 60/10$) units per week and on Y is 400 ($40 \times 60/6$) units per week.

The five basic interactions are as follows:

1. Y to X: flow from nonbottle neck to bottleneck.

$$\longrightarrow \boxed{Y} \longrightarrow \boxed{X} \longrightarrow$$

2. X to Y: flow from bottleneck to nonbottleneck.

$$\longrightarrow \boxed{X} \longrightarrow \boxed{Y} \longrightarrow$$

3. Y1 to Y2: flow from one nonbottleneck to another.

$$\longrightarrow \boxed{Y1} \longrightarrow \boxed{Y2} \longrightarrow$$

4. X1 to X2: flow from one bottleneck to another.

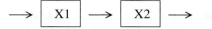

5. X and Y to assembly: a bottleneck and nonbottleneck feed into an assembly operation.

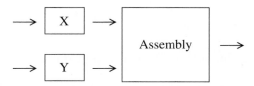

In Case 1, if the first resource is scheduled in traditional fashion, it would produce 400 per week and these would feed to X. Note that X can only move 240 per week through, leaving an inventory buildup of 160 per week. This does not increase the throughput of the system because that is determined by the bottleneck to be 240 per week. If Y produces only 240 per week, it will be utilized for twenty-four hours per week. Note that if this machine is down for an hour due to a mechanical failure, there is no effect on the throughput of 240 units per week (see Rule 5). On the other hand, if X is down for one hour, then the throughput now becomes 234 units; this cannot be made up and is lost (see rule 4). This situation illustrates that the Y resource must be synchronized with the X resource to balance the flow.

In Case 2, the X resource produces 240 units per week that are fed to the Y resource, which can process those 240 in twenty-four hours. As long as there is material available, X can continue to produce at maximum throughput. This case also illustrates the concept of transfer batch size. Define the process batch as the 240 units per week that are to be processed through this system. If all 240 are first processed at X and then moved as a batch to Y, the total lead time to process all 240 will be sixty-four hours (forty hours on X plus twenty-four hours on Y), as shown in Figure 16.1.

A transfer batch is the quantity that will be transferred to the next operation. Assume that the transfer batch is ten units. The first ten units will take 100 minutes to move through X; these can then be moved to Y where the processing can start and be completed in sixty minutes. The second transfer batch of ten units will be done on X in 200 minutes, transferred to Y, and completed on Y at 260 minutes. The total lead time to process the 240 units will be forty-one hours (forty hours on X plus sixty minutes to

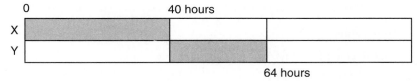

FIGURE 16.1 Total lead time when transfer batch equals process batch.

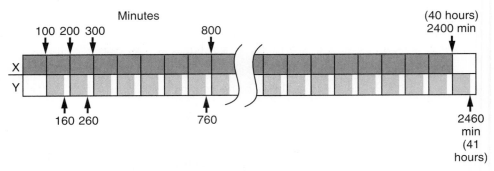

FIGURE 16.2 Total lead time when transfer batch equals ten.

process the last transfer batch of ten on Y), as shown in Figure 16.2. This is a twenty-three hour reduction in throughput time!

Case 3 illustrates the situation when both resources are nonbottlenecks. If the demand for this product is only 240 units per week, there is no need to produce at a rate of 400 per week because there will be an inventory buildup of 160 per week. In this case it is important to have the 2Y resources synchronized with the demand.

Case 4 illustrates the situation where the capacities are both 240 units per week or when the demand is greater than the capacity of either, even if the throughput rates are different for X1 and X2. For example, say that X1 has a throughput rate of 240 units per week while X2 has a throughput rate of 200 per week. If the demand is 260 per week, then both are bottlenecks because their capacity is less than the demand placed on them. If the demand were 240 per week, then X2 would be a Y nonbottleneck resource.

Case 5 shows why these two resources must be synchronized when feeding into an assembly operation. Keep the throughput at 240 for X and 400 for Y. Obviously, the assembly operation will not be able to produce more than 240 per week because the X resource is the constraint. Note that the Y resource thus must be synchronized at 240 per week, or inventory will build up after Y and before the assembly operation.

Capacity Constraint Resource

A capacity constraint resource (CCR) is a resource that, if not properly scheduled and managed, is likely to prevent the product flow to deviate from the planned flow. Note that a bottleneck can be a CCR, but so could a nonbottleneck if not properly scheduled. For example, in case 1 above say that another product uses sixteen hours of capacity on the Y resource. If the Y resource is processed first during the week and sent to another work center, then the twenty-four hours of processing for the product to flow to X would be done later in the week. It may arrive too late at the X resource to be processed in sufficient time to reach the customer at the desired time. Thus Y can be considered a CCR.

Constraints

A CCR is not the only type of constraint that can be holding back performance. Market constraints may prevent the full utilization of the manufacturing resources available. Increased market will increase throughput and net profit. Material constraints may prevent the utilization of the resources. If capacity is greater than the current throughput with a material constraint, more materials would increase throughput and profit.

Logistical constraints would include planning and control functions such as order entry or material control systems. Long lead times in order entry and scheduling may place a constraint on the company's ability to win orders in the market through fast customer service.

Managerial constraints are those strategies and policies that prevent the system from improving performance. Consider the process of batch sizing using EOQ or other mathematical processes. These may provide batches that are not the right size for synchronous product flow.

Behavioral constraints may be the most difficult constraints to eliminate. In earlier discussions of utilization of nonbottleneck resources, it was apparent that it is better to underutilize the resource because inventory will build up. A significant behavioral problem is that of keeping machinery busy so that it looks good to management. Idle workers and machines are perceived to be wastes of labor time and valuable assets, when in reality they may indicate that product flow is being synchronized.

SYNCHRONOUS MANUFACTURING

In his groundbreaking novel *The Goal,* Goldratt used the concept of a troop of Boy Scouts on a hike through a forest to illustrate the idea of synchronization. Consider a troop of Boy Scouts that is to march in a line as being analogous to process flow. During the march, some hikers are slower than others. If each is allowed to set his own pace (produce at his own rate), then the troop will become spread out (as work-in-process inventory increases), as shown in Figure 16.3. Our goal is to keep the troop together be-

Finishing Goods   Raw Material

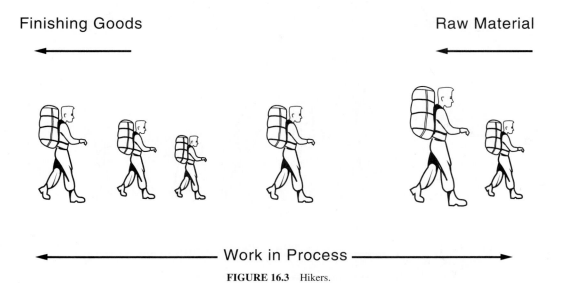

Work in Process

FIGURE 16.3   Hikers.

cause the slowest person in the group is the one who determines when all the hikers arrive at their destination. Thus the bottleneck (slowest person) is the resource that determines the throughput.

How do we keep the troop together (keep work in progress down) and reach our destination in the least amount of time (lowest lead time, maximize throughput)? One possibility is to place the slowest hikers in the front and fastest in the rear, as shown in Figure 16.4. Then the process will proceed at the rate of the first, slowest hiker. This is fine if you have the ability for a process flow with the bottleneck at the raw material level and the resources with the greatest capacity at the finished goods level. This is unlikely, however, without significant capacity purchases.

A second possibility, seen in Figure 16.5, is to leave everyone in the original order and tie them all together with a rope to ensure that they do not spread out (the paced assembly line!). This strategy will work in systems with products that can be economically produced on paced lines, but it is useless in a job shop.

A third possibility is to have a drummer set the pace at the gating operation (raw materials), see Figure 16.6. The other hikers would have to listen to the drum and keep pace or be urged to make up space if they become spread out. If the slowest hiker does not keep pace with the drummer, then the slowest, and all behind, will be separated from the front part of the troop.

Drum-Buffer-Rope Approach

The way to get all hikers synchronized is to combine the drummer with the rope. If the slowest hiker is tied by rope to the front and the drummer sets the pace of the slowest hiker, than all hikers will be forced to march at the same pace; see Figure 16.7. The hiker at the front is forced to march at the pace of the slowest because of the rope. Those be-

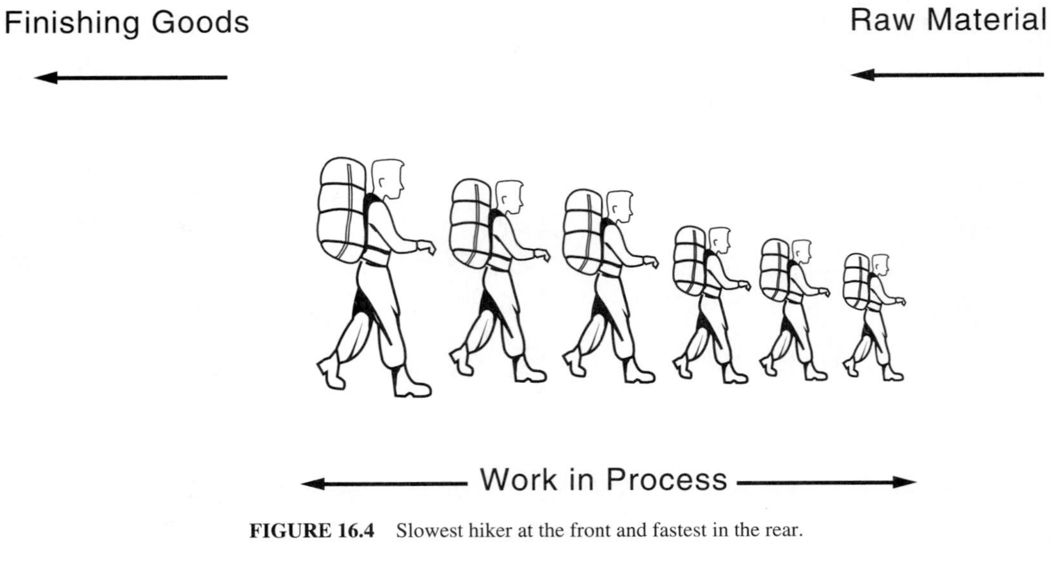

Finishing Goods   Raw Material

Work in Process

FIGURE 16.4   Slowest hiker at the front and fastest in the rear.

Finishing Goods   Raw Material

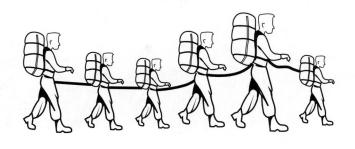

———— Work in Process ————

FIGURE 16.5   Hikers tied together with a rope.

hind the first in line will be forced to march at the same rate. Because the slowest hiker sets the pace, those behind that hiker will be forced to march at that rate. All are then marching at the same rate. If there is some variation in the rate of the front section, then some slack is provided in the rope so that the hikers may be able to slow down and speed up without interfering with the slowest hiker. This is known as the drum-buffer-rope approach.

To see how this will work with a product flow, see Figure 16.8. The CCR (black circle) will determine the throughput; this is the drummer that determines the throughput

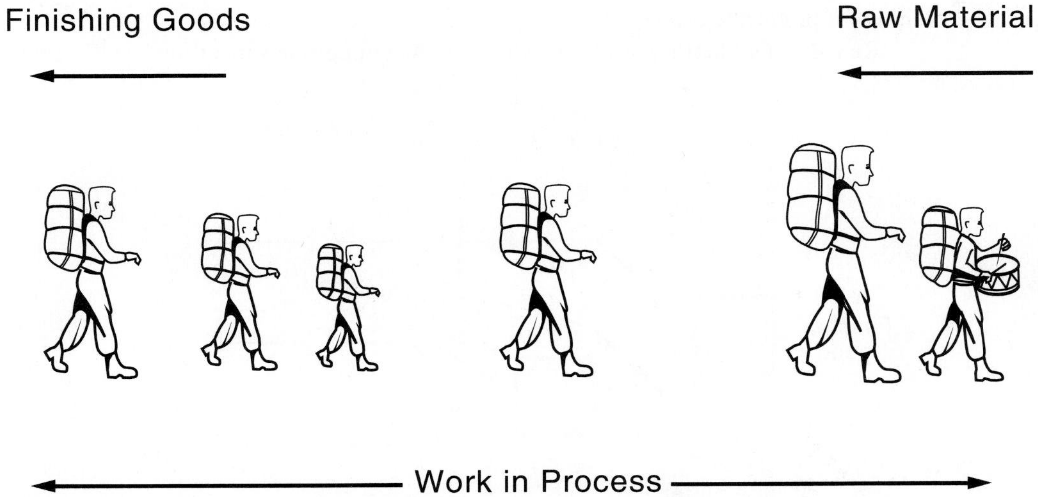

Finishing Goods   Raw Material

———— Work in Process ————

FIGURE 16.6   Drummer at the front.

Finishing Goods

Raw Material

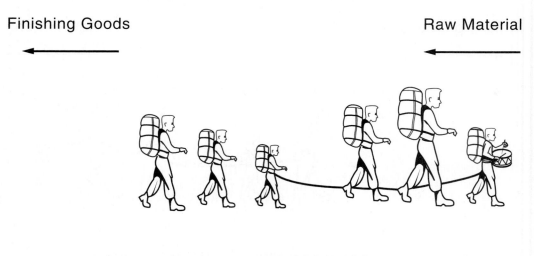

FIGURE 16.7 Drum-buffer-rope approach ties slowest hiker to the front.

of the entire operation. All operations behind the CCR will be scheduled according to the schedule of the CCR. A rope, represented by the dotted line, is tied back to the gating operation at the raw material level.

In our hiker example, the slack in the rope took care of variation in hiking speed by those hikers between the front and the slowest hiker. In the manufacturing situation, we have the same problem with variation in processing time of each operation in the process between the raw material and the CCR. If there is variation in the process times, it is possible that product flow will not be smooth into the CCR and that the CCR may sit idle waiting for product to arrive.

Rule 4 of Goldratt's rules for production scheduling states that time lost at a bottle-

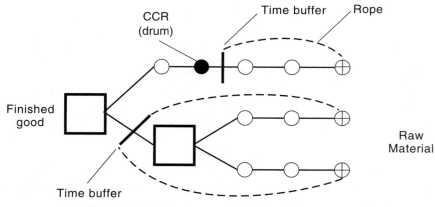

FIGURE 16.8 Manufacturing process with drum-buffer-rope approach.

neck is time lost for the entire system. It is absolutely essential that the bottleneck and CCR not lose production time. There should be a buffer just before the CCR to ensure that the CCR has no lost production time. This is known as a time buffer because it is determined by how much time needs to be buffered. For example, if disruptions occur before the CCR that could last as long as two days, it would be a good idea to have a three-day time buffer of inventory just before the CCR. Then, if there is a disruption lasting two days, the CCR has three-day's worth of inventory that it can work on and not lose any production time.

In Figure 16.8, notice the time buffers just before the CCR and just before the final assembly operation. The CCR schedule will provide the flow into this final assembly. The branch supplying the other parts into the final assembly should also be protected to ensure that the final assembly schedule is not disrupted by problems in this right branch. A time buffer should be established in front of an assembly operation requiring a part from a CCR. This time buffer then provides the start of the ropes back to the gating operations on that branch. That branch is then scheduled according to the final assembly schedule, and the final assembly schedule is protected against disruptions possible in that branch.

Example 16.1

Consider the following process flow in Figure 16.9 with the numbers indicating the capacity in units per week. (1) What is the bottleneck in this process? (2) What is the throughput rate for the process? (3) Where should the time buffers be located? (4) Where do the ropes go? (5) What will be the production rates at each work center?

Solution: (1) The lowest capacity is in process C with a rate of 240 units per week. This bottleneck will determine the throughput for the entire process. A time buffer will be located just before C with a rope leading back to A. (2) Work centers A and B will be scheduled at a rate of 240 units per week, with A the gating operation. (3) There should

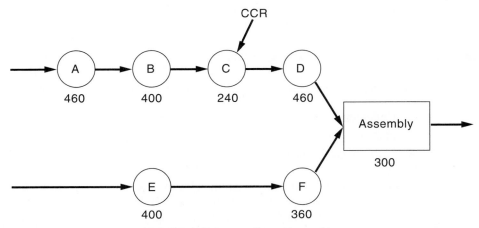

FIGURE 16.9   Process flow with capacities.

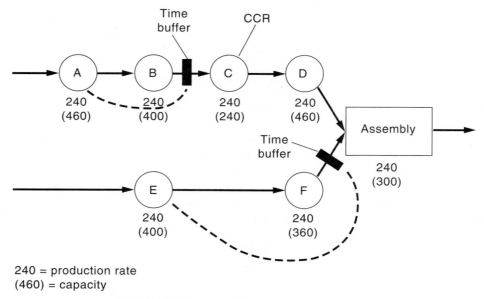

240 = production rate
(460) = capacity

FIGURE 16.10   Process flow the drum-buffer-rope way.

be sufficient time buffer inventory just before C to ensure that the variability that may occur in operations A and B will not hamper the maximum utilization of resource C. Resource D will also be scheduled at a rate of 240 per week. Because the bottleneck resource feeds into final assembly, there should be a second time buffer in the lower branch just before the final assembly operation to ensure that the final assembly will not be disrupted by the variability in the processes E and F. (A) A rope will go back to process E to ensure that both E and F are producing at a rate of 240 units per week. (5) The final assembly will also be conducted at a rate of 240 units per week. Now all flows are balanced, not the capacities. Some of the work centers will have excess capacity, but these should not be used up to the maximum to build up inventory. The drum-buffer-rope approach is shown in Figure 16.10.

MARKETING AND PRODUCTION

Example 16.2 illustrates some of the synchronous manufacturing concepts.

Example 16.2

Figure 16.11 shows two products, M and N, selling for $100 and $110, respectively, with market demands of 90 and 100 per week, respectively. Assume that we can sell that demand each week if we desired. Product M is produced from RM1, which costs $40 per unit and first requires fifteen minutes of processing on resource A and another ten minutes of processing on resource A. Product N requires RM1 processed through resource A at fifteen minutes and RM2 (cost of $20 per unit) through resource C with a time required of five minutes per unit. The two are then moved to final assembly, which

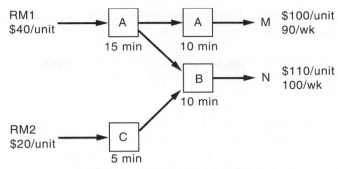

FIGURE 16.11   Market and production.

requires ten minutes on resource B. There are 2400 minutes per week available for each resource. Assume that one unit of RM1 is used to produce one unit of M and that one unit of RM1 and one unit of RM2 are required to produce one unit of N. What production schedule will maximize profit?

Solution: The first step would be to determine the net profit per unit. M would have a net profit of $100 − $40 = $60. Product N has a net profit of $110 − $40 − $20 = $50. Conventional accounting profit figures indicate that M has the greatest profit margin. Thus we would maximize production of M and use the remaining time for N.

If ninety units of M are produced, then 2250 minutes of resource A are used and a net profit of $5400 is realized from M. Resource A has 150 minutes of time left that may be used to produce ten units of N. This results in a profit of $500 from N, for a total of $5900 net profit. Resource B is used for 100 minutes and resource C is used for fifty minutes per week. Marketing would be induced to sell M and reduce the emphasis on N.

The problem with this analysis is that it does not consider the utilization of the bottleneck resource A, which is the constraint preventing increased profit. Note that M uses twenty-five minutes per unit of A, while N uses only fifteen minutes. Remember rule 4, which states that an hour lost on the bottleneck is an hour lost for the entire system. The throughput of the entire system is dependent on the bottleneck and suggests that the bottleneck should be investigated as the point to improve profits.

Because N uses less of bottleneck resource A, perhaps N should be scheduled first, leaving the remaining time for M. Production of 100 units of N would use 1500 minutes of A, 1000 minutes of B, and 500 minutes of C and would result in a net profit of $5000. That leaves 900 minutes on A for product M, and thirty-six units of M could be produced. This results in a net profit of $2160 and a total net profit of $7160, an increase of $1260 in net profit.

Applying the Theory of Constraints

At this point we can apply the five-step process to improve performance. Review the five steps given at the beginning of this chapter. Step 1 is to identify the system's constraints. Example 16.2 indicated that resource A is obviously a constraint because it is a bottleneck constraining throughput. Note, however, that the market is also a constraint because

100 units per week of N are fully produced and sold. An increase in demand for N would also increase profit.

Increase Resource A Capacity. We continue with step 1 to identify the most important constraint to study. If we could increase the demand for N, we would reach another limit at a demand of 160, which would totally consume the A resource. The net profit would be $8000, all from N.

If we focus on the resource A constraint, we will be able to increase profit and allow the production and sale of both M and N. Say that we plan to focus on resource A as the first constraint to exploit. Step 2 indicates that we need to describe how to exploit this constraint. Assume that through process improvement and setup time reductions we are able to reduce the processing times significantly on A to those shown in Figure 16.12. Now we can produce all ninety units of M and 100 units of N each week and can use 2350 minutes for production of the 2400 available. This results in a profit of $5000 from N and $5400 from M for a total of $10,400.

Steps 3 and 4 have been completed. Resource A is now using only 2350 minutes of the 2400 available, so that constraint has been broken and we can identify the next constraint to attack. The demand constraint is now the one that is preventing increased profits, so our effort will now focus on increasing demand for M or N.

Increase Demand. If we could increase demand for M or N by 1, would we increase profit the most? If we look at just the net profit for each, M with a profit of $60 is still more appealing than N with a profit of $50. If M were increased to ninety-one, then the resource A would be used 2365 minutes per week, still within the 2400 minutes available. If we increased the sales of M to the maximum allowed by the resource A available time, we could produce ninety-three units per week of M and see a profit increase of $180.

We may again be misled because we have not considered the impact of the resource A constraint. What would happen if we increased N instead, to the limit allowed by resource A? We could produce 105 units of N and increase profit by $250. If the tradeoff is between an increase of five units of N or three units of M, then N is preferred because it increases net profit the most due to its lower utilization of resource A.

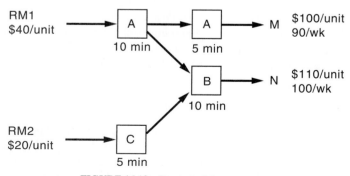

FIGURE 16.12 Resource A improvement.

In addition, if sales could be increased beyond those figures above, then profit may increase even more. Say that demand for N can be increased to 120 units per week. If ninety units of M were also sold, the time required on resource A would be 2550, which is greater than the available time. If we sell and produce 120 units of N, that would require 1200 minutes on A, leaving 1200 minutes for M. Since M requires fifteen minutes on A, eighty units of M could be produced and sold. The net profit for N would be $6000, the net profit for M would be $4800, and the total profit would be $10,800.

Say that we could increase demand for M by twenty units also. The 110 units per week of M would require 1650 minutes on A, leaving 750 for N. Then seventy-five units of N could be produced and sold for a net profit of $3750. The 110 units of M would have a net profit of $6600, for a total net profit of $10,350. This is lower than increasing N by twenty units per week and, in fact, is $50 less than if we did nothing to increase sales!

Now that sales of N have been increased to 120 per week, resource A is again a constraint. The five-step process continues to break the capacity constraint or increase sales.

This situation illustrates that the importance of identifying the constraints and determining which improvement will result in the greatest gains. The net profit margin per unit turned out to be the wrong indicator of which item to increase sales or to take the first piece of the resource constraint.

SUMMARY

Goldratt's ideas regarding production scheduling and the use of capacity have made many production and inventory professionals think differently about how to utilize resources properly. The goal of every firm is to make a profit. Constraints that prevent an increase in profits are present in every organization. The theory of constraints presents a framework for identifying constraints and improving profit performance.

Capacity constraint resources (CCR) are one of the most important constraints. The theory of constraints allow us to investigate and find other constraints, such as the market, that may be more important than the capacity to increasing profit. An important lesson with capacity is that the most important capacity constraint, a bottleneck or CCR, requires the most effort to schedule and manage, because time lost on that resource is time lost for the entire process.

PROBLEMS

1. There are forty hours per week available on each work center on the production line shown below. The times indicate the minutes required per unit at each work center.

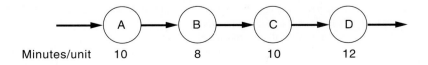

    a. Which is the bottleneck?

    b. What is the throughput per week for this line?

    c. If each work center is utilized at the rate of the bottleneck, what would be the utilization times for each work center?

    d. What would happen if we tried to run each work center at full utilization?

    e. Where should the time buffer be located?

2. The following problem uses Examples 16.1 and 16.2 (Figures 16.9 and 16.10). Since resource C is the CCR, that constraint should be broken to increase profit.

    a. If the capacity of C is increased to 400, what is the new output of this system?

    b. What resource is the new CCR?

    c. Where should the time buffers be located? Where do the ropes go?

    d. Increase this new CCR's capacity by 100. Now what is the new CCR?

    e. What is the output rate?

    f. Now where would the time buffers be located?

3. Consider the situation illustrated below. The two final products have weekly demand at the selling price as indicated. One unit of each raw material is needed for one unit of final product. Each resource has 2400 minutes of time available per week. Raw material costs are indicated.

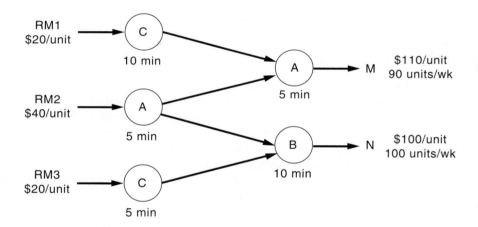

    a. What should the production volumes be to maximize profit?

    b. What are the constraints?

    c. If we had the choice of increasing demand by 100 for either product but not both, which would you select? Why?

    d. Increase the demand for both products by 100. Now what is the mix of production for maximum profit?

    e. Now what are the constraints?

    f. Increase the time available on resources A and C by 600 each. What production volumes will maximize profits?

    g. Now what are the constraints?

REFERENCES AND BIBLIOGRAPHY

1. M. Baudin, *Manufacturing Systems Analysis* (Englewood Cliffs, N.J.: Prentice Hall–Yourdon Press, 1990).
2. J. H. Blackstone, Jr., *Capacity Management* (Cincinnati: South-Western, 1989).
3. J. Brown, J. Harhen, and J. Shivnan, *Production Management Systems, a CIM Perspective* (Reading, Mass.: Addison-Wesley, 1988).
4. R. B. Chase, and N. J. Aquilano, *Production Operations Management,* 6th ed. (Homewood, Ill.: Richard D. Irwin, 1992).
5. D. W. Fogarty, J. H. Blackstone, Jr., and T. R. Hoffmann, *Production and Inventory Management,* 2nd ed. (Cincinnati: South-Western, 1991).
6. E. M. Goldratt and J. Cox, *The Goal* (Croton-on-Hudson, N.Y.: North River Press, 1984).
7. E. M. Goldratt and J. Cox, *The Race* (Croton-on-Hudson, N.Y.: North River Press, 1986).
8. E. M. Goldratt, "Computerized Shop Floor Scheduling," *International Journal of Production Research,* Vol. 26, No. 3 (1988), pp. 443–455.
9. E. M. Goldratt, *What Is This Thing Called Theory of Constraints* (Croton-on-Hudson, N.Y.: North River Press, 1990).
10. M. L. Srikanth and H. E. Cavallaro, Jr., *Regaining Competitiveness—Putting the Goal to Work* (New Haven, Conn.: Spectrum Publishing, 1987).
11. M. M. Umble and M. L. Srikanth, *Synchronous Manufacturing* (Cincinnati: South-Western, 1990).

APPENDIX 16A

How the Constraint Theory Improved a Job-Shop Operation
David S. Koziol

The benefits of Just-in-Time have been documented in many articles. However, the only companies to profit from JIT are those in repetitive manufacturing. For example, in a job-shop operation interactions among resources are more complex than in an assembly line or process operation. Consequently, any interruption in the flow of material through a JIT shop as the result of "Murphy's Law" will cause the entire system to crash to a halt until the problem is corrected. Also, implementing a true JIT system requires considerable lead time that is likely to be measured in years and at great expense to a company.

I have been plant controller of Valmont/ALS, a job-shop steel fabricator in Brenham, Texas, for the past five years. When our plant was faced with mediocre inventory turnovers, increasing amounts of overtime, and a recession in our industry, we believed the solution to improving our operations would be to adapt the JIT formula, rather than introduce a true JIT system.

Reprinted from *Management Accounting.* (Vol. 69, No. 11 May 1988), pp. 44–49. Copyright by Institute of Management Accountants, Montvale, N.J.

THE MONTH END SYNDROME

Our plant manufactures tapered steel poles in heights from 20 to more than 200 feet according to customer specification. Our product lines can be split into the two categories: large pole and small pole. Small pole consists of smooth round tapered tubes that are purchased from our home site and fabricated into street, highway, area lighting, and traffic signal poles.

Large pole products range from sports lighting structures for stadiums to those used for power line transmission, signs, high mast lighting for interstate highways, and cellular communications.

Most large pole type orders require extensive engineering to customer specification and there is very little parts standardization. Small pole designs have some degree of standardization and repeat orders are frequent but there are still hundreds of possible finished pole designs due to varying height, strength, and finish requirements. Our industry would not be considered particularly "high-tech," but it is capital intensive.

Our business systems including standard cost, MRP scheduling, and labor reporting systems have been centralized on a mainframe computer system at the home office in Nebraska. The standard cost system averages 6,500 active part numbers and these are categorized as raw material, purchased parts, manufactured parts, and finished goods. There are nearly 40 manufacturing cost centers, each with burden rates set for labor, variable expense, and fixed expense. Although somewhat cumbersome for a company of our size, the system provides an accurate valuation of inventory for our financial reports.

In 1986, Lew Hays, our division president, distributed copies of *The Goal* by Eli Goldratt and Jeff Cox to his staff. Jeff Wood, our general manager, gave copies to his staff. As we read about plant manager Max Rogos' struggle to keep his plant open, we realized that the problems he faced were similar to our own.

For example, I always had been at a loss to explain the "month-end syndrome" that seemed to occur every fiscal period. The first week of the month would start fine. Shop activity levels were high and everyone seemed to be busy, although few products were shipped and invoices sent.

During the next few weeks shipments would improve but the shop would become chaotic. Production control personnel devoted much of their time to resolving scheduling conflicts and expediting manufactured parts to the final production stages. At the end of the month we had to schedule a significant amount of overtime in the final assembly, finishing, and shipping departments. Shipments on the last days of the fiscal period often represented 40% of the total month. It had become increasingly difficult to forecast earnings from the production control schedules. Our reports to the home office often required last minute downward revisions.

This scenario resulted in tense emotions and finger-pointing. Low productivity, inadequate machinery, equipment breakdowns, bad scheduling, employee turnover, and absenteeism were blamed for the problems.

Inventory levels had been gradually increasing over the last few years. Write-offs of obsolete inventory were always a problem to be heavily accrued for. Inventory turns were mediocre, but it was pointed out that they were near the average for our industry.

Capital projects that would improve productivity by decreasing direct labor requirements in our products usually were approved if they met the target hurdle rates. Requisitions for increased inventory storage space often were discussed as well as increasing capacity in some expensive asset areas.

Early in 1987, the geographic area served by our plant suddenly felt the effect of the 1986 "oil bust" and a recession in the construction industry. New orders for the year were coming in at lower than the optimistic levels of our operating plan. As a result, in the spring of 1987 we laid off one-third of the shop employees and administrative staff. Our future was not very promising at this point. The traditional management principles for operating in a recession were calling for further cost control measures, improvements in shop productivity, and aggressive marketing efforts.

The ideas suggested in *The Goal* and its sequel, *The Race*, made a lot of sense. However, they seemed to contradict what we had been taught in college and our experiences in the manufacturing industry. Although our administrative staff was relatively young and small (nine, including shop supervisors), it represented many years of manufacturing experience.

We spent a great deal of time discussing the books and how we could improve our own situation. A hot topic was determining the constraints (bottlenecks). We agreed that the company faced an external constraint in our market. We had the capacity to produce far more than we could sell. The engineering function at the home office also was identified as an external constraint. In our plant, most agreed that bottlenecks occurred in our weld assembly areas, although there appeared to be other "floating" bottlenecks that moved through the plant during the month.

BREAKING WITH TRADITION

When we first began to change the way our plant was managed, we were not quite ready to throw out all of our long-held beliefs. As a result, early steps were not bold.

The first item we looked at was batch sizing for our small pole parts production. Traditional scheduling using our MRP system encouraged large batch sizes to save setups and maximize efficiencies for each of our manufacturing resources. While we didn't know exactly what size to set our batches at the time, Jeff decided to reduce all small parts batches by aprroximately 40%. This scheduling change caused a subtle but noticeable improvement. The product flow through the shop experienced less disruption as a result of stockouts for small parts, and the general chaos of expediting orders seemed to diminish.

Meanwhile, we were learning as much as possible about the Theory of Constraints from the American Production Inventory and Control Society publications. I was embarrassed to discover these concepts had been around for a number of years but I'd never been exposed to them in any of my accounting publications.

Many of the APICS articles discussed a growing dissatisfaction with MRP-based scheduling systems. A vigorous debate was being waged between proponents of MRP and various "zero inventory" groups. One side argued that traditional MRP systems were the antithesis of the successful Japanese JIT systems. The other side defended MRP systems as fundamentally sound, but that too many companies were misusing them. The proponents of MRP systems charged that a lack of accuracy in companies' databases was a major obstacle to the systems' success. Other experts viewed constraint theory as a valid management tool for the future but too complex to implement immediately.

Reading these articles prompted us to reevaluate our MRP system. At this time we were totally dependent on the MRP system as were our other plants. The MRP system had undergone continuous changes over the last 15 years. It was basically a three-level system that batched production requirements into weekly "buckets." The system would access and schedule our customer backlog of orders, and planning bills also would be scheduled for future customer requirements based on assumptions drawn from historic data. New customer orders were being quoted with predetermined lead times based roughly on the industry standard. Our database had a high degree of accuracy, but of course it wasn't 100% correct.

During the second quarter of last year, the divisional controller and I attended seminars conducted by the Goldratt Institute on "Logistics of Constraint and Finance." Jeff Wood and Operations Manager Bob Evans attended a similar seminar on engineering and quality.

At the seminar, Bob and Jeff discussed the nature of our business and suspected constraints with Eli Goldratt and Rob Fox. The question was raised during the conversation: "Why produce inventory to stock rather than only to customer order?" Our old school of thought had been that it was not possible to manufacture a new order on a consistent basis in less than the standard lead time, so a large inventory of manufactured parts and finished goods were required.

Our new challenge now would be to produce only to customer order while substantially reducing the amount of time it took to produce an order. It would be necessary to identify our constraints(s) in our operation and to use it as the heartbeat of our entire plant. Our scheduling system would allow for strategically sized inventory buffers placed directly in front of our constraints. The size of the inventory buffer required would depend on an analysis of the average downtime of the nonconstraint resources (cost centers with overcapacity) that feed the constraint (bottleneck).

An important difference between the theory of constraints and a pure JIT system is the allowance of these inventory buffers. A JIT job-shop would have small buffers throughout the plant rather than just a few strategically sized and placed ones.

The constraint would become the focal point of our organization's attention. Any new program, policy, asset addition, etc., would have to be viewed in light of how the constraint would be affected. It would be critical that flow of product through our constraint remain constant with no interruptions. Preventative maintenance and quality efforts would need to be intensified on the constraint.

We decided a three-day buffer was sufficient in front of our constraints. It would allow enough time to rectify a problem in the flow of product from our overcapacity resources without interruption in the flow through the bottlenecks. If our buffer was too small or nonexistent, the constraint could become starved for work and throughput (sales) for the period would be lost.

The new system appeared to allow our job-shop to operate as a modified Kanban (JIT) system-inventory would be PULLED through the shop at a rate dictated by our constraints. Our existing MRP system was trying to PUSH inventory through the shop without acknowledging the existence of our constraint. By ignoring the effects of statistical fluctuations and dependent events as described in detail in *The Race*, our MRP system had caused work to flow through the shop in large "waves." These "waves" would peak at the end of the month and drown our weld assembly, finishing, and shipping resources.

With the elimination of manufactured parts and finished goods from inventory, the buffer concept could be expanded to include certain raw material stock that had been difficult to maintain because of unreliable suppliers.

Equally important to the buffer concept would be the requirement that our overcapacity resources be strictly limted to produce only what was required for the buffer. The entire system is known as the "Drum-Buffer-Rope" and is examined in detail in *The Race*.

Our decision to build products only to customer order rather than to stock was based on a careful analysis of our total system constraints and markets. Every company is unique in the products produced and markets served. A company that performs a similar analysis may decide to pursue a different plan based on its own situation.

OBSTACLES AND OPPORTUNITIES

Jeff and I saw a tremendous opportunity in the new system to exploit the reduced production lead time that was expected. Our market share could be increased for orders in which short lead time was important to the customer. Our employment levels after the lay-off were at their lowest point ever. By increasing throughput (sales, not inventory produced), and without increasing operating expenses, our actual unit cost of production would be reduced.

There were two formidable obstacles to be hurdled before successfully implementing the new strategy. Although our plant was responsible for its own bottom line, marketing and engineering both reported directly to the home office. Both groups would become key players in exploiting the new opportunities but neither was totally aware of what our plant was trying to accomplish. The reeducation process that had been taking place in our small operation would have to be expanded to our entire organization.

The second major obstacle concerned how to properly schedule the shop in the most effective manner. The seminars we had attended used simulations of functional plants on personal computers. We had learned that by running these relatively simple plants in a traditional manner, bankruptcy was the usual result. By applying the principles we learned, we saw it was possible manually to schedule the various simulations with profitable results. We now had to apply these concepts to a real world setting—our more complicated shop floor.

The option to purchase our own mainframe computer and any required software would be prohibitively expensive and time-consuming. The MRP system could still be used but major modifications would be required. These modifications, however, would not be possible because of the shared nature of the MRP programs throughout the division. Over the last few years, our company had seen the cost of a mainframe transaction drop considerably, but not without increasing the demands placed on our corporate programmers to make system changes. We realized that changing an MRP system that had served the company so well in the past would be a difficult, often political struggle.

Part of the solution grew out of our ability to use a program to query our inventory system database. Our corporate data processing department had completed the necessary program installation and training, but until now our usage had been limited to maintenance of the standard cost system and excess inventory analysis.

Jeff has extensive knowledge of our MRP system and was able to develop suitable queries on our database that could be used to schedule the shop floor on a daily basis. (The weekly "bucket" scheduling of the existing MRP would be far too imprecise for our new needs.) Machine loads for each of the shop resources could now be updated on a daily basis. By modifying transit days on the master production schedule (assumed days between routing sequences), it was possible to incorporate the planned buffers into the new query programs. Transit days were eliminated entirely wherever buffers were not required.

Glenn Reimer, our production control manager, would now manually schedule the constraints using the query reports. The MRP and additional queries would be used to schedule all supporting resources (nonconstraints).

From what we had learned in *The Goal,* we now had a terminology that was understandable throughout our entire plant. The concepts of throughput, inventory and operating expenses were much easier to convey than the formal and often abstract accounting terms that had been used in the past. By identifying our operating constraint area of the weld assembly areas, we also had a focal point to concentrate on. Improving the flow of product through the constraint while lowering inventory and maintaining operating expenses were now our primary objectives.

Of course, my definition of excess inventory immediately changed. All existing manufactured parts and finished goods were now considered excess inventory. We realized that the excess inventory in the storage lots would provide a cushion of time until these large buffers were consumed through the constraints. By mid-summer, the majority of the excess was consumed and much of what remained was scrapped. A program was set up to reduce remaining slow-moving finished goods. Industrial engineers began analyzing incoming orders for possible substitutions and modifications of existing items, and sales reps were given listings to be sold at reduced prices.

August was the critical month. The success or failure of our efforts would be revealed, and we would know whether or not the shop could function effectively in a low inventory environment.

The results were better than expected. The invisible "rope" of the scheduled constraint was pulling small parts through the shop just before they were needed. Shipments for the period and the months to follow hit record highs. Overtime was held to low levels and customers were being routinely contacted to accept orders on earlier dates.

The Pantograph machine was an important indicator of how well we were doing. This machine cuts small parts from steel plate. Just a few months before, it had to be run almost nonstop yet it still had trouble keeping up with requirements in the weld assembly areas for parts. Today this asset often is idle for a day or more per week. Producing only what is needed, just before it is required, and in much smaller batch sizes has improved the flow of product to the customer because small part stockouts in the weld assembly areas have been eliminated.

Originally we had thought that the inventory cushion produced by the Pantograph when it ran nonstop was supposed to help throughput for the period. Instead it actually had caused the opposite effect and reduced throughput.

   Our high inventories had actually been costing the company far more than the traditional carrying charge calculation. As described by Professor Goldratt and Mr. Cox, the masking effect of high inventories and large batch sizes can have detrimental effects on quality assurance programs and the speed with which product improvements can be implemented. By increasing the product throughput without increasing operating expenses, real gains in lower unit costs are obtained. In addition, improving due-date performance and shortened quoted lead times makes a company far more responsive to the marketplace.

GAIN BUT NOT WITHOUT PAIN

By the end of the year, the improvements as a result of the drum-buffer-rope method of operation were plainly visible. Quality efforts were being redirected to improve the process, rather than merely inspecting the product. Engineering change orders by the customer were less disruptive because less parts were affected. Smaller batch sizes were making it possible to change the processes more quickly.

   The actual lead time for certain product times had been cut to half the assumed industry standard lead time. Due date shipment performance had improved to the mid-to-upper 90% range. Shipment levels increased to their highest levels in our company history, yet personnel were still near the lowest layoffs levels. Shop overtime had been kept at reasonable levels and shipping costs as a percentage of sales had fallen.

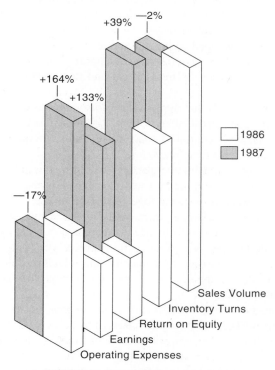

FIGURE 1   Changes in key measurements

Throughput had risen without a proportionate rise in operating expenses, and inventories had been rejected. Our new set of measurements all indicated favorable results.

What about the more traditional accounting measurements? Net earnings were 40% better than our operating plan for the period. Return on equity was up, and cash flow was 60% greater than the operating plan. These favorable results occurred despite a 9% decrease in actual volume as called for in our aggressive annual operating plan (see Figure 1).

On a personal note, 1987 was particularly difficult for me. Many of the cost accounting concepts I had learned in college and experienced over the last 15 years were being challenged as no longer valid.

I realized that by hounding our manufacturing department about efficiencies and standard costs variances, I often moved the company away from rather than towards our goal. Because I had ignored our system constraints and relied on traditional standard cost calculations, past decisions made to out-source production and pricing our product were often wrong.

The standard cost system remains the same as well as my month-end financial reporting requirements. Little has been changed in regard to the external reporting system. But the internal reports that receive our prime attention for operations have undergone significant changes.

Our labor reporting system still generates thousands of pages each month. They give detailed information on efficiencies by part number, order number, employee, cost centers, and departments. Our prime emphasis is now on the rate of product flow through our constraint areas and adherence to daily machine load schedules.

Jeff has developed additional queries to better assist in the management of the inventory buffers. Three queries provide production control with a detailed analysis of parts missing from the desired buffers and their current location and process time remaining.

We have found that applying the constraint theory to our operation offers greater opportunities and benefits than just having another cost control or scheduling system. To ensure the system's continued success we are expanding the concepts to all operations, with particular attention to marketing.

In strategic planning the question should be raised that with any specific manufacturing operation what is the best way to compete in the marketplace? Without manufacturing and marketing working together, Srikanth and Cavallaro point out in *Regaining Competitiveness*, marketing could easily pursue a mix of product orders that severely hampers the operating performance of a company.

In addition to marketing, administrative departments should be aware of the strengths and weaknesses of the manufacturing facilities. For example, if engineering designs products with the system's constraint in mind, the success of the operation will be greatly enhanced.

Instead of being defensive about our systems, we should take a hard look at how to improve them. The recent surge in research by academe and practitioners has been encouraging. Another positive sign is that more of us are getting out on the shop floor. I believe that if management accountants keep an open mind and share what is learned, companies' efforts to compete in today's global marketplace will be enhanced and not hindered.

Chapter Seventeen

Project Management Techniques

INTRODUCTION

Operations managers must plan, organize, and control a variety of operations. Some of them are repetitive activities, and others are one-shot deals. Routine activities might be continuous or intermittent. The continuous production functions can be managed by assembly line balancing, whereas intermittent or job shop production requires scheduling, loading, and control techniques. The one-shot deals are generally one-time projects, such as constructing a hospital, research and development of a Boeing 767, or building a Polaris missile. A project approach is frequently used to develop and market new products and services. The type of technique required to manage these activities depends on the complexity of the project. For small projects, Gantt charts are adequate, whereas for large and complex projects, the critical path method (CPM) or the program evaluation and review technique (PERT) would be more effective. The problem becomes more complicated when common resources are used to execute multiple projects. For example, NASA launches several space flights from the same facility. Flights must be scheduled so that the capacity of various equipment is not exceeded. Such problems require resource-constrained multiple project scheduling techniques. Other situations may require limited quantities of repetitive production. Line of balance techniques can be used to manage such situations economically. In this chapter we discuss a variety of techniques for managing many types of project functions.

GANTT CHARTS

For small projects, Gantt charts are sufficient. Preparation of Gantt charts was discussed in detail in Chapter 14. A Gantt chart is simply a bar chart that time-phases activities. Every activity should have a start date and a finish date. Figure 17.1 exhibits a typical Gantt chart for a building construction project. All major activities are described.

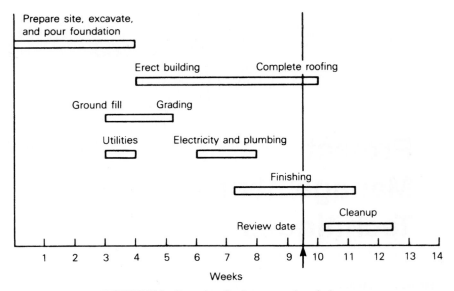

FIGURE 17.1 Gantt chart for the construction of a house.

Gantt charts serve as a record keeping and monitoring tool for projects. The major drawback is that the chart does not show the interrelationships of activities in the project. Gantt charts are frequently used to supplement CPM results.

THE USE OF CPM AND PERT

Large projects present many problems to the manager because of the nonroutine, one-time nature of activities. Planning and coordinating all activities can be very complex for large projects. For example, ship construction involves design, fabrication, purchasing, and building. The total number of activities could easily exceed 5000. Such projects have a limited time framework and require a host of internal and external approval procedures. Again, in building a Boeing 767, a Polaris missile, or a space shuttle, many technical problems arise. In many instances, new technology emerges. In general, the manager considers all activities of a project and the individual tasks that must be accomplished and relates tasks to one another and to the calendar. For planning and monitoring large projects, network techniques such as CPM and PERT are widely used.

CPM was originally developed by J. E. Kelly of Remington Rand Corporation and M. R. Walker of DuPont for planning and coordinating maintenance projects in the chemical industries. PERT evolved from Gantt charts in the late 1950s through the joint efforts of Lockheed Aircraft, the U.S. Navy Special Projects Office, and Booz, Allen and Hamilton, a consulting firm. It was first applied to the U.S. Navy's Polaris submarine project. Although CPM and PERT evolved independently, they have a great deal in common. As users borrowed certain features from each other, the gap narrowed further. These techniques can be used for planning, monitoring, and controlling project activities.

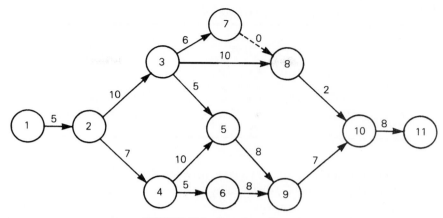

FIGURE 17.2   Precedence diagram.

CPM and PERT are used to find the minimum expected time to complete a project. By identifying the bottleneck operations, we can monitor the progress of the project to assure project completion on time. In addition, PERT can estimate the probability of meeting certain due dates. As delays occur, certain critical activities may need to be hastened. These techniques are most useful in evaluating the time and cost trade-offs of specific project activities.

Activity Networks

The main feature of activity networks is the use of a precedence diagram to depict all or major project activities and their sequential relationships to other activities. Each activity must have an associated time estimate and precedence relationship, as shown by the example in Figure 17.2. The arrows in the diagram represent activities, and the circles represent nodes. The nodes depict both the completion of one activity and the start of the next activity. The activities may represent steps in building a house, milestones in the construction of a ship, or major steps in an research and development project. One important aspect of a network diagram is that two or more activities cannot start and finish at the same two nodes. If this did happen in the real world, we would add a dummy activity. For example, see activity 7–8 in Figure 17.2; dummy nodes are generally connected by dotted lines. Let us now try to understand what the sequential or precedence relationship means. Activity 5–9, for instance, cannot be started until activities 3–5 and 4–5 are completed. Also, neither activity 2–3 nor activity 2–4 can be started prior to completion of activity 1–2, and so forth.

THE CRITICAL PATH
METHOD (CPM)

Once the precedence diagram is drawn and the duration of each activity is estimated, the next step involves tracing all paths. A path is defined as the sequence of activities leading from the starting node to the completion node. Table 17.1 lists all possible paths for

TABLE 17.1 List of All Paths for the Precedence Diagram

Path Identifier	Path	Length (weeks)	Slack (weeks)
A	1, 2, 3, 8, 10, 11	35	45 − 35 = 10
B	1, 2, 3, 7, 8, 10, 11	31	45 − 31 = 14
C	1, 2, 3, 5, 9, 10, 11	43	45 − 43 = 2
D	1, 2, 4, 5, 9, 10, 11	45	45 − 45 = 0
E	1, 2, 4, 6, 9, 10, 11	40	45 − 40 = 5

the network in Figure 17.2. In a large network, there will be several paths. The lengths of individual paths can be calculated by adding the durations of all activities on each path. Path C, for example, consists of activities 1–2, 2–3, 3–5, 5–9, 9–10, and 10–11, with durations 5, 10, 5, 8, 7, and 8, respectively. Summing the durations, we obtain the path length or duration of path C as forty-three weeks. We can see from the table that the longest path is forty-five weeks. Since all other paths are less than forty-five weeks, we know that we need at least forty-five weeks from the start to complete the entire project. The longest path is known as the *critical path.* If there are any delays along the longest path, there will be corresponding delays in the project completion time. Similarly, if the completion time is not acceptable, the management should attempt to shorten the activities on the critical path, which are called the *critical activities.*

Activities that are not on the critical path can experience some delays, to the extent known as *slack,* without affecting the completion time of the project. Slack is the maximum amount of slippage allowed in each path. The slack of any path is obtained by subtracting the length of that path from the critical path length. For example, path C has 45 − 43 = 2 weeks of slack. Obviously, the critical path has zero slack, as exhibited in Table 17.1. It is important to understand the difference between the slack of a path and the slack of an activity. For example, activity 2–3 is on paths A, B, and C, where the slacks are ten, fourteen, and two weeks respectively. Even though path A has ten weeks of slack, activity 2–3 has only two weeks of slack. Thus the amount of slack for activities on the same path can vary.

Determining the Critical Path and Slacks

In the foregoing small problem, we were easily able to list all paths and to identify the critical path and the slacks associated with all paths. As the number and the precedence complexity of activities increase, the identification of individual paths by enumeration becomes impractical. A systematic approach is essential. First, we will define the following variables for activity $i–j$. Let t_{i-j} represent the duration of activity $i–j$.

ES_{i-j} = earliest possible start time for activity $i–j$

EF_{i-j} = earliest possible finish time if the activity is started at ES_{i-j}

$= ES_{i-j} + t_{i-j}$

LF_{i-j} = latest feasible finish time for activity i–j

LS_{i-j} = latest feasible start time if activity i–j needs to be completed by LF_{i-j}

$\quad\;\; = LF_{i-j} - t_{i-j}$

Next we illustrate how these four variables are generated for each activity, using the example illustrated in Figure 17.2.

Calculation of *ES* and *EF*

All activities that do not have precedence can start at time zero. Thus activity 1–2 in Figure 17.2 can start at time zero, or $ES_{1-2} = 0$. Therefore, the earliest finishing time $EF_{1-2} = ES_{1-2} + t_{1-2} = 0 + 5 = 5$ weeks. Activities 2–3 and 2–4 can be started as soon as activity 1–2 is completed. Therefore,

$$ES_{2-3} = 5$$
$$ES_{2-4} = 5$$
$$EF_{2-3} = ES_{2-3} + t_{2-3} = 5 + 10 = 15$$
$$EF_{2-4} = ES_{2-4} + t_{2-4} = 5 + 7 = 12$$

Similarly,

$$ES_{3-5} = ES_{3-7} = ES_{3-8} = EF_{2-3} = 15$$
$$EF_{3-5} = 15 + 5 = 20$$
$$EF_{3-7} = 15 + 6 = 21$$
$$EF_{3-8} = 15 + 10 = 25$$
$$EF_{7-8} = 21 + 0 = 21$$

and

$$ES_{4-5} = ES_{4-6} = EF_{2-4} = 12$$
$$EF_{4-5} = 12 + 10 = 22$$
$$EF_{4-6} = 12 + 5 = 17$$

The activities at node 5 can be started only after all activities ending in node 5 are completed. We have

$$EF_{3-5} = 20$$
$$EF_{4-5} = 22$$

Therefore, the earliest time that activity 5–9 can be started is week 22; that is, $\max(EF_{3-5}, EF_{4-5})$. Then

$$ES_{5-9} = 22$$
$$EF_{5-9} = 22 + 8 = 30$$

Similarly,

$$ES_{6-9} = EF_{4-6} = 17$$
$$EF_{6-9} = 17 + 8 = 25$$

and

$$ES_{8-10} = \max[EF_{3-8}, EF_{7-8}]$$
$$= \max[25, 21]$$
$$= 25$$
$$EF_{8-10} = 25 + 2 = 27$$
$$ES_{9-10} = \max[EF_{5-9}, EF_{6-9}]$$
$$= \max[30, 25]$$
$$= 30$$
$$EF_{9-10} = 30 + 7 = 37$$

Proceeding in the same way, we obtain

$$ES_{10-11} = \max[27, 37] = 37$$
$$EF_{10-11} = 37 + 8 = 45$$

The calculations of *ES* and *EF* are illustrated in Figure 17.3 and summarized in Table 17.2.

Calculation of *LF* and *LS*

Determining the latest finish time (*LF*) and latest start time (*LS*) follows a similar procedure, but the calculations must start from the end of the network diagram. The procedure is also referred to as the *backward pass*. Once we know the earliest possible completion time of the critical path, we can calculate the latest possible times at which we can start each activity without delaying the project. For example, activity 8–10 can be started as late as week 35 and completed by week 37 without affecting the completion time of the project. That is,

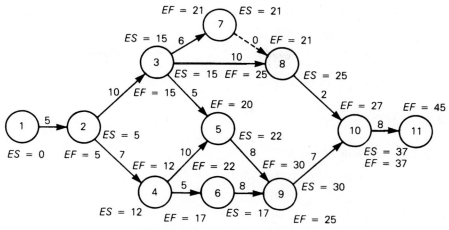

FIGURE 17.3 Calculation of *ES* and *EF.*

$$LF_{8-10} = 37$$

$$LS_{8-10} = LF_{8-10} - t_{8-10} = 37 - 2 = 35$$

By subtracting the duration t_{i-j} of an activity from the latest feasible time LF_{i-j}, we obtain the latest feasible start time LS_{i-j}. The LF times for the immediately preceding activities are equal to the smallest LS times for all immediate successor activities. That is,

$$LF_{3-8} = LS_{8-10} = 35$$

$$LF_{7-8} = LS_{8-10} = 35$$

TABLE 17.2 Summary of *ES, EF, LS, LF,* and Slack and Identification of Critical Path

Activity	Duration	ES	EF	LS	LF	Slack	Critical Path
1–2	5	0	5	0	5	0	*
2–3	10	5	15	7	17	2	
2–4	7	5	12	5	12	0	*
3–5	5	15	20	17	22	2	
3–7	6	15	21	29	35	14	
3–8	10	15	25	25	35	10	
4–5	10	12	22	12	22	0	*
4–6	5	12	17	17	22	5	
5–9	8	22	30	22	30	0	*
6–9	8	17	25	22	30	5	
7–8	Dummy	—	—	—	—	—	
8–10	2	25	27	35	37	10	
9–10	7	30	37	30	37	0	*
10–11	8	37	45	37	45	0	*

$$LS_{3-8} = 35 - 10 = 25$$
$$LS_{7-8} = 35 - 0 = 35$$
$$LS_{3-7} = LS_{7-8} = 35$$
$$LS_{3-7} = 35 - 6 = 29$$

Similarly,

$$LF_{9-10} = LS_{10-11} = 37$$
$$LS_{9-10} = LF_{9-10} - t_{9-10} = 37 - 7 = 30$$
$$LF_{5-9} = LS_{9-10} = 30$$
$$LS_{5-9} = 30 - 8 = 22$$

and

$$LF_{3-5} = LS_{5-9} = 22$$
$$LS_{3-5} = 22 - 5 = 17$$

Now we have the latest starts of all activities starting at node 3. They are 25, 29, and 17 respectively, for activities 3–8, 3–7, and 3–5. Therefore, the LF for all activities ending at node 3 is 17—that is, min(LS_{3-8}, LS_{3-7}, LS_{3-5}). Then,

$$LF_{2-3} = 17$$
$$LS_{2-3} = 17 - 10 = 7$$

Proceeding in the same way, we calculate

$$LF_{6-9} = LS_{9-10} = 30$$
$$LS_{6-9} = 30 - 8 = 22$$
$$LF_{4-6} = LS_{6-9}$$
$$LS_{4-6} = 22 - 5 = 17$$

and

$$LS_{4-5} = 22 - 10 = 12$$

We now have the latest start times for all activities starting at node 4. They are 12 and 17 for activities 4–5 and 4–6, respectively. Therefore, the LF time for all activities ending at node 4 is 12. Now we can complete the calculations for the rest of the activities:

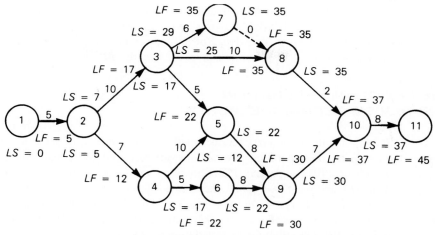

FIGURE 17.4   Calculation of *LF* and *LS*.

$$LS_{2-4} = 12 - 7 = 5$$

$$LF_{1-2} = \min[LS_{2-3}, LS_{2-4}]$$

$$= \min[5, 7] = 5$$

$$LS_{1-2} = 5 - 5 = 0$$

The calculations are illustrated in Figure 17.4 and summarized in Table 17.2.

The slack for each activity is obtained by subtracting the *ES* from the *LS* or the *EF* from the *LF* for that activity. Figure 17.5 exhibits the slack (*S*) for all activities. All activities with zero slack are the critical activities, and they make up the critical path.

Any activity with a positive slack can be delayed by that amount of time without delaying completion of the project. Once an activity is delayed by some amount of time,

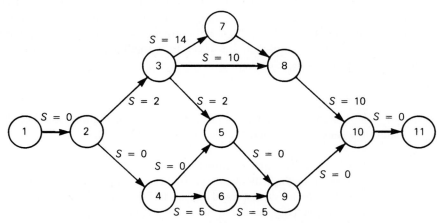

FIGURE 17.5   Calculation of slack and critical path.

the slack is changed not only for that activity but for all activities on the path. For example, if activity 4–6 is delayed by five weeks, then the slack for activity 4–6 and for activity 6–9 becomes zero. A knowledge of individual activity slack times provides the managers with flexibility for planning and monitoring the allocation of scarce resources.

THE PROGRAM EVALUATION AND REVIEW TECHNIQUE (PERT)

The foregoing discussion of CPM assumed that the durations of activities are known. For most projects, these durations are random variables, and hence many situations require a probabilistic PERT approach. PERT involves three time estimates: optimistic (a), pessimistic (b), and most likely (m). These time estimates are obtained either from past data or from the experience of those who are responsible for complete a specific activity. Once these time estimates are available, the β (beta) probability distribution becomes very handy for computing the expected time t_{i-j} and the variance σ^2_{i-j} of each activity:

$$t_{i-j} = \frac{a + 4m + b}{6}$$

$$\sigma^2_{i-j} = \left(\frac{b - a}{6}\right)^2 = \frac{(b - a)^2}{36}$$

The magnitude of the variance reflects the uncertainty involved in the activity. As the variance gets larger, the uncertainty becomes greater. A knowledge of the variance of the activities on the critical path, and hence the total variance associated with the critical path, facilitates making probability estimates for the project completion dates. The variance of the critical path is given by the sum of the variances of individual activities on the critical path. For example, top management or a government official might like to know the probability of completing a research and development project in forty-two weeks or in a year. Unless we have the probability estimates associated with all activities, we cannot answer these questions. Once we have the probabilistic estimates, the normal probability distribution generally provides a reasonable approximation to the distribution of project completion time, even for a small number of activities. The following example illustrates these concepts.

Example 17.1

The network diagram for a project is shown in Figure 17.6. The three time estimates, in weeks, for each activity are shown in Table 17.3.

1. Compute the expected time for each activity.
2. Determine the critical path and the expected project duration.
3. Compute the variance and standard deviation of the critical path.
4. Which activity has the most precise time estimate?
5. What is the probability that the project will be completed in twenty weeks?

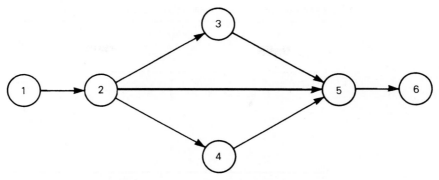

FIGURE 17.6   Precedence diagram for Example 17.1.

Solution: (1) The expected durations and variances of all activities are shown in Table 17.4.

(2) We plot the expected times on the network and find *ES, EF, LS,* and *LF,* for all activities, as shown in Figure 17.7. The critical path is depicted by a sequence of activities with slack $S = 0$. Therefore, the critical path in our case is 1, 2, 5, 6. In addition, activities 2–4 and 4–5 are also critical. In other words, we have dual critical paths: 1, 2, 4, 5, 6 and 1, 2, 5, 6. The expected duration T of the critical path is nineteen weeks.

(3) The total variance of the critical path is obtained by summing individual activity variances on the critical path:

$$\sigma^2 = \sigma^2_{1-2} + \sigma^2_{2-4} + \sigma^2_{4-5} + \sigma^2_{5-6}$$
$$= 1.78 + 1 + 0.44 + 1.78 = 5.0.$$

or

$$\sigma^2 = \sigma^2_{1-2} + \sigma^2_{2-5} + \sigma^2_{5-6}$$
$$= 1.78 + .44 + 1.78 = 4.0$$

TABLE 17.3   Time Estimates for Example 17.1

Activity	Duration (weeks)		
	a	*m*	*b*
1–2	5	6	13
2–3	2	2	2
2–4	2	5	8
2–5	6	8	10
3–5	3	5	7
4–5	1	3	5
5–6	2	3	10

TABLE 17.4   Expected Durations and Variances, Example 17.1

Activity	Expected Time $\dfrac{a + 4m + b}{6}$	Variance $\left(\dfrac{b - a}{6}\right)^2$
1–2	7	1.78
2–3	2	0
2–4	5	1.00
2–5	8	0.44
3–5	5	0.44
4–5	3	0.44
5–6	4	1.78

The variance we use is the larger of the two—that is, 5.0—and the standard deviation is the square root of the variance:

$$\sigma = \sqrt{5} = 2.24$$

(4) Activity 2–3 has the least variance, since its variance is zero.

(5) The normal probability distribution will generally provide a reasonable approximation to the distribution of project durations, and hence it can be used to determine the probabilities for various project completion times:

$$z = \frac{y - T}{\sigma} = \frac{20 - 19}{2.24} = 0.4464$$

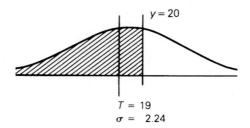

$$y = 20$$
$$T = 19$$
$$\sigma = 2.24$$

Referring to the normal probability tables, we determine that the shaded area is 0.6736. Therefore, we can state that the probability of completing the project in twenty weeks is 0.6736, or 67.36%.

Crashing

In many instances, a manager would like to shorten the total duration of a project. The project manager may have some options, such as hiring additional workers and machines or increasing control with supervision on critical projects to expedite their completion time. In these instances the goal is to identify the time versus cost trade-offs and to eval-

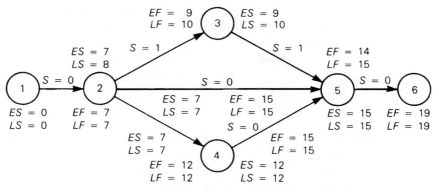

FIGURE 17.7   *ES, EF, LS* and *LF* calculations for Example 17.1.

uate alternative plans for minimizing the sum of the indirect and direct project costs. Versions of CPM and PERT can be used to evaluate how best to shorten, or *crash,* the project completion time.

Example 17.2

Suppose that the manager wishes to complete the project shown in Figure 17.2 in forty weeks. The crash time estimates—the amount of time it would take to complete an activity if additional resources were allocated—and the costs of allocating resources to corresponding tasks are given in Table 17.5. All possible paths are listed in Table 17.1. It is evident that we need to expedite activities in paths C and D or activities common to both C and D, since they exceed forty weeks. From Table 17.5, we find that the cost of crashing activity 2–4 is the cheapest and 2–3 is next cheapest. By crashing these activities, we can reduce the path by three weeks. By crashing activity 10–11, we gain two

TABLE 17.5   Time versus Cost Trade-off Data, Example 17.2

Activity	Expected Time (weeks)	Crash Time (weeks)	Cost of Crashing ($)	Incremental Cost per Week ($)
1–2	5	3	10,000	5,000
2–3	10	7	6,000	2,000
2–4	7	4	3,000	1,000
3–5	5	3	4,000	2,000
4–5	10	8	8,000	4,000
5–9	8	5	9,000	3,000
9–10	7	6	10,000	10,000
10–11	8	6	7,000	3,500

more weeks. The total cost associated with this crashing is $16,000. This crashing and an alternative crashing—and their corresponding costs—are as follows:

Alternative	Activities Crashed	Weeks Gained	Cost of Crash ($)
1	2–3, 2–4, 10–11	5	16,000
2	5–9, 10–11	5	16,000

Any other combination yields a higher cost, unless partial crashing is permitted. For example, if we were allowed to reduce two weeks of activity 5–9 for $6,000 in addition to crashing activities 2–3 and 2–4, our total cost of crashing would be only $15,000. In a large network, it is a formidable job to test all combinations and to find all paths that exceed the desired length of duration. Most applications of PERT and CPM use canned computer programs that have been developed to perform the necessary network analysis.

Cash Flow Analysis

In preceding sections we described many techniques for planning and monitoring projects. CPM and PERT can be very useful, not only for scheduling activities but also for monitoring and controlling cash flows. Referring to Figures 17.3 and 17.4, we find that CPM and PERT provide a feasible time-phased schedule of activities of the project in either an early start and finish or a late start and finish configuration. Using this information, we can draw Gantt charts. If we were given the cost of completing individual activities, we could find the total cash flow needs for each period according to Gantt chart schedules. For simplicity of calculations, we assume that the cost of an activity is linear.

TABLE 17.6 Project Activity Costs, Example 17.3

Activity	Duration (weeks)	Cost ($1000s)	Cost per Period ($1000s/week)
1–2	5	20	4
2–3	10	10	1
2–4	7	35	5
3–5	5	10	2
3–7	6	18	3
3–8	10	20	2
4–5	10	30	3
4–6	5	40	8
5–9	8	48	6
6–9	8	48	6
8–10	2	20	10
9–10	7	35	5
10–11	8	40	5

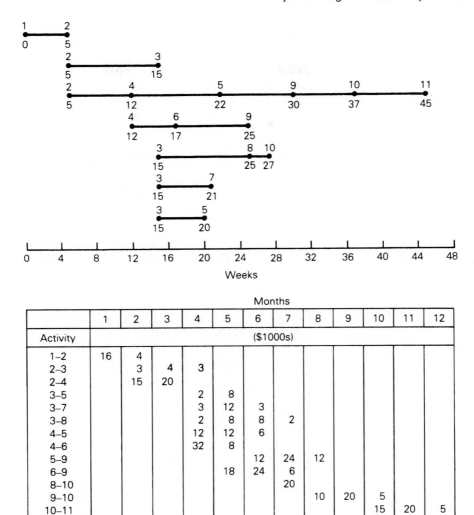

FIGURE 17.8   Gantt chart and cash flow analysis based on the earliest start–earliest finish schedule.

For example, if a project costs $10,000 and it takes four weeks to complete, then we as-sume that the cost per period is $2500. This assumption is not necessary when working with many CPM/PERT software packages, however.

Monitoring and controlling project costs involve comparing the expected range of cash flows with the actual costs incurred. If discrepancies are encountered, the project manager can take appropriate action to alleviate the problem.

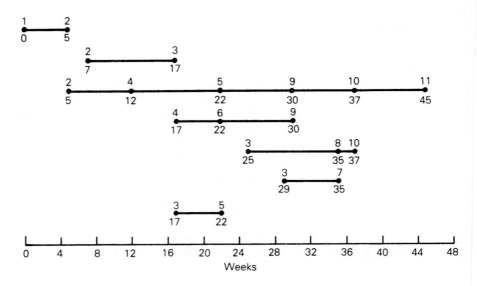

	Months											
	1	2	3	4	5	6	7	8	9	10	11	12
Activity					($1000s)							
1–2	16	4										
2–3		1	4	4	1							
2–4		15	20									
3–5					6	4						
3–7									9	9		
3–8							6	8	6			
4–5				12	12	6						
4–6					24	16						
5–9						12	24	12				
6–9						12	24	12				
8–10										10	10	
9–10								10	20	5		
10–11										15	20	5
Total	16	20	24	16	43	50	54	51	45	30	20	5

FIGURE 17.9   Gantt chart and cash flow analysis based on the latest feasible start–latest feasible finish schedule.

Example 17.3

The cost of completing each activity in the project represented by Figure 17.2 is given in Table 17.6. Perform a cash flow analysis. State your assumptions.

Solution: Based on the data in Figure 17.3, we draw a Gantt chart for the earliest possible start (*ES*) and the earliest possible finish (*EF*). Since the duration is expressed in weeks,

TABLE 17.7   Range of Expected Cash Needs per Month, Example 17.3

Month	Based on LS Schedule ($1000s)		Based on ES Schedule ($1000s)	
	Per Period	Cumulative	Per Period	Cumulative
1	16	16	16	16
2	20	36	22	38
3	24	60	27	65
4	16	76	51	116
5	43	119	66	182
6	50	169	53	235
7	54	223	52	287
8	51	274	22	309
9	45	319	20	329
10	30	349	20	349
11	20	369	20	369
12	5	374	5	374

we assume that a month consists of four weeks. Furthermore, for simplicity of calculations, we assume that activity expenditures are linear. The cash flow for each activity per week is exhibited in Table 17.6. For example, activity 1–2 takes five weeks and consumes $20,000. Based on a four-week month, we show expenses of $16,000 and $4000 for months 1 and 2, respectively, in Figure 17.8. Total cash flows per month are also exhibited. Similar calculations are shown in Figure 17.9 for the latest feasible start (*LS*) and the latest feasible finish (*LF*), as exhibited in Figure 17.4. The final step consists of predicting the monthly cash requirements for the project. The ranges are given by the *ES–EF* and *LS–LF* schedule cash flows. Table 17.7 exhibits both the per period and the cumulative expected cash flows for the project.

LINE OF BALANCE (LOB)

An MRP system is very useful when an item containing several production stages and long lead times is mass-produced. In many situations, however, the quantity of items produced is limited, and an MRP system may not be economically feasible. Some examples are the production of aircrafts, missiles, and heavy machinery, where different batches of production could contain different items and there might not be many common parts among products. In such instances, the line of balance (LOB) technique may be more suitable. LOB is a charting and computational technique for monitoring and controlling products and services that are made to meet certain specific delivery dates.

The concept of LOB is similar to the time-phased order-point (TPOP) system and the material requirements planning (MRP) system. We start with the final production date and quantity. Then the product structure tree is laid on the horizontal scale, offsetting lead time on a time scale that reflects the stages of production. The stages could also represent purchased parts, as well as subassembly and assembly operations to support delivery on schedule. The LOB chart depicts the quantity of components, subassemblies,

assemblies, and end products produced at every stage and at any given time. It indicates the quantity of goods or services that should be completed at every stage and at any given time so as to meet the delivery date. Therefore, the LOB is the scale against which the progress is measured at any given time. Obviously, actual progress is represented by the LOB chart. We will explain the technique with the following example.

Example 17.4

Amity, Inc., has received orders to deliver beamers, for which the product structure tree and delivery schedule are given here. The test and delivery procedure, which is not shown here, takes a month.

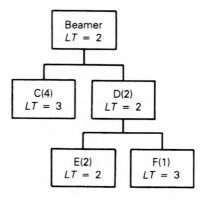

Month	Quantity	Cumulative Quantity
6	50	50
7	100	150
8	200	350
9	300	650
10	300	950
11	300	1250
12	300	1550

Four months into the shipping schedule, the following cumulative quantities of units have passed through corresponding steps in the production process:

Process Steps	Cumulative Production
I	1220
II	1185
III	1150
IV	1125
V	850
VI	720
VII	700

Develop an LOB chart and evaluate the status of production at each production stage, using the line of balance technique.

Solution: As the first step, draw the process plan for producing one unit of beamer (Figure 17.10). Second, construct a cumulative delivery schedule and the progress chart on a graph sheet (Figure 17.11). Then draw vertical bars on the progress chart to indicate the actual cumulative number of units produced for each production step.

Next, enter the review period—that is, the fourth month—on the graph. We know from the delivery schedule that we should have completed 650 units, and we can also obtain this information by drawing a vertical line from the review period to the cumulative delivery schedule curve. Now draw a horizontal line from that point on the curve to stage VII of the progress chart, representing the line of balance for stage VII of production. For other stages, we follow a different procedure. We know, for example, that final assembly and inspection takes three weeks. The cumulative end item production quantity three weeks from the review period will be equal to or less than the cumulative production of part C and subassembly D at stage V now. Therefore, we advance a period of three weeks from the review period—that is, four months and three weeks—and draw a vertical line to the cumulative delivery schedule curve. Now we draw a horizontal line to obtain the line of balance for production at stage V. This means that we should have enough units of C and D to complete 875 units by there weeks from now. In a similar fashion, we can draw the line of balance for every production stage at any given review period. Comparing cumulative production to date and the line of balance, we find that we are behind at stages I and V, whereas we are ahead in stages III, IV, and VII.

A new LOB can be drawn on the progress chart at any given time. Thus, a snap-

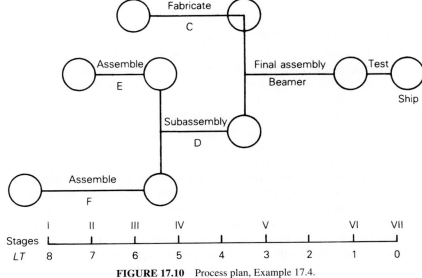

FIGURE 17.10   Process plan, Example 17.4.

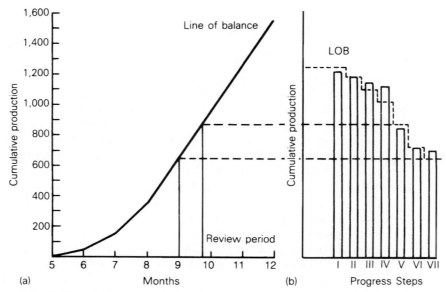

FIGURE 17.11 LOB chart, Example 17.3: (a) cumulative delivery schedule: (b) progress chart.

shot evaluation is made of each production stage at regular intervals of weeks or months. These periodic evaluations provide the manager with information about the performance at each stage of the schedule.

PROJECT SCHEDULING WITH CPM/MRP

The use of CPM and PERT for cash flow and budgeting decisions were illustrated in an earlier section. The total cost was spread uniformly over the duration of the activity for budgeting purposes. CPM and PERT assume that workers, machines, and materials are available in the right quantity, at the right time, and in the right place. In the real world, however, material has to be ordered in advance, and adequate capacities of equipment and machinery must be made available when necessary. Suppose that we prepare a master project schedule, list all material requirements, and use an MRP system for procuring them. Essentially, we would have interfaced the CPM and MRP systems. The MRP system can aid in planning and ordering materials according to the current demand obtained from the master project schedule, the amount of inventory on hand, and the quantity of material on order. Inventory records also help in replanning and rescheduling project activities when they are restricted by the capacities of equipment and machinery. Therefore, if we included the procurement of material as part of project planning, the total time necessary to procure materials and complete the activities will be greater than the critical path. Aquilano and Smith [1] stress the importance of an integrated approach in developing a CPM/MRP system. Gessner [9] reports that such a system has been developed by IBM and is used in the shipbuilding industry.

The CPM/MRP Model

The CPM/MRP model will be described with the aid of the sample project given in Figure 17.6. The project can also be tabulated in the form of an MRP-type indented bill of materials. See Table 17.8 for a list of activities, labor requirements, facilities, and equipment, in addition to the materials usually listed by MRP. Table 17.9 describes the contents of the inventory status file, including all activities, materials, labor, machines, and equipment and their associated codes. The file also includes the durations of activities and the lead times required to acquire these resources. Figure 17.12 shows the project in the form of a project structure tree. The solid lines connect activities of a path in the CPM diagram. Also shown are the codes for the resources, such as personnel, labor, equipment, and machinery, necessary to complete the task. Activity durations (t) or lead times (LT) are given along with the codes. Since an activity may be common to one or more paths, the project structure tree may show an activity more than once. For example, activity 100–200 is part of three paths, and hence it is listed three times. This may lead to the duplication of resource inputs. To prevent this, an activity is allowed to occur only one time, when it is first needed in the schedule. Duplications of requirements should be ignored by the system as they occur [1].

TABLE 17.8   Project Bill of Material

Parent	Child	Quantity	Description
500–600			
	200–500		
	300–500		
	400–500		
	451	1	Machine
	051	2	Material
300–500	200–300		
	452	1	Equipment
	052	4	Material
400–500	200–400		
	452	1	Equipment
	051	2	Material
200–500	100–200		
	451	1	Machine
	052	2	Material
200–300	100–200		
	451	1	Machine
	052	2	Material
200–400	100–200		
	451	1	Machine
	052	2	Material
100–200	055	3	Labor

TABLE 17.9 Inventory Status File Description

Description	Code	On Hand	On Order	Lead Time	Activity Duration
Activity	500–600	0	0		4
Activity	400–500	0	0		3
Activity	300–500	0	0		5
Activity	200–500	0	0		8
Activity	200–400	0	0		5
Activity	200–300	0	0		2
Activity	100–200	0	0		7
Material	051	0	0	2	
Material	052	0	0	5	
Labor	055	0	0	4	
Machine	451	0	0	2	
Equipment	452	0	0	4	

Using the procedure described earlier, we compute *ES, EF, LS,* and *LF* for the project structure tree, as exhibited in Figure 17.12. The numbers at the corners of the project structure rectangles represent these variables. For example, activity 200–300, which duration $t = 2$, is exhibited as follows:

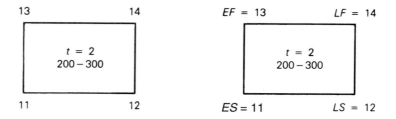

Once the computations are made for all activities and resources, the information can be used in the MRP-like system or in an actual MRP environment. Table 17.10 portrays MRP tables for the project schedule with latest possible starts and Table 17.11 gives them with earliest feasible starts. Thus these two tables provide the lower and upper limits of dates for starting activities or acquiring resources. We see from Figure 17.12 that the project needs twenty-three weeks, instead of nineteen weeks, as originally planned.

The CPM/MRP schedules assume that adequate capacities of resources exist. Suppose that we do not have adequate capacity for activity 200–300 either by latest or earliest starts. Then we might have to start the activity one or more periods earlier to complete the project on time. If this is infeasible, the project may have to be prolonged beyond the dates given by the CPM/MRP schedules. In that case we might end up revising several other activity schedule dates to comply with the new project completion date. Steinberg, Lee, and Khumawala [13] report that they have developed an MRP-type resource-constrained CPM system, which was adopted by NASA for flight operations planning and scheduling.

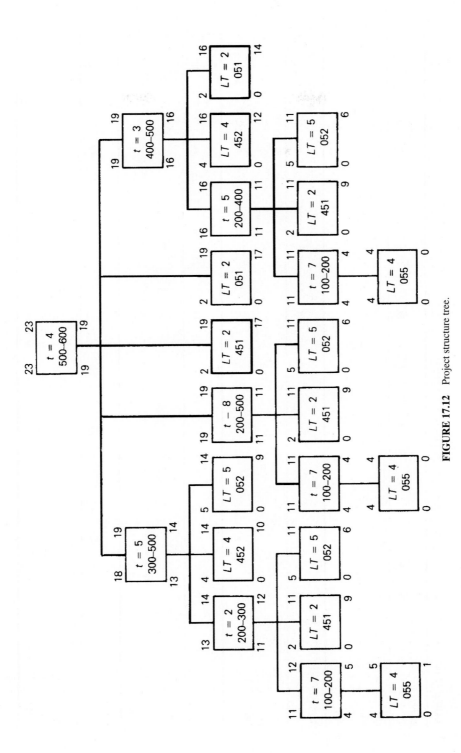

FIGURE 17.12 Project structure tree.

TABLE 17.10 CPM/MRP Latest Start Schedule for Branch 055-100-200-300-500-600

Activity 500–600 (LT = 4)

	0	1	2	3	4	5	6	7	8	9	10	11	12	13	14	15	16	17	18	19	20	21	22	23
GR																								1
PR																								1
OH	0																							0
POR																				1				

Activity 300–500 (LT = 5)

	0	1	2	3	4	5	6	7	8	9	10	11	12	13	14	15	16	17	18	19	20	21	22	23
GR																				1				
PR																				1				
OH	0																			0				
POR															1									

Activity 200–300 (LT = 2)

	0	1	2	3	4	5	6	7	8	9	10	11	12	13	14	15	16	17	18	19	20	21	22	23
GR															1									
PR															1									
OH	0														0									
POR													1											

Activity 100–200 (LT = 7)

	0	1	2	3	4	5	6	7	8	9	10	11	12	13	14	15	16	17	18	19	20	21	22	23
GR													1											
PR													1											
OH	0												0											
POR																								

TABLE 17.10 Continued

Activity 055 (LT = 4)

	0	1	2	3	4	5	6	7	8	9	10	11	12	13	14	15	16	17	18	19	20	21	22	23
GR						3																		
PR						3																		
OH	0					0																		
POR		3																						

Equipment 452 (LT = 4)

	0	1	2	3	4	5	6	7	8	9	10	11	12	13	14	15	16	17	18	19	20	21	22	23
GR															1									
PR															1									
OH	0														0									
POR											1													

Machine 451 (LT = 2)

	0	1	2	3	4	5	6	7	8	9	10	11	12	13	14	15	16	17	18	19	20	21	22	23
GR												1												
PR												1												
OH	0											0												
POR										1														

Material 052 (LT = 5)

	0	1	2	3	4	5	6	7	8	9	10	11	12	13	14	15	16	17	18	19	20	21	22	23
GR															1									
PR															1									
OH	0														0									
POR							1			1														

TABLE 17.11 CPM/MRP Earliest Start Schedule for Branch 055-100-200-300-500-600

Activity 500–600 (LT = 4)

	0	1	2	3	4	5	6	7	8	9	10	11	12	13	14	15	16	17	18	19	20	21	22	23
GR																								1
PR																								1
OH	0																							0
POR																				1				

Activity 300–500 (LT = 5)

	0	1	2	3	4	5	6	7	8	9	10	11	12	13	14	15	16	17	18	19	20	21	22	23
GR																			1					
PR																			1					
OH	0																		0					
POR														1										

Activity 200–300 (LT = 2)

	0	1	2	3	4	5	6	7	8	9	10	11	12	13	14	15	16	17	18	19	20	21	22	23
GR														1										
PR														1										
OH	0													0										
POR												1												

Activity 100–200 (LT = 7)

	0	1	2	3	4	5	6	7	8	9	10	11	12	13	14	15	16	17	18	19	20	21	22	23
GR												1												
PR												1												
OH	0											0												
POR					1																			

TABLE 17.11 Continued

Activity 055 (LT = 4)

	0	1	2	3	4	5	6	7	8	9	10	11	12	13	14	15	16	17	18	19	20	21	22	23
GR					1																			
PR					1																			
OH	0				0																			
POR	1																							

Equipment 452 (LT = 4)

	0	1	2	3	4	5	6	7	8	9	10	11	12	13	14	15	16	17	18	19	20	21	22	23
GR					1																			
PR					1																			
OH	0				0																			
POR	1																							

Machine 451 (LT = 2)

	0	1	2	3	4	5	6	7	8	9	10	11	12	13	14	15	16	17	18	19	20	21	22	23
GR			1																					
PR			1																					
OH	0		0																					
POR	1																							

Material 052 (LT = 5)

	0	1	2	3	4	5	6	7	8	9	10	11	12	13	14	15	16	17	18	19	20	21	22	23
GR						2																		
PR						2																		
OH	0					0																		
POR	2																							

PROJECT MANAGEMENT USING MICROCOMPUTERS

Project planning systems have been utilized successfully in managing large projects such as building construction, shipbuilding, and new-product development. Previously these software packages were available to only managers of large project. With the advent of powerful microcomputers and user-friendly project management software, however, they are now quite accessible. Among the popular software packages on microcomputers are Harvard Project Manager, Super Project Expert, Microsoft Project, Project Workbench, Timeline, and Primevera. A survey by Wassil and Assad [14] indicates that over 500 such packages are available today. To increase the productivity in all levels of management and support staff, project management tools are being used to plan and control myriad smaller as well as short-term activities common to both corporate and governmental offices [7]. In addition, such software facilitates the planning and control of multiproject staff scheduling to (1) minimize multiproject makespan, (2) minimize total weighted project lateness, (3) minimize total overtime costs, and (4) minimize the workload variability. Dean [6] summarizes the advantages of such a system:

1. It acts as a cost effective tool in producing, reporting, and communicating requirements.
2. It provides response to direct inquiries including "what if" questions.
3. It develops clearly defined specifications of activities and schedules.
4. It derives activity networks and resource schedules.
5. It satisfies management needs for frequent project reporting.

SUMMARY

Many project scheduling techniques, including CPM and PERT, are valuable tools for controlling project times and costs.

In addition to the popular project planning tools described in this chapter, mathematical programming techniques have also been widely used for project planning and control. Askin et al. [2] have designed an integrated information system and resource allocation model for implementing concurrent engineering projects for Motorola's semiconductor industry. Dean et al. [7] formulated a large-scale integer programming model for scheduling multiple-project staffing of toxic substance program management where the resource constraints could vary; they then solved the problem using PC-based software and heuristic procedures. Several CPM/PERT commercial software packages are available in the market for use in large computers, minicomputers, microcomputers and personal computers.. It must be realized, however, that CPM/PERT is not a panacea for all the ills of industry and business. If management is poor, if planning efforts are slight, or if the estimates for the project are unrealistic, then CPM/PERT will be of little help. Wheelright and Clark [15] emphasize the importance of creating an aggregate project plan by mapping an eight-step procedure to the overall product development process. It is important to realize that these are only tools to aid good management in making better decisions.

PROBLEMS

1. The following table lists all activities necessary to procure and build a soda bottling facility:

Activity	Description	Predecessors	Duration (weeks)
1–2	Market research and design building		12
2–3	Site selection	1–2	8
2–4	Supplier selection	1–2	6
2–6	Personnel selection	1–2	6
3–5	Construct building	2–3	8
4–5	Procure machinery	2–4	12
4–6	Prepare procedures	2–4	4
5–7	Install machinery	3–5, 4–5	2
6–7	Personnel training	2–6, 4–6	2
7–8	Launch	5–7, 6–7	4

Construct a CPM diagram and identify the critical path. What is the minimum time required to get the project rolling? Make a table of *ES, EF, LS, LF,* and slack.

2. The following table provides the costs of each activity in problem 1 for normal time and crash time:

Activity	Normal Time (weeks)	Crash Time (weeks)	Normal Cost ($1000s)	Crash Cost ($1000s)
1–2	12	8	80	120
2–3	8	6	20	25
2–4	6	4	10	15
2–6	6	4	10	15
3–5	8	5	150	180
4–5	12	8	250	300
4–6	4	3	10	15
5–7	2	2	60	—
6–7	2	2	50	—
7–8	4	2	50	70

The owner feels that the project should be completed in thirty-two weeks. Estimate the additional cost that will be incurred due to crashing.

3. The owner of the soda bottling firm is financing the project through a bank loan. The bank requires a monthly (four-week) cash flow statement for the project. Using the *ES* and *LS* approaches, calculate the range of cumulative cash flows required for completing the project using normal time durations.

4. A project has the following schedule:

Activity	Predecessors	Duration (days)
A	—	10
B	—	3
C	A	7
D	B	10
E	B	12
F	C, D	6
G	E	5
H	E	6
I	G	7
J	H	5
K	I, J	2
L	F	4
M	L, K	10

Construct a CPM diagram and find the critical path. Tabulate *ES, EF, LS, LF,* and slack for each activity. (*Hint:* In this problem, we have specified activity labels instead of node numbers. Make up your own node numbers when you draw the diagram. We have started the diagram; you should not have much trouble in completing it.)

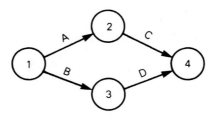

5. The manager feels that the durations given in problem 4 are very tentative. Therefore, he would like to use probabilistic time estimates, as follows:

Activity	Optimistic Duration (days)	Pessimistic Duration (days)	Probable Duration (days)
A	8	15	10
B	2	7	3
C	5	10	7
D	8	15	10
E	8	20	12
F	5	9	6
G	4	6	5
H	5	5	5
I	5	9	7
J	4	8	5
K	2	2	2
L	3	6	4
M	8	12	10

Based on these estimates, calculate *ES, EF, LS, LF,* and minimum duration for completion of the project.

6. What is the probability of completing the project described in problem 5:
   a. In thirty-five days?
   b. In forty-five days?
   c. In the time frame specified by PERT?
   d. Which activity in problem 5 has the most uncertainty?
   e. Which activity has the least uncertainty? Why?
7. The following activities should be accomplished for completing an R&D project:

Activity	Time months	Predecessors	Cost ($1000s)
A	3	—	30
B	6	—	40
C	4	A, B	40
D	2	B	30
E	8	A	10
F	6	C	15
G	9	E, F	10
H	9	D, F	20
I	2	G, H	15
J	13	B	30
K	5	I, J	15

Construct a CPM diagram and determine the project duration. Construct a table containing the slack for each activity.

8. Prepare a quarterly cash flow analysis for problem 7, using *ES* dates.
9. Holland Electric Company installs wiring and electrical fixtures for residential and commercial buildings. The manager generally likes to organize all activities so that she can complete the project on time. The activities and their optimistic completion times (*a*), most likely completion times (*m*), and pessimistic completion times (*b*) are as follows:

Activity	Days a	m	b	Immediate Predecessor
A	7	9	11	—
B	5	6	9	—
C	3	5	8	—
D	10	20	35	A
E	6	7	8	C
F	7	9	11	B, D, E
G	6	8	10	B, D, E
H	5	7	12	F
I	14	15	17	F
J	6	8	12	G, H
K	4	5	6	I, J
L	9	9	9	G, H

a. Construct a network for this problem.
b. Determine the expected times and variances for each activity.

    c. Calculate *ES, EF, LS, LF,* and slack for each activity.

    d. Determine the critical path and project the completion time.

    e. Determine the probability that the project will require more than sixty-five days.

10. The costs of completing each activity in problem 9 are as follows:

Activity	Cost ($)
A	200
B	300
C	500
D	100
E	200
F	700
G	1500
H	2500
I	1700
J	2000
K	500
L	1000

Assuming a five-day work week and LS schedule, provide a cash flow analysis for problem 9.

11. A manufacturer of stereo equipment has the following information related to activities necessary to procure and build a local facility:

Activity	Description	Predecessor	Duration (weeks)
1–2	Market research/design	—	10
2–3	Site selection	1–2	4
2–6	Personnel selection	1–2	4
2–5	Legal/environmental	1–2	6
2–4	Select vendors	1–2, 2–6	8
3–5	Building construction	2	8
4–5	Establish procedures	2–4	4
4–6	Install test equipment	4–5, 2–5, 3–5	4
5–7	Establish operations	2–5, 3–5, 4–5	4
6–7	Train personnel	2–6, 4–6	2
7–8	Begin operations	6–7, 5–7	4

Construct a CPM diagram and identify the critical path. What is the minimum time required to get the business started? Construct a table of *ES, FS, LS, LF,* and slack.

12. The following are the costs of each activity in problem 11 for normal time and crash time:

Activity	Normal Time (weeks)	Crash Time (weeks)	Normal Cost ($1000s)	Crash Cost ($1000s)
1–2	10	8	75	100
2–3	4	2	20	30
2–6	4	2	20	30

2–5	6	6	10	10
2–4	8	6	50	60
3–5	8	6	180	220
4–5	6	4	160	200
4–6	4	3	20	30
5–7	4	3	40	—
6–7	2	2	40	40
7–8	4	2	60	90

The owner feels that the project should be completed in thirty weeks. Estimate the additional cost that will be incurred due to crashing.

13. For financing requirements, and using the normal time durations and *LS* approaches, calculate the range of cumulative cash flows required for completing the project in problem 11. (Cash flow statements are due every four weeks.)

14. A manager of a construction company is reviewing the activities in a current project. The schedule is as follows:

Activity	Predecessors	Duration (weeks)
A	—	10
B	—	5
C	A	8
D	A	12
E	B	13
F	B, D	8
G	C	6
H	E, F	4
I	B, D	8
J	G, I	10
K	H, J	6

Construct a CPM diagram and identify the critical path. Tabulate *ES, EF, LS, LF,* and slack for each activity. For this problem, make up your own node numbers.

15. Probabilistic time estimates for the project in problem 14 are as follows:

Activity	Optimistic Duration (weeks)	Pessimistic Duration (weeks)	Probable Duration (weeks)
A	8	14	10
B	2	7	5
C	6	10	8
D	10	15	12
E	10	15	13
F	6	15	8
G	6	6	6
H	4	6	4
I	4	9	8
J	8	15	10
K	4	8	6

Based on these estimates, calculate *ES, EF, LS, LF,* and minimum duration for completion of the project.

16. What is the probability of completing the project in problem 15:
    a. In forty days?
    b. In the time frame supplied by PERT?
    c. Which activity in problem 15 has the most uncertainty?
    d. Which activity has the least uncertainty? Why?

17. Mason Company has received orders for M-120 terminals, for which the product structure tree and delivery schedule are given. The test and delivery procedure (not shown) takes one month. Lead times (LT) is specified in weeks.

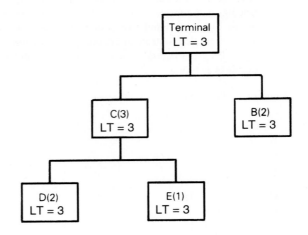

Month	Quantity	Cumulative Quantity
2	100	100
3	150	250
4	200	450
5	200	650
6	350	1000
7	350	1350
8	350	1700

Five months into the shipping schedule, the following cumulative quantities of units have passed through corresponding steps in the production process.

Process Steps	Cumulative Production
1	1320
2	1260
3	1220
4	1000
5	800

Develop an LOB chart and evaluate the status of production at each production stage, using the line of balance technique.

18. An oil exploration project has been proposed. The following table shows the activities involved in completing the project:

Activity	Time (months)	Predecessors	Cost ($10,000s)
1	4	—	20
2	6	—	30
3	3	1, 2	30
4	9	2	10
5	8	1, 2	10
6	7	3, 4	20
7	8	3, 4	40
8	2	5, 7	40
9	12	8	15
10	4	6, 9	10

Construct a CPM diagram and determine the project duration. Construct a table containing the slack for each activity.

19. Prepare a quarterly cash flow analysis for problem 18, using *LS* dates.

20. For Carter Company, an electrical engineering firm, the project activities and their optimistic (*a*) most likely (*m*), and pessimistic (*b*) completion times are as follows:

Activity	a	m	b	Immediate Predecessor
A	5	6	8	—
B	7	9	12	—
C	3	5	7	—
D	6	7	8	C
E	10	15	18	A
F	5	7	12	B, E
G	14	20	24	A
H	4	6	8	D, F
I	8	9	10	G, H
J	4	5	7	I

a. Construct a network for this problem.
b. Determine the expected times and variances for each activity.
c. Determine the critical path and project the completion time.
d. Determine the probability that the project will not be completed in fifty days.

21. A U.S. firm has decided to locate a plant in France. The firm's bills of material are shown in Table 17.12.
a. Based on the parent activities, find the shortest amount of time necessary to complete the project.
b. Based on the combined parent-child relationship, find the critical path and its duration.

TABLE 17.12 Project Bills of Material, Problem 21

Parent	Child	Activity Duration (weeks)	Quantity (Lead Time)	Description
500–600		6		
	200–500			
	300–500			
	400–500			
	351		1 (6)	Equipment
	051		2 (2)	Material
400–500		8		
	100–400			
	352		2 (8)	Equipment
	052		4 (3)	Material
300–500		3		
	100–300			
	353		1 (4)	Equipment
	052		2 (1)	Material
200–500		4		
	100–200			
	354		4 (5)	Equipment
100–400		2		
	355		3 (6)	Equipment
	053		2 (1)	Material
100–300		2		
	054		2 (5)	Material
100–200		2		
	055		2 (2)	Material

REFERENCES AND BIBLIOGRAPHY

1. N. J. Aquilano and D. E. Smith, "A Formal Set of Algorithms for Project Scheduling with Critical Path Scheduling/Material Requirements Planning," *Journal of Operations Management,* Vol. 1, No. 2 (November 1980), pp. 57–67.
2. R. G. Askin, M. Sodhi, O. R. Liu Shenh, and J. S. Ramberg, "Information System Design and Resource Allocation for Concurrent Engineering in Semiconductor Industry." Paper presented at the TIMS/ORSA Joint National Meeting, in Orlando, Fl., April 26–29, 1992.
3. T. M. Cook and R. A. Russell, *Introduction to Management Science* (Englewood Cliffs, N.J.: Prentice Hall, 1981).
4. I. Curtalus and E. W. Davis, "Multiproject Scheduling: Catagorization of Heuristic Rules Performances," *Management Science,* Vol. 28, No. 2 (1982), pp. 161–172.
5. E. W. Davis, "Project Scheduling under Resource Constraints—Historical Review and Categorization of Procedures," *AIIE Transactions,* Vol. 5, No. 4 (December 1973), pp. 297–313.
6. B. V. Dean, "Use of Project Management Microcomputer Software for the Effective Management of Technology Start-Up Firms." In *Impact of Microcomputers on Operations Research,* S. Gass, ed. (New York: Elsevier Science, 1986), pp. 139–149.
7. B. V. Dean, D. Denzler, and J. Watkins, "Multiple Staff Scheduling with Variable Resource

Constraints," *IEEE Transactions on Engineering Management,* Vol. 39, No. 1 (February 1992), pp. 59–72.

8. H. F. Evarts, *Introduction to PERT* (Boston: Allyn and Bacon, 1964).

9. R. Gessner, "Use Networking, MRP, or Both?" *Production and Inventory Management Review and APICS News,* December 1981, pp. 22–23.

10. R. I. Levin and C. A. Kirkpatrick, *Planning and Control with PERT/CPM,* (New York: McGraw-Hill, 1966).

11. J. Riis, U. Thorsteinsson, and H. Mikkaelsen, "Project Management Aided by Microcomputers," *International Journal of Project Management,* Vol. 5, No. 2 (May 1987), pp. 91–96.

12. L. A. Smith and P. Mahler, "Comparing Commercially Available CPM/PERT Computer Programs," *Industrial Engineering,* Vol. 10, No. 4 (April 1978), pp. 37–39.

13. E. Steinberg, W. B. Lee, and B. M. Khumawala, "A Requirements Planning System for the Space Shuttle Operations Schedule," *Journal of Operations Management,* Vol. 1, No. 2 (November 1980), pp. 69–76.

14. E. A. Wasil and A. A. Assad, "Project Management on the PC: Software, Applications, and Trends," *Interfaces,* Vol. 18, No. 2 (March–April 1988), pp. 75–84.

15. S. Wheelright and K. B. Clark, Creating Project Plans to Focus," *Harvard Business Review,* Vol. 70, No. 2 (March–April 1992), pp. 70–82.

PART FIVE

MANUFACTURING STRATEGY AND TECHNOLOGY

Important advances have been made in strategic thinking about getting products to the marketplace sooner than the competition. Successful manufacturing and service strategies must be integrated into the overall strategy of a firm. In Part V we discuss time-based competition and innovations in technology that are driving the rapid changes in the manufacturing environment. We emphasize how these strategies and changes can effect production planning and control activities.

Figure V.1 is the Arthur Young's framework for competitive advantage. Strategic vision (discussed in Chapter 1) provides the overall guide for world-class manufacturing. Chapter 18 introduces the importance of speed from the new-product and process-design phase to the production and delivery of products to customers. Here we address such vital concepts such as benchmarking, simultaneous engineering, reverse engineering, technology audits, and strategic alliances, all of which are essential to the productivity and in some cases to the survival of the firm itself. It is not hard for anyone in manufacturing today to understand the importance of speed in meeting customer demands and delivering products.

Chapter 19 discusses technology and chapter 20 discusses quality, two other key components of the world-class manufacturing resources shown in Figure V.1. Technological innovations will continue to change the way products are manufactured dramatically. Computer-aided design, group technology, flexible manufacturing systems, and automated storage and retrieval systems are a few of the concepts that have changed manufacturing recently.

Quality management has become one key to success in business today. The basics of total quality management and the quality improvement process are discussed in Chapter 20. We conclude with a view toward the future (Chapter 21); rapid change will have dramatic effects on future production planning and inventory control techniques.

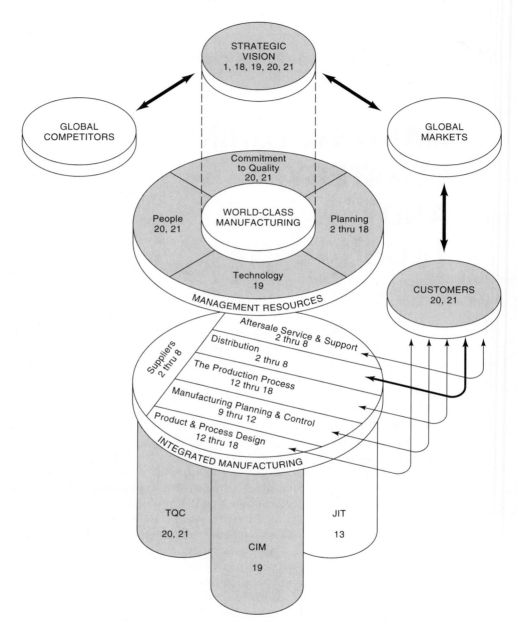

FIGURE V.1 Arthur Young's manufacturing for competitive advantage framework. *Source:* Copyright ® 1987 by Ernst & Young (formerly Arthur Young, a member of Arthur Young International). Reprinted by permission. *Note:* numbers on this figure indicates chapter numbers covering these subjects.

The following chapters are in Part V:

Chapter 18, Speed to Market
Chapter 19, Technological Innovations in Manufacturing
Chapter 20, Total Quality Management
Chapter 21, Factory of the Future

Chapter Eighteen

Speed
to
Market

INTRODUCTION

Executives have always known that time is money. Successful publishing companies thrive only by getting their "hot" news to the public before their competitors. The concept of the one-minute manager and facsimile machines have invaded the workplace. Time pressure was initially felt only in markets driven by style, fashion, and fad. But as competition among international companies intensifies and as improved products are continually introduced, speed has taken on added significance. For example, new models of Japanese cars in the past have reached the market faster than new models by U.S. manufacturers. Multinational firms are forced to shorten their product life cycles, to minimize the lead times, and to adopt a new approach in the process of launching new products. Ford Motor Company, for example, established Team Taurus to launch the Taurus and Sable line of automobiles [14]. The team, which included designers, engineers, production specialists, and market researchers, addressed important factors such as feasibility, profitability, competitiveness, and consumer satisfaction in the design and the production stages. By working together the team knew how exactly the car would be built even before the design team made the clay model. Ford's 1980s rebirth is largely attributed to these actions.

As speed to the market has become the norm for the 1990s, the product introduction process has been further complicated by ever-changing technology and customer niches. Products and processees must be developed simultaneously by teams. Products include both goods and services. Processes include manufacturing, marketing, procurement, quality, and after-sale operations that keep the customers satisfied. Although most of the technologies have come from external sources, successful companies have used incremental advances in product and process technologies to gain market share. They have also aggressively pursued the transfer of technology from elsewhere in the world and integrated it in their own businesses. An organization's ability to manage the product life cycle more effectively by using technology in the product and process has become an important issue. Japanese companies maintain an ongoing assessment of technological changes. They keep close touch with the marketplace and with research communities around the world by translating technical publications and finding useful ideas to increase productivity.

Cooperative research and development, joint ventures, alliances, reverse engineering, benchmarking, project management, technical audit, and product design teams are some ways multinational firms plan to meet the challenges of the future. These methods aid management in organizing, analyzing, and filtering information to obtain the best information for their next product or process. In today's rapidly changing technological environment, failure to act quickly in investing products and processes could open the doors for the competition. In fact, the cost of catching up may be prohibitive [11]. Therefore, for a firm to survive, design teams, manufacturing organizations, or material managers can no longer isolate themselves from other functions of the corporation. Material managers and production planners must understand these concepts and be able to work with others within the firm to increase quality and productivity. Hence, this chapter addresses the vital issues successfully used by many companies to speed their product to the marketplace while improving their productivity and quality.

BENCHMARKING

Benchmarking includes identifying the performance measures of competitors, measuring their performance levels, comparing their performances, and then taking appropriate steps to close the performance gaps [2]. While Xerox Corporation was being trounced by foreign competitors during the late 1970s and early 1980s in its own copier market, the corporation found that benchmarking was its savior; thus Xerox popularized the concept. In the past, corporations examined the business practices and products of their direct competitors. Increasingly, this concept is being extended to operational processes and practices such as engineering, purchasing, manufacturing, sales, research and development, accounting, finance, marketing, and office operations. Competitive benchmarking can shed light on determining which company has the best or proven practice of performance to a particular function within an enterprise. By understanding where "best in-the-class companies" have achieved top performance for a given function, a firm can target clear and understandable goals.

Today, multinational corporations not only study their competitors but also study specific functions of other well-managed companies. For example, Xerox studied the order processing system of L. L. Bean in detail because it was considered to be best in-the-class. Biesada [2] points out that learning new approaches by examining other organizations and then implementing those systems to improve a firm's own business is a formalized way of managing change. In addition, identifying a competitor's strengths and weaknesses can help a company achieve superior performance. Camp [6] identifies ten major steps involved in the benchmarking process (see Figure 18.1).

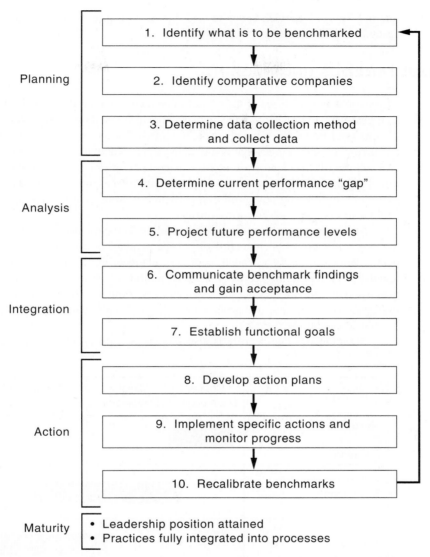

Planning

1. Identify what is to be benchmarked

2. Identify comparative companies

3. Determine data collection method and collect data

Analysis

4. Determine current performance "gap"

5. Project future performance levels

Integration

6. Communicate benchmark findings and gain acceptance

7. Establish functional goals

Action

8. Develop action plans

9. Implement specific actions and monitor progress

10. Recalibrate benchmarks

Maturity

- Leadership position attained
- Practices fully integrated into processes

FIGURE 18.1 Ten steps in benchmarking. From R. C. Camp, *Benchmarking: The Search for Industry Best Practices that Lead to Superior Performance* (Milwaukee: ASQC Press, 1989), p. 259. Reprinted with the permission of ASQC.

Examples of benchmarking abound in the literature. ALCOA has studied Dow Chemical, DuPont, and Hercules Inc., for safety procedures and General Electric for management processes. When Motorola was trying to speed up its delivery process of cellular telephones, it benchmarked Domino's Pizza and Federal Express. MBNA America, a credit-card company, bench marked Xerox, IBM, L. L. Bean, and U.S. Sprint to study customer service practices. To learn the best practices of the industry, the

Internal Revenue Service benchmarked American Express for billing and Motorola for accounting procedures.

SIMULTANEOUS ENGINEERING

Approximately two-thirds of the cost of manufacturing a product is determined in the design phase [7]. Poor designs can cause manufacturing and quality problems. Complicated designs can lead to excessive delays in the introduction of the product while creating frequent engineering changes that could have been avoided in the design phase. As technology and industry standards are rapidly changing, consumer expectations are also changing. Newer tools in automation undoubtedly place more demands on employees. Therefore, developing new products, particularly those involving advanced technology, could result in cost and schedule overruns.

Multinational corporations are scrambling to meet customer needs by introducing higher-quality products in record time. The product development environment brings people from various fields, such as design, engineering, manufacturing and marketing, together. In the past, these cross-functional multidisciplinary teams communicated infrequently. A traditional product development and commercialization model [11] is shown in Figure 18.2. The approval process takes place serially and hence the total lead time is prolonged needlessly.

Simultaneous engineering is also known as team design, concurrent engineering,

Activity	Concept Development	Design Development	Design Validation	Production Development
Marketing Product Planning	███			
Engineering		████		
Testing			████	
Manufacturing				████
Conventional Engineering				

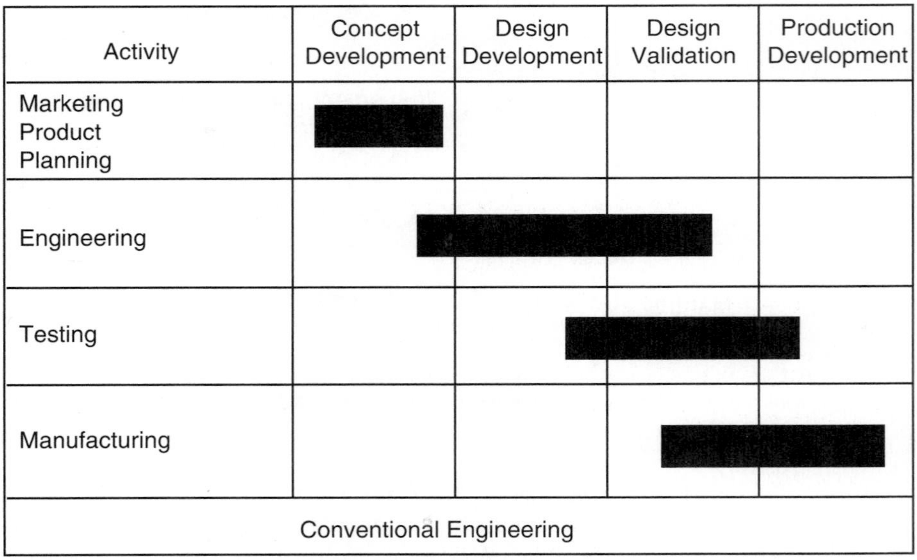

FIGURE 18.2 Traditional product commercialization model using conventional engineering. From John R. Hartley, *Concurrent Engineering: Shortening Lead Times, Raising Quality, and Lowering Costs* (Mass.: Productivity Press, 1992), p. 17. Reprinted with the permission of Hawtal Whiting Design & Engineering Co. Ltd, Warwickshire, England.

integrated product development, or parallel release. By fostering cultural considerations and emphasizing communication among team members, this process can help firms to reduce the total lead time necessary to introduce new products. Simultaneous engineering is a process by which coordinating disciplines work together to conceive, develop, and implement new-product and process decisions. The synergy developed between the business and the technology units of the company can create a link among research and development, manufacturing engineering, production, marketing, and other supporting organizations and hence can get the product or service to the market faster than under the old system. In Japanese manufacturing, simultaneous engineering has been a key factor for reducing the lead time and the cost of launching products while increasing the quality.

The importance of getting a product to the market cannot be overemphasized. Although the total development time is a function of the complexity of tasks and the number of persons involved, among other factors, the overall time frame has to respond to marketplace needs. Delayed introduction of the product to the marketplace can lead to significant loss in revenues for each month the product is late [7, 21]. The following formula illustrates the relationship between revenue loss as a function of delay and market window:

$$\frac{D\,(3W - D)}{2W^2}$$

where D and W represent the delay to the market and market window, respectively, in months. Table 18.1 exhibits the cost of arriving late to the market [21].

Successful completion of projects require effective action from all major functions of the business. Engineering should provide manufacturable designs, well-executed tests, and a high-quality prototype. Marketing should thoughtfully consider product position-

TABLE 18.1   The Cost of Arriving Late to Market (and still on budget)

| If your company is late to market by: | | | | | |
6 Mo.	5 Mo.	4 Mo.	3 Mo.	2 Mo.	1 Mo.
Your gross profit potential is reduced by:					
−33%	−25%	−18%	−12%	−7%	−3%
Improve time to market by only 1 mo., profits improve:					
+11.9%	+9.3%	+7.3%	+5.7%	+4.3%	+3.1%
For revenues of $25 Million, annual gross profit increases:					
+$400K	+$350K	+$300K	+$250K	+$200K	+$150K
For revenues of $100 Million, annual gross profit increases:					
+$1600K	+$1400K	+$1200K	+$1000K	+$800K	+$600K

Source: McKinsey & Company.

Reprinted with the permission of APICS, Inc. The New Competitors: "They Think in Terms of Speed to Market," Production and Inventory Management by J.T. Vesey, Vol 33, No. 1, 1992, pp. 71–77.

ing and have solid consumer analysis, and thought-out product plans. Manufacturing should initiate capable processes, forecast cost estimates, and skillfully pilot production and rapid manufacturing. An effective project team commercialization model [11] is portrayed in Figure 18.3. Application of simultaneous engineering in the product design phase could result in 40% savings in the product introduction time. Comprehensive frameworks for cross-functional integration have been recommended by Wheelwright and Clark [22] as well as by Carter and Baker [5].

Not all development projects need to deploy cross-functional integration. Especially when the product designs are stable, customer requirements are well defined, and the life cycle and lead times are long, cross-functional groups can develop new products with modest coordination through established rules and procedures and occasional meetings. But when the manufacturing and the technology is more dynamic and the lead time is critical element of competition, cross-functional interface takes an important dimension. Nevens, Summe, and Uttal [12] found many common traits among leading companies. Compared with their competition, these firms successfully communicate two

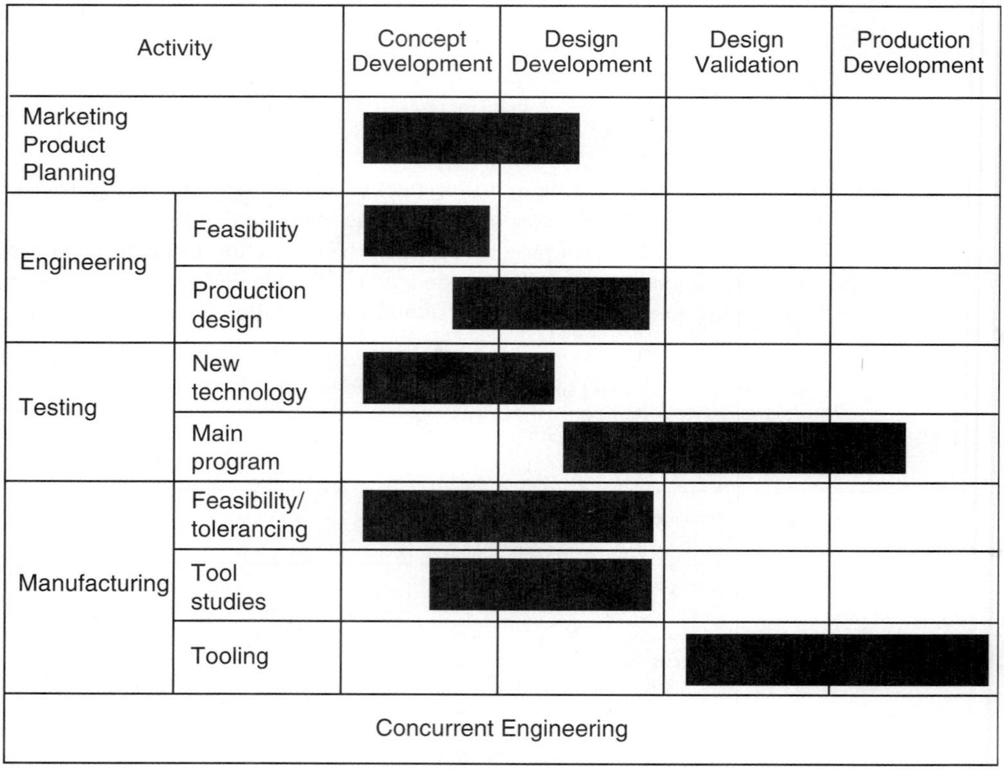

Activity		Concept Development	Design Development	Design Validation	Production Development
Marketing Product Planning		██████			
Engineering	Feasibility	███			
	Production design		████		
Testing	New technology	████			
	Main program			████████	
Manufacturing	Feasibility/ tolerancing	██████			
	Tool studies	████			
	Tooling				██████
Concurrent Engineering					

FIGURE 18.3 Project team commercialization model using concurrent engineering. From John R. Hartley, *Concurrent Engineering: Shortening Lead Times, Raising Quality, and Lowering Costs* (Cambridge, Mass.: Productivity Press, 1992), p. 17. Reprinted with the permission of Hawtal Whiting Design & Engineering Co. Ltd. Warwickshire, England.

to three times faster among the team members and bring two to three times the number of products to market. They incorporate technology two to five times more than their competitors. In addition, they introduce their products in less than half the time and compete in twice as many products and geographical areas as their competitors.

REVERSE ENGINEERING

Reverse engineering (see Figure 18.4) has always been practiced by most multinational corporations and, in fact, has become a popular and preferred practice. Companies buy their competitor's products and then literally strip them apart and learn to improve the functional capability of their own products. In essence, reverse engineering recovers the physical and logistical functions of the original design specifications of component parts, items, and products. The design and manufacture of IBM Proprinter cloning Epson printers has been touted in the literature as an excellent example of reverse engineering. Using powerful concepts such as the design for manufacturability (DFM), companies can improve their designs and reduce their costs while increasing the quality of their products. Design for manufacturability and design for assembly (DFA) are further discussed in Chapter 19.

If the product is software, reverse engineering can extract the essential and complex structures of entities and relationships of computer software and systems. Increasingly, corporations use reverse engineering to upgrade their aging applications software. Here the existing system is analyzed, understood, and refined and the system is rewritten, thus taking advantage of the new technology while maintaining all functions and features that were useful in the original system [15]. Once the systems' functions are understood in simplified form, the system can be revised by removing redundancies and updated to create new products. Computer-assisted software engineering (CASE) tools are available to aid the reverse engineering process.

Another interesting application of reverse engineering is expediting products from the conceptual level to the design stage quickly. Suppose that a clay model of an item is

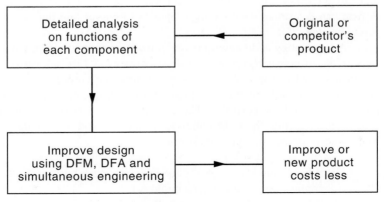

FIGURE 18.4   Reverse engineering.

available. Modern digitizing equipment and computer-assisted design tools can transform the clay model to a computer-aided drawing (CAD) model of the part. Specialized software can be used to test the integrity of the design and to develop production tooling. Through modern technology, reverse engineering can mechanize the tedious manual dimensioning or tracing process and thus cut product introduction time significantly. In the early 1980s, by using the reverse engineering process, Xerox found that they could buy copier components from vendors much more cheaply than they could manufacture them internally. The success of Advanced Micro Devices Corporation in using reverse engineering to clone Intel computer chips is also well documented.

TECHNOLOGY AUDITS

The process of organizing and using technical knowledge and skills that are acquired and used from internal and external sources is part of the management of technology (MOT). The objectives of technology audits are to determine (1) the innovative capabilities of a firm, (2) the merits of existing technological strategy of a firm, and (3) a firm's ability to transfer technology efficiently. A firm's audit could cover such factors as its research and development record, customer needs, competition, the position of its technology in the life cycle of the product, and how the technology is utilized in the business strategies [8]. First, by comparing technology audits of a firm with those of the competitor and then by comparing them with the state-of-the-art technology, a firm can plot its course of technology with its customers. The technology policy, in general, is formulated in relation to the life cycle of the product and the effectiveness with which technology is used in the given business strategy. By identifying suitable policies and practices, a firm can develop performance standards for the management of technology. A study by van de Meer and Calori [19] reveals that several key factors—including goal clarity, clear definition of the business, simultaneous engineering, and an organizational climate conducive to openness—lead to success in the technology audit area. Japanese companies have found benchmarking a viable tool for technology audits.

During a technology audit a firm can find products and processes that are obsolete and can determine what is hot now and what areas are expected to flourish in the future. A firm can determine what strategic options to pursue and thus set a direction for their future business.

A technology audit should reveal a firm's innovation in product and service offerings, in the fit between corporate business strategy and its innovative capabilities, in and the firm's need to support an ongoing business and corporate strategy [5]. Petrov [14] and Burgleman and Maidque [5] provide a list of questions a firm should answer as a part of a technology portfolio. These questions can provide a rational categorization process for selecting the right policy to technology and to provide a strategic direction. These questions are summarized as follows:

1. What technologies exist in the firm or in each business unit, and what are other applications of existing technology?
2. Which technologies are currently used in each product, item, or part, and what

are the competing technologies in each application? Should we obtain these technologies, and if so, from where?

3. How critical is each technology to an individual product or service, and what should the priorities of the technology development or purchase be?
4. Which external technologies seem critical in the future, and what is the likely evolution of these technologies?
5. What are the implications of technology as business portfolios for corporate strategy? How critical is a firm's technology in each application? When should technology be introduced to the market?

A document stating a firm's major mission areas and key technological needs to support each activity should be developed, thus integrating the objectives with strategic technologies of the firm.

Audit of Competitors' Technology

A technology audit covers the determination of (a) the current technological and business position of competitors' products, processes, and services; (b) an overall evaluation of their management of technology; (c) a comparison of competition in each activity; and (d) the mechanisms the competitor used for technology transfer. The audit should include the following activities:

1. Bibliographical reviews
2. Competitive analysis
3. Comparison of technology position for each product, process, or service
4. Reverse engineering
5. Comparison of the competitive market position
6. Use of strategic technical areas for positioning
7. Assessment of critical success factors to maintain market position
8. Evaluation of uniqueness of firm's products, processes, and services
9. Evaluation of elements that allow a customer to evaluate the desirability of the product or service

Assessing competitors' technology is difficult, and discretion should be exercised. Remember that your competitor is also trying to do the same. A folder for each competitor should be maintained and updated periodically. These files should be reviewed by key professionals and midlevel managers at regular intervals.

Audit of a Company's Own Technology

An audit determines the probability of success of a firm's products, processes, and service development programs, which include both business objectives and strategies. Major new products, services, processes, and other development projects besides existing products and services can be included. Project audits give an insight into the overall technological health of an institution.

Auditing of State-of-the-Art Technology

The purpose of auditing state-of-the-art technology is to compare the firm's own technology with the competitor's as well as with state-of-the-art technology. Briggs and Coleman [4] suggest the following steps:

1. Identify potential information from sources such as journals, periodicals, government studies, research institutions, databases, patent filings, and universities regarding technical issues in which you are currently or potentially interested.
2. Assess potential breakthroughs and incremental improvements in those technologies.
3. Compare the position of the competitor with state-of-the-art technologies and then with your technological strengths.

Briggs and Coleman also recommend that companies maintain a research database on environmental and competitive information and closely monitor their competitors.

Project Evaluation

A technology audit should allow the company to identify basic technologies, distinctive technologies, and external technologies. Basic technologies offer a foundation for their existence, distinctive technologies provide a potential competitive advantage, and external technologies refer to those provided by suppliers and other associated sources. A firm should be able to answer questions such as, Which technologies should be used as a basis for the future business? and Where should the firm become proficient given the distinctive competence? Such questions can help a firm set future technological strategies.

Technological audits can provide several benefits to the firm (1) by better predicting the probable success rate of individual products, processes and services, (2) by independently assessing efforts in technology transfer and MOT within the firm, (3) by identifying problems and opportunities where assistance is needed, and (4) by improving communication among professionals and creating awareness. The process helps a firm know how other firms, private and government organizations, and universities are doing. These audits also provide an overall understanding of each technology's contribution to the strategic direction of the firm and identify technologies that require external assistance through collaboration and joint venture. The technology audits, especially the company's own audits, should be conducted regularly.

STRATEGIC ALLIANCES

Strategic alliances in business and industry range from formal licensing contracts, transfers of technology, research consortia, technology acquisition, joint ventures, and spin-off agreements to informal agreements to work together and conserve resources. Strategic alliances are formed with long-term mutual benefit of entering parties. The benefit is derived from sharing technologies, markets, facilities, or the development of products and processes. It is very difficult for any single company to learn fully the varying customer markets and changing technologies around the globe. By forming long-term relationships or alliances, all parties stand to gain overall competitive advantage by sharing escalating costs. Working together also increases innovation. In essence, acquired key competitive technologies improve flexibility, provide access to new markets, reduce the total cost of products and processes by sharing fixed costs, and give a competitive weapon to respond to changes in the current market. It also decreases product development costs and product life cycles to satisfy changing customer demands more rapidly.

Forms of Alliances

General Motors, Ford, and Chrysler have formed a strategic alliance to conduct research for developing long-life batteries to power future cars. GM also has alliances with several firms working on machine vision. Bethlehem Steel and USX Corporation cooperate on conducting research on continuous casting of thin steel sections. AT&T and Sun Microsystems formed an alliance to develop a standard version of AT&T's Unix operating system. These companies will share results and license any invention or any data derived from the research.

Consortia Formed
to Encourage Research

Consortia are formed to conduct long-range research. For example, SEMATEC, by about fifty computer and communications companies, conducts research on computer chip manufacturing technology. When the cost of developing processes become too prohibitive for any single company, consortiums become an efficient vehicle.

Acquisition, Mergers, and Licensing

Acquisitions and mergers are another way to gain access to technology that are not developed internally. Acquisition of Apollo computers facilitated Hewlett-Packard Corporation to get into the computer workstation market. By acquiring NCR Corporation, AT&T was not only able to enlarge its computer business but also prevent huge losses incurred in its original computer business operations. The success of such agreements depends on the ability of the merging companies to meld into one monolithic organization.

Joint Ventures

Joint ventures facilitate firms to share in other firms' technologies, new markets, or other advantages. Joint ventures provide an avenue for effectively sharing technology or new markets, especially in foreign countries where the distribution system and market presence of a firm may be limited. The success of a joint venture depends on the partners learning to work with and trust one another. GM and Toyota formed NUMMI in Freemont, California, to produce Chevy Novas and Toyota Corollas. GM wanted to learn Toyota's manufacturing systems while Toyota wanted to learn more about GM's manufacturing technology and supplier arrangements. Similar alliances were formed between

Chrysler and Mitsubishi, Apple Computer and Sony for developing notebook laptop computers, IBM and Toshiba for developing laptop computer lightweight screens, Xerox and Fuji for developing copiers, and General Electric and Tungsram for manufacturing lighting products [16].

A success story encompassing many these aspects is illustrated by the Ford-Mazda alliance. Haigh [10] indicates that a new body of organizational skills and learning is emerging to manage their relationship. Seeds of a joint project by Ford and Mazda emerged in early 1980s when a $500 million stamping and assembly plant was conceived. Mazda wanted to learn about Ford's cost-control practices and also to experience the process of starting a major manufacturing facility in a foreign country. Ford wanted to learn Mazda's product design, tooling and equipment, quality deployment system, and employee training methods. Mazda's quality deployment system, in which the technicians, designers, engineers and customer service representatives work together, is a way of translating customer inputs as well as employee inputs to improve the quality of cars built. Mazda designed Mercury Tracer using the Mazda 323, Ford provided the styling, the project was completed on schedule. Ford videotaped Mazda's car sequence of operations, work locations, and required tools and learned that discipline and timeliness are very important. Mazda learned smart marketing and cost-control procedures from the joint venture. They have each come to respect the other, and this alliance has led to further cooperation. Ford and Mazda have since joined Kia Motors of South Korea to produce Ford Festiva and market the car in North America and Taiwan.

SUMMARY

In this chapter we dealt with many important concepts for increasing the productivity and quality of products and services that a firm produces and markets. Benchmarking aids a firm in evaluating their current operations and in setting goals for the future. Other concepts and techniques described in this chapter help a firm speed their products to the customer. Although the benefits of applying these concepts appear simple, it is important to realize that they contribute individually and cumulatively toward decreasing product introduction time, and hence they reduce the total cost of products and services. Speed to the market creates market leadership, opportunities for higher market share, and profits. Management should foster an organizational environment where change and innovation are a part of life and should provide employees with the most current and proven tools to perform their job [14]. The 1990s has been called the *value decade* [18] as customers become value-conscious and demand for products with highest quality at lowest prices. General Electric successfully increased productivity and quality by creating a concept called *workout program* where speed and simplicity were given prime importance, and employees at all levels teamed up to better their systems. For a program to be successful, employees must have open minds and must listen, debate, trust, and proceed with their best ideas. When multiple projects are undertaken simultaneously, a firm must coordinate these activities effectively using project management techniques, as described in Chapter 17.

PROBLEMS

1. Explain the purpose and major steps of benchmarking in manufacturing and service operations.
2. What is the difference between concurrent or simultaneous engineering and reverse engineering? In what situations can they be applied either individually or collectively?
3. Can reverse engineering and/or simultaneous engineering be applied in a nonmanufacturing environment? Explain where and how.
4. Explain the importance of technology management in manufacturing planning and control.
5. Can we apply the management of technology concepts to service operations? Give an example where these concepts were successfully applied.
6. In your opinion, what are some important steps we should take to increase the quality and productivity of products and services in the United States?
7. In your opinion, is the United States making any progress in regaining competitiveness in any sector of industry? If so, what sectors? How was the industry able to achieve the breakthroughs?
8. Would you agree that the acquisition of technology is important for the growth of a firm? If so, explain why. If not, explain why.
9. What are some avenues that firms follow to acquire technology? Is there a best way to do that? Explain.
10. What is the purpose of technology audits in a firm? Explain some major steps involved in the process.

REFERENCES AND BIBLIOGRAPHY

1. R. Abella, J. Daschbach, and L. Pawlicki, "Human Skill Interface in Reverse Engineering," *Computers and Industrial Engineering,* Vol. 24, No. 1–4 (1991) pp. 495–499.
2. A. Biesada, "Benchmarking," *Financial World,* Vol. 160, No. 19 (September 17, 1991), pp. 28–32.
3. G. Bootroyd, "Design for Assembly—The Key to Design for Manufacturing," *International Journal of Advanced Manufacturing Technology,* Vol. 2, No. 3 (August, 1987), pp. 3–11.
4. W. Briggs and J. Coleman, "Compact Discs: New Tool for Competitive Analysis," *Planning Review,* Vol. 15, No. 6 (November–December 1987), pp. 32–48.
5. R. A. Burgelman and M. A. Maidque, *Strategic Management of Technology and Innovation* (Homewood, Ill.: Richard D. Irwin, 1988).
6. R. C. Camp, *Benchmarking: The Search for Industry Best Practices that Lead to Superior Performance* (Milwaukee: ASQC Press, 1989).
7. D. E. Carter and B. S. Baker, *Concurrent Engineering* (New York: Addison-Wesley, 1992).
8. D. I. Cleland and K. M. Bursic, *Strategic Technology Management* (New York: AMACOM, a division of American Management Association, 1992).
9. D. Ford, "Develop Your Technology Strategy," *Long-Range Planning,* (October 1988), pp. 85–89.
10. R. W. Haigh, "Building a Strategic Alliance: The Hermisillo Experience as a Ford-Mazda Proving Ground," *Columbia Journal of World Business,* Vol. 27, No. 1 (1991), pp. 61–74.
11. W. G. Howard, Jr., and B. R. Guile, Ed. *Profiting from Innovation* (New York: Free Press, 1992).

12. M. T. Nevens, G. L. Summe, and B. Uttal, "Commercializing Technology: What the Best Companies Do" *Harvard Business Review,* Vol. 68, No. 3 (May–June 1990), pp. 154–163.

13. T. J. Peters, Liberation Management: necessary disorganization for the nanosecond nineties, (New York: A. A. Knopf, 1992).

14. B. Petrov, "The Advent of Technology Portfolio," *Journal of Business Strategy,* Vol. 3, No. 2 (1982), pp. 70–75.

15. J. B. Rochester, D. P. Douglas, and E. J. Chikofsky, "Re-Engineering Existing Systems, Commentary," *I/S Analyzer,* Vol. 29, No. 10 (October 1991), pp. 1–16.

16. M. K. Starr, *Global Corporate Alliances and the Competitive Edge* (New York: Quorum Books, 1991).

17. R. S. Teitelbaum, "The New Rage of Intelligence," *Fortune,* November 2, 1992, pp. 104–110.

18. N. M. Tichy and S. Sherman, *Control Your Destiny or Someone Else Will* (New York: Doubleday, 1993); excerpts in *Fortune,* January 25, 1993, pp. 86–93.

19. J. B. H. van de Meer and R. Calori, "Strategic Management in Technology Intensive Industries," *International Journal of Technology Management,* Vol. 4, No. 2, (1989), pp. 127–139.

20. G. S. Vasilash "Defining the Unknown Part," *Production,* Vol. 101, No. 2 (February 1989), pp. 57–59.

21. J. T. Vesey, "The New Competitors: They Think in Terms of Speed to the Market," *Production and Inventory Management Journal,* Vol. 33, No. 1 (1992), pp. 71–77.

22. S. C. Wheelwright and K. B. Clark, *Revolutionizing Product Development* (New York: Free Press, 1992).

APPENDIX 18A

The Benchmarking Bonanza
Beth Enslow

Jeanette Frick was frustrated. As head of the benchmarking program for GTE Directories, the Dallas-based subsidiary of GTE Corporation and the world's largest publisher of Yellow Pages, she kept running into the same stumbling block: The telephone industry's notorious secretiveness made it exceedingly difficult and time-consuming to get comparative information, an essential for benchmarking. "We had benchmarked against companies in other industries," says Frick, "but I felt that benchmarking against our competitors was also essential—after all, that's who your customers compare you with."

Frick proposed that GTE Directories start a Yellow Pages benchmarking consortium so that all the companies could have access to data on the best practices both inside and outside the industry. GTE Directories' senior managers liked the idea, so in late 1990 the company approached the Yellow Pages Publishers Association (YPPA), a trade association of 130 directory publishers.

YPPA formed a task force to investigate the consortium concept. The task force decided that YPPA would act as coordinator of the consortium, but that a third party—

Reprinted with permission from *Across the Board,* Vol. XXIX, No. 4 (April 1992), pp. 16–22.

Ernst & Young—would be in charge of compiling and overseeing the benchmarking data base, an arrangement that would calm companies' fears of losing proprietary information. All the information collected, the companies decided, would be disguised, so that no one would know which company provided which data.

The YPPA benchmarking consortium completed a 12-company ministudy last fall that benchmarked such areas as cycle time (the time it takes for a directory to be completed from start to finish) and the time needed to resolve customer complaints. The results were promising, says Frick, and the consortium began a full-blown study last month.

The telephone directory industry isn't alone in its new cooperative spirit. Throughout corporate America, companies are throwing off their veils of secrecy and placing their faith in the competitive power of cooperation. "Getting companies to share information readily is a significant directional change in the corporate culture in this country," says Robert C. Camp, manager of benchmarking competency at Xerox Corporation and author of the seminal book *Benchmarking: The Search for Industry Best Practices That Lead to Superior Performance.* "We're beginning to realize that sharing benchmarking data benefits everyone." Indeed, it may not be overstating the case to say that the future of U.S. business lies in its ability to effect this culture change.

Benchmarking, according to David T. Kearns, former CEO of Xerox, "is the continuous process of measuring products, services, and practices against the toughest competitors or those companies recognized as industry leaders." Actually, says John Scharlacken, a member of Arthur D. Little's operations management section who frequently helps clients benchmark, the principle of benchmarking has been around for a long time—in the form of corporate espionage. But today, with more companies freely sharing information, it's much more up front.

Benchmarking also differs from corporate espionage in that it typically involves gathering data from disparate industries. "Benchmarking is so powerful because you can apply processes outside your industry to your own company," says Harry L.M. Artinian, vice president of corporate quality at Colgate-Palmolive Company and a protégé of W. Edwards Deming, the best-known guru of the quality movement. "It gives you ideas and helps clarify universal principles that you can then apply to your own processes."

Six or seven years ago, only half a dozen *Fortune*-500 companies were benchmarking, reports Kaiser Associates, a benchmarking firm. Today, the majority are doing so. Why the exponential growth? Lawrence S. Pryor, a Kaiser vice president, says it's because U.S. companies consider benchmarking to be a powerful way to take action against competition from Japan and Europe. A contributing factor is the new corporate obsession with the Malcolm Baldrige National Quality Award, which has a strong benchmarking program as one of its criteria. Pryor says that Kaiser alone has done 500 to 600 benchmarking projects in the past six years.

"But executives must realize that benchmarking is not as simple as it may sound," warns Pryor. "Telling an employee to jump on a plane and go see what Xerox is doing will not produce results." Artinian agrees: "It's important to realize that you can't just copy another company's process, because your own system is unique. Even if the company you benchmarked has the same setup as you do, it's still different. You have a different environment and different people. This is a mistake that many companies starting out in benchmarking make. You need to understand how to apply benchmarking data to your own company."

As companies become more sophisticated at benchmarking, they're starting to realize that its free-for-all nature, with hundreds of companies running around collecting data on thousands of processes, is anarchic. As a result, leaders like Jeanette Frick are trying to establish some type of order. What are emerging are three clusters of benchmarking organizations: consultancies, consortiums, and clearinghouses.

CONSULTANCIES

"Management consultants have had their toes in the water for a long time on this subject," says Christopher Bogan, president of Best Practices Benchmarking in Lexington, Massachusetts. "For years they've been conducting competitive studies and then selling the results to clients. Benchmarking is just a new way to market what they do." That may be so, but benchmarking also requires a fundamentally different relationship between consultant and client—one of mutual involvement. "You realize the true value of benchmarking only when the people involved in the day-to-day work process—the line people—do the actual benchmarking," explains Xerox's Camp. "Otherwise, you don't know how to apply the results. The line people have to be learning the best practices firsthand."

Most consultants agree with this view. "There's a limited role for consultants in benchmarking, despite the great numbers of them out there," says James E. Staker, a vice president at the Strategic Planning Institute, a Cambridge, Massachusetts, management consulting firm. "Benchmarking is a do-it-yourself project."

According to Saul Berman, a partner in Price Waterhouse's strategic consulting group, consultants are useful in helping a company establish a system and methodology of benchmarking so that it can benchmark on a self-sustaining basis, using consultants only for advisory matters. Kaiser's Pryor concurs: "A company should not have to spend big bucks hiring an outside firm every time it wants to do a benchmarking project." To this end, many consultants offer benchmarking training programs and seminars to teach a client's managers how to find the best practices, measure them, and apply them to their own company.

Even with such altruistic principles, consultants haven't been left twiddling their thumbs. One of the most difficult aspects of benchmarking is finding out whom to benchmark; digging up best-practice companies, especially those in radically different industries, is time-consuming. "Companies whose benchmarking projects fail to deliver significant improvements have often not been diligent in finding out who is best in class," says Pryor. "Perhaps they'll see a name of a company listed in a magazine article as best of class, so they'll choose it. And one of the managers has a brother-in-law in some other company, so they'll decide to benchmark that company also. You need to do solid research to be sure that you're really benchmarking the best; benchmarking an average company is beside the point. As they say in the computer industry, 'Garbage in, garbage out.'" Many consultants have massive data bases that are ideal for helping companies pinpoint best practices, thus speeding up the benchmarking process.

John Scharlacken says that Arthur D. Little is often contacted to help manage particularly complicated or broadly based benchmarking programs. If time is a significant factor, the added manpower and expertise of consultants will often be worthwhile. Scharlacken points out that consultants also offer an objective assessment of both the data collected and how the results are being applied. "Most people have a bias to rate themselves higher than they are," he says.

One of the greatest attributes of a consultant, says Saul Berman, is the ability to put benchmarking in its proper perspective. "At Price Waterhouse, we believe in using benchmarking as a step in competitive analysis," he says. "Benchmarking one process and making it world-class won't help you if your business is structurally uncompetitive as a whole."

When a consulting firm is involved in setting up a benchmarking project, it typically puts together a survey group of three to eight companies that are the best in the process to be measured. The consultants and representatives from the client company then conduct field studies to gather data. The final report, with companies' identification removed, is given to the survey group as well as to the client.

How difficult is it to get companies to participate in a benchmarking study? "You'd be surprised how open and willing to participate companies are, once they realize the benefits of benchmarking," says Scharlacken. Companies that are reluctant to let themselves be benchmarked often fear that they will reveal proprietary information—but that concern is usually overblown. "If there was an area in which we felt Xerox had a competitive advantage, and we were approached about benchmarking that area, we'd probably say no thanks," says Camp. "But when you focus on business processes, as most benchmarking does, you rarely run into proprietary problems."

CONSORTIUMS

As useful as consultancies are when used properly, there is no getting around the fact that they're expensive. And here's where consortiums enter the picture. "Sharing the cost of a study lets a company benchmark many more processes than it could afford by contracting with a consultant individually," says Orval Brown, director of business process architecture and benchmarking for Ameritech, and the guiding force of the Telecommunications Industry Benchmarking Consortium. "Having to spend only a 17th the amount to get a comprehensive benchmarking study is a real plus."

Ameritech had been tossing around the idea of a benchmarking consortium for a while, says Brown. But a presentation he heard by Jeanette Frick on her experience setting up the Yellow Pages consortium triggered the company into action last July. It sent out a letter of interest to various telecommunications firms. Bell South, Bell Canada, and Cincinnati Bell, among others, said they would like to join. In fact, says Brown, *all* the companies contacted expressed interest, though some haven't joined the consortium yet because they don't have the funds available.

The consortium contracted with a third party—A.T. Kearney—for antitrust and regulatory reasons and to ease the bookkeeping load. The consultancy will help to put together the benchmarking surveys, visit sites to collect data, and produce final reports using disguised data. "The consortium has moved forward much more quickly than we expected," Brown says. "We were thinking that we'd still be in the planning stages at this point. Instead, we've already started our first study." The biggest struggle so far, he says, has been working out the legal agreements between A.T. Kearney's counsel and the 17 lawyers from the individual companies.

The consortium's first project involves benchmarking maintenance for special services, such as WATS lines and 800 numbers. The study will analyze the process in each of the 12 member companies participating in the project and find the industry's best. Seven external companies—prospects include Xerox, Honda, and Federal Express—will also be benchmarked for that process; a "best of best" company will then be chosen. Ideally, says Brown, the consortium will be able to benchmark three or four processes a year. "Due to antitrust restrictions," he says, "we found that we cannot benchmark how cost relates to price. Pricing product services, for instance, is a no-no; process comparisons, however, are fine."

In addition to saving time and money, a consortium gives you the right contacts in each company and builds camaraderie. "That was missing in our industry before," says Brown. "We've been able to knock down the secrecy barrier." The consortium should also improve the competitiveness of the entire telecommunications industry by raising its collective skill level. According to Brown, more than half the consortium's members have never before done any benchmarking.

Individual companies appear to be taking the lead in forming benchmarking consortiums, often approaching an industry association to be the spearhead of the project. "Associations are the logical place for consortiums to form," says Lawrence Pryor of Kaiser. "Companies already have a relationship and the group is knowledgeable about the area to be benchmarked. All they need to learn is how to set up a benchmarking data base." Says Frick: "By having an industry association head the project there's less suspicion that one company will be gaining at another's expense. This is especially important if you're in a secretive industry like ours. The whole point of an industry consortium is to get everyone in the industry performing better, not to help just a few companies."

A variant of the industry consortium that's evolving is the subject-specific consortium. The Readiness to Compete Consortium, started this past summer by United Technologies Corporation, will focus on human resources processes, such as service quality and employee involvement. Members include major computer, consumer-products, and defense companies. The Navy's best manufacturing practices consortium is one of the many other subject consortiums that have been formed.

For-profit groups are also getting into the consortium act. Best Practices Benchmarking, founded in 1991, is developing a data base of best practices in the areas of human resources management, customer satisfaction, and strategic planning. Information will be available on a subscription basis, with companies that share their own data receiving more comprehensive information and additional services, such as benchmarking training and, perhaps, consulting services. Other for-profit benchmarking groups include Real Decisions Corporation's information-technology association. (See box on page 19.)

It's clear that the consortium is an efficient way to share information within an industry. For smaller companies with limited staff and tight budgets, a consortium may also be the only way to afford comprehensive benchmarking. Another asset: A consortium can help define industry benchmarking standards. "Before the Yellow Pages consortium," says Frick, "we never knew if our benchmarking partner was measuring its processes in the same way we were."

Consultants agree that consortiums make getting data a lot easier, but they're quick to add that companies shouldn't rely solely on that method of benchmarking. "If a client has specific needs, I believe it should develop its own tailored benchmarking program rather than give money to support a benchmarking consortium in the hopes that the data it needs will magically come to it," says Arthur D. Little's Scharlacken.

Price Waterhouse's Berman voices a related concern: "In working with consortiums, the problem is that you can't be sure of the quality of the information you're getting. I've sent three different people into a plant on the same day to collect benchmarking data, and they've each come out with different results. Your people need the opportunity to work through the data and cross-check and verify them. Only by doing this can they put the pieces of the puzzle together and understand how to apply the insights to your own system." Harry Artinian agrees: "The main failure of companies when benchmarking is trying to apply a benchmark without knowing how to apply it to their particular case."

A consortium can also run into the pitfall of becoming too insular. "If you only benchmark direct competitors there is a risk," says Artinian. "General Motors, Chrysler, and Ford have historically done a lot of competitive benchmarking among themselves. Over time, their processes became similar. But then came the import competition, which had totally different processes and which blew the Big Three away." Continues Artinian: "It was like three club tennis players who all had similar levels of skill and who knew each other's games inside and out—and then Bjorn Borg walked on the court."

Others defend the practice of benchmarking direct competitors. "We tell clients to benchmark against world-class companies outside of their industry but also to benchmark against the industry leader," says Berman. "After all, it's nice to be better than your nonindustry competition, but it's critical to be better than your industry competition."

CLEARINGHOUSES

The third—and most recent—cluster of benchmarking groups to emerge are clearinghouses. By far the most ambitious clearinghouse is the one recently set up by the American Productivity & Quality Center, a nonprofit organization in Houston. After eight months of planning and collecting data, the APQC International Benchmarking Clearinghouse opened its doors in early February. Its data base contains not only benchmarking data across industries on everything from customer service to flexible manufacturing but information on the processes behind the numbers. The clearinghouse also offers training and assistance in benchmarking.

"The APQC clearinghouse's real benefit will be that you can go to a central location to get information on benchmarking," says Camp. "In the past, there's been no place like that." Camp hopes that the clearinghouse will also take some of the load off Xerox: "When companies want to know about benchmarking, they all do the same thing: call Xerox," says Camp. "As Baldrige winners, we have an obligation to help people, but we've really been overwhelmed with requests. The APQC clearinghouse may alleviate the number of calls we get."

Xerox is a member of the clearinghouse's steering committee; other members include AT&T, Honeywell, IBM, and Browning Ferris Industries. Response to the clearinghouse concept has been phenomenal, says Charlotte Scroggins, a senior vice president at the APQC. "When we first started the project, we felt we would be doing well to attract 15 or 20 companies. We now have 52 member companies and are still going strong," Scroggins says. "We've had to rethink the entire scope of the project."

Some observers think that scope is the problem with the APQC's project; they characterize the task of building a data base for all functions as monumental, even grandiose. Many expect the benchmark information in the data base to be of limited use to companies. "The data will no doubt be useful in supporting market intelligence and research," says John Scharlacken, "but there's a limit to the usefulness that such generalized data can provide. You really need data that are tailored to the specific company's needs."

Another concern is that the clearinghouse's data will be relatively static: What's best in class today isn't necessarily best in class tomorrow. "It's an enormous task to create an ongoing, living data base for such a spectrum of functions," says Pryor. "It's also difficult to make sure that you're comparing apples with apples." Scroggins acknowledges that these are concerns, but says that the clearinghouse is using standard survey forms so that data will be comparable. All information entered into the data base will also be dated, so companies will know how current the benchmarks are.

But the benchmarking data are probably the least important aspect of the APQC clearinghouse. Far more valuable, say companies, are the networks it will create, particularly among industries. "There's literally hundreds of professional and trade associations that have data on industry practices," says Camp. "But these practices never get shared between industries. We need a better way of using the information that exists. The APQC is going to give us that way." As part of its services, the APQC clearinghouse has set up directories of benchmarking trainers and organizations. It also offers abstracts of benchmarking studies, with names of people to contact for more information. Companies can use the clearinghouse services even if they are not members, but members receive special discounts.

In addition to serving as a matchmaker for establishing relationships between companies that wish to benchmark, the clearinghouse is helping the corporate community establish benchmarking guidelines. "Currently, there is no established protocol for benchmarking, no guidelines as to what is permissible to do with the data once they have been gathered," says Camp. "Once the APQC guidelines are set, a company that's approached to be benchmarked can say, 'We go by the APQC guidelines. If you do so also, then we can probably make a go of it.'"

The APQC clearinghouse isn't the only game in town. Ernst & Young and the American Quality Foundation have joined up to create a benchmarking data base of management practices aimed at improving quality. The data base is an outgrowth of the pair's International Quality Study, which examined more than 1,000 quality-focused management practices within the automotive, retail banking, computer, and health-care industries in the United States, Canada, Germany, and Japan. Preliminary results suggest that there are marked differences in management practices and the intensity of their use among countries, says Stephen L. Yearout, national director of operations and quality management for Ernst & Young's management consulting group in Cleveland.

The data base will let companies benchmark their quality-focused management practices against industry and country standards and best practices. To use the data base, a company fills out the same questionnaire as the original survey participants, a task that takes an average of 20 hours. The information, with the company's name disguised, will then be made part of the data base, helping to keep it current.

The Conference Board's Total Quality Management Center also has a clearinghouse in the works. Its data base will be a repository of information and resources on benchmarking and quality literature, awards, training courses, conferences, and techniques, and will contain a national calendar of quality-related events. "The clearinghouse will be an affordable way for firms to educate themselves on what's happening in the quality movement," says Lawrence Schein, director of the quality center. Only publicly available material on benchmarking will be included, says Schein. He expects the data base to be ready in 1993.

A number of companies, including Xerox, Ameritech, and IBM, have or are in the process of setting up their own internal benchmarking repositories. Ameritech's repository, Information Central, will contain information not only on all the company's benchmarking projects, but on all its quality efforts and business process and engineering projects, says Orval Brown. Xerox's data base will let a manager find out about any of the company's benchmarking projects through a computer terminal. Of course, says Robert Camp, there will be levels of security built into the system; some managers will be able to call up more information than others.

The real value of many of the current benchmarking efforts, particularly the consortiums and clearinghouses, lies in their networking potential. More than one company has shelled out millions of dollars for a consultant to conduct a benchmarking study, only to find that the company learned more from a few networking sessions with its peers. To that end, the Strategic Planning Institute set up its benchmarking council two years ago to help companies share benchmarking experiences and tools. The council began with seven companies but through word of mouth has grown to 40, including Xerox, Procter & Gamble, and Florida Power & Light. The SPI council is now investigating particular techniques, says James Staker, who runs the council. Owens Corning, for instance, recently told members about its "Hidden Gems" program, which seeks out low-profile midsize firms that are good in a particular function to benchmark; Johnson & Johnson explained how it is getting its middle managers to buy in to benchmarking.

Ameritech's Brown particularly likes the council's informal combination of presentations and discussions. "We can take back our peers' experiences and ideas and tell our management, 'Hey, look at what they were able to do. Let's do that here,'" he says.

Benchmarking's effects on business have been dramatic; Xerox, for instance, credits it with being one of the main factors behind the company's revival in the 1980s. But consultants warn that benchmarking can have unwanted side effects if findings are applied blindly. "Our view is that there is no one best in anything; it's folly to think there is," says Christopher Bogan of Best Practices Benchmarking. "We look for a group of companies that achieve high results in a particular area and that have developed approaches to the task that will be appropriate for your own company."

"You also have to remember that you're competing against a moving target," says Jeanette Frick. "You have to keep in mind that the benchmark information is from last year; you can't assume that you're doing good if your business stacks up well against that data. It's likely that other companies in your industry are improving their processes beyond last year's benchmarks."

"Meeting best-of-class standards is not the end-all it's simply a milepost for where you're going," says Harry Artinian. "The real target needs to be. 'Does this give the customers what they want? Does it anticipate their needs?'" Companies are just now realizing that they need to focus on benchmarking those processes that are critical to their success, says SPI's Staker. "A year ago, the hot benchmark subject was machine maintenance; now our council members are looking to focus on things that touch customers."

Benchmarking in America is still experiencing growing pains, but more and more companies are turning to it as a treatment for their competitiveness problems. "Companies that perceive a competitive threat from Japan or Europe are the most likely to benchmark," says Lawrence Pryor. "Those that feel that their only competition is U.S.-based are less likely to share benchmarking data."

Many think that the emerging benchmarking structure of consultancies, consortiums, and clearinghouses will make the United States more competitive as a whole. "The ultimate benefits of such a system would be absolutely tremendous," says Artinian. "It would lead to higher-quality products, lower costs, more jobs, and a higher standard of living. Let me give you just one example: The Big Three automakers recently agreed to common standards for suppliers. Prior to this, each company had its own rules and specifications for suppliers, which added tremendous cost to the suppliers' work—cost that led to higher-priced cars for consumers. This is one example of how U.S. industry will be more competitive on a global level through cooperative activity.

"But cooperation won't come easy for U.S. managers," Artinian warns, "because they've been taught how to manage only in win-lose situations. Successful benchmarking is about managing so that all sides win."

Chapter Nineteen

Technological Innovations in Manufacturing

INTRODUCTION

Competition in the international manufacturing segment is now at an all-time high. Although the United States has lead the way in the innovation of newer technologies, new product design, manufacturing processes, and automation, it is now being challenged by Germany, Japan, and other industrialized nations. During the post–World War II years, manufacturing organizations concentrated on producing and delivering similar products economically. But today consumer demand, characterized by short product life cycles and high product diversity, has become increasingly complex and diverse. These changing consumer expectations challenge manufacturing organizations to reduce the lead time and meet quality and reliability standards. In many instances, however, the existing organizational structure is not flexible enough to respond to customer needs in a timely manner. Some manufacturers are responding to this challenge by increasing the manufacturing process flexibility, that is, by using robots, manufacturing cells, computer-assisted design/computer-assisted manufacturing, computer-integrated manufacturing, and other innovative techniques. These topics are discussed briefly in this chapter.

COMPUTER-INTEGRATED MANUFACTURING

Taylorism has been a dominating guideline for structuring and managing organizations during most of the twentieth century. Each department controlled all aspects of business pertaining to its activities. Information traveled vertically from one department to another. Because of the lead time associated with each data transfer, the total time needed to complete a project is at least the sum of the departmental lead times participating in the data transfer. Obviously, any notion of an integrated database would reduce the total lead time; hence, the idea of computer-integrated manufacturing (CIM) was born. Although it is difficult to find a commonly acceptable definition for CIM, the term refers to the integrated information processing requirements for the technical and operational tasks of industrial enterprise [22]. The operational side is represented by the production planning and control systems (PPC), whereas the more technical and engineering activities are characterized by computer-aided design and computer-aided manufacturing (CAD/CAM). The integration of PPC and CAD/CAM not only poses a challenge to management but also to hardware and software producers to coordinate their individual development of systems for the technical and operational sides of manufacturing and business. Although the term *manufacturing* is used as a part of CIM, it is much broader than just that industry; hence, this concept not only misleads but also confuses many. In summary, CIM encompasses a spectrum of coordinated activities from design to delivery of a product.

The activities involved in traditional order handling are compared with the computer-integrated order handling environment in Figure 19.1. At first is a departmentally determined data organization (Figure 19.1a) in which every department has its own database. In addition to the data transfer time is a lead-in period associated with every depart-

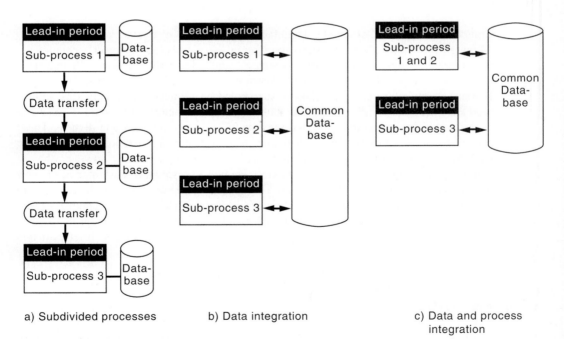

a) Subdivided processes   b) Data integration   c) Data and process integration

FIGURE 19.1 Reintegration of functionally driven operations. *Source:* A. W. Scheer, *CIM, Computer Integrated Manufacturing: computed steered industry* (New York: Springer-Verglag, 1988). Reprinted with the permission of Springer-Verlag.

Improved customer service

Improved product quality

Shorter customer lead time

Shorter flow time

Shorter vendor lead time

Reduced inventory levels

Improved schedule performance

Greater flexibility and responsiveness

Improved competitiveness

Lower total cost

Greater long-term profitability

Shorter time to market with new product

Increased manufacturing productivity

Decrease in work-in-process inventory

FIGURE 19.2   Potential Benefits of CIM.

ment. If a common database is established among the participating departments (Figure 19.1b), information transfer times can be eliminated and thus the process can be considerably accelerated.

Scheer [22] indicates that it is often possible to reduce the total administration and order handling time from three weeks to three days. As the human ability to control complex tasks grows, specialization takes a backseat; in terms of Figure 19.1, subprocesses 1 and 2, for example, can be combined. Consequently, the lead in time occurs just once as shown in Figure 19.1c, instead of twice. Clearly, the additional manual handling of data can be virtually eliminated as data integration and process integration takes place. When we apply the same concept to traditional and CIM systems, the savings in lead time becomes even more evident.

The potential benefits that could be derived from CIM as described by Bedworth et al. [2] are shown in Figure 19.2. The challenge is in the integration of information collected from various sources. In this chapter we briefly cover some popular topics in CIM including various tools for manufacturing planning, automation, material control, and integration, that are not covered elsewhere in this book.

TOOLS FOR MANUFACTURING PLANNING

This section deals with several computerized tools and techniques that could be extremely helpful in manufacturing planning. They include computer-assisted design (CAD), computer-assisted manufacturing (CAM), group technology (GT), computer-assisted process planning (CAPP), decision support systems (DSS), and expert systems (ES).

CAD/CAM

Groover and Zimmers [9] define CAD as the application of computers to assist in the creation, modification, analysis, or optimization of the design of a product. CAD hardware typically encompasses computers, data entry and graphics display terminals, and peripheral equipment. CAD software consists of computer graphics and application programs for use in the design of products. CAD was initially used in drafting and was found to increase the productivity in drafting and generating accurate drawings. A fourfold increase in productivity was not uncommon. Its use quickly spread to geometric modeling and analysis of components on the graphics terminal. Once the drawings are created they can be dimensioned, stored, retrieved, and drawn on the plotter automatically.

CAM refers to the use of computer systems to plan, monitor, and control the manufacturing operations. CAM consists of computer-assisted process planning (CAPP), numerical control (NC) machining, operations planning, robotics, assembly, testing, and manufacturing management. Obviously, manufacturing management consists of production planning and control, total quality management, and other decision support programs.

Group Technology/Computer-Assisted Process Planning (GT/CAPP)

Group technology is a technique by which similar or related components are identified so that design and manufacturing processes can minimize duplication of components. The concept of group technology (GT) was invented by Sokolowski. Although the concept has been in existence for a long time, it was formalized only in 1958 when Russian engineer S. P. Mitrofanov published *The Scientific Principles of Group Technology* [4]. For the design engineer, part similarities exist in shape and dimensions. For the manufacturing engineer, similarities are based on process routings. Traditionally, engineers used a cumbersome manual production flow analysis chart [23]. As a first step, the chart lists all component items to be manufactured and the associated machines or processes in a matrix. As a second step, engineers manually group them into families of items having a common process. This is illustrated in Tables 19.1a and b.

Using similar logic, group analysis subdivides the major families and machine groupings into smaller families such that each family can be processed entirely in one machine group without intermediate visits to other groups of machines. Based on the process flow of the parts within the family, line analysis develops the best arrangement of machines for the components with in a group to flow. Similarly, tooling analysis finds the tooling families for parts processed on each machine and the optimum sequence of loading of the tools. Gallagher and Knight [7] developed a procedure so that part families and machines groupings can be done simultaneously. Because the details of the procedure is beyond the scope of this book, the interested reader is directed to their work.

With the advent of modern computers, several commercial GT classification and coding systems are available for design and manufacturing applications. Coding establishes symbols for the product's attributes that aid the software logic to group processes or machines, whereas classification creates groups based on the existing attributes.

TABLE 19.1   Production Flow Analysis Chart
(a) unsorted

		Part Number							
		1	*2*	*3*	*4*	*5*	*6*	*7*	*8*
	1	X					X		
	2		X	X		X			X
Machine	3			X		X			X
number	4				X				
	5	X					X		
	6		X	X		X			
	7	X			X		X		

(b) sorted

		Part Number							
		1	*6*	*2*	*5*	*3*	*4*	*8*	*7*
	1	X	X						
	5	X	X						
	7	X	X				X		
Machine	2			X	X	X		X	
number	6			X	X	X			
	3				X	X		X	
	4						X		

Popular commercial systems include Optiz developed by Optiz; Brisch developed by Brisch; Code MDSI by Schlumberger Technologies Inc.; MICLASS, MULTICLASS, and MULTI-II developed by the Organization for Industrial Research (OIR); and DCLASS by the CAM Research Center of Brigham Young University. The MULTI-II system is capable of addressing engineering and production as well as business applications. The application of group technology can increase productivity in design and manufacturing processes while decreasing waste associated with component inventories. It is obvious that GT is a viable tool for improving productivity in CAD, CAM, and process layout.

Computer-assisted process planning software, a subset of CAM, is yet another useful tool to automate the process routings. Once a GT code exists for a part, CAPP analyzes the routings and standardizes them. GT/CAPP software can also be applied to group a firm's most often used machines into FMS cells so that productivity can be boosted. The process engineer can then review the routings and determine their adequacy.

Decision Support System and Expert Systems (DSS/ES)

Keen and Morton [12] define DSS as follows: "Decision Support Systems couple the intellectual resources of individuals with the capabilities of the computer to improve the quality of decisions. They are computer-based support for management decision makers who deal with semistructured problems." In short, a DSS is designed to aid (not replace) the effectiveness (not efficiency) of the managerial decision-making process, especially in semistructured complex tasks, objectives, or problems. The DSS helps decision makers to respond to unexpected situations in a logical (analytic), timely manner. DSS/ES can be useful in all phases of CAD, CAM, and PPC for improving productivity and for more effective decision making.

All organizations cannot afford to have an in-house expert to solve complex problems. The rarer and complex the situation, the more expensive the advice. An expert system (ES), in many instances, can provide comparable or even better solutions. An ES

TABLE 19.2 DSS/ES Framework for Flexible Manufacturing Systems (FMS)

Type of Decision	Operational Control	Managerial Control	Strategic Planning	Support Needed
Structured	Inventory control Material replenishment Routing and part movement Computer-aided manufacturing Time standards Workplace handling devices Tool management	Resource planning Cost estimation Reporting systems Material planning Software development	Systems budgeting Equipment analysis Financial analysis	EDP MIS Operations research Models
Semistructured	Operational planning Computer-aided inspec. Computer-aided design Method analysis Data collection and encoding Failure analysis Operation sequencing Work order dispatching	Master prod. scheduling Part and tool transport. Material handling Maintenance planning Line balancing Plant layout Cost-benefit of equipment Data system design Project planning and schedul. Machine utilization plng.	Facility planning Man-machine interface Quality assurance planning Systems modification Part-mix selection and changes Capacity expansion Economic justif. of FMS	DSS
Unstructured	Failure diagnosis Corrective actions User interfaces Exceptions tracing and actions Operations and tool scheduling Equipment reallocation	Level and timing of selection and integration of robots Testing programs Group technology Coding Ergonomics analysis Safety and health planning Computer-aided instruction	Level and application of FMS R&D planning Long-term business planning Feasibility study	DSS ES

Source: E. Turban and M. Sepehri, "Applications of Decision Support and Expert System in Flexible Manufacturing Systems," *Journal of Operations Management,* Vol. 6, No. 4 (August 1986), p. 438. Reprinted with the permission of Elsevier Science Publishers BV, Academic Publishing Division.

TABLE 19.3   An Applications Map for Expert Systems in Production Management

	INT	DIA	MON	PRD	PLA	DES	CON	TEA
					Expert Task			
Decision Area								
Technological:								
(ME)					1	2		1, 2
(IE)						4	3	
(M)		5	5	5	5		5	
(QC)	6		6		6			
Logistical:								
(PC)					7, 11		7	
(MC)						9	8	
(P)							10	
(IM)								
(F)				12				

Source: P. Mertens and J. J. Kanet, "Expert Systems in Production Management," *Journal of Operations Management,* Vol. 6, No. 4 (August 1986), p. 395. Reprinted with the permission of Elsevier Science Publishers BV, Academic Publishing Division.

is an accumulation of hardware or software systems that hold a body of task-specific knowledge accumulated by human. When specific advice is needed, the system can help the decision maker make inferences and arrive at specific recommendations or conclusions. The ES can also be built to explain the reasoning behind the decision.

The DSS/ES can be of immense help in the design of FMS. Turban and Sepehri [26] summarize the applicability of DSS/ES for FMS situations as exhibited by Table 19.2.

Another survey by Mertens and Kanet [16] finds production/operations management (POM) to be a fertile ground for applications of ES. They summarize that nine production management functions (consisting of technological and logistical activities) and eight type of expert tasks can be combined to form an application map (see Table 19.3) consisting of twelve situations for ES in the field of production management. The nine

> Technological activities:
>   Manufacturing engineering
>   Industrial engineering
>   Maintenance
>   Quality control
> Logistical activities:
>   Production planning and control
>   Material control
>   Purchasing
>   Inventory management
>   Forecasting

FIGURE 19.3   Nine production management functions.

Interpreting
Diagnosing
Monitoring
Predicting
Planning
Designing
Consulting
Teaching

FIGURE 19.4   Expert decision areas.

production management functions, eight expert tasks, and twelve situations are exhibited by Figures 19.3, 19.4, and 19.5, respectively.

Mertens and Kanet [16] conclude from their survey that engineering process design and maintenance activities would be the most promising areas suitable for ES application. Production and operation scheduling as well as forecasting would be next. A combined task of production scheduling and equipment maintenance would be a fruitful area for further investigation. From the history of applications, they also foresee that future successful ES applications most likely hinge on skillfully blending technological and logistical knowledge.

MANUFACTURING AUTOMATION

This section describes automated facilities and flexible manufacturing systems and explains how various components of a manufacturing facility can be tied together for increased productivity.

Process selection
Process design
Facility location
Facility layout
Maintenance
Quality control
Production planning and control
Material selection
Storeroom design
Vendor selection
Capacity planning
Forecasting

FIGURE 19.5   Activities common to nine decision areas.

Automated Facility

An automated manufacturing facility consists of CAD, CAM, numerical control (NC) machines, storage mechanisms, and conveyors. NC machines help part programmers generate programs easily. The programmers simply retrieve the part number, description, and dimensions directly from the engineering drawing and then determine the manufacturing processes, tools to be used, and operations sequences. The programs can be stored on tapes or disks and loaded on a computer. These computer programs can also be stored in a host computer and downloaded on individual machine centers as necessary. Host computers and distributed processing systems can also incorporate other decision support systems to facilitate efficient manufacturing and delivery of products.

Flexible Manufacturing Systems (FMS)

An flexible manufacturing system is an automated factory. Although these systems have been around since the 1970s, they have been primarily used in the manufacture of tools and complex parts. An FMS consists of one or more cells. A sophisticated FMS cell consists of a set of machines with multiple interchangeable tools and fixtures, an automated material handling system possibly with robots, and an automated identification system for a computer system to scan, coordinate, and control all activities. The flexibility of an FMS permits several different parts to be produced at the same workstation because it can instantly receive information regarding the materials, processes, and sequence of operations from the databank.

When a part is scheduled for production in a sophisticated FMS environment, the control system (computer-assisted process planning system) determines the proper quantities of raw materials, appropriate tools, fixtures, and machines. The control system schedules the production time at these machines. It subsequently instructs the automated material handling systems about where, when, and how much of these raw materials are available and when to deliver them to any specific machine center. Thus the control system automatically routes the completed parts to the next machine center for subsequent processes.

Brown et. al. [3] classify the characteristics of FMS as (1) machine flexibility, (2) product flexibility, (3) process flexibility, (4) operation flexibility, (5) routing flexibility, (6) volume flexibility, (7) expansion flexibility, and (8) production flexibility. A survey by Turban and Sepehri [26] summarizes the characteristics of FMS into five major categories: (1) It is flexible, (2) it is fast learning, (3) it can be automated, (4) it provides continuous production, and (5) it is mostly underutilized.

Advantages of FMS

As markets become global and consumers ask for niches (more specialized products), manufacturers are reexploring new process technologies and manufacturing techniques. The competing environment has decreased the life cycles of products and increased customization of parts. Yet there is a desire to reduce the total cost and the product delivery lead time. Because FMS has the flexibility of a job shop operation and the production capability of medium- to large-lot mimicking a flowshop while reducing labor costs, manufacturers now believe that it can help achieve these goals. The viable uses of FMS in industrialized nations such as the United States, Europe, and Japan are steadily increasing.

The prime advantage of FMS is its ability to incorporate the changes in design automatically and to download the data on machines. Another advantage of the system is its ability to handle a variety of product mixes. Because these machines are capable of performing more than one operation, FMS permits scheduling flexibility as well as possibly increased utilization with consistent quality of products. In addition, because a specified order can be incorporated in the production schedule earlier, work-in-process is reduced and the lead time is slashed. It is interesting to note that a survey by Turban and Sepehri [26] indicates that FMS systems are now chronically under utilized and expert systems could be a potential vehicle to improve the situation.

Implementation Problems with FMS

Yet FMS is not for everybody. It is not only complex and very expensive, but it also requires specialized training of operators and rigid maintenance requirements. Flexibility comes at a cost. With existing cost accounting procedures, it is difficult to justify an FMS because the present system is built on labor costs. Labor costs, however, are no longer a major component of a product in an FMS environment. To further complicate the cost justification procedure, in many instances an FMS is a part of the total system and hence it is difficult to separate its benefits from those of the total system. An FMS needs higher-level planning and more disciplined environment than other systems. Another complication is not knowing the necessary level of flexibility when designing an FMS. More flexibility means added cost, whereas less flexibility leads to limitation in its use.

In many instances the benefits of an FMS are hard to quantify. While the reduction in inventory is possible to quantify, it is difficult to calculate the effects of decreased lead time, easier design changes, flexible products scheduling, and a more consistent quality. Adaptation of simultaneous engineering and the design for manufacturing (DFM) concepts reduce the number of parts in a product and simplifies the processes that, in turn, can reduce the cost of a product. DFM was often found to eliminate the need for FMS. As JIT becomes more popular and as employees are trained in multifunctional activities, the tendency in the industry has been a decrease in automation. Research in manufacturing has lead to delegation of more responsibility to individuals and groups rather than to machines and, again, automation is reduced.

MATERIAL CONTROL

This section deals with efficient movement of materials in a manufacturing facility. The systems described here also increase the productivity of a firm.

Automated Storage and Retrieval Systems (AR/AS)

Vertical storage racks, narrow aisles, and cranes generally symbolize computerized warehousing systems. Material handling systems for CIM need to move materials between points with an emphasis on speed, accuracy, and uniformity of product flow. They comprise conveyers, feed mechanisms, rotary rack storage systems, small monorails, and automated guided vehicles. Although these systems are called an automated storage and retrieval systems (AR/AS) they serve as automated inventory and material handling systems as well. Comprehensive automated AR/AS systems track incoming parts, storage, and retrieval locations and delivers required items at specified locations. Thus they are designed to replace manual and remote-controlled systems.

Frequently, automation occurs on the shop floor assembly line and then extends to other manufacturing operations. With the popularity of the CIM philosophy, the AR/AS system offers increased capabilities such as preparing kits for assembly, storing work-in-process inventories, and acting as delivery systems to flexible manufacturing cells (FMS). Thus AR/AS in conjunction with automated guided vehicle systems (AGVS) are becoming an integral part of factory automation. Hence, these systems can also serve to meet the demands of JIT manufacturing environment.

If all parts are made at the same rate and are used, in-process inventory can be reduced and the manufacturing productivity can be increased. Therefore, effective integration and communication among AR/AS, equipment, AGVS, manufacturing processes, and scheduling functions becomes essential in a CIM environment.

Automatic Identification Systems

Bar coding is the most popular automated data entry system. Bar coding and other automatic identification systems can furnish information regarding the status of a customer order in the system accurately and instantly. These traits are essential for inventory status, shop floor scheduling, and tracking systems. An automatic identification system consists of a scanner's light source that passes across bar code symbols. A photo detector in the scanner converts the pattern of bar codes into an analog signal that is decoded by a microprocessor in the scanner. The information is then transmitted into a computer. Due to their speed, flexibility, and accuracy, automatic identification systems have the potential of integrating shop floor, inventory, material tracking, scheduling, receiving, and other major functions in a manufacturing environment. In addition, bar coding applications are economical to implement from start to finish in many service and manufacturing operations [6, 18].

INTEGRATION ISSUES IN CIM

The success of any CIM project will depend on the effectiveness of transformation of data among the components of the total CIM system.

Data Exchange
and Network Communications

Business computer users have been successful in bridging computer operations and data-processing equipment for their own applications because compatibility among systems is not a significant problem. These users' successes is because they are primarily dealing with at most two or three systems and they can be connected to communicate among themselves with the help of bridging equipment. In a manufacturing environment, however, each piece of equipment is highly specialized and specific pieces are chosen based on their superior performance. It is very difficult to find a vendor who can supply an entire array of superior equipment; hence, the manufacturer ends up buying equipment from different vendors. In addition, most manufacturing facilities have been slowly automated over time. The result is the creation of islands of automation. Because these machines were not originally built for this purpose, it may be not only be very expensive but also impossible to bridge them technically. In some instances it may even be cheaper to replace the entire system than attempt to bridge the operations. Unfortunately, existing governmental and corporate investment policies may not justify such an expense.

CIM and Standardization

Industry leaders such as General Motors, in cooperation with large computer manufacturers, are spearheading open system architecture (OSA) standards known as manufacturing automation protocol (MAP) and technical and office protocol (TOP) for the CIM, based on computer-integrated manufacturing open systems architecture (CIMOSA), a model established by the International Standards Organization. MAP, based on OSA, employs a standard set of protocols for exchanging data sets between different types of microprocessors (computers). The protocols are built into devices such as computer-controlled machine tools, engineering workstations, process controllers, factory floor terminals, and control rooms. The OSA has seven layers, each of which deals with a different aspect of linking computer systems [2].

Europeans have embarked on developing and standardizing modular approach for various components of CIM under the terms of the European Strategic Programme for Research and Development in Information Technology (ESPRIT) Project 688, based on CIMOSA. The project's object is to provide clearly defined mechanisms for analyzing the manufacturing requirements of a business and translate the requirements into specific CIM architecture. In fact, CIMOSA is a methodology that helps the parts of the system to work together to satisfy the specifications. Because the consortium lead by GM has been coordinating with European counterparts, which have been experiencing difficulties in developing standards for CIM, the creation of standards for the seven layers of MAP has been impeded. The standards being established under the initiatives of the United States and Europe and the MAP 3.0 version were introduced during the 1988 Enterprise Networking Event in Baltimore, Maryland. These efforts are further solidified by the Computer-Aided Manufacturing International (CAMI), a nonprofit organization that researches developing integrated production methodology such as solid modeling, planning of production processes, application of AI in scheduling, and the basic architecture of CIM itself. The shortcomings in the development of universal standards are addressed by Initial Graphics Exchange Specification (IGES) and by the Product Data Exchange Standard (PDES) programs in the United States and by CAD-I, an European ESPRIT program. Further, Electronic Data Interchange Standards for Administration, Commerce and Transport (EDIFACT) program has been established for exchanging data, independent of their organization, communication system used and the type of data processing equipment used. These standards have been approved by the ISO.

The U.S. government through the Department of Defense sponsors research activities in manufacturing automation under its manufacturing technology program (MAN-TECH). Through its International Modernization Incentives Program (IMIP), the Department of Defense has also been assisting U.S. manufacturing organizations to implement new technologies. The National Science Foundation sponsors automation research at universities, and the National Bureau of Standards Center for Manufacturing Engineering conducts research to manufacturing automation. The development these standards should bear fruit on other efforts for automating all functions of an enterprise.

SUMMARY

In this chapter we have dealt with many important topics associated with the manufacturing technologies. In the middle of the 1980s there was an initial burst of enthusiasm regarding the potential use of CIM in manufacturing. Successful applications include IBM's University Research Park Proprinter Plant, Westinghouse's College Station Electronic Assembly Plant, Texas Instrument's Johnson City Circuit Board Facility, and Fujitsu's Robot Factory. Existing cost accounting procedures, however, do not justify the use of CIM and interest in CIM slowed considerably. Now, though, there is a renewed interest now in CIM as technological problems are being solved, and management education and user training aspects are being dealt with [2, 26]. Before embarking on a CIM journey, firms are also learning more about the concepts and application of simultaneous engineering, just-in-time, and design for manufacturability. The net result is that more firms will be using new technologies in selective areas of manufacturing. We can only be sure of one thing about the future: constantly changing environment and technology.

QUESTIONS

1. What are the major components of CIM?
2. What are some perceived benefits of CIM?
3. How does CIM aid in manufacturing planning?
4. How does CIM aid in manufacturing automation?
5. How does CIM aid in material control?
6. Discuss ways in which CIM could help in increasing productivity of organizations.
7. Discuss various impediments that hinder the implementation of CIM.
8. Discuss the importance of cost justification in implementing CIM.

REFERENCES

1. N. Adams, New Warehouse Automation Improves Customer Service at Kinney Drugs, *Industrial Engineering,* Vol. 25, No. 10 (October 1993), pp. 21–23.
2. D. Bedworth, M. Henderson, and P. Wolfe, *Computer-Integrated Manufacturing* (New York: McGraw-Hill, 1991).
3. J. Brown et. al., "Classification of Flexible Manufacturing Systems," *FMS Magazine,* Vol. 2, No. 2 (April 1984), pp. 114–117.

4. T. Chang and R. A. Wysk, *An Introduction to Automated Process Planning Systems* (Englewood Cliffs, N.J.: Prentice Hall, 1985).

5. A. Diehl, Understanding IGES, *Manufacturing Engineering,* Vol. 112, No. 6 (June 1994), pp. 55–56.

6. J. Fetter, "The Use of Bar Code Technology in Production-Oriented Applications," *Production and Inventory Management,* Vol. 25, No. 4 (1984), pp. 1–20.

7. C. C. Gallagher and W. A. Knight, *Group Technology* (London: Butterworth, 1973).

8. M. P. Groover, *Automation, Production Systems, and Computer-Aided Manufacturing* (Englewood Cliffs, N.J.: Prentice Hall, 1980).

9. M. P. Groover and E. W. Zimmers, Jr., *CAD/CAM: Computer-Aided Design and Manufacturing* (Englewood Cliffs, N.J., 1984).

10. T. G. Gunn, *Computer Applications in Manufacturing* (New York: Industrial Press, 1981).

11. K. Hitomi, Moving Toward Manufacturing Excellence for Future Production Perspectives, *Industrial Engineering,* Vol. 26, No. 6 (June 1994), pp. 48–50.

12. P. G. W. Keen and M. S. S. Morton, *Decision Support Systems, An Organizational Perspective* (Reading, Mass.: Addison-Wesley, 1978).

13. J. R. Koelsch, A New Look at Transfer Lines, *Manufacturing Engineering,* Vol. 112, No. 5 (May 1994), pp. 73–78.

14. C. LeMaistre and A. El-Sawy, *Computer-Integrated Manufacturing: A Systems Approach* (White Plains, N.Y.: NUIPUB/Kraus International Publications, 1987).

15. J. McCloud, McDonnell Douglas Saves Over $1,000,000 per Plane with Reengineering Effort, *Industrial Engineering,* Vol. 25, No. 10 (October, 1993), pp. 27–30.

16. P. Mertens and J. J. Kanet, "Expert Systems in Production Management: An Assessment," *Journal of Operations Management,* Vol. 6, No. 4 (August 1986), pp. 393–404.

17. P. W. Moir, *CIM and Its Applications for Today's Industry* Efficient beyond Imagining, (New York: Halsted Press, 1989).

18. S. L. Narasimhan and R. C. Koza, "Automatic Identification Systems Serve as Integrators in the Factory of the Future," *Industrial Engineering,* Vol. 17, No. 2 (February 1985), pp. 58–66.

19. J. V. Owen and E. E. Sprow, The Challenge of Change, *Manufacturing Engineering,* Vol. 112, No. 3 (March 1994), pp. 33–46.

20. J. V. Owen and E. E. Sprow, The New World of Work, *Manufacturing Engineering,* Vol. 112, No. 5 (May 1994), pp. 37–44.

21. P. G. Ranky, *Computer-Integrated Manufacturing* (Englewood Cliffs, N.J.: Prentice Hall, 1986).

22. A. W. Scheer, *CIM: Computer-Integrated Manufacturing: Computer Steered Industry* (New York: Springer-Verlag, 1988).

23. C. S. Snead, *Group Technology* (New York: Van Nostrand Reinhold, 1989).

24. E. E. Sprow, Machine in the Dark, *Manufacturing Engineering,* Vol. 112, No. 5 (May 1994), pp. 61–65.

25. R. Suri, and R. Desiraju, Design for Analysis Leads to A New Concept in Plating Lines, *Industrial Engineering,* Vol. 25, No. 8 (August 1993), pp. 54–60.

26. E. Turban and M. Sepehri, "Applications of Decision Support and Expert Systems in Flexible Manufacturing Systems," *Journal of Operations Management,* Vol. 6, No. 4 (August 1986), pp. 433–448.

APPENDIX 19A

The Age of Re-Engineering
A. J. Vogl

Serious diseases require serious medicine, and the United States is suffering from various plagues that conspire to make it a poor competitor. Promises of cure, of course, are never further than a consultant away, but probably no cure is as radical as reengineering, which bids to become for the '90s what Total Quality Management was for the '80s.

Two leading exponents of reengineering are Dr. Michael Hammer, a former professor at the Massachusetts Institute of Technology and president of his own consulting firm in Cambridge, Mass., and James Champy, chairman and CEO of CSC Index Inc., a management-consulting firm also based in Cambridge. Together they have written *Reengineering the Corporation: A Manifesto for Business Revolution,* which was published last month by HarperBusiness.

The dust jacket of their book proclaims, "Forget what you know about how business should work. Most of it is wrong." The zealotry of that statement was apparent during a recent visit to New York, when the two revolutionaries expounded upon their manifesto to A.J. Vogl, editor of *Across the Board.*

Gentlemen, let me start by reading from a press release that accompanied your book: *"Reengineering the Corporation* **sets aside much of the received wisdom of the last 200 years of industrial management, and in its place presents a new set of organizing principles by which managers can rebuild their businesses." Now, even allowing for the excesses of press releases, that seems a rather strong statement; it's akin to saying you've found the Magic Bullet or the Holy Grail. I think many executives would regard your claim with skepticism.**

Champy: We believe we've found something fundamental about the way organizations are designed and the way they do their work that needs to be radically changed. In that sense, there is some Holy Grail in the book. But it's not a Magic Bullet in the sense that it's an easy fix.

If anybody asks for a quick definition of reengineering, you put it simply: "starting over."

Champy: Yes, starting over—meaning, first of all, taking the clean-sheet approach to the design of how work is done and how organizations are structured. By and large, what people do is take what they've inherited and find ways to make it better. We tell them to forget what they've inherited; that's the problem, not the solution. Rather, we say, pretend you're starting a new company from scratch: How would you design it, organize it, and have it run? That's the mindset of reengineering.

Reprinted from *Across the Board,* Vol. XXX, No. 6 (June 1993), pp. 26–33.

Where did reengineering come from? It didn't come fullblown from the brow of Zeus, did it?

Champy: In one sense, reengineering is a discovery that work based on fragmentation and specialization—and organizations based on fragmentation and specialization—cannot respond to a fast-changing business context that we're in. Another discovery came from observing a few years ago the way a small number of successful organizations changed by starting to look at processes rather than tasks, getting dramatic business improvement by operating on the processes.

We also recognized three or four years ago that information technology, as it's been typically and traditionally applied, automates the way work is done currently, rather than rethinks it. So some of reengineering's roots lie in the discovery that, frankly, most information technology hasn't added value to what we do, given what we spend on it, and a realization that if we rethought the business processes first and then applied technology, we'd get a lot more leverage out of technology.

Hammer: At no point do we claim to have invented reengineering. What we've tried to do is organize, systematize, label, and put in a conceptual framework bits and pieces that many organizations have been doing to get it to the point where other companies can understand what's going on, the motivations for it, and decide to do it themselves. We started using the word *reengineering* about '87, '88, but the ideas have been swimming around for close to 20 years.

Champy: I think the fastest way for people to understand reengineering is to look at it from a results perspective rather than from a conceptual one. Think of an organization that used to take 30 days to fill a customer order that now does it on demand on the same day, or a pharmaceutical company that had taken eight years to develop a new drug and now does it in four years' time, or think of a new-product-development cycle that went from two years to three months.

Impressive-sounding, certainly, but I suspect many of our readers might react by saying, "Yeah, but my business is different." Is that a typical reaction?

Hammer: Maybe at first, but very quickly they see that the issues have nothing to do with a particular industry. The deep issue is one of functional specialization and fragmentation—the way we've taken work, broken it into pieces, and proceeded with the assumption that if every individual worried about his or her piece of the work, the whole would take care of itself. In fact, everyone does worry about his or her piece, but the whole is going to hell. That seems to be essentially a universal truth.

I don't know if you saw *Absence of Malice* with Paul Newman. In this movie, the Justice Department thinks Newman has information about a crime. He doesn't, but the Justice Department officials believe he does, so they pressure him to release it, ultimately leaking some very embarrassing information about him and a friend. The press prints it, and the friend is so embarrassed that she commits suicide. Last scene in the movie: Paul Newman's there with the Justice Department and the press, and he has a memorable line. He looks around the room and says, "Everybody here is smart. Everybody here is doing their job, and my friend is dead." I tell executives that by changing one word of that sentence, we have something we can engrave over the entrances of their firms. It becomes, "Everybody here is smart. Everybody here is doing their job, and my company is dead."

Because everybody doing his or her piece leads to disaster. And that's a fundamental premise that is wired into any enterprise—public or private sector, large company or medium or small. It's a universally held truth, and it's simply a lie.

Again putting myself in readers' shoes, another reaction you'd likely get is: "Reengineering looks as if it'd be terribly disruptive to my company. Do I need that on top of all my other problems?"

Hammer: First of all, you don't need it on top of all your other problems; you need it *instead* of all your other problems. But your assertion is exactly correct: Reengineering is enormously disruptive because it gets to the heart of the way work is organized. We say: You've got to do your work differently. There's nothing more fundamental than that.

Secondly, as our book argues, if you make significant changes in the way in which work is formed, it has fantastic ripple effects on every other aspect of the business. It changes people's jobs. It changes the kind of people working at those jobs. It changes the way people are organized. The focus is on teams instead of departments. It changes how people are measured—they're measured on end-to-end processes rather than on end tasks. It changes compensation—the emphasis is less on position in the hierarchy and more on contribution to business performance. It changes how people regard each other, the values, the attitudes they have. So it basically says, "We'll leave the walls standing and we'll nuke everything on the inside."

Does one have to reengineer every function across the board, every department?

Champy: No. In most cases, that's impractical. It's too disruptive. Aetna Life & Casualty, for instance, may reengineer three of its processes at any given time. So it's reengineering claims for property casualty and it's reengineering its health-insurance business, and then it's going on to others. So only the part that's being reengineered is in chaos; the rest are relatively stable until their turn comes. It's only for companies that are in the deepest trouble that we'd ever say, "Reengineer the whole place."

I think the most common misunderstanding is that reengineering is a change program of normal proportions that can be managed in what I would call normal management styles and normal management processes. So you'll see people attempting it without the degree of intensity and focus it demands, and without the management tools and techniques that they need in order to accomplish transformational change. Some of that, I think, is a rationalization. It's easier to think of your organization as being one that doesn't need radical change, so you rationalize back to a level that you see is manageable.

Hammer: One common misperception that I run across is that reengineering is about building new computer systems. It is true that reengineering is enabled by technology, and that when you reengineer a process, it usually requires a new computer system to support it. The process horse, of course, belongs in front of the technology cart. People sometimes get that mixed up. Another common misperception is that it's the same as Total Quality Management because it has some points in common with TQM—specifically, the emphasis on customers and the emphasis on process. But after that, they diverge radically. TQM is basically about improving something that is basically okay, and reengineering is about taking something that is irrelevant and starting over.

Are you saying TQM and reengineering are mutually exclusive?

Hammer: No, they're complementary. For a given process at any point in time, you should be doing one or the other, but not both. If it's basically sound, you look to improve it. If it has hit the wall, you look to nuke it.

How do you decide whether to use TQM or reengineer?

Hammer: It's a combination of how far off performance is from what the customer requires, where your big cost centers are, and where you see the opportunity for greatest competitive leverage in the marketplace.

Champy: The other important factor is timing. If you need big change fast, you can't get there with total-quality programs, which are built around bottom-up, heavy-participation, large-scale organizational change. For big change, the program has to be very laser-like, very intense, very focused, and has to be driven top-style for a couple of years.

Now, reengineering may be as wonderful as you say, but it can't be invariably successful. What causes reengineering programs to fail?

Hammer: I often respond in metaphor. I say that reengineering is like chess, not like roulette. People lose often in both, but when you lose at roulette, it's fate, chance, odds, bad luck. You don't lose at reengineering because of bad luck. You lose at reengineering out of stupidity. It's sitting down to play the game and not knowing the rules, and missing the queen—that's what causes people to lose at reengineering. It's a combination of not understanding what it is, getting caught up by the term, and having a vague and fuzzy sense, but not having any disciplined technique or tools, and/or trying to make it happen without the queen, which is an intensely committed executive leader. Those are the two biggest causes of reengineering failure: ignorance and a lack of strong leadership.

Workers on the line—when they hear of reengineering, do they translate that to mean downsizing and redundancies, and do they ask: "If I'm going to lose my job, why should I participate in my own execution?"

Hammer: Actually, it's worse than you said because even people who *aren't* going to be losing their jobs are scared to death of reengineering because, even if they keep their jobs, it ain't gonna be the same job. The jobs are different, and they have great anxiety. "Gee," they ask, "will I be able to do the job?" Or, "What will my future be? The career path doesn't look the same. The pay system is different."

Now, what we find is that reengineering usually results in a smaller number of better, higher-paying jobs. But people are afraid, and when folks are afraid, their initial reaction is to resist and put up barriers. That's a huge implementation problem in reengineering, and the difference between the winners and losers is who manages to overcome those problems. How do you do that? One piece of the answer is to create a sense of inevitability. At one manufacturer I know, the union was about to go on strike over reengineering. On the eve of the strike, management went to the union and said, "Listen, we understand your concerns, but you have to understand, this company is customer-driven, not union-driven. We have to do this. We're going to do it. We'll do it with you, we'll do it without you, but we are doing it." The union called off the strike.

Champy: Let me describe to you what I consider to be the condition of three different audiences in companies today. First, there are the people on the factory floor, the workers. Among them, there is, on most occasions, a fear about reengineering. Some of that fear comes from cynicism because they've gone through the "program of the month." Is this downsizing under a different name, they wonder, or another manifestation of the lack of leadership in getting the job done that needs to be done? I find that deep in the organization, people recognize the actual need for change, but they are increasingly cynical about their leaders, who haven't been able to produce some change. The people on the factory floor are the bravest people because they recognize that their jobs are at risk, but they also recognize that the company is at risk if something doesn't happen.

Second, sitting at the top of the organization are the leaders, who increasingly recognize that unless they can take the company from Place A to Place B, there may not be a company in five years. So they're ready to move, but they're also frustrated and privately voice a lot of worry about whether they'll be able to get their organization to change. Whenever I'm alone with a CEO, I hear a deep expression of concern about how difficult the ship is to turn. "Gee, my managers have only been in the business of adding capacity," he'll say. "Never have we had to exercise any muscle around creatively restructuring this place."

The last audience is composed of so-called middle managers. These people have gotten to a place in their careers under a set of rules that have been well-understood. Now we come in and we say, "All the rules are going to change. There may not even be the place you thought you were moving towards when this reengineering job is done." It's these people who are the most intransigent to change. And the truth is that maybe 75 percent of them aren't going to be there—at least not in their current capacity. And this group also has the most difficult time transitioning into a new model of work. What you're saying to those people is, "Managerial work as you knew it is over. Nothing personal in this, but in your work you have not brought value added to what we do. So what we want you to do is get back out there on the street, on the route, back out there with the customer, where we really need the value added." Now, that's a new game for some of these people, who, quite frankly, thought that what a career was about was getting away from the real work.

Hammer: In a reengineered organization, most people are workers, which is where all the value is added. Managers by definition are overhead, so you get by with a lot fewer of them. The managerial roles that remain primarily are coaching roles—cheerleading, helping, supporting. That kind of role is very foreign to what most managers have been trained to do, and that's a terribly hard transition for them to make.

So top management then will have to deal with covert resistance from middle managers?

Hammer: Yes, it usually manifests itself as covert, ranging from, "Gee, that idea will never work" to, "We tried that idea and it didn't work" to, "I really think that's a great idea, but this is our busy time of the year, and then we have the holidays and the vacations, so come back a year from Wednesday" to, "That's a very good idea, but we should really do it slowly and carefully to get it right." Like, let's take about 10 years.

Champy: Other times, it's ignoring the effort entirely. Just pay no attention and believe this, too, will pass.

Hammer: To put it bluntly, the way that you have to deal with this resistance is a combination of relentless communication, support incentives, and a bloody ax. Al Capone once said, "You get a lot further with a gun and a kind word than with a kind word alone."

Jim's company did a lot of work at Hallmark, which has been very successful in reengineering. Bob Stark, who was president of the card business, had a famous line when he started reengineering at Hallmark. He said, "We are going on a journey. On this journey, we will carry the wounded and shoot the stragglers." That says it perfectly.

Your mention of Hallmark raises a question in my mind. To me, reengineering involves an engineering mindset, while Hallmark seems perhaps the epitome of a nonengineering company; in fact, I wouldn't be surprised if the people who produce greeting cards—designers, artists, writers—had an anti-engineering mindset.

Hammer: Your insight is not off-track. Reengineering is about the engineering of business processes, which is an engineering-like discipline. It's not about physical science, but it does require an engineering mentality. Some organizations are comfortable with that—some electronics companies, even some insurance companies, because insurance is a form of financial engineering. On the other hand, there are some consumer-products companies that are brilliant marketers, but I'm not sure they could organize my 9-year-old's class to go to the zoo together. I mean, they're just not oriented to a process and that kind of discipline.

Champy: The image of Hallmark hides a characteristic about the firms helped by reengineering, and that is that they have some proclivity around operations, or a need for an operations infrastructure. To be sure, Hallmark has floor after floor of creative people, but it also has all the people who do the manufacturing and the distribution and the selling and the marketing. Hallmark is a company that has 75,000 different products out there at any point in time, and 25,000 of them change each year. Just think of the kind of infrastructure that's required for a company to do that. Hallmark used to take two years to get one of those new products to market, and we got it down to a few months. That was critical, because it's a company that must catch consumer tastes.

Actually, from an operations perspective, Hallmark was the best in the industry. But it recognized that niche players were taking bits and pieces of its markets. There were also outfits like Wal-Mart that were saying, "All right, we're now your single largest customers. We're going to dictate the terms of the deal. And so, Hallmark, you'd better have a cost infrastructure that allows you to deal with us at a point where we can both make a lot of money from this relationship." Bob Stark was visionary enough to see these market shifts, and Hallmark began reengineering when it was having the best year in its history. I remember one of the early meetings, when one of the managers raised his hand—there were about 60 key managers present—and said, "Well, don't we all have to agree that we're going to do this because that's the way we've always done it here in

Kansas City?" There was a very pregnant pause in the room, and Stark, who was up in front, just turned and said, "No longer. No longer."

Hammer: There are other companies like Hallmark that start reengineering when they're in good shape. One that I've spent some time with recently is the Trane Co., the heating and air-conditioning people; they're part of American Standard. The company just came off its best year ever. It is No. 1 in market share, and it is also in the midst of a dramatic amount of reengineering of its basic processes. Its business model was based on new construction. But you look around today, there's not a lot of new construction. People see the idea of rehabilitation of existing properties as both a threat and an enormous opportunity, and so it requires a fundamental shift in the way they operate.

They used an interesting technique to create a sense of urgency. Jim Schultz, the executive vice president, called together the top 150 people in the company, from all over the country. At the beginning of the meeting, Schultz stood up, thanked them for how well the year had gone, then said, "That's the past. We're here to talk about the future." Then he turned on a video that started with funereal music and a black screen with white letters that said, "What follows could happen to us." What followed was a bogus documentary about how Trane went out of business, based on interviews with people purporting to be customers, distributors, and employees. A customer was heard to say, "Trane used to be No. 1, but our needs changed and they were too slow to respond, so we switched to somebody else." Or an employee might say, "That system was too complicated, and though we tried hard, we just couldn't make it work." The video concludes that Trane has gone bankrupt. Then Schultz stood up and said, "Look, that's a possibility. What really happens is going to be up to us." That got their attention and commitment.

Your book charges that so-called classical business structures often are characterized by what you call the "Humpty Dumpty school of management." Explain.

Hammer: What happens in traditional organizations is that we take a process—let's say new-product development—and we break it into pieces. We put each piece in a separate department, then we have to hire all the king's horses and all the king's men to put those pieces back together again—liaison expediters, comptrollers, auditors, supervisors, managers, vice presidents. It's all overhead. The real work is scattered, and to make it fit together we need this managerial overhead. If you look at a typical manufacturing company, direct labor is 10 percent of product costs. It's overhead that's killing us. In many cases, the glue is more than the paper.

Champy: The Humpty Dumpty school of management has been dominant for the last 50 years. It's what Alfred Sloan taught us about how to run companies that basically were fragmented, and it worked for some period of time. But times have changed. Sloan was right for the 1920s, not for the 1990s.

You mean reengineering wouldn't have worked in the 1920s?

Hammer: You didn't need it. In those days, you had a fantastically growing economy with increasing demand and unsophisticated customers. What you needed was to ramp up mass production, and the ability to predict and forecast in a very limited fashion. And that's what the traditional hierarchical organization provided.

Champy: Just produce and it will sell.

Hammer: That's it. It's like the movie *Field of Dreams:* If we make it, they will come. You can't operate that way anymore. Virtually every industry I know is in a global condition of overcapacity. You've got sophisticated customers, you've got global markets, global production, global movement of capital. You need to focus on a different set of issues that didn't worry Sloan and Henry Ford.

Traditionalists might also be taken aback by something else you say in your book: "Paying people based on their position in the organization—the higher up they are the more money they make—is inconsistent with the principles of reengineering."

Champy: It's what we should have been doing for the last hundred years. It's where the real value added is. And, by the way, there already exist compensation systems like that: Most sales-compensation systems pay good salespeople more than sales managers.

But what about the creative side? To take your example of Hallmark, what about all those talented people who come up with the clever blurbs on the cards and draw the pictures? How do you evaluate that kind of creativity?

Hammer: You absolutely have to be able to measure things, but creativity is not what really counts. What counts is getting the product to market, and having products to sell. So if I'm a graphic artist and I'm part of the team that's developing a new set of cards, we measure how long it takes for us to go from concept to realization in the marketplace, and then how well those cards are received in the marketplace.

In general, the idea of solo contribution becomes the exception rather than the rule. By and large, business value is created by groups of people, so you look at the group of people who created some value, measure that value, and reward them in proportion to how much value they created. In fact, most productivity-measurement systems in organizations today are useless because all they measure are pieces of the work, tasks.

You say: "A company that looks for problems first and then seeks technology solutions for them cannot reengineer." This speaks to the core of reengineering, does it not?

Hammer: Right. The traditional approach, which we've all been taught since we were children, is first understand your problem and then go look for a solution. Trouble is, when you're trying to break out of existing models, the definition of your problem is shaped in conventional terms. And what you need to do is think creatively about your problem. The power of the new technologies is that they allow you to redefine what your problem is. An example we cite in the book: Until the Xerox 914 copier came along, nobody thought they had a convenience/copy problem. If you asked anybody, "Do you have a problem with copying?", they would have said no because they didn't think instant copying was possible. Once they saw that they could have instant copies made, everybody said, "I've been waiting for this all my life."

Champy: We could argue that most information technology in the last 50 years has been misapplied, that it has been used to automate current processes and not used in an inductive way, and that you look at the new technology and say: Well, how might this allow me to do my work? It's so fundamentally different.

If there's no use simply asking people how they would use a technology in their business, since they inevitably will reply in terms of how that technology might improve a task they do already, this suggests to me that most marketing research is of very little value.

Hammer: That's *exactly* our assertion. Most traditional market research is only good for extensions to what you already have—say you're considering putting the same toothpaste in a larger tube, or in a box with a slightly different color, or giving it a slightly different flavor. Akio Morita at Sony has a famous line: "You can't do market research on a product that doesn't exist." It's not until you see how people adapt to the capabilities of the product that you start to see what's possible.

If you look at the history of innovation, you'll see people always think about the new in terms of the old, and they miss what's really important about it. Thomas Edison, for instance, thought that the future of the phonograph record was in recording the deathbed wishes of dying gentlemen. And Guglielmo Marconi, who invented radio, literally saw it as a wireless telegraph. He didn't understand the idea of broadcasting at all.

Champy: And most people who made PCs didn't understand the application of spreadsheets until Lotus came around and started making spreadsheets. They didn't understand what in the hell they were going to use all that stuff for. Now, there was a suspicion that there was an application, but they didn't predict where the applications were really going to be.

All right, let's say a CEO reads this interview in *Across the Board* and becomes intrigued with the idea of reengineering. Let's also say he's not only skeptical but a little scared about its implications. Is it reasonable for him to try reengineering on a pilot basis—or should it be pervasive throughout his company?

Hammer: Absolutely, you pilot it, but you can't do it quietly, because reengineering always entails such big change that to do it quietly with low energy means it'll die. So you need to motivate it very strongly, with a big fanfare, but you don't do it everywhere at once.

Champy: I would also ask that CEO about the condition of his business. There are some businesses where there isn't time to pilot: The need to create the change is so immediate that you have to get on with it. That means moving into higher-risk territory, but frankly, I think you'll be in a higher-risk area just staying where you are.

Now, if you're going to start with a pilot program, you've got to make sure it produces a meaningful business-performance improvement. We could, for instance, reengineer the coffee service at The Conference Board.

Marvelous Idea.
Champy: And the coffee would be better and it would cost 50 percent less, but will that really improve the business performance of The Conference Board? Not likely. Not only

that, it will cause us as much pain to redo that coffee service as it does to redo what managers do who sit around this table.

A curious thing: The word "teamwork" recurs in your book, but you also say reengineering demands a czar. A self-managed team concept on the one hand; strong top-down leadership on the other. Isn't there a conflict?

Hammer: The short answer is it's one of the conundrums of reengineering that when you're done with it you've created an environment in which people are largely self-managed, self-directed—but getting there is a very top-down, very autocratic, very nondemocratic process. You either get on the train or we'll run over you with the train, but when you get on the other side of the river it becomes a totally different environment.

To conclude, you say in your book: "None of the management fads of the last 20 years—not MBO, diversification, Theory Z, zero-based budgeting, value chain analysis, decentralization, quality circles, 'excellence,' restructuring, portfolio management, management by walking around, matrix management, intrapreneuring, or one-minute managing—has reversed the deterioration of America's corporate competitive performance. They have only enhanced the manager's sense of self-importance." My question: Won't reengineering also enhance the manager's sense of self-importance?

Hammer: The last thing in the world that reengineering does is enhance the manager's sense of self-importance, because one of the things that reengineering says is that managing isn't so important. All those fads we mentioned are about managing. Reengineering is about working, which is a different emphasis.

You don't have any fear, then, that reengineering might end up on a list of fads in some future book?

Champy: I'll tell you why I don't have that fear: I genuinely believe that within the next five to 10 years, every organization must fundamentally rethink its infrastructure and the way the employees do their work, or else it will be noncompetitive. It's possible that reengineering will be called something else a few years from now, but it will never be a fad. It's just too important.

Chapter Twenty

Total
Quality
Management

INTRODUCTION

Quality improvement as a critical management process began to converge on the concept and title of total quality management (TQM) during the late 1980s and early 1990s. The earliest quality concepts relied on quality control and inspection sampling as a means to "inspect in" quality at the end of a production process. During the 1970s and 1980s, the concept of zero defects became popular, with an increased use of control charts to reduce the variation in production processes.

These two approaches to quality left out the most important component of all: the customer. The TQM movement places the customer, whether internal or external, as the most important definer of quality. The TQM improvement process is drive by improved products and services for the customer, with the result that production planning and control will be significantly impacted through the process of quality improvement.

TOTAL QUALITY MANAGEMENT (TQM)

A Brief History

In modern industrial society, the first quality processes relied on inspection. The inspection approach implies that manufacturing produces some level of poor quality and that the only way to get good quality to the customer is to inspect out the poor quality. This approach was justified by assuming that the cost of high-quality production on the first manufacturing pass was much higher than inspection. Few manufacturing managers now believe this is to be true. As can be seen in Figure 20.1, the process loops until sufficient acceptable product is delivered.

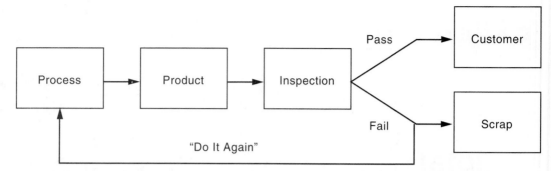

FIGURE 20.1 The process in a typical inspection approach. From *Total Quality Management* (Houston: American Productivity and Quality Center, 1990).

The second major era of quality control could be called the statistical process control (SPC) era. Process control involves measuring the variation of a process, setting limits to the variation, and allowing the process to be quickly adjusted toward target means and standard deviations. Effective use relies on frequent measurement, quick feedback, and the operator's ability to adjust the process. The advantages of this method are that little defective material will be passed to the subsequent operations and the final product will have a significantly higher pass rate. If done properly, the final inspection can be eliminated. It is now realized that although SPC is a valuable tool, it has done little to identify the customer's definition of quality. A product that conforms to specification but does not meet customer expectations is still not high quality.

The TQM era has evolved as a natural result of marketplace influences. Customers are now more vocal in their dislike for low quality and are willing to spend more for higher quality, either in the design or delivery of the product or in reliability. Dissatisfied customers will switch to higher-quality competitors, and the result has been disastrous for many domestic manufacturers. In this approach, proper analysis and action taken during the design stage ensures a product that will satisfy the customer. Figure 20.2 shows how this approach differs from the inspection approach.

There is evidence of a new era emerging that has been tentatively dubbed the WOW era, so named because customers are so happy at the quality of a product or service that they are heard to spontaneously exclaim "Wow." In this approach the process in Figure 20.2 is continued to an extreme so that customers are surprised and delighted; customer expectation is exceeded.

TQM Defined

Total quality management (TQM) is a philosophy of continuous improvement of quality. Customers are the key to the definition of quality, and TQM relies on a set of tools and concepts, teamwork, and employee empowerment.

Total. Everyone in an organization should be involved in the improvement process. All facets of the organization are focused on continuous quality improvement to the customer in a process that continues all the time, every day. The result is a continu-

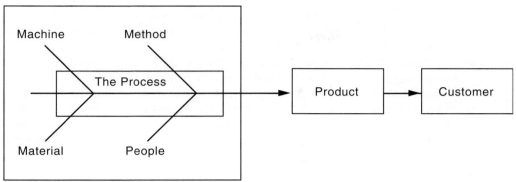

FIGURE 20.2   The process in a TQM approach. From *Total Quality Management* (Houston: American Productivity and Quality Center, 1990).

ous process of quality improvement, through both small incremental improvements and major breakthrough improvements in quality.

Quality.   The focus of TQM is on the customer of the product or service. Customers can be both internal and external, although the latter are usually considered first because they may be the buyer of the product or service. The internal customer should be given a significant role, however, because each person or process may eventually have an impact on the product delivered to the external customer. For example, an operator on an assembly line passes a product to the next person on the line, who is a customer of the product the operator has just worked on. That next customer expects a high-quality product, or they may not be able to do their job properly.

The customer defines the quality of the product or service. This is why strict adherence to design standards may still result in unhappy customers. Consider the electric automobile, which is now technically feasible to build, yet does not have a mass market. Why not? Because several significant design limitations to electric automobiles exist, most notably, a limited mileage range before recharging is required. Even if the electric car were manufactured to exact specifications and had high reliability, very few people would consider it a high-quality car because of the nuisance of frequent recharges. Thus customers of cars have defined the electric car as low quality at this time and will not buy them.

In today's competitive marketplace, it is critically important that the product or service meet or exceed the customer's expectations. Customer input into design of the product or service is crucial.

Management.   TQM is a system of management, not just a program that is put in place and then remotely managed. Systems need to be established that accomplish the quality improvement process. Management of the resources within an organization takes on a different approach when the issue of customer focus on quality is paramount. Employees are empowered to suggest and make improvements in processes and products.

Goals of TQM

The primary goal of TQM is total customer satisfaction. Many organizations are now suggesting that the goal is to delight and surprise customers to make them so satisfied that they will become loyal, long-term customers.

Second, goals of TQM include the ability to "do it right the first time," that is, to ensure that the product or service is delivered correctly at the first attempt. This can be accomplished by designing quality into the product or service, not trying to inspect it in. Proper design will prevent defects that may not surface until the product is in the customers hands. The TQM process should involve and empower all employees to place customer satisfaction foremost and to make continuous improvements to the process and product.

TQM Pioneers

The quality movement has seen significant contributions from individuals now revered as "gurus" in the business. One of the earliest proponents was Walter Shewart, who is credited with developing statistical process control while at AT&T Bell Labs in the 1930s. W. Edwards Deming refined these concepts and is probably the most famous quality pioneer. He has been credited with leading the recent quality revolution. Deming was invited to Japan after World War II to help reconstruction of industry there. His basic message that quality was a key competitive advantage. It is ironic that many years later, American businesses are starting to realize this same message. The Japanese were so grateful for Deming's influence that they created a quality award now known as the Deming Prize.

Joseph M. Juran is another pioneer whose career paralleled Deming's. He also spent time in Japan after World War II and has had a distinguished career in quality in the United States. Juran is credited with, among other things, creating the idea that quality is defined by the customer as "fitness for use." That is, the customer of the product or service defines whether it is of high quality by their use of it.

Philip B. Crosby made significant contributions to quality at ITT as vice-president of quality before he left in 1979 to found the Crosby Quality College. He is credited with creating the concept of zero defects. His message of consistency, known as conformance to requirements, is directed toward top management, who have the ability to improve the system so that the workforce can produce zero defects.

These quality pioneers have set the stage for the recent quality movement within American businesses.

TQM AS A KEY COMPONENT
OF COMPETITIVE INITIATIVES

TQM has become the most important part of competitive initiatives of organizations in the 1990s. The key reason is that it has customer satisfaction as the primary focus.

JIT and TQM Compared

Figure 20.3 illustrates the key difference between JIT and TQM. Both represent philosophical methods of reducing waste, increasing productivity, and ultimately improving quality. The key difference is that JIT focuses internally on the process and TQM focuses externally on the customer.

This significant difference explains why TQM has caught on so rapidly. JIT has been implemented in many companies, with many success stories, but has never been hailed as the great approach that made a difference in the competitive marketplace. The reason is that it focuses on the internal, leaving the customer wondering if the low quality of the product will ever be improved, even though the company was announcing great reductions in inventory levels and making more small lot deliveries. Because it became important to have defect-free product move directly to the shop floor, JIT did force organizations to improve quality in the small lot deliveries. This still did not attack a problem of poor quality design.

As illustrated in Figure 20.3, JIT starts with inventory and waste reduction, with subsequent improvements in cycle time and productivity. This leads to improved quality. TQM starts with customer defined quality, which forces employees to find ways to improve the quality of the product and process. This may lead to cycle time, inventory, and waste reductions if necessary for customer satisfaction.

Because of its customer focus, TQM is now the major initiative in organizations. Upper management sees the advantages of TQM because it can result in more sales through customer satisfaction. This is not to say that TQM is the only initiative, but that

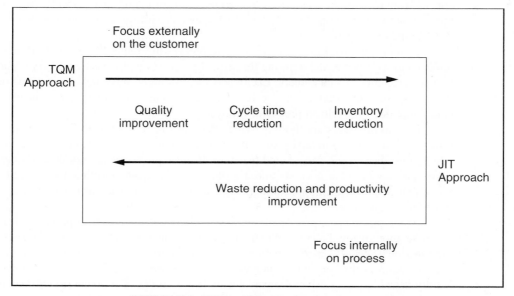

FIGURE 20.3   TQM and JIT approaches to improvement.

it is the primary initiative. JIT and other methodologies will be employed within the framework of TQM to improve quality for the customer. There will be increased emphasis on process variation reduction, setup time reduction, and design for manufacturability. These initiatives will all be occurring simultaneously within organizations. The impact of TQM on the production planning and inventory control functions will be profound.

THE TQM PROCESS
OF IMPROVEMENT

TQM involves several processes. Top management is involved in starting the process through a full planning process. A ten-step process is employed to solve problems and to make quality improvements. Within that process, Deming's plan-do-check-act cycle is employed.

Top-Management Leadership

The implementation of TQM requires commitment from top management. Most of the known success stories indicate that the full support, commitment, and involvement of top management is critical to success. Top management must take the initiatives to involve and empower employees to make improvements. Leadership is more important than management.

Teams are formed throughout an organization to attack quality problems. At the top-management level, teams are formed to guide the process and to work on major "breakthrough" activities, activities that could result in major changes in the product or the way the company operates. Figure 20.4 shows the full planning process that should occur at the top-management level.

Two of the most important steps in the process in Figure 20.4 are identifying customers and their needs and defining the critical processes and measures. Every organization has a number of customers. Each class of customers must have their needs determined so that the product or service can be designed to satisfy those needs. The determination of the critical processes follows.

Critical processes are those activities that must be accomplished to satisfy the customers' needs. For a university or college, some important critical processes would be admissions, curriculum design, and teaching. If these processes are not done properly, a university would either have no students (through inadequate enrollment) or would provide such a poor education (through poor curriculum or teaching) that its existence would be in jeopardy.

At the top-management level, the determination of these critical processes leads to a combination of top-management strategy changes and of strategic initiatives passed to the functional areas. Consider again the university's critical processes. For example, the determination of the curriculum, degrees, and majors is a complicated management and faculty decision that is driven, in part, by student response. Top management may make decisions about the degree programs, and individual colleges and departments make determination of courses that are critical to a state-of-the-art degree program.

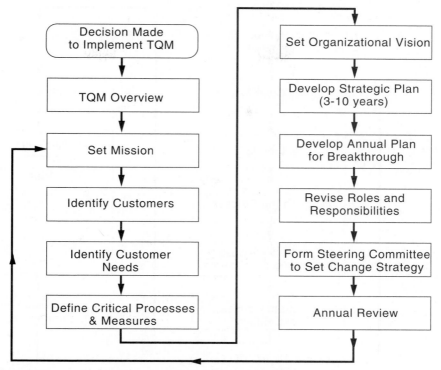

FIGURE 20.4 Top management "breakthrough" planning process. From L. Edwin Coate, *Implementing Total Quality Management in a University Setting.* Used with permission. Copyright Oregon State University, Corvallis, Oregon, 1990. All rights reserved.

The determination of the critical process that should be passed to functional areas forms the basis of problem solving at those levels in the organization.

Problem-Solving Process

Within an organization's functional areas, departments and divisions need to determine their own customers and the critical processes involved. When a team or individual finds an activity that could use improvement, the ten-step process shown in Figure 20.5 can be followed.

> *Step 1: Critical process selected:* This step involves the selection of the most important opportunity for improvement.
>
> *Step 2: Survey customers:* Identify the customer, either external or internal. Determine how the customer feels about the problem.
>
> *Step 3: Select the issue:* Determine what the issue is that needs to be improved to increase customer satisfaction.
>
> *Step 4: Diagram the process:* Draw a flowchart of the process as it is actually happening. Do not draw it as some believe it should be happening, but

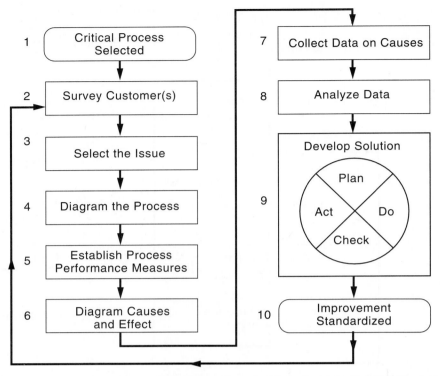

FIGURE 20.5   Ten-step problem solving process. From L. Edwin Coate, *Implementing Total Quality Management in a University Setting.* Used with permission. Copyright Oregon State University, Corvallis, Oregon, 1990. All rights reserved.

as the people involved in the process actually see it. This is important because many processes that may be written down as policy may, in fact, not be followed by the workforce. Perhaps they have found a better way or are they skipping important parts of the process.

Step 5:   *Establish process performance measures:* Determine what should be measured in the process that is a good indicator of the quality aspects to be improved.

Step 6:   *Diagram causes and effects:* Find out some causes of the identified problems.

Step 7:   *Collect data on causes:* Gather data on the process so that problems can be isolated based on fact, not rumor or speculation.

Step 8:   *Analyze data:* Use graphs and charts to illustrate what the data are telling.

Step 9:   *Develop solution:* Use brainstorming or other techniques such as the Shewhart cycle to find a solution.

Step 10:   *Improvement standardized:* Find a method to ensure that the solution to the problem has been made a standard part of the operating procedure.

The Shewhart Cycle

(Deming, 1986)

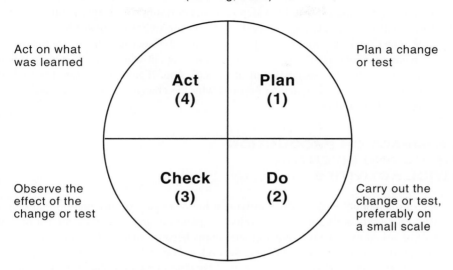

5. Repeat Step 1, with new knowledge
6. Repeat Step 2, and onward

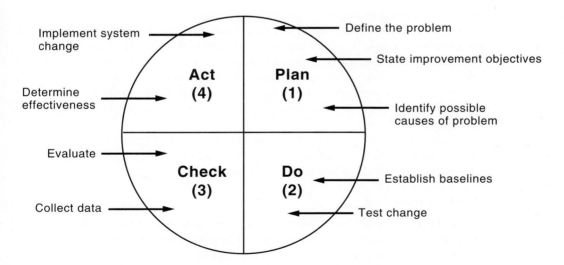

5. Repeat Step 1, with new knowledge
6. Repeat Step 2, and onward

FIGURE 20.6 The Shewart cycle: plan-do-check-act. From *Total Quality Management,* (Houston: American Productivity and Quality Center, 1990).

The Shewhart Cycle: Plan-Do-Check-Act

Deming [2] discusses the Shewart cycle (see Figure 20.6) as a continuous method for finding improvements. The Plan step involves defining the problem and identifying possible solutions. The Do step suggests that the improvement should be tested on a small scale. The Check step involves collecting data and observating to ensure that the changes did have a positive result. The Act step is the stage at which the change is implemented at the system level if successful at the Check step. If not successful at the Check step, then proceed to the Plan step to find out why the change did not work and move to another possible solution.

TQM'S IMPACT ON PRODUCTION PLANNING AND INVENTORY CONTROL ACTIVITIES

The implementation of TQM will have a major impact on production planning and inventory control activities. First, production planning and inventory control professionals will be asked to serve on improvement teams because many issues of customer satisfaction involve inventory and delivery. Second, actions taken to improve customer satisfaction in other areas of an organization will have an impact on planning and control activities.

Inventory

The focus of TQM will continue to be reduction in inventory levels and waste. JIT techniques will be employed to assist in this area. Customers will demand faster delivery of more customized products, which will force a change in the location of important inventories. To speed the throughput of product, changes in production planning methods will be needed. Determining which planning heuristics to use in each situation will be critical to ensure that customers are satisfied. Organizations that determine how to speed the flow of product through the production steps will have a competitive advantage in the marketplace.

Internally, defects will decrease throughout the production process, and less safety stock will be required at each stage to cover these defects. The reduction in variation at each process step will also allow reduction in safety stock.

Setup Time Reduction

If reduced cycle time and faster throughput are required to improve quality to customers, then setup time will continue to be an important item for study. Shorter setup times allow smaller economic lot sizes, reduced inventory, and faster throughput. Smaller lots will allow more flexibility in scheduling and more variety in product features. The result could be customized product that customers define as high quality.

Single-minute exchange of die (SMED) is a method of setup time reduction that relies on quick die exchange. The goal is to reach setup of less than ten minutes (a single minute) through technology changes and rapid changeover techniques.

Material Requirements Planning (MRP)

As yields improve, the use of material requirements planning (MRP) as a tool will become easier. Nervousness can be the result of changes in yields in previous stages, thus changing the gross requirements. As yields improve through the quality improvement process, nervousness will be reduced. In actual practice, lot-for-lot and other small lot procedures will become more popular, a smaller lot size can allow for more customer satisfaction through small lot delivery and customization. Lead times will need to be reduced to allow faster customer delivery.

Distribution Requirements Planning (DRP)

Faster delivery and distribution requirements planning (DRP) will become a key customer requirements in some industries. DRP could also be used by companies as a competitive advantage. Methods of distribution, location of inventory, and transshipment issues will need to be addressed to ensure that customers receive product rapidly.

Capacity Planning

Reduction in defects during the production process means less rework and reduction in capacity dedicated to rework. Because rework will not sporadically take away the capacity, such reductions will free capacity for first-time production and will allow more accurate scheduling of capacity.

MALCOLM BALDRIDGE NATIONAL QUALITY AWARD

The Malcolm Baldridge National Quality Award was created by the U.S. Congress and signed into law by President Ronald Reagan in August 1987. The award was created as a means to encourage competitiveness through recognition and commendation of quality in U.S. businesses. Since its inception, the award and its criteria have had a major impact on the way many businesses operate. Thousands of copies of the award's guidelines are requested each year, although a much smaller number of business actually apply.

The award's criteria (shown in Figure 20.7) focus on a business's total quality management system and the improvements the system generates. These criteria are evaluated on three dimensions: (1) the soundness of the approach, (2) the deployment of the system and (3) the results provided by the system.

Many businesses are now using the criteria as a method of self-analysis, even if they do not apply for the award. Such self-analysis provides a basis for the quality improvement process.

Figure 20.8 shows the winners in the award's three categories—large manufacturer, small business, and service—since its inception. Notice that there does not need to be a winner in each category. The maximum number of winners in any category in any year is two.

	Points Possible
1.0 Leadership	
1.1 Senior executive leadership	45
1.2 Management for quality	25
1.3 Public responsibility	20
Section 1.0 total	90
2.0 Information and analysis	
2.1 Scope and management of quality and performance date and information	15
2.2 Competitive comparisons and benchmarks	25
2.3 Analysis and uses of company-level data	40
Section 2.0 total	80
3.0 Strategic quality planning	
3.1 Strategic quality and company performance planning process	35
3.2 Quality and performance plans	25
Section 3.0 total	60
4.0 Human resource development and management	
4.1 Human resource management	20
4.2 Employee involvement	40
4.3 Employee education and training	40
4.4 Employee performance and recognition	25
4.5 Employee well-being and morale	25
Section 4.0 total	150
5.0 Management of process quality	
5.1 Design and introduction of quality products and services	40
5.2 Process management—product and service production and delivery processes	35
5.3 Process management—business processes and support services	30
5.4 Supplier quality	20
5.5 Quality assessment	15
Section 5.0 total	140
6.0 Quality and operational results	
6.1 Product and service quality results	75
6.2 Company and operations results	45
6.3 Business process and support service results	25
6.4 Supplier quality results	35
Section 6.0 total	180
7.0 Customer focus and satisfaction	
7.1 Customer relationship management	65
7.2 Commitment to customers	15
7.3 Customer satisfaction determination	35
7.4 Customer satisfaction results	75
7.5 Customer satisfaction comparison	75
7.6 Future requirements and expectations of customers	35
Section 7.0 total	300
Grand total possible = 1000	

FIGURE 20.7   Baldridge point values and categories.

	Large Manufacturer	Small Business	Service
1993	Eastman Chemical Company	Ames Rubber Corporation	
1992	AT&T Network Systems Group, Transmission Systems Business Unit and Texas Instruments Defense Systems and Electronics Group	Granite Rock Company	Ritz-Carlton Hotel Company and AT&T Universal Card Services
1991	Solectron Corporation and Zytec Corporation	Marlow Industries	
1990	Cadillac Motor Car Company and IBM Rochester	Wallace Company, Inc.	Federal Express Corporation
1989	Milliken and Co. and Xerox Business Products and Systems		
1988	Motorola, Inc. and Westinghouse Commercial Nuclear Fuel Division	Globe Metallurgical, Inc.	

FIGURE 20.8   Baldridge award winners.

SUMMARY

Total quality management (TQM) has quickly become the most important management initiative to improve the global competitiveness of domestic manufacturers and service providers. TQM will have a major impact on production planning and inventory systems. New ways of planning and control will need to be developed to satisfy customers with faster delivery of customized products. Just-in-time (JIT) methodologies will continue to be an important part of solutions. Production planning and inventory control professionals will be called on to work on TQM improvement projects. This can have a positive effect on the production planning and inventory control profession as improved customer service from faster cycle time and customized products gives an organization a competitive advantage.

PROBLEMS

1. What does TQM mean?
2. Give two examples, one of customer satisfaction and one of dissatisfaction, that you re-

cently experienced. What concepts of TQM applied to each situation? In the satisfaction situation, what did the organization or person do that made you satisfied? In the dissatisfaction situation, what would the organization or person have to do differently to make you satisfied?

3. Place yourself in the role of the instructor of a course that uses this textbook. You are to work on improving this course. Refer to Figure 20.4.
   a. What is the mission of the instructor in delivering this course? That is, what should be accomplished in this course?
   b. Who are the customers of this course?
   c. What are the customer needs of this course?
   d. Define some critical processes for this course. That is, what must occur for this course to complete its mission successfully?

4. Refer to Problem 3. Consider what needs to be done to make this course better. Refer to the ten-step process in Figure 20.5.
   a. Select one critical process, such as exams, grading of exams, research papers, or homework assignments.
   b. Survey customers to get their reaction to identify problems in that critical process.
   c. Select an issue to study.
   d. Diagram the process selected. What have you learned from this exercise?
   e. Establish process performance measures.
   f. Determine some causes and effects.
   g. If possible, collect data on causes.

5. Select a chapter from this text that you have studied.
   a. Survey other students about the quality of the selected chapter. What are the key issues that students mentioned as low quality?
   b. Select an issue that should be improved.
   c. Collect information from other students about that issue. What could be improved and how?
   d. Write that information down and send it to one of the authors of this text.

6. This following questions refer to Appendix 20A.
   a. An important factor in success of quality improvement is top management support. In what ways has IBM implemented this factor?
   b. The issue of cycle time reduction is important to IBM. Why are they interested in this?
   c. In new product design, how does IBM get customer information?
   d. Figure 20.4 presents that many steps that top management should take in the planning process. For each step in Figure 20.4, what are some of the activities that IBM has taken?

7. The following questions refer to Appendix 20B.
   a. How does the IBM Rochester site maintain customer contact?
   b. Discuss how the concept of just-in-time was used to improve the quality.

REFERENCES AND BIBLIOGRAPHY

1. L. E. Coate, *Implementing Total Quality Management in a University Setting* (Corvallis: Oregon State University 1990).
2. W. E. Deming, *Out of the Crisis* (Cambridge, Mass.: MIT Center for Advanced Engineering Study, 1986).

3. E. C. Huge, *Total Quality, An Executive's Guide for the 1990's* (Homewood, Ill.: Richard D. Irwin, 1990).

4. K. D. Lam, F. D. Watson, and S. R. Schmidt, *Total Quality, A Textbook of Strategic Quality and Leadership and Planning* (Colorado Springs, Colo.: Air Academy Press, 1991).

5. R. J. Schonberger, *World-Class Manufacturing: The Lessons of Simplicity Applied* (New York: Free Press, 1986).

6. *Total Quality Management* (Houston: American Productivity and Quality Center, 1990).

7. *Total Quality Management Master Plan, An Implementation Strategy,* GOAL/QPC Research Committee 1990 Research Report No. 90–12–02 (Methuen, Mass.: GOAL/QPC 1990).

8. M. Walton, *The Deming Management Method* (New York: Perigee Books, 1986).

APPENDIX 20A

Market-Driven Quality: IBM's Six Sigma Crusade
Bruce C. P. Rayner

Since taking the helm of IBM in 1986, chairman John Akers has been restructuring the $62 billion company in an attempt to restore its image and regain its preeminence as a technology and product leader. So far, results have been mixed at best. But starting this year, Akers is redoubling his efforts with a focus on quality.

Most of Akers' efforts to date have been in the people department. He removed two layers of management, cut the U.S. work force by 37,000—including 7,000 managers—and increased IBM's field force of sales and service personnel by nearly 20%. In the fourth quarter of 1989, IBM wrote off $2.4 billion in job-reduction and plant-closing costs. "Morale has been down since 1986," comments one ex-IBMer.

More distressing to IBM executives is the decline in market share. Since 1986, the company has lost market share among major customers in both mainframes and personal computers and will probably keep losing market share in software and peripherals, according to Gartner Group Inc., a research firm in Stamford, Conn. "IBM is trying to come back, but it's a tough fight," notes Randall Brophy, a Gartner senior research analyst.

Big Blue is plagued by manufacturing problems as well. In 1989, difficulties with the production of the high-end PS/2 personal-computer line delayed sales while semiconductor snags slowed mainframe production. Manufacturing glitches also held up production of the recently introduced 3390 disk drive, while the RS/6000 workstation was delayed because of reported problems with the operating system. "Our business as a whole did not live up to our expectations," Akers told stockholders at the company's annual meeting in April.

In response, IBM's chairman has launched a full frontal assault on what he now believes is the root cause of the company's problems: poor quality. At last January's

Reprinted from Electronic Business, October 15, 1990. Vol. 16, No. 19 pp. 26–30. Reproduction with permission from *Electronic Business Buyer* ® 1990, Reed Elsevier Inc.

IBM senior management meeting, Akers rolled out his heavy artillery: the Market-Driven Quality program, or MDQ. The objective is to make IBM more customer-responsive and more productive—two changes management hopes will strengthen sales and boost profits.

The MDQ program has three components: a set of quality initiatives, a system of process review and a system of quality measurement. All are aimed at cutting defects to near zero in everything IBM does and at shortening product cycle time. Beginning this year, MDQ will affect every customer, every vendor and every IBM employee world-wide.

"This really is a survival issue," Akers told his senior managers in January. Echoed Jack Kuehler, IBM's president, "The hard fact is that we're not doing well enough."

No one expects IBM to go belly-up in the near term. In the first half of 1990, the company reported a healthy revenue growth of 9.8% and a net income growth of 6.8% over the same period a year ago. Still, over the long term, IBM has been less than successful. Five-year figures compiled by ELECTRONIC BUSINESS show Big Blue's revenue growth at 6.4% and net income growth at a *negative* 10.6%. Because of this track record, senior executives are getting serious about improving both product quality and customer satisfaction. After all, it is high-profile, quality-conscious competitors, like Compaq Computer Corp. and Hewlett-Packard Co., that have been stealing market share from IBM in recent years.

"Every year the customer sets higher and higher standards," says Stephen Schwartz, IBM's senior vice president for market-driven quality. "If they don't get the quality they need from us, they will get it from someone else." Schwartz moved into the new position in April from the top job at IBM's Application Business Systems (ABS) division in Rochester, Minn., one of this year's finalists in the Malcolm Baldridge National Quality Award.

IBM sources refuse to disclose the cost of the company's quality program, but Schwartz admits that it will be high. Just training IBM's managers for two days will require about 800,000 hours, or at least $20 million. "It's costing us some money, but we are already seeing a payback," says Schwartz.

To drive the MDQ message home, Akers and his management committee—Schwartz, Kuehler and a few other senior vice presidents—are touring all of IBM's major U.S. facilities and many overseas sites. Akers sits down with the rank and file to impress upon them that MDQ is the top priority at the highest levels of the company. "Market-driven quality is the first thing on our agenda at every site," says Akers.

Employee empowerment is an important quality theme, and IBM's chairman drove that message home during a recent speech. "Market-driven quality starts with making customer satisfaction an obsession and empowering our people to use their creative energy to satisfy and delight their customers," Akers said.

Employees are pleasantly surprised by Akers' recommendations: "He told us to take more risks . . . and spend less time double-checking everything we do," says Jim McDonald, manager of systems assurance at IBM's Entry Systems Division in Boca Raton, Fla. "I liked that."

So what drove IBM's chairman to take such a radical course? After all, MDQ involves a sweeping cultural change at Big Blue. For one thing, IBM gained a dictatorial reputation during the 1980s as a company that often ignored customers' demands. Today, improved customer satisfaction is seen as absolutely vital to IBM's future.

The argument for implementing a customer-focused corporate quality program has been gaining momentum as the result of successes at Schwartz's old ABS division in Rochester. The division, which was a Baldrige finalist in 1989 and 1990, manufactures AS/400, S/36 and S/38 mid-range computers and data-storage devices. Since the early 1980s, the division also has run a quality-improvement program that has been used in development of the AS/400 and other new products.

The AS/400 was an important factor in the corporate adoption of MDQ. In developing the computer, management broke with IBM tradition by involving customers at the earliest stage of product definition. The reason was that the AS/400 was to replace the S/36 and S/38 models already owned by 300,000 customers. The AS/400 program also was a testing ground for shortening product cycle: 28 months for the AS/400 compared with 5 years for the S/38, says Larry Osterwise, ABS director and site general manager.

Akers' commitment to quality got another push last November, when a group of 31 top IBMers visited Motorola Inc., the winner of the 1988 Baldridge award, for a two-and-a-half day session on quality. The delegation was headed by Heinz Fridrich, vice president and manufacturing; Bob Talbot, formerly director of quality and now assistant general manager for special bids; and Bob Friesen, vice president of IBM U.S.'s development and manufacturing. Back at IBM, the group urged top management to adopt quality initiatives.

They succeeded. But Akers' chances of pulling off a corporate cultural change is tempered by the fact that he retires at age 60 at the end of 1994. Whether such a mammoth turnaround is possible with such a short tenure is unclear. Still, Akers told ELECTRONIC BUSINESS he wants "MDQ installed throughout IBM to the greatest degree possible" by the time he retires. Schwartz concurs: "MDQ is part of the legacy John wants to leave."

NO EASY TARGETS

Akers does not believe in setting easy targets. He expects IBM's entire worldwide organization to slash defects by a mind-boggling factor of 20,000 over the next five years and cut average product cycle time in half. One of the central initiatives IBM has adopted is Motorola's "six sigma" approach to eliminating defects. Six sigma is a statistical term denoting about 3.4 defects per million operations.

Kuehler has nicknamed IBM's defect reduction initiative "Excellence in Execution." Whatever name the program takes, IBM's quality executives clearly have their work cut out for them. The corporate goal is to reach six sigma by 1994. In January 1990, the company was running at an average level of three sigma (66,800 defects per million operations), according to Kuehler. Most experts rate U.S. manufacturing on the whole at below four sigma.

Achieving six sigma within five years will involve a sea change throughout IBM. Incremental goals have been set at a number of divisions. One division, for example, plans a tenfold reduction in defects each year through 1993 and a twentyfold reduction in 1994.

But while a number of senior manufacturing and marketing executives are optimistic about their goals, many doubt that six sigma is achievable by 1994. Even Schwartz is hesitant when asked if IBM will make its six-sigma deadline. "Some say yes, but time will tell," he says. He prefers to stress the positive: "If you don't set ambitious goals, then you don't change the thinking of the people."

Cycle time reduction is IBM's other primary concern. IBM wants to compress the process that starts when a customer expresses a need and that ends when a customer pays the bill. That includes everything from revamping intelligence-gathering—"market information capture," as IBM calls it—to speeding up design, development, manufacture, ordering, shipping and billing.

Management training involves a two-day MDQ session and is conducted by IBM staff, including top executives such as Schwartz. Both Akers and Terry Lautenbach, senior vice president and IBM U.S. general manager, maker video presentations. There also are contributions from outsiders, such as Paul Noakes, vice president and director of external quality programs at Motorola, who discuss their companies' quality.

The two-day course includes most of the quality basics—defining initiatives, deploying a quality-based process-management system, applying the Baldridge criteria, implementing six sigma, benchmarking and measuring quality. These managers then can take training material back to their employees. Managers at IBM Rochester, for example, were trained in February and were expected to have all 8,000 plant employees trained in MDQ by August, says site manager Osterwise.

Market-driven quality also applies to non-manufacturing sectors of the company. Ken Thornton, general manager of marketing in the mid-Atlantic region, attended a managers' MDQ session in February and has already set up of a quality program. All 7,400 of his employees will have received MDQ training by September 30, the end of the last fiscal quarter.

Sweeping changes already have occurred, notes Thornton. Each branch office is required to set five measurable quality parameters for improving customer service. The parameters can be as simple as the time it takes to return a phone call or as complex as improving employee skills. To stress his commitment, Thornton has appointed a regional MDQ manager responsible for implementing MDQ, who reports directly to him.

In all branch offices in the mid-Atlantic region, Thornton has installed a new suggestion system. Within the first three weeks, the Hagerstown, Md., branch was deluged with 128 suggestions from its 54 employees. The spectrum of ideas was broad. One employee recommended the accounts-payable staff visit customer sites to solve certain problems while another looked to improve the efficiency of the branch library, says Howard Rockwell, branch manager and 37-year IBM veteran. Many suggestions have already been implemented, he adds.

At IBM's Santa Teresa Laboratory for the development of software in San Jose, Calif., management is using Baldridge criteria to measure each function within the division, says Tom Furey, assistant general manager of programming systems and site general manager. Measurements include information-gathering, customer satisfaction measurements and work force use for all 2,000 employees. In one product development group, the first-quarter score was a disappointing 390 out of a possible 1,000. After some process changes, says Furey, the second-quarter score was improved to 570.

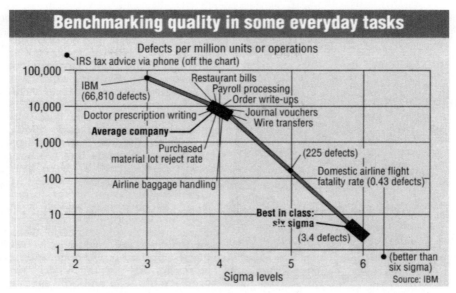

Benchmarking quality in some everyday tasks

Defects per million units or operations

IRS tax advice via phone (off the chart)

100,000 —

IBM
(66,810 defects)

Restaurant bills
Payroll processing
Order write-ups

10,000 —

Doctor prescription writing

Journal vouchers
Wire transfers

Average company

1,000 —

Purchased
material lot reject rate

(225 defects)

100 —

Airline baggage handling

Domestic airline flight
fatality rate (0.43 defects)

10 —

**Best in class:
six sigma**

(3.4 defects)

1 —

2   3   4   5   6   • (better than
six sigma)

Sigma levels

Source: IBM

At the beginning of 1990, IBM admitted its overall defect rate was around three sigma, or 66,310 defects per million operations. By 1994, IBM wants to reach six sigma.

Making MDQ work—and in particular, making six sigma a reality—must involve suppliers. But Schwartz is still unsure about how much pressure will be necessary to convince suppliers to adopt their own MDQ principles. "We will ensure that quality is up to scratch, but at this time there is no mandatory MDQ supplier program," he says.

For now, the pressure is relatively subtle. Over the past few years, IBM has been paring its supplier base and establishing closer relationships with vendors, or "business partners" as the company calls them. The number of IBM's significant suppliers dropped from 4,000 in 1988 to 3,000 this year, due in part to higher quality standards. Given the threat of further cuts, the pressure on suppliers to perform is sure to drive up vendor quality.

One major IBM supplier has already responded. National Semiconductor Corp. in Santa Clara, Calif., recently unveiled its own quality initiative, which is compatible with most of IBM's goals, claims Tim Thorsteinson, National's director of quality performance. Thorsteinson is responding to the needs of all its customers, not just IBM. "I think all computer vendors are redoubling their efforts, including IBM," he says.

But the task could demand more than subtle pressure. At IBM's Rochester facility. Osterwise says he "demands" supplier compliance in terms of focusing on quality, understanding and implementing the Baldridge application criteria and sharing suppliers' quality approaches with IBM. The alternative, says Osterwise, is to lose IBM as a customer. One West Coast supplier, who asked not to be identified, confirms IBM's demands, saying that Big Blue requires adoption of both the Baldridge application criteria *and* six sigma. "IBM may have the most extensive quality requirements [in the industry]," he says.

While MDQ is still in the early stages, there is little doubt that as it progresses in other IBM divisions around the world, they too will be forced to place more and more stringent demands on suppliers. "We expect them to strive for constant improvement," says Schwartz.

APPENDIX 20B

IBM's Rochester Facility Strives for a Perfect 10
Robert Haavind

When Olympic gymnast Peter Vidmar visited IBM's Rochester, Minn., plant recently, he told the 8,000 employees that performing well in a gymnastic routine rates about 9.2 points out of a possible 10.0. Technical excellence adds 0.2 to the score, risk-taking another 0.2 and innovation 0.2 more. To rate a perfect 10, Vidmar added, the gymnast must earn a final 0.2 points for "virtuosity."

His audience got the point. Ever since IBM chairman John Akers kicked off the company's Market-Driven Quality program early this year, IBMers have known that their own virtuosity is being rated every day. And although the Rochester unit—a producer of mid-range computer systems and data-storage products—hasn't yet scored a 10, improvement efforts at the site serve as a model for IBM's sweeping new drive for total corporate quality.

Following an unsuccessful bid in 1989 for the Malcolm Baldrige National Quality Award, IBM-Rochester is a finalist in this year's competition. The plant is upgrading nearly every one of its operations, with the emphasis on customer satisfaction. Product design procedures are being streamlined, product cycles shortened, inventories trimmed and waste and rework reduced. The Rochester plant's philosophy could be summed up in seven words, says Larry Osterwise, site general manager and director of IBM's Application Business Systems group: "If it's not perfect, make it better."

As with many other Baldrige hopefuls, IBM-Rochester learned that applying for the Award is itself an exhausting exercise. In this case, the process required some 20 man-years of effort, according to Roy Bauer, manager of engineering planning and operations at Rochester. Bauer and his team uncovered quality problems at practically every turn. The result: 71 "wart reports" that were divided into 34 top-priority tasks and 37 less urgent problems. "We're working on problems other companies don't even know they have," he says. People were asked to develop strategies and action plans to work on each problem. Rework levels were too high in some production operations, for example, so efforts were made to gain better control.

Rather than going for exotic, complex solutions, the quality team is concentrating on "basic blocking and tackling," says Osterwise, That requires coaching, and Rochester managers often exchange visits with other U.S. companies known for their strong quality efforts. David Kearns, CEO of Xerox Corp.—one of last year's two Baldrige winners—spent six hours touring the Rochester facility. IBMers themselves have visited such innovative companies as 3M Corp., American Express Co. and Disney Co. "Although Disney is in the entertainment business, their quality procedures were not that different from ours," according to Richard Lueck, Rochester's director of site services.

The customer satisfaction campaign is paying off. Rochester reportedly has made significantly progress in improving its customer service, particularly in its software sector. IBM-Rochester even changed the name of its Software Development Support Center to Software Partners Lab in order to stress the importance of customer participation. Customer teams come to the lab from a wide assortment of businesses—financial services, oil exploration, workstations, health maintenance, even bowling-alley construction—and work with IBM employees in well-equipped offices to perfect new applications.

TEST BED FOR QUALITY

Largely because of the Baldrige competition's heavy emphasis on customer satisfaction. Rochester is constantly pushing its involvement with buyers upstream. Before the AS/400 mid-range computer was introduced in mid-1988, customers were asked to join round table talks in which IBM discussed its plans, laid out various options and encouraged customers to help set priorities. They discussed over 155 key issues related to attaching PCs to the new IBM systems and, of these, 14 were identified as high-priority. Acting on the customers' recommendations, IBM agreed to share software codes with other manufacturers of equipment that was used by the customers, according to Ray Harney, manager of Software Partners Lab.

The Rochester plant also has worked to speed products to market. While the development cycle for the S/38 took a full 5 years to complete, the cycle for the AS/400 was cut to just 28 months by using cross-functional teams that merged software and hardware design with manufacturing.

IBM has called the introduction of the AS/400 the most successful product launch in its history. Within six months of introduction, more than 25,000 AS/400s were installed worldwide, with two-thirds of those sales overseas, according to Osterwise. Exports from the Rochester facility were estimated to total $575 million in 1989.

The fundamental architecture of the AS/400 family resulted from customer input. Because of the popularity of object-oriented programming, the system was geared to running this type of code. To get still more information, thousands of machines were shipped early for evaluation by customers.

Even now, the customer link does not stop once a system is shipped. After 90 days, a "customer partner" call is made to check on whether there are any problems with the new installation. IBM retirees, working part time, make many of the calls because of their familiarity with the company and its product line. "These are not survey calls," says Robert Tremain, a market-support representative. "We train our people to listen." Research and development staffers also get a chance to participate in these calls, reports Osterwise, in order to receive direct feedback from users.

Customer problems are referred to the local servicing group for action, and then a 30-day follow-up insures that the problem has been solved.

If customer complaints happen to come through chairman Akers' office, company policy requires a response within 10 days, according to Margie Spohn, manager of customer partnerships and telemarketing. And what if the field-service representative can't handle the problem or they don't have the right parts? "Someone gets on an airplane," Spohn says. "[Surveys show that] we have less than 5% dissatisfied customers, and that is going down."

IBM-Rochester favors continuous-flow manufacturing and just in time methods to shorten production cycles and reduce inventories. IBM's management studied the techniques of Shigeo Shingo, the Japanese consultant who helped set up Toyota Motor Corp.'s system, but they decided to take a gradual approach toward just in time deliveries, says Osterwise. Indeed, as they began operating with less and less inventory, a shortage arose at their Guadalajara, Mexico, plant—a source for actuators for their thin-film disk drives. Once the reason for the shortage was located and eliminated, the Rochester plant was able then to gradually reduce inventories even further.

To provide flexibility on the disk-drive production floor, strips of colored tape designating locations for the rolling five-layer carts are used to deliver parts and subassemblies to workstations. This is known as a *kanban*—or "pull-when-needed"—system. Blue-taped areas indicate items in preparation for production; green tape identifies disk drives that have gone through testing, and black-taped areas are rework stations.

Efforts to empower workers are evident throughout the disk-drive production area. Lists of worker committees, dealing with issues that range from "safety" to "six-sigma quality," are posted in the office of George Thompson, manager of Rochester storage products operations. Six sigma is a statistical measure that indicates that are about 3.4 defects per million operations.

The worker committees are producing results. Thompson explains that incremental goals for a new disk drive had been exceeded well before the deadline. The initial move to 4.0 sigma—the equivalent of 6,210 defects per million—from 3.85 sigma happened so quickly that a new goal of 4.5 sigma had to be substituted, and that, in turn, was within reach nearly two months early.

Osterwise stresses the need for creating an environment of "high morale" so employees want to do close to perfect work. Improvements must come both incrementally and in quantum leaps, he says; only then will virtuosity be achievable and the perfect 10 be possible.

Chapter Twenty-one

Factory
of
the
Future

INTRODUCTION

Forecasting the future is difficult, particularly in regard to the direction of manufacturing and technology. At least the economist has a fifty-fifty chance of forecasting interest rates if the only choices are "up" or "down." The objective of this chapter is not to forecast the future, but to present a brief introduction to the trends in industry today that will have an impact on production planning and inventory control professionals. The number of initiatives currently under way in many companies (not all at the same time) is truly staggering: total quality management, Baldrige Award analysis, reengineering, computer-aided drafting and computer-aided manufacturing, computer-integrated manufacturing, flexible manufacturing systems, group technology, setup time reduction, just-in-time, total preventive maintenance, activity-based costing, quality function deployment, concurrent engineering, electronic data interchange, and benchmarking (to name a *few*).

At present, the two key competitive factors in manufacturing are time-based competition and quality. Manufacturers are recognizing that either or both of these factors will result in a significant competitive advantage. These two factors are driving today's change in manufacturing methods that will result in the factory of the future. We are already seeing significant changes in the size of, structure of, and management approaches to manufacturing.

Manufacturing has generally been thought to possess four basic capabilities: cost efficiency, quality, dependability, and flexibility. The most common theory of trade-off manufacturing suggested that we could be excellent in one or perhaps two of those factors at the expense of the others. If we tried to improve on one, it would have to be at the expense of another. Now we are seeing companies that have higher quality, have lower costs, are more dependable, and respond faster to changing market conditions. Ferdows and De Meyer [3] have theorized that rather than a trade-off in capabilities, under the right situation improvement programs are cumulative and result in capability improvement on all dimensions. A quality program is the most important to start with; dependability is second. If a quality program is successful, dependability can be improved without a loss of quality.

Stalk [9] has theorized the movement toward flexible, innovative manufacturing facilities based on his observations of leading Japanese companies. The strategic focus has shifted from low labor costs to scale-based economies to focused factories. These were all cost-based strategies. Today's strategies are based on time; they start with flexible manufacturing, followed by rapid response, expanding variety and increased innovation. If this is indeed the trend, then the concepts of quality improvement, just-in-time, and flexible manufacturing will be key to successful time-based strategies.

Trends in manufacturing today and the implications for production planning and inventory control activities are now discussed briefly.

TRENDS

Total Quality Management

The quality revolution of the late 1980s will continue throughout the 1990s and beyond due to the necessity of competitive forces. Rather than a separate quality improvement program, quality improvement processes in the future will be part of everyone's day-to-day activities. Large, major companies, such as Motorola and Procter and Gamble, have made commitments to quality improvement activities over the long run. Companies that do not make such efforts will fall behind. There is no longer any doubt that quality is an important competitive force.

Product and process design has become a more critical component of quality; proper product design ensures the manufacturability of the product. Design and manufacturing engineers are merging into single or closely related working groups. The result is faster time to market and products that have a better record of design for manufacturability.

One of the most important quality improvement initiatives has been and will continue to be cycle time (and lead time) reduction. This will benefit customers, most importantly, but will also result in faster throughput and reduced inventory. Lead time has been an important parameter throughout this text; as lead times are reduced, many inventory control procedures will need to be altered to account for rapid delivery.

These changes will have significant impact on production planning and inventory control professionals and procedures. Lead times will be cut, and this reduction will need to be reflected in control procedures. Consider the lead time in MRP systems (see Chapter 11). If lead times are shorter than the time bucket of one week, say, then how do we reflect this: as a lead time of zero? There may be a need to move time buckets to days to ensure that the cumulative effect of daily time buckets is gathered. What happens when lead times are less than one day? Questions such as these will need to be addressed by both practitioners and theorists.

Companies that implement zero-defect and process control programs see not only improvement in the quality of the product but also reduced costs (less scrap and rework labor) and faster delivery (fewer delays in rework). Production and inventory professionals will be called upon to be active in quality improvement.

Time-Based Competition

Time-based competitive strategies have started to emerge in some industries. The ability to deliver faster than your competitor will be an important competitive advantage to be exploited. MRP systems will be used to help determine the total lead time to delivery. Shortening that total lead time will be of great importance. Input from planning and control to determine where lead time can be reduced will be required. It is not as simple as reducing the lead times in control procedures, however. Setup time reduction is one important initiative that will result in shorter lead times due to smaller lots and quicker response to machine changeovers.

Ferdows and De Meyer [3] show that organizations that implement quality improvement programs are better able to implement time-based improvements and that the effects of improvement programs will be cumulative if they are done in the correct sequence. It appears that the best sequence of implementation should be quality, dependability, speed, and cost efficiency. In fact, the cost efficiencies will come about because of the success of the other programs, not because of any particular cost efficiency activities undertaken.

Just-In-Time (JIT)

Just-in-time made significant gains throughout the 1980s. Additional implementation will continue, but many gains will be smaller and more difficult on the margin. One area that will see significant gain will be JIT purchasing. Planning and control procedures will be integrated with suppliers to foster a faster linkage for delivery. If suppliers are considered another link in a chain of manufacturing processes, then they can be included in planning and control activities.

Electronic data interchange (EDI) promises to reduce the supplier information time link even further. In EDI, purchase orders can be transmitted to suppliers electronically using a standardized data transfer protocol. Paperwork is omitted, and the time to transmit the purchase order is reduced to the time it takes to enter the data into the EDI format.

Setup time reduction is an area where improvement is necessary to gain advantages in quality, flexibility, and dependability. Lot-sizing procedures and inventory control methods require some input in setup time or cost. As the cost or time decreases, there is a corresponding decrease in the lot size. Again, planning and control professionals will be called upon to develop analysis to assist engineers in improvement projects.

As quality is improved in the production process and as shorter setup times allow smaller lot sizes, flexibility is introduced into the product line and costs can decrease.

Focused Factory (Revisited)

The trend toward smaller, focused factories (and the factory within a factory) continues to intensify. For many years the focused factory has been prescribed as the way to improve efficiency and effectiveness. Several new developments have accelerated this transition.

The first is the move toward fewer levels of management, more decision making by the workforce at the line level, and fewer supervisors. The result is that planning and control output will no longer be directed at supervisors but at workers. There has also been a trend away from the centralized philosophy typical of production planning and control. The new approach is to have planning and control activities determined at the local level. In fact, simplification of many procedures may completely eliminate the need for planning and control. If a lot size is one for a particular operation and the setup time is so short that this is feasible, then all lot-sizing procedures would be useless at this level.

A second major change is in the simplification of product design, which will lead to simplification of process design. Simpler design may result in more common components and parts, with a resulting simplification of the planning and control process.

A third major change is in the move toward multiplant clusters. Rather than large plants, more focused plants will be clustered and linked. Most planning and control procedures discussed in this text are for single plants and departments. Heuristics and procedures necessary for clusters of small plants will need to be developed.

Flexible Manufacturing Systems (FMS)

A flexible manufacturing system (FMS) is a group of programmable pieces of equipment, arranged sequentially, with automated transport systems. They are typically used for batch production of families of similar parts.

The movement toward focused factories will require a movement toward flexible manufacturing and group technology. No longer will similar production processes be grouped together, but cells will be formed with the manufacturing equipment to manufacture a set of similar products. Planning and control procedures for this type of process will need to be further developed and introduced. Flexibility and the success of cells will be dependent on other factors such as quality and just-in-time processes. Shorter setup times will allow machines within cells to run small lot sizes and will offer greater flexibility and dependability to customers. A cell would have no trouble deciding to change over for a small production run if the cost is minimal to do so and the quality is guaranteed. Less rework and reduced inventory result in lower cost while providing higher quality and faster delivery to customers.

Computer-Integrated Manufacturing (CIM)

Computer-integrated manufacturing (CIM) is a major advancement from FMS. CIM suggests a completely automated facility that can produce a mix of low-volume products. This would allow an increase in product mix with little cost of the flexibility inherent in typical manufacturing systems. In a typical system an increase in mix is usually accompanied by an increase in per unit cost. With CIM, the per unit cost would remain the same regardless of the mix or volume. Another important benefit is the consistency of quality gained through a CIM system.

CIM has only had limited success at this time due to the complexities of controlling such a system.

The Virtual Corporation

One intriguing new company model is known as the virtual corporation. In this model, partnerships are formed with other companies to tap the expertise of each. In some arrangements a new company is formed that utilizes the resources of the partnering companies and then disbands when its mission is done.

The virtual corporation introduces a new element into Young's model of manufacturing for competitive advantage. In his model, all the key functions are housed in one company. In the virtual company, product design may be done by one company, process design by a second, actual production for a third, and distribution by another. Young's concept of integrated manufacturing as the glue that holds all these together will not necessarily work in the virtual company.

Manufacturers within a virtual company must be able to connect with the marketing organization for customer orders and forecasts and then connect with the distribution organization on the other end of the process. This maya be accomplished through electronic data interchange (EDI), a faster way to transmit information.

IMPLICATIONS FOR PRODUCTION PLANNING AND INVENTORY CONTROL

Some important implication for planning and control professionals have been discussed in the previous sections of this chapter. It is obvious that we will need to adapt quickly to changes, regardless of the direction of change. Quality improvements will require planning and control system changes to adapt to reduced lead times and improved yields. New scheduling procedures for FMS and CIM applications will need to be developed and implemented. Setup costs will be driven down, resulting in changes to planning and control procedures. Opportunities for faster cycle times will need to be found. Integration of systems through EDI will also allow faster cycle times. Focus can reduce complexity of procedures and foster a return to less complex systems.

It will be necessary to change our thinking regarding many of the procedures discussed in this text. Consider basic inventory models: We have analyzed the trade-off of inventory carrying costs and setup cost for many years. New thinking suggests that if we attack the setup cost, then overall costs may actually decrease while we enjoy reduced lot sizes and inventory levels.

SUMMARY

The manufacturing setting is rapidly changing in response to global competitive changes. To remain competitive, manufacturing will need to continue this change. There will be a need for different methodologies in planning and control in response to time-based competition, quality improvements, and increased use of cells and flexible manufacturing systems.

REFERENCES AND BIBLIOGRAPHY

1. R. U. Ayres and D. C. Butcher, "The Flexible Factory Revisited," *American Scientist,* Vol. 81 (September–October 1993), pp. 448–459.
2. J. D. Blackburn, ed., *Time-Based Competition—The Next Battle Ground in American Manufacturing* (Homewood, Ill.: Business One–Irwin, 1991).
3. K. Ferdows, and A. De Meyer, "Lasting Improvements in Manufacturing Performance: In Search of a New Theory," *Journal of Operations Management,* Vol. 9, No. 2 (April 1990), pp. 168–184.
4. M. Hammer and J. Champy, *Reengineering the Corporation* (New York: Harper Business, 1993).
5. R. L. Harmon and L. D. Peterson, *Reinventing the Factory* (New York: Free Press, 1990).
6. R. L. Harmon, *Reinventing the Factory II.* (New York: Free Press, 1992).
7. M. Jelinek, and J. D. Goldhar, "The Strategic Implications of the Factory of the Future," *Sloan Management Review,* Summer 1984; pp. 58–60, 63–64, 66.
8. D. W. Richardson, "A Call for Action: Integrating CIM and MRP II," *Production and Inventory Management,* Vol. 29, No. 2 (Second Quarter 1988), pp. 32–35.
9. G. Stalk, Jr., "Time—The Next Source of Competitive Advantage," *Harvard Business Review,* (July–August 1988), pp. 41–51.
10. T. A. Stewart, "Reengineering—The Hot New Management Tool," *Fortune,* August 23, 1993, pp. 41–48.
11. "The Virtual Corporation," *Business Week,* February 8, 1993, pp. 98–103.

APPENDIX 21A

Tools to Become World Class Are Available. It's Up to You to Use Them.
Glenn A. Fishman

In today's globally competitive business climate, if your company is not exploring how to use quality and process improvement strategies to improve business performance, you virtually guarantee your organization of a competitive disadvantage. Most firms today realize that, just to stay competitive, they must implement at least some of the now popular tools and strategies that are available.

Federal Express chairman Fred Smith says, "If companies are not aggressively pursuing total quality today, they are following a self liquidating strategy." Global competition means no business can afford to stand still when it comes to quality or process improvement.

Source: APICS—The Performance Advantage, August, 12993, pp. 28–31. Vol. 3, No. 8

NEW TOOLS FOR IMPROVEMENT

Total quality management (TQM) programs, in addition to management innovations like business process re-engineering or business process innovation, are dramatically changing the way competitive advantage is realized and work is performed. This new thinking is inspiring managers to challenge conventional wisdom and traditional ideas on how businesses should operate or be organized. What's more, a host of new tools, techniques and strategies have recently evolved that help businesses reach world-class status.

Our new tool chest includes some of the following buzzwords that have become associated with quality management or process improvement strategies: activity based costing (ABC), bench-marking, total employee empowerment, design for manufacturability, design for competitive advantage, manufacturing resource planning (MRP II), distribution resource planning, Just-in-Time (JIT), total quality control, business process re-engineering, quality function deployment, concurrent engineering, electronic data interchange, Theory of Constraints, Malcolm Baldrige and ISO 9000. The list could go on and on.

While there is general agreement that improvement programs create business advantages, there are many different views on which programs are most effective. Some companies champion TQM programs, some focus on MRP II or JIT, some pursue re-engineering, some pursue benchmarking, and still others have attempted to integrate two or more of these strategies simultaneously.

Although there are many tools to choose from, the essence of becoming a world-class company requires companies to choose not just one philosophy or tool, but rather an integrated set of these strategies in the quest for world-class membership. Each tool and strategy makes its own contribution in our search for competitive advantage. For example, if it were not for TQM, quality and customer focus might not be considered a strategic variable. If it were not for JIT, continuously reducing lead times and batch sizes would not be at the top of our agendas. Furthermore, if it were not for benchmarking, we might never learn how to be "best in class."

Unfortunately, there is much confusion as to what many of these techniques stand for and how they strategically fit with the organization's overall business goals. Indeed, the preponderance of so many different tools and strategies, often implemented simultaneously, can lead to frustration, conflicting goals and even organizational paralysis. Adding to this confusion is the fact that many of these tools are still evolving as the collective experience and understanding of industry change the way we utilize them. For example, MRP, originally considered a tool to plan and schedule production requirements, has evolved into MRP II reflecting its now much broader role in managing all of a firm's production resources[2]. Even today, the application of MRP II continues to evolve towards an "extended enterprise" or "business network" concept whereby com-panies try to manage business processes beyond conventional organizational boundaries by integrating and linking their operations with those of their suppliers and customers.

Complicating matters, many of these new approaches to doing business are misunderstood or considered only in their narrowest sense. For example, if JIT is considered simply as an "inventory-reduction project," as it sometimes is, and not as a broad process that helps expose waste, continuously improving our ability to economically respond to change[3], it can offer little strategic value. Similarly, if MRP II is though of as just a "systems project" rather than as a game plan for planning and monitoring all the resources of a manufacturing company[4], then it can have little impact on competitive positioning.

SHIFTING PARADIGMS

The new thinking that accompanies such strategies as JIT, TQM and business process re-engineering, reflects nothing less than a complete paradigm shift in perspective from traditional ways of doing business. Consider, for example, how manufacturing has embraced a fundamentally new paradigm as a direct result of the total quality process (Table 1).

Table 1 illustrates the profound impact that the new manufacturing paradigm has had on the way manufacturing strategy is understood. New paradigms for conducting business aren't restricted to manufacturing either. New paradigms for customer service, marketing, distribution, research and development, and information technology, are emerging that are radically redefining the roles these functions play in contributing to the corporation's competitive positioning. Clearly, breaking with old paradigms and embracing new tools and strategies has many strategic implications that can lead to an entirely different competitive positioning for the organization.

TABLE 1: Old and New Manufacturing Paradigms Compared

Old Rules of Manufacturing	*New Paradigm of Manufacturing*
• Line personnel shouldn't challenge current practices	• The person closest to the problem is the world's best expert
• Large lot sizes are better because we amortize setup and changeover times over more uits	• Constantly try to economically reduce lot size and setup times
• Layout the factory by function (e.g., drillers on one side, sanders on the other)	• Cellular layout
• Always keep people busy and equipment humming	• Make only as much as you need and only when you need it
• Inventory is an asset	• Inventory is a liability
• Traditional performance measures such as labor and machine efficiencies, purchase price variance and overhead absorption rates	• Measurements focus on improvement rates in cost, quality, flexibility, value-added activities and customer satisfaction
• Quality is inspected at end of line	• Build quality in throughout entire process

A BUSINESS ADVANTAGE

To understand why quality of process improvement is a strategic factor, imagine you are a custom office furniture manufacturer with a normal industry lead time of three weeks from order entry to customer delivery of product. If competitors are integrating TQM, MRP II, JIT, EDI and other strategies that serve to reduce cycle time to three days, and your customers value this faster service, you might quickly see your market share erode. On the other hand, if you are already implementing these strategies and your competitors aren't, it could provide you with a very real source of product or service differentiation and, consequently, a competitive advantage.

Similarly, forward-thinking companies realize that by using tools and approaches like quality function deployment, design for manufacturability and concurrent engineering, they can involve suppliers and customers early on in the design, engineering and product development stages of the product life cycle. This not only shortens lead times to allow market introduction ahead of the competition, but also vastly improves product quality and lowers cost for both customer and supplier alike. These new approaches to doing business are increasingly becoming key sources of competitive advantage—even more effective and sustainable than such traditional sources of competitive advantage that stem from market positioning or cost leadership. Consequently, business advantage today is derived more from doing business differently than from crafting grandiose strategic plans.

More and more companies are jumping on the TQM or re-engineering band-wagon nowadays, yet few take the extra step to define, from a strategic perspective, how the improvement activities they undertake link to the firm's overall strategy and business goals. Instead, most companies adopt improvement initiatives on an ad hoc basis or in response to a competitor's initiative. For competitive advantage to be realized, however, there must be a fit between the business strategy and the focus of improvement. Consequently, managers and employees alike must make the link between the firm's overall strategy, its business goals and the various internal improvement efforts that may be going on. This necessary strategic dimension is often lacking in improvement activities. Moreover, the problem becomes even more pronounced when hundreds of people are organized around different improvement efforts that may span many functional areas of an organization.

STRATEGISTS' ROLE

Because of the inherent relationship between quality or process improvement on the one hand and competitive strategy on the other, the strategist can make a valuable contribution in helping align improvement activities with the firm's overall business strategy. But to do so, the traditional definition of strategic planning needs to be broadened. This new definition needs to reflect its much broader role in helping the company plan for the integration of new quality and process improvement tools and strategies to realize business advantage. The implications of this new role are that future strategists will be as concerned with improvement and re-engineering strategies as they used to be with the more traditional areas of strategic planning (competitive intelligence, product/market strategies, alliances and so on). This new role for the strategist embodies a shift from thinking about the external environment (competitive analysis, etc.) towards internally directed activities as an additional source of competitive advantage.

HOLISTIC APPROACH NEEDED

Because companies are generally organized functionally rather than by business process, improvement activities typically are managed with a functional view of the world. Thus, there is an engineering group, a manufacturing group and a marketing group, but there is no group that is responsible for the process of obtaining a customer order and seeing it through to delivery to the customer. Similarly, product development work generally cuts across many functional areas without any one person or group responsible for the overall process of developing a new product. Moreover, people in functional areas are typically separated by divisions, unwritten rules, floors, buildings and even continents, which act as barriers to the kind of coordination that real improvement demands.

With this structure, it is obviously difficult to see things from the customer's perspective and focus improvement efforts on the right things. Moreover, improvement in this environment often results in "island of excellence" that are not linked across functional boundaries and do not provide the firm with a strategically significant advantage. Improvements in manufacturing cycle time, for example, may be squandered as product languishes on the loading dock waiting to be shipped due to poor coordination in the shipping area. Even worse, a functional orientation may engender smaller segments of the firm to "improve" themselves at the expense of some other portion of the business. Organizations therefore must take a holistic view of the enterprise and manage improvement activities across business processes that put an end to the throw-it-over-the-wall-it's-someone-else's-problem approach. Bell Atlantic, for example, recognized the problems inherent in the functional structure and created process-oriented teams to respond to customer requests for installation of high-speed digital service links. By organizing work across business processes, they were able to reduce installation time from 30 days to only three[5].

Not only do business processes transcend functional boundaries, they often cross into supplier and customer boundaries as well. Consequently, many companies have included customers and suppliers as part of multifunctional teams that tackle improvement objectives across business processes. Adopting a process view of the business enables people to think outside their functional roles and see things from the customer's vantage point. IBM Credit Corporation, for example, which is in the business of financing the computers IBM sells, dramatically slashed the time needed to process a request for financing from seven days to just four hours by adopting a process view of their operations and then fundamentally redesigning the whole credit issuance process[6]. Likewise, Kodak re-engineered its product development process and cut the time required to move its 35mm single-use camera from concept to production nearly in half, to just 38 weeks[7].

Kodak was able to achieve dramatic improvements by looking at entire processes that cut across organizational boundaries such as credit issuance and product development. To accomplish the kinds of dramatic improvements IBM or Kodak realized, however, requires people to think in terms of the overall business goals of the organization—even many layers down the corporate ladder. Strategic thinking and a process orientation must therefore permeate all levels of the organization and cross all functional boundaries. A key tool which can provide the framework to set all improvement strategies into perspective and towards a common agenda is the process of creating a strategic vision.

WHAT IS A STRATEGIC VISION?

A strategic vision is an overarching expression and shared understanding of what the firm would like to become in the long run. It essentially defines a future desired state for the organization to strive for. Authors Collins and Porras describe a vision as a "guiding star" that gives people a context to make decisions, thus increasing the likelihood that actions taken by dispersed parties will be mutually compatible[8]. Many synonyms have been used interchangeably with the term "vision," including mission, goals, purpose, strategic, intent, philosophy and so on. Some of these terms are used synonymously and sometimes to mean different things. Regardless of the term your organization uses, what you are essentially seeking is a compelling, clear, gut-grabbing, inspirational statement that reflects what kind of organization you would like to become.

Doing the "vision thing" nowadays is very popular, and the process of formulating a vision is certainly a healthy one even if you are not currently sponsoring any improvement or change efforts. If your organization is, however, launching improvement or change efforts, the need to formulate a strategic vision is even more compelling. Vision is the link between strategy and action and serves as the overarching purpose that drives these activities.

A strategic vision helps an organization focus on a singular purpose. When purpose and goals combine with team commitment, they become a powerful engine for organizational change—change that can span all functional boundaries and link up those "islands of excellence." Indeed, the very process of working to shape a common and meaningful vision often helps teams develop direction, momentum and motivation.

At Pfizer, our Corporate Information Services (CIS) division has launched a process to develop a shared vision that will express how information technology shall be managed as a source of business advantage. Because our vision building process has enrolled members throughout the division to participate, the vision developed will be one that all members are committed to. Our vision will serve not only as our "guiding star," but also as the foundation from which to manage a variety of re-engineering and improvement initiatives. These initiatives will continuously improve Pfizer's ability to thrive in the dynamically changing health care marketplace.

CIS's efforts are only one component of an even broader re-engineering initiative encompassing the operations of Pfizer's 800-person finance division. In addition to redefining the role of CIS, other improvement projects under way include the redesign of tax, accounts payable, general ledger and payroll areas. While the tasks are different, these individual initiatives share a common theme. They all stem from a shared vision of our goal of becoming a world-class organization.

The time has come to recognize improvement strategies as a clear source of competitive advantage and as a lever in directing strategic action. The strategic value your company derives from these new strategies will depend largely on how closely you align improvement activities towards your overall strategy and business goals.

In an increasingly competitive business environment, strategic focus of improvement efforts will be what provides your firm with a competitive edge. The use of an organizational vision is one tool to help people relate the improvement activities they engage in to the overall goals of the organization and provide strategic direction. The tools to become world class are finally available. Now it's up to us to use them—the way they were intended.

REFERENCES

1. Richard J. Schonberger, "Total Quality Management Cuts A Broad Swath Through Manufacturing And Beyond," *Organizational Dynamics Magazine,* Nov. 1992.
2. Oliver Wight, *The Executive's Guide To Successful MRP II,* Oliver Wright Ltd. Publications, 1985.
3. Bill Sandras, *Just In Time: Making It Happen,* Oliver Wight Ltd. Publications, 1985.
4. Oliver Wight, *The Executive's Guide To Successful MRP II,* Oliver Wright Ltd. Publications, 1985.
5. Michael Hammer and James Champy, *Reengineering The Corporation,* Harper Business, 1993.
6. Ibid.
7. Ibid.
8. James Collins and Jerry Porras, "Organizational Vision and Visionary Organizations," *California Management Review,* Fall 1991.

Index